Lecture Notes in Artificial Intelligence 9436

Subseries of Lecture Notes in Computer Science

Davide Ciucci · Guoyin Wang
Sushmita Mitra · Wei-Zhi Wu (Eds.)

Rough Sets
and Knowledge Technology

10th International Conference, RSKT 2015
Held as Part of the International Joint Conference
on Rough Sets, IJCRS 2015
Tianjin, China, November 20–23, 2015
Proceedings

Springer

Editors
Davide Ciucci (iD)
University of Milano-Bicocca
Milano
Italy

Sushmita Mitra
Indian Statistical Institute
Kolkata
India

Guoyin Wang
Chongqing University of Posts
   and Telecommunications
Chongqing
China

Wei-Zhi Wu
Zhejiang Ocean University
Zhejiang
China

ISSN 0302-9743   ISSN 1611-3349   (electronic)
Lecture Notes in Artificial Intelligence
ISBN 978-3-319-25753-2   ISBN 978-3-319-25754-9   (eBook)
DOI 10.1007/978-3-319-25754-9

Library of Congress Control Number: 2015951773

LNCS Sublibrary: SL7 – Artificial Intelligence

Springer Cham Heidelberg New York Dordrecht London

Printed on acid-free paper

This Springer imprint is published by the registered company Springer Nature Switzerland AG
The registered company address is: Gewerbestrasse 11, 6330 Cham, Switzerland

Preface

This volume comprises papers accepted for presentation at the 10th International Conference on Rough Sets and Knowledge Technology (RSKT) conference, which along with the 15th Rough Sets, Fuzzy Sets, Data Mining and Granular Computing (RSFDGrC) International conference was held as a major part of the 2015 International Joint Conference on Rough Sets (IJCRS 2015) during November 20–23, 2015, in Tianjin, China. IJCRS is the main annual conference on rough sets. It follows the Joint Rough Set Symposium series, previously organized in 2007 in Toronto, Canada; in 2012 in Chengdu, China; in 2013 in Halifax, Canada; and in 2014 in Granada and Madrid, Spain. In addition to RSFDGrC and RSKT, IJCRS 2015 also hosted the Third International Workshop on Three-way Decisions, Uncertainty, and Granular Computing (TWDUG).

IJCRS 2015 received 97 submissions that were carefully reviewed by two to five Program Committee (PC) members or additional reviewers. After a rigorous process, 42 regular papers (acceptance rate 43.3) were accepted for presentation at the conference and publication in the two volumes of proceedings. Subsequently, some authors were selected to submit a revised version of their paper and after a second round of review, 24 more papers were accepted.

At IJCRS 2015 a ceremony was held to honor the new fellows of the International Rough Set Society. The fellows had the possibility to present their contributions in the rough set domain and to propose open questions. This volume contains six of the 12 contributions authored by the fellows Jerzy Grzymala-Busse, Tianrui Li, Duoqian Miao, Piero Pagliani, Andrzej Skowron, and Yiyu Yao.

Moreover, this volume contains the contribution by Hong Yu, invited speaker of the International Workshop on Three-Way Decisions, Uncertainty, and Granular Computing, and the tutorial by Andrzej Janusz, Sebastian Stawicki, Marcin Szczuka, and Dominik Ślęzak.

It is a pleasure to thank all those people who helped this volume to come into being and made IJCRS 2015 a successful and exciting event. First of all, we express our thanks to the authors without whom nothing would have been possible. We deeply appreciate the work of the PC members and additional reviewers who assured the high standards of accepted papers.

We deeply acknowledge the precious help of all the IJCRS chairs (Yiyu Yao, Qinghua Hu, Hong Yu, Jerzy Grzymala-Busse) as well as of the Steering Committee members (Duoqian Miao, Andrzej Skowron, Shusaku Tsumoto) for their work and suggestions with respect to the process of proceedings preparation and conference organization. Also, we want to pay tribute to the honorary chairs, Lotfi Zadeh, Roman Słowiński, Bo Zhang, and Jianwu Dang, whom we deeply respect for their countless contributions to the field.

We also gratefully thank our sponsors: the Tianjin Key Laboratory of Cognitive Computing and Application, School of Computer Science and Technology at Tianjin

University, for organizing and hosting the conference, the Institute of Innovative Technologies EMAG (Poland) for providing the data and funding the awards for the IJCRS 2015 Data Challenge, and Springer for sponsorship of the best paper awards.

Our immense gratitude goes once again to Qinghua Hu for taking charge of the organization of IJCRS 2015.

We are very grateful to Alfred Hofmann and the excellent LNCS team at Springer for their help and cooperation. We would also like to acknowledge the use of Easy-Chair, a great conference management system.

Finally, it is our sincere hope that the papers in the proceedings may be of interest to the readers and inspire them in their scientific activities.

November 2015

Davide Ciucci
Guoyin Wang
Sushmita Mitra
Wei-Zhi Wu

Organization

Honorary General Co-chairs

Jianwu Dang   Tianjin University, Tianjin, China
Roman Słowiński   Poznan University of Technology, Poznań, Poland
Lotfi Zadeh   University of California, Berkeley, CA, USA
Bo Zhang   Tsinghua University, Beijing, China

General Chairs

Qinghua Hu   Tianjin University, Tianjin, China
Guoyin Wang   Chongqing University of Posts and
                           Telecommunications, Chongqing, China

Steering Committee Chairs

Duoqian Miao   Tongji University, Shanghai, China
Andrzej Skowron   University of Warsaw, Warsaw, Poland and Polish
                           Academy of Sciences, Warsaw, Poland
Shusaku Tsumoto   Shimane University, Izumo, Japan

Program Chairs

Davide Ciucci   University of Milano-Bicocca, Milano, Italy
Yiyu Yao   University of Regina, Regina, SK, Canada

Program Co-chairs for RSFDGrC 2015

Sushmita Mitra   Indian Statistical Institute, Kolkata, India
Wei-Zhi Wu   Zhejiang Ocean University, Zhejiang, China

Program Co-chairs for RSKT 2015

Jerzy Grzymala-Busse   University of Kansas, Lawrence, KS, USA
Hong Yu   Chongqing University of Posts and
                           Telecommunications, Chongqing, China

Workshop Co-chairs for TWDUG 2015

Davide Ciucci   University of Milano-Bicocca, Milano, Italy
Jerzy Grzymala-Busse   University of Kansas, Lawrence, KS, USA
Tianrui Li   Southwest Jiaotong University, Chengdu, China

Program Co-chairs for TWDUG 2015

Nouman Azam   National University of Computer and Emerging
                               Sciences, Peshawar, Pakistan
Fan Min   Southwest Petroleum University, Chengdu, China
Hong Yu   Chongqing University of Posts and
                               Telecommunications, Chongqing, China
Bing Zhou   Sam Houston State University, Huntsville, TX, USA

Data Mining Contest Organizing Committee

Andrzej Janusz   University of Warsaw, Warsaw, Poland
Marek Sikora   Silesian University of Technology, Gliwice, Poland
Łukasz Wróbel   Silesian University of Technology, Gliwice, Poland
Sebastian Stawicki   University of Warsaw, Warsaw, Poland
Marek Grzegorowski   University of Warsaw, Warsaw, Poland
Dominik Ślęzak   University of Warsaw, Warsaw, Poland

Tutorial Co-chairs

Dominik Ślęzak   University of Warsaw, Warsaw, Poland
Yoo-Sung Kim   Inha University, Incheon, Korea

Special Session Co-chairs

JingTao Yao   University of Regina, Regina, SK, Canada
Tianrui Li   Southwest Jiaotong University, Chengdu, China
Zbigniew Suraj   University of Rzeszow, Rzeszow, Poland

Publicity Co-chairs

Zbigniew Ras   University of North Carolina at Charlotte, Charlotte,
                               NC, USA and Warsaw University of Technology,
                               Warsaw, Poland
Georg Peters   Munich University of Applied Sciences, Munich,
                               Germany and Australian Catholic University,
                               Sydney, Australia
Aboul Ella Hassanien   Cairo University, Cairo, Egypt

Azizah Abdul Manaf   Universiti Teknologi Malaysia, Kuala Lumpur,
    Malaysia
Gabriella Pasi   University of Milano-Bicocca, Milano, Italy
Chris Cornelis   University of Granada, Granada, Spain

Local Co-chairs

Jianhua Dai   Tianjin University, Tianjin, China
Pengfei Zhu   Tianjin University, Tianjin, China
Hong Shi   Tianjin University, Tianjin, China

Program Committee

Arun Agarwal
Adel Alimi
Simon Andrews
Nidhi Arora
Piotr Artiemjew
S. Asharaf
Ahmad Azar
Nakhoon Baek
Sanghamitra
   Bandyopadhyay
Mohua Banerjee
Nizar Banu
Andrzej Bargiela
Alan Barton
Jan Bazan
Theresa Beaubouef
Rafael Bello
Rabi Bhaumik
Jurek Błaszczyński
Nizar Bouguila
Yongzhi Cao
Gianpiero Cattaneo
Salem Chakhar
Mihir Chakraborty
Chien-Chung Chan
Chiao-Chen Chang
Santanu Chaudhury
Degang Chen
Hongmei Chen
Mu-Chen Chen
Mu-Yen Chen
Igor Chikalov
Wan-Sup Cho

Soon Ae Chun
Martine de Cock
Chris Cornelis
Zoltán Csajbók
Alfredo Cuzzocrea
Krzysztof Cyran
Jianhua Dai
Bijan Davvaz
Dayong Deng
Thierry Denoeux
Jitender Deogun
Lipika Dey
Fernando Diaz
Ivo Düntsch
Zied Elouedi
Francisco Fernandez
Jinan Fiaidhi
Wojciech Froelich
G. Ganesan
Yang Gao
Günther Gediga
Neveen Ghali
Anna Gomolińska
Salvatore Greco
Shen-Ming Gu
Yanyong Guan
Jianchao Han
Wang Hao
Aboul Hassanien
Jun He
Christopher Henry
Daryl Hepting
Joseph Herbert

Francisco Herrera
Chris Hinde
Shoji Hirano
Władysław Homenda
Feng Hu
Xiaohua Hu
Shahid Hussain
Namvan Huynh
Dmitry Ignatov
Hannah Inbaran
Masahiro Inuiguchi
Yun Jang
Ryszard Janicki
Andrzej Jankowski
Andrzej Janusz
Jouni Järvinen
Richard Jensen
Xiuyi Jia
Chaozhe Jiang
Na Jiao
Tae-Chang Jo
Manish Joshi
Hanmin Jung
Janusz Kacprzyk
Byeongho Kang
C. Maria Keet
Md. Aquil Khan
Deok-Hwan Kim
Soohyung Kim
Young-Bong Kim
Keiko Kitagawa
Michiro Kondo
Beata Konikowska

Jacek Koronacki
Bożena Kostek
Adam Krasuski
Vladik Kreinovich
Rudolf Kruse
Marzena Kryszkiewicz
Yasuo Kudo
Yoshifumi Kusunoki
Sergei Kuznetsov
Wookey Lee
Young-Koo Lee
Carson K. Leung
Huaxiong Li
Longshu Li
Wen Li
Decui Liang
Jiye Liang
Churn-Jung Liau
Diego Liberati
Antoni Ligęza
T.Y. Lin
Pawan Lingras
Kathy Liszka
Caihui Liu
Dun Liu
Guilong Liu
Qing Liu
Xiaodong Liu
Dickson Lukose
Neil Mac Parthaláin
Seung-Ryol Maeng
Pradipta Maji
A. Mani
Victor Marek
Barbara Marszał-Paszek
Tshilidzi Marwala
Benedetto Matarazzo
Nikolaos Matsatsinis
Stan Matwin
Jesús Medina-Moreno
Ernestina Menasalvas
Jusheng Mi
Alicja
   Mieszkowicz-Rolka
Tamás Mihálydeák
Pabitra Mitra

Sadaaki Miyamoto
Sabah Mohammed
Mikhail Moshkov
Tetsuya Murai
Kazumi Nakamatsu
Michinori Nakata
Amedeo Napoli
Kanlaya Naruedomkul
Hungson Nguyen
Linhanh Nguyen
Maria Nicoletti
Vilém Novák
Mariusz Nowostawski
Hannu Nurmi
Hala Own
Piero Pagliani
Krzysztof Pancerz
Taeho Park
Piotr Paszek
Alberto Guillén Perales
Georg Peters
James F. Peters
Frederick Petry
Jonas Poelmans
Lech Polkowski
Henri Prade
Jianjun Qi
Jin Qian
Yuhua Qian
Keyun Qin
Guo-Fang Qiu
Taorong Qiu
Mohamed Quafafou
Annamaria Radzikowska
Vijay V. Raghavan
Sheela Ramanna
Zbigniew Raś
Kenneth Revett
Leszek Rolka
Leszek Rutkowski
Henryk Rybiński
Wojciech Rząsa
Hiroshi Sakai
Abdel-Badeeh Salem
Miguel Ángel Sanz-Bobi
Gerald Schaefer

Jeong Seonphil
Noor Setiawan
Sitimariyam Shamsuddin
Lin Shang
Ming-Wen Shao
Marek Sikora
Arul Siromoney
Vaclav Snasel
Urszula Stańczyk
Jarosław Stepaniuk
Kazutoshi Sumiya
Lijun Sun
Piotr Synak
Andrzej Szałas
Marcin Szczuka
Tomasz Szmuc
Marcin Szpyrka
Li-Shiang Tsay
Gwo-Hshiung Tzeng
Changzhong Wang
Chaokun Wang
Hai Wang
Junhong Wang
Ruizhi Wang
Wendyhui Wang
Xin Wang
Yongquan Wang
Piotr Wasilewski
Junzo Watada
Ling Wei
Zhihua Wei
Paul Wen
Arkadiusz Wojna
Karlerich Wolff
Marcin Wolski
Michał Woźniak
Gang Xie
Feifei Xu
Jiucheng Xu
Weihua Xu
Zhan-Ao Xue
Ronald Yager
Hyung-Jeong Yang
Jaedong Yang
Xibei Yang
Yan Yang

Yingjie Yang
Yong Yang
Yubin Yang
Nadezhda G. Yarushkina
Dongyi Ye
Ye Yin
Byounghyun Yoo

Kwan-Hee Yoo
Xiaodong Yue
Sławomir Zadrożny
Hongyun Zhang
Nan Zhang
Qinghua Zhang
Xiaohong Zhang

Yan-Ping Zhang
Cairong Zhao
Shu Zhao
Xian-Zhong Zhou
William Zhu
Wojciech Ziarko
Beata Zielosko

Additional Reviewers

Banerjee, Abhirup
Benítez Caballero, María José
Błaszczyński, Jerzy
Chiaselotti, Giampiero
Czołombitko, Michał
D'Eer, Lynn
De Clercq, Sofie
Garai, Partha
Hu, Jie
Jie, Chen
Li, Feijiang
Liang, Xinyan
Wang, Jianxin
Wang, Jieting
Xu, Xinying
Zhang, Junbo

Contents

Generalized Rough Sets

Three-Way Decision

Logic and Algebra

Clustering

Rough Sets: The Experts Speak

A Rough Set Approach to Incomplete Data

Jerzy W. Grzymala-Busse[1,2]([⊠])

[1] Department of Electrical Engineering and Computer Science,
University of Kansas, Lawrence, KS 66045, USA
[2] Department of Expert Systems and Artificial Intelligence,
University of Information Technology and Management,
35-225 Rzeszow, Poland
jerzy@ku.edu

Abstract. This paper presents main directions of research on a rough set approach to incomplete data. First, three different types of lower and upper approximations, based on the characteristic relation, are defined. Then an idea of the probabilistic approximation, an extension of lower and upper approximations, is presented. Local probabilistic approximations are also discussed. Finally, some special topics such as consistency of incomplete data and a problem of increasing data set incompleteness to improve rule set quality, in terms of an error rate, are discussed.

Keywords: Incomplete data · Characteristic relation · Singleton concept and subset approximations · Probabilistic approximations · Local probabilistic approximations

1 Introduction

It is well-known that many real-life data sets are incomplete, i.e., are affected by missing attribute values. Recently many papers presenting a rough set approach in research on incomplete data were published, see, e.g., [4,7,9–17,21–27,31–34,37,38,40–42,44,47–62,68–72,74–78,80,81].

Most of the rough set activity in research on incomplete data is conducted in data mining. Using a rough set approach to incomplete data, we may distinguish between different interpretations of missing attribute values.

If an attribute value was accidentally erased or is unreadable, we may use the most cautious approach to missing attribute values and mine data using only specified attribute values. The corresponding type of missing attribute values is called *lost* and is denoted by "?". Mining incomplete data affected by lost values was studied for the first time in [44], where two algorithms for rule induction from such data were presented. The same data sets were studied later, see, e.g., [76,77].

Another type of missing attribute values happens when a respondent refuses to answer a question that seems to be irrelevant. For example, a patient is tested for a disease and one of the questions is a color of hair. The respondent may consider the color of hair to be irrelevant. This type of missing attribute values is called a *"do not care" condition* and is denoted by "*". The first study of "do

© Springer International Publishing Switzerland 2015
D. Ciucci et al. (Eds.): RSKT 2015, LNAI 9436, pp. 3–14, 2015.
DOI: 10.1007/978-3-319-25754-9_1

not care" conditions, again using rough set theory, was presented in [17], where a method for rule induction in which missing attribute values were replaced by all values from the domain of the attribute was introduced. "Do not care" conditions were also studied later, see, e.g. [50,51].

In yet another interpretation of missing attribute values, called an *attribute-concept* value, and denoted by "−", we assume that we know that the corresponding case belongs to a specific concept X and, as a result, we may replace the missing attribute value by attribute values for all cases from the same concept X. A *concept* (class) is a set of all cases classified (or diagnosed) the same way. For example, if for a patient the value of an attribute *Temperature* is missing, this patient is sick with *Flu*, and all remaining patients sick with *Flu* have *Temperature* values *high* then using the interpretation of the missing attribute value as the attribute-concept value, we will replace the missing attribute value by the value *high*. This approach was introduced in [24].

An approach to mining incomplete data presented in this paper is based on the idea of an attribute-value block. A characteristic set, defined as an intersection of such blocks, is a generalization of the elementary set, well-known in rough set theory [63–65]. A characteristic relation, defined by characteristic sets, is, in turn, a generalization of the indiscernibility relation. As it was shown in [21], incomplete data are described by three different types of approximations: singleton, subset and concept.

For rule induction from incomplete data it is the most natural to use the MLEM2 (Modified Learning form Examples Module, version 2) [2,18–20] since this algorithm is also based on attribute-value pair blocks. A number of extensions of this algorithm were developed in order to process incomplete data sets using different definitions of approximations, see, e.g., [5,31,43,45].

One of the fundamental concepts of rough set theory is lower and upper approximations. A generalization of such approximations, a probabilistic approximation, introduced in [79], was applied in variable precision rough set models, Bayesian rough sets and decision-theoretic rough set models [46,66,67,73,82–86]. These probabilistic approximations are defined using the indiscernibility relation. For incomplete data, probabilistic approximations were extended to characteristic relation in [30]. The probabilistic approximation is associated with some parameter α (interpreted as a probability). If α is very small, say $1/|U|$, where U is the set of all cases, the probabilistic approximation is reduced to the upper approximation; if α is equal to 1.0, the probabilistic approximation is equal to the lower approximation. Local probabilistic approximations, based on attribute-value blocks instead of characteristic sets, were defined in [7], see also [31].

2   Fundamental Concepts

A basic tool to analyze incomplete data sets is a *block of an attribute-value pair*. Let (a, v) be an attribute-value pair. For *complete* data sets, i.e., data sets in which every attribute value is specified, a block of (a, v), denoted by $[(a, v)]$, is the set of all cases x for which $a(x) = v$, where $a(x)$ denotes the value of the attribute a for the case x. For incomplete data sets the definition of a block of an attribute-value pair is modified.

- If for an attribute a there exists a case x such that $a(x) = ?$, i.e., the corresponding value is lost, then the case x should not be included in any blocks $[(a, v)]$ for all values v of attribute a,
- If for an attribute a there exists a case x such that the corresponding value is a "do not care" condition, i.e., $a(x) = *$, then the case x should be included in blocks $[(a, v)]$ for all specified values v of attribute a.
- If for an attribute a there exists a case x such that the corresponding value is an attribute-concept value, i.e., $a(x) = -$, then the corresponding case x should be included in blocks $[(a, v)]$ for all specified values $v \in V(x, a)$ of attribute a, where

$$V(x, a) = \{a(y) \mid a(y) \text{ is specified}, y \in U, \ d(y) = d(x)\}.$$

For a case $x \in U$ the *characteristic set* $K_B(x)$ is defined as the intersection of the sets $K(x, a)$, for all $a \in B$, where the set $K(x, a)$ is defined in the following way:

- If $a(x)$ is specified, then $K(x, a)$ is the block $[(a, a(x))]$ of attribute a and its value $a(x)$,
- If $a(x)) =?$ or $a(x) = *$, then the set $K(x, a) = U$.
- If $a(x) = -$, then the corresponding case x should be included in blocks $[(a, v)]$ for all known values $v \in V(x, a)$ of attribute a. If $V(x, a)$ is empty, $K(x, a) = U$.

The *characteristic relation* $R(B)$ is a relation on U defined for $x, y \in U$ as follows

$$(x, y) \in R(B) \text{ if and only if } y \in K_B(x).$$

The characteristic relation $R(B)$ is reflexive but—in general—does not need to be symmetric or transitive.

3 Lower and Upper Approximations

For incomplete data sets there is a few possible ways to define approximations [24,43]. Let X be a concept, let B be a subset of the set A of all attributes, and let $R(B)$ be the characteristic relation of the incomplete decision table with characteristic sets $K_B(x)$, where $x \in U$. A *singleton* B-lower approximation of X is defined as follows:

$$\underline{B}X = \{x \in U \mid K_B(x) \subseteq X\}.$$

A *singleton* B-upper approximation of X is

$$\overline{B}X = \{x \in U \mid K_B(x) \cap X \neq \emptyset\}.$$

We may define lower and upper approximations for incomplete data sets as unions of characteristic sets. There are two possibilities. Using the first way, a *subset B*-lower approximation of X is defined as follows:

$$\underline{B}X = \cup\{K_B(x) \mid x \in U, K_B(x) \subseteq X\}.$$

A *subset B*-upper approximation of X is

$$\overline{B}X = \cup\{K_B(x) \mid x \in U, K_B(x) \cap X \neq \emptyset\}.$$

The second possibility is to modify the subset definition of lower and upper approximation by replacing the universe U from the subset definition by a concept X. A *concept B*-lower approximation of the concept X is defined as follows:

$$\underline{B}X = \cup\{K_B(x) \mid x \in X, K_B(x) \subseteq X\}.$$

The subset B-lower approximation of X is the same set as the concept B-lower approximation of X. A *concept B*-upper approximation of the concept X is defined as follows:

$$\overline{B}X = \cup\{K_B(x) \mid x \in X, K_B(x) \cap X \neq \emptyset\} = \cup\{K_B(x) \mid x \in X\}.$$

Two traditional methods of handling missing attribute values: Most Common Value for symbolic attributes and Average Values for numeric attributes (MCV-AV) and Concept Most Common Values for symbolic attributes and Concept Average Values for numeric attributes (CMCV-CAV), for details see [36], were compared with three rough-set interpretations of missing attribute values: lost values, attribute-concept values and "do not care" conditions combined with concept lower and upper approximations in [27]. In turned out that there is no significant difference in performance in terms of an error rate measured by ten-fold cross validation between the traditional and rough-set approaches to missing attribute values.

In [26] the same two traditional methods, MCV-AV and CMCV-CAV, and other two traditional methods (Closest Fit and Concept Closest Fit), for details see [36], were compared with the same three rough set interpretations of missing attribute values combined with concept approximations. The best methodology was based on the Concept Closest Fit combined with rough set interpretation of missing attribute values as lost values and concept lower and upper approximations.

Additionally, in [28], a CART approach to missing attribute values [1] was compared with missing attribute values interpreted as lost values combined with concept lower and upper approximations. In two cases CART was better, in two cases rough set approach was better, and in one case the difference was insignificant. Hence both approaches are comparable in terms of an error rate.

In [29,39], the method CMCV-CAV was compared with rough set approaches to missing attribute values, the conclusion was that CMCV-CAV was either worse or not better, depending on a data set, than rough-set approaches.

4   Probabilistic Approximations

In this section we will extend definitions of singleton, subset and concept approximations to corresponding probabilistic approximations. The problem is how useful are *proper* probabilistic approximations (with α larger than $1/|U|$ but smaller than 1.0). We studied usefulness of proper probabilistic approximations for incomplete data sets [3], where we concluded that proper probabilistic approximations are not frequently better than ordinary lower and upper approximations.

A B-singleton probabilistic approximation of X with the threshold α, $0 < \alpha \leq 1$, denoted by $appr_{\alpha,B}^{singleton}(X)$, is defined by

$$\{x \mid x \in U, \ Pr(X \mid K_B(x)) \geq \alpha\},$$

where $Pr(X \mid K_B(x)) = \frac{|X \cap K_B(x)|}{|K_B(x)|}$ is the conditional probability of X given $K_B(x)$ and $|Y|$ denotes the cardinality of set Y.

A B-subset probabilistic approximation of the set X with the threshold α, $0 < \alpha \leq 1$, denoted by $appr_{\alpha,B}^{subset}(X)$, is defined by

$$\cup\{K_B(x) \mid x \in U, \ Pr(X \mid K_B(x)) \geq \alpha\}$$

A B-concept probabilistic approximation of the set X with the threshold α, $0 < \alpha \leq 1$, denoted by $appr_{\alpha,B}^{concept}(X)$, is defined by

$$\cup\{K_B(x) \mid x \in X, \ Pr(X \mid K_B(x)) \geq \alpha\}.$$

In [6] *ordinary* lower and upper approximations (singleton, subset and concept), special cases of singleton, subset and concept probabilistic approximations, were compared with *proper* probabilistic approximations (singleton, subset and concept) on six data sets with missing attribute values interpreted as lost values and "do not care" conditions, in terms of an error rate. Since we used six data sets, two interpretations of missing attribute values and three types of probabilistic approximations, there were 36 combinations. Among these 36 combinations, for five combinations the error rate was smaller for proper probabilistic approximations than for ordinary (lower and upper) approximations, for other four combinations, the error rate for proper probabilistic approximations was larger than for ordinary approximations, for the remaining 27 combinations, the difference between these two types of approximations was not statistically significant.

Results of experiments presented in [9] show that among all probabilistic approximations (singleton, subset and concept) and two interpretations of missing attribute values (lost values and "do not care" conditions) there is not much difference in terms of an error rate measured by ten-fold cross validation. On the other hand, complexity of induced rule sets differs significantly. The simplest rule sets (in terms of the number of rules and the total number of conditions in the rule set) we accomplished by using subset probabilistic approximations combined with "do not care" conditions.

In [8] results of experiments using all three probabilistic approximations (singleton, subset and concept) and two interpretations of missing attribute values: lost values and "do not care" conditions were compared with the MCV-AV and CMCV-CAV methods in terms of an error rate. For every data set, the best among six rough-set methods (combining three kinds of probabilistic approximations with two types of interpretations of missing attribute vales) were compared with the better results of MCV-AV and CMCV-CAV. Rough-set methods were better for five (out of six) data sets.

5   Local Approximations

An idea of the local approximation was introduced in [41]. A local probabilistic approximation was defined in [7]. A set T of attribute-value pairs, where all attributes belong to the set B and are distinct, is called a B-*complex*. In the most general definition of a local probabilistic definition we assume only an existence of a family $\mathcal{T}$ of B-complexes T with the conditional probability $Pr(X|[T])$ of $X \geq \alpha$, where $Pr(X|[T]) = \frac{|X \cap [T]|}{|[T]|}$.

A B-*local probabilistic approximation* of the set X with the parameter α, $0 < \alpha \leq 1$, denoted by $appr_\alpha^{local}(X)$, is defined as follows

$$\cup \{[T] \mid \exists \text{ a family } \mathcal{T} \text{ of } B - complexes \ T \text{ of } X \text{ with } \forall \ T \in \mathcal{T}, \ Pr(X|[T]) \geq \alpha\}.$$

In general, for given set X and α, there exists more than one A-local probabilistic approximation. However, for given set X and parameter α, a B-local probabilistic approximation given by the next definition is unique.

A *complete B-local probabilistic approximation* of the set X with the parameter α, $0 < \alpha \leq 1$, denoted by $appr_\alpha^{complete}(X)$, is defined as follows

$$\cup \{[T] \mid T \text{ is a } B - complex \text{ of } X, \ Pr(X|[T]) \geq \alpha\}.$$

Due to computational complexity of determining complete local probabilistic approximations, yet another local probabilistic approximation, called a *MLEM2 local probabilistic approximation* and denoted by $appr_\alpha^{mlem2}(X)$, is defined by using A-complexes Y that are the most relevant to X, i.e., with $|X \cap Y|$ as large as possible, etc., following the MLEM2 algorithm.

In [31] concept probabilistic approximations were compared with complete local probabilistic approximations and with MLEM2 local probabilistic approximations for eight data sets, using two interpretations of missing attribute values: lost vales and "do not care" conditions, in terms of an error rate. Since two interpretations of missing attribute values and eight data sets were used, there were 16 combinations. There was not clear winner among three kinds of probabilistic approximations. In four combinations the best was the concept probabilistic approximation, in three cases the best was the complete local probabilistic approximation, and in four cases the best was the MLEM2 local probabilistic approximation. For remaining five combinations the difference in performance between all three approximations was insignificant.

6 Special Topics

When replacing existing, specified attribute values by symbols of missing attribute values, e.g., by "?"s, an error rate computed by ten-fold cross validation may be smaller than for the original, complete data set. Thus, increasing incompleteness of the data set may improve accuracy. Results of experiments showing this phenomenon were published, e.g., in [34, 35].

Yet another problem is associated with consistency. For complete data sets, a data set is consistent if any two cases, indistinguishable by all attributes, belong to the same concept. The idea of consistency is more complicated for incomplete data. This problem was discussed in [4] and also in [54, 60, 72].

7 Conclusions

Research on incomplete data is very active and promising, with many open problems and potential for additional progress.

References

1. Breiman, L., Friedman, J.H., Olshen, R.A., Stone, C.J.: Classification and Regression Trees. Wadsworth & Brooks, Monterey (1984)
2. Chan, C.C., Grzymala-Busse, J.W.: On the attribute redundancy and the learning programs ID3, PRISM, and LEM2. Technical report, Department of Computer Science, University of Kansas (1991)
3. Clark, P.G., Grzymala-Busse, J.W.: Experiments on probabilistic approximations. In: Proceedings of the 2011 IEEE International Conference on Granular Computing, pp. 144–149 (2011)
4. Clark, P.G., Grzymala-Busse, J.W.: Consistency of incomplete data. In: Proceedings of the Second International Conference on Data Technologies and Applications, pp. 80–87 (2013)
5. Clark, P.G., Grzymala-Busse, J.W.: A comparison of two versions of the MLEM2 rule induction algorithm extended to probabilistic approximations. In: Cornelis, C., Kryszkiewicz, M., Ślęzak, D., Ruiz, E.M., Bello, R., Shang, L. (eds.) RSCTC 2014. LNCS, vol. 8536, pp. 109–119. Springer, Heidelberg (2014)
6. Clark, P.G., Grzymala-Busse, J.W., Hippe, Z.S.: An analysis of probabilistic approximations for rule induction from incomplete data sets. Fundam. Informaticae 55, 365–379 (2014)
7. Clark, P.G., Grzymala-Busse, J.W., Kuehnhausen, M.: Local probabilistic approximations for incomplete data. In: Chen, L., Felfernig, A., Liu, J., Raś, Z.W. (eds.) ISMIS 2012. LNCS, vol. 7661, pp. 93–98. Springer, Heidelberg (2012)
8. Clark, P.G., Grzymala-Busse, J.W., Kuehnhausen, M.: Mining incomplete data with many missing attribute values. A comparison of probabilistic and rough set approaches. In: Proceedings of the Second International Conference on Intelligent Systems and Applications, pp. 12–17 (2013)
9. Clark, P.G., Grzymala-Busse, J.W., Rzasa, W.: Mining incomplete data with singleton, subset and concept approximations. Inf. Sci. 280, 368–384 (2014)

10. Cyran, K.A.: Modified indiscernibility relation in the theory of rough sets with real-valued attributes: application to recognition of fraunhofer diffraction patterns. Trans. Rough Sets 9, 14–34 (2008)
11. Dai, J.: Rough set approach to incomplete numerical data. Inf. Sci. 241, 43–57 (2013)
12. Dai, J., Xu, Q.: Approximations and uncertainty measures in incomplete information systems. Inf. Sci. 198, 62–80 (2012)
13. Dai, J., Xu, Q., Wang, W.: A comparative study on strategies of rule induction for incomplete data based on rough set approach. Int. J. Adv. Comput. Technol. 3, 176–183 (2011)
14. Dardzinska, A., Ras, Z.W.: Chasing unknown values in incomplete information systems. In: Workshop Notes, Foundations and New Directions of Data Mining, in Conjunction with the 3-rd International Conference on Data Mining, pp. 24–30 (2003)
15. Dardzinska, A., Ras, Z.W.: On rule discovery from incomplete information systems. In: Workshop Notes, Foundations and New Directions of Data Mining, in Conjunction with the 3-rd International Conference on Data Mining, pp. 24–30 (2003)
16. Greco, S., Matarazzo, B., Slowinski, R.: Dealing with missing data in rough set analysis of multi-attribute and multi-criteria decision problems. In: Zanakis, H., Doukidis, G., Zopounidis, Z. (eds.) Decision Making: Recent developments and Worldwide Applications, pp. 295–316. Kluwer Academic Publishers, Dordrecht (2000)
17. Grzymala-Busse, J.W.: On the unknown attribute values in learning from examples. In: Raś, Z.W., Zemankova, M. (eds.) ISMIS 1991. LNCS, vol. 542, pp. 368–377. Springer, Heidelberg (1991)
18. Grzymala-Busse, J.W.: LERS–a system for learning from examples based on rough sets. In: Slowinski, R. (ed.) Intelligent Decision Support, pp. 3–18. Handbook of Applications and Advances of the Rough Set Theory. Kluwer Academic Publishers, Dordrecht (1992)
19. Grzymala-Busse, J.W.: A new version of the rule induction system LERS. Fundamenta Informaticae 31, 27–39 (1997)
20. Grzymala-Busse, J.W.: MLEM2: A new algorithm for rule induction from imperfect data. In: Proceedings of the 9th International Conference on Information Processing and Management of Uncertainty in Knowledge-Based Systems, pp. 243–250 (2002)
21. Grzymala-Busse, J.W.: Rough set strategies to data with missing attribute values. In: Notes of the Workshop on Foundations and New Directions of Data Mining, in Conjunction with the Third International Conference on Data Mining, pp. 56–63 (2003)
22. Grzymała-Busse, J.W.: Characteristic relations for incomplete data: a generalization of the indiscernibility relation. In: Tsumoto, S., Słowiński, R., Komorowski, J., Grzymała-Busse, J.W. (eds.) RSCTC 2004. LNCS (LNAI), vol. 3066, pp. 244–253. Springer, Heidelberg (2004)
23. Grzymala-Busse, J.W.: Data with missing attribute values: generalization of indiscernibility relation and rule induction. Trans. Rough Sets 1, 78–95 (2004)
24. Grzymala-Busse, J.W.: Three approaches to missing attribute values–a rough set perspective. In: Proceedings of the Workshop on Foundation of Data Mining, in Conjunction with the Fourth IEEE International Conference on Data Mining, pp. 55–62 (2004)

25. Grzymała-Busse, J.W.: Incomplete data and generalization of indiscernibility relation, definability, and approximations. In: Slezak, D., Wang, G., Szczuka, M.S., Düntsch, I., Yao, Y. (eds.) RSFDGrC 2005. LNCS (LNAI), vol. 3641, pp. 244–253. Springer, Heidelberg (2005)

26. Grzymala-Busse, J.W.: A comparison of traditional and rough set approaches to missing attribute values in data mining. In: Proceedings of the 10-th International Conference on Data Mining, Detection, Protection and Security, Royal Mare Village, Crete, pp. 155–163 (2009)

27. Grzymala-Busse, J.W.: Mining data with missing attribute values: A comparison of probabilistic and rough set approaches. In: Proceedings of the 4-th International Conference on Intelligent Systems and Knowledge Engineering, pp. 153–158 (2009)

28. Grzymala-Busse, J.W.: Rough set and CART approaches to mining incomplete data. In: Proceedings of the International Conference on Soft Computing and Pattern Recognition, IEEE Computer Society, pp. 214–219 (2010)

29. Grzymala-Busse, J.W.: A comparison of some rough set approaches to mining symbolic data with missing attribute values. In: Kryszkiewicz, M., Rybinski, H., Skowron, A., Raś, Z.W. (eds.) ISMIS 2011. LNCS, vol. 6804, pp. 52–61. Springer, Heidelberg (2011)

30. Grzymała-Busse, J.W.: Generalized parameterized approximations. In: Yao, J.T., Ramanna, S., Wang, G., Suraj, Z. (eds.) RSKT 2011. LNCS, vol. 6954, pp. 136–145. Springer, Heidelberg (2011)

31. Grzymala-Busse, J.W., Clark, P.G., Kuehnhausen, M.: Generalized probabilistic approximations of incomplete data. Int. J. Approximate Reasoning **132**, 180–196 (2014)

32. Grzymala-Busse, J.W., Grzymala-Busse, W.J.: Handling missing attribute values. In: Maimon, O., Rokach, L. (eds.) Data Mining and Knowledge Discovery Handbook, pp. 37–57. Springer, Heidelberg (2005)

33. Grzymala-Busse, J.W., Grzymala-Busse, W.J.: An experimental comparison of three rough set approaches to missing attribute values. Trans. Rough Sets **6**, 31–50 (2007)

34. Grzymala-Busse, J.W., Grzymala-Busse, W.J.: Improving quality of rule sets by increasing incompleteness of data sets. In: Proceedings of the Third International Conference on Software and Data Technologies, pp. 241–248 (2008)

35. Grzymala-Busse, J.W., Grzymala-Busse, W.J.: Inducing better rule sets by adding missing attribute values. In: Chan, C.-C., Grzymala-Busse, J.W., Ziarko, W.P. (eds.) RSCTC 2008. LNCS (LNAI), vol. 5306, pp. 160–169. Springer, Heidelberg (2008)

36. Grzymala-Busse, J.W., Grzymala-Busse, W.J.: Handling missing attribute values. In: Maimon, O., Rokach, L. (eds.) Data Mining and Knowledge Discovery Handbook, 2nd edn, pp. 33–51. Springer, Heidelberg (2010)

37. Grzymala-Busse, J.W., Grzymala-Busse, W.J., Goodwin, L.K.: A comparison of three closest fit approaches to missing attribute values in preterm birth data. Int. J. Intell. Syst. **17**(2), 125–134 (2002)

38. Grzymala-Busse, J.W., Grzymala-Busse, W.J., Hippe, Z.S., Rzasa, W.: An improved comparison of three rough set approaches to missing attribute values. In: Proceedings of the 16-th International Conference on Intelligent Information Systems, pp. 141–150 (2008)

39. Grzymala-Busse, J.W., Hippe, Z.S.: Mining data with numerical attributes and missing attribute values–a rough set approach. In: Proceedings of the IEEE International Conference on Granular Computing, pp. 144–149 (2011)

40. Grzymała-Busse, J.W., Hu, M.: A comparison of several approaches to missing attribute values in data mining. In: Ziarko, W.P., Yao, Y. (eds.) RSCTC 2000. LNCS (LNAI), vol. 2005, p. 378. Springer, Heidelberg (2001)
41. Grzymala-Busse, J.W., Rzasa, W.: Local and global approximations for incomplete data. In: Greco, S., Hata, Y., Hirano, S., Inuiguchi, M., Miyamoto, S., Nguyen, H.S., Słowiński, R. (eds.) RSCTC 2006. LNCS (LNAI), vol. 4259, pp. 244–253. Springer, Heidelberg (2006)
42. Grzymala-Busse, J.W., Rzasa, W.: Local and global approximations for incomplete data. Trans. Rough Sets **8**, 21–34 (2008)
43. Grzymala-Busse, J.W., Rzasa, W.: A local version of the MLEM2 algorithm for rule induction. Fundamenta Informaticae **100**, 99–116 (2010)
44. Grzymala-Busse, J.W., Wang, A.Y.: Modified algorithms LEM1 and LEM2 for rule induction from data with missing attribute values. In: Proceedings of the 5-th International Workshop on Rough Sets and Soft Computing in conjunction with the Third Joint Conference on Information Sciences, pp. 69–72 (1997)
45. Grzymala-Busse, J.W., Yao, Y.: Probabilistic rule induction with the LERS data mining system. Int. J. Intell. Syst. **26**, 518–539 (2011)
46. Grzymala-Busse, J.W., Ziarko, W.: Data mining based on rough sets. In: Wang, J. (ed.) Data Mining: Opportunities and Challenges, pp. 142–173. Idea Group Publ, Hershey (2003)
47. Guan, L., Wang, G.: Generalized approximations defined by non-equivalence relations. Inf. Sci. **193**, 163–179 (2012)
48. Hong, T.P., Tseng, L.H., Chien, B.C.: Learning coverage rules from incomplete data based on rough sets. In: Proceedings of the IEEE International Conference on Systems, Man and Cybernetics, pp. 3226–3231 (2004)
49. Hong, T.P., Tseng, L.H., Wang, S.L.: Learning rules from incomplete training examples by rough sets. Expert Syst. Appl. **22**, 285–293 (2002)
50. Kryszkiewicz, M.: Rough set approach to incomplete information systems. In: Proceedings of the Second Annual Joint Conference on Information Sciences, pp. 194–197 (1995)
51. Kryszkiewicz, M.: Rules in incomplete information systems. Inf. Sci. **113**(3–4), 271–292 (1999)
52. Latkowski, R.: On decomposition for incomplete data. Fundamenta Informaticae **54**, 1–16 (2003)
53. Latkowski, R., Mikołajczyk, M.: Data decomposition and decision rule joining for classification of data with missing values. In: Tsumoto, S., Słowiński, R., Komorowski, J., Grzymała-Busse, J.W. (eds.) RSCTC 2004. LNCS (LNAI), vol. 3066, pp. 254–263. Springer, Heidelberg (2004)
54. Leung, Y., Wu, W., Zhang, W.: Knowledge acquisition in incomplete information systems: a rough set approach. Eur. J. Ope. Res. **168**, 164–180 (2006)
55. Li, D., Deogun, I., Spaulding, W., Shuart, B.: Dealing with missing data: algorithms based on fuzzy set and rough set theories. Trans. Rough Sets **4**, 37–57 (2005)
56. Li, H., Yao, Y., Zhou, X., Huang, B.: Two-phase rule induction from incomplete data. In: Wang, G., Li, T., Grzymala-Busse, J.W., Miao, D., Skowron, A., Yao, Y. (eds.) RSKT 2008. LNCS (LNAI), vol. 5009, pp. 47–54. Springer, Heidelberg (2008)
57. Li, T., Ruan, D., Geert, W., Song, J., Xu, Y.: A rough sets based characteristic relation approach for dynamic attribute generalization in data mining. Knowl. Based Syst. **20**(5), 485–494 (2007)

58. Li, T., Ruan, D., Song, J.: Dynamic maintenance of decision rules with rough set under characteristic relation. In: Proceedings of the International Conference on Wireless Communications, Networking and Mobile Computing, pp. 3713–3716 (2007)
59. Meng, Z., Shi, Z.: A fast approach to attribute reduction in incomplete decision systems with tolerance relation-based rough sets. Inf. Sci. **179**, 2774–2793 (2009)
60. Meng, Z., Shi, Z.: Extended rough set-based attribute reduction in inconsistent incomplete decision systems. Inf. Sci. **204**, 44–69 (2012)
61. Nakata, M., Sakai, H.: Rough sets handling missing values probabilistically interpreted. In: Slezak, D., Wang, G., Szczuka, M.S., Düntsch, I., Yao, Y. (eds.) RSFDGrC 2005. LNCS (LNAI), vol. 3641, pp. 325–334. Springer, Heidelberg (2005)
62. Nakata, M., Sakai, H.: Applying rough sets to information tables containing missing values. In: Proceedings of the 39-th International Symposium on Multiple-Valued Logic, pp. 286–291 (2009)
63. Pawlak, Z.: Rough sets. Int. J. Comput. Inf. Sci. **11**, 341–356 (1982)
64. Pawlak, Z.: Rough Sets: Theoretical Aspects of Reasoning about Data. Kluwer Academic Publishers, Dordrecht (1991)
65. Pawlak, Z., Grzymala-Busse, J.W., Slowinski, R., Ziarko, W.: Rough sets. Commun. ACM **38**, 89–95 (1995)
66. Pawlak, Z., Skowron, A.: Rough sets: some extensions. Inf. Sci. **177**, 28–40 (2007)
67. Pawlak, Z., Wong, S.K.M., Ziarko, W.: Rough sets: probabilistic versus deterministic approach. Int. J. Man-Mach. Stud. **29**, 81–95 (1988)
68. Peng, H., Zhu, S.: Handling of incomplete data sets using ICA and SOM in data mining. Neural Comput. Appl. **16**, 167–172 (2007)
69. Qi, Y.S., Wei, L., Sun, H.J., Song, Y.Q., Sun, Q.S.: Characteristic relations in generalized incomplete information systems. In: International Workshop on Knowledge Discovery and Data Mining, pp. 519–523 (2008)
70. Qi, Y.S., Sun, H., Yang, X.B., Song, Y., Sun, Q.: Approach to approximate distribution reduct in incomplete ordered decision system. J. Inf. Comput. Sci. **3**, 189–198 (2008)
71. Qian, Y., Dang, C., Liang, J., Zhang, H., Ma, J.: On the evaluation of the decision performance of an incomplete decision table. Data Knowl. Eng. **65**, 373–400 (2008)
72. Qian, Y., Li, D., Wang, F., Ma, N.: Approximation reduction in inconsistent incomplete decision tables. Knowl. Based Syst. **23**, 427–433 (2010)
73. Ślęzak, D., Ziarko, W.: The investigation of the bayesian rough set model. Int. J. Approx. Reason. **40**, 81–91 (2005)
74. Song, J., Li, T., Ruan, D.: A new decision tree construction using the cloud transform and rough sets. In: Wang, G., Li, T., Grzymala-Busse, J.W., Miao, D., Skowron, A., Yao, Y. (eds.) RSKT 2008. LNCS (LNAI), vol. 5009, pp. 524–531. Springer, Heidelberg (2008)
75. Song, J., Li, T., Wang, Y., Qi, J.: Decision tree construction based on rough set theory under characteristic relation. In: Proceedings of the ISKE 2007, the 2-nd International Conference on Intelligent Systems and Knowledge Engineering Conference, pp. 788–792 (2007)
76. Stefanowski, J., Tsoukiàs, A.: On the extension of rough sets under incomplete information. In: Zhong, N., Skowron, A., Ohsuga, S. (eds.) RSFDGrC 1999. LNCS (LNAI), vol. 1711, pp. 73–82. Springer, Heidelberg (1999)
77. Stefanowski, J., Tsoukias, A.: Incomplete information tables and rough classification. Computat. Intell. **17**(3), 545–566 (2001)

78. Wang, G.: Extension of rough set under incomplete information systems. In: Proceedings of the IEEE International Conference on Fuzzy Systems, pp. 1098–1103 (2002)
79. Wong, S.K.M., Ziarko, W.: INFER–an adaptive decision support system based on the probabilistic approximate classification. In: Proceedings of the 6-th International Workshop on Expert Systems and their Applications, pp. 713–726 (1986)
80. Yang, X., Yang, J.: Incomplete Information System and Rough Set Theory: Model and Attribute Reduction. Springer, Heidelberg (2012)
81. Yang, X., Zhang, M., Dou, H., Yang, J.: Neighborhood systems-based rough sets in incomplete information systems. Knowl. Based Syst. **24**, 858–867 (2011)
82. Yao, Y.Y.: Probabilistic rough set approximations. Int. J. Approx. Reason. **49**, 255–271 (2008)
83. Yao, Y.Y., Wong, S.K.M.: A decision theoretic framework for approximate concepts. Int. J. Man Mach. Stud. **37**, 793–809 (1992)
84. Yao, Y.Y., Wong, S.K.M., Lingras, P.: A decision-theoretic rough set model. In: Ras, Z.W., Zemankova, M., Emrich, M.L. (eds.) Methodologies for Intelligent Systems, North-Holland, pp. 388–395 (1990)
85. Ziarko, W.: Variable precision rough set model. J. Comput. Syst. Sci. **46**(1), 39–59 (1993)
86. Ziarko, W.: Probabilistic approach to rough sets. Int. J. Approx. Reason. **49**, 272–284 (2008)

PICKT: A Solution for Big Data Analysis

Tianrui Li[✉], Chuan Luo, Hongmei Chen, and Junbo Zhang

School of Information Science and Technology, Southwest Jiaotong University,
Chengdu 610031, China
{trli,hmchen}@swjtu.edu.cn, luochuan@my.swjtu.edu.cn,
JunboZhang86@163.com

Abstract. Emerging information technologies and application patterns in modern information society, e.g., Internet, Internet of Things, Cloud Computing and Tri-network Convergence, are growing in an amazing speed which causes the advent of the era of Big Data. Big Data is often described by using five V's: Volume, Velocity, Variety, Value and Veracity. Exploring efficient and effective data mining and knowledge discovery methods to handle Big Data with rich information has become an important research topic in the area of information science. This paper focuses on the introduction of our solution, PICKT, on big data analysis based on the theories of granular computing and rough sets, where P refers to parallel/cloud computing for the Volume, I refers to incremental learning for the Velocity, C refers to composite rough set model for the Variety, K refers to knowledge discovery for the Value and T refers to three-way decisions for the Veracity of Big Data.

Keywords: Big data · Rough set · Granular computing · Incremental learning

1 Introduction

Enormous amounts of data are generated every day with the amazing spread of computers and sensors in a range of domains, *e.g.*, search engines, social media, health care organizations, insurance companies, financial industry, retail, and many others [1,2]. Now we are in the era of Big Data, which is characterized by 5Vs, *i.e.*, Volume, Velocity, Variety, Value and Variety. Volume means the amount of data that needs to be handled is very huge. Velocity means that the speed of data processing is very high. Variety means that the data is varied in nature and there are many different types of data that need to be properly combined to make the most of the analysis. Value means high yield will be achieved if the big data is processed correctly and accurately. Veracity means the inherent trustworthiness of data, namely, the uncertainty about the consistency or completeness of data and other ambiguities [3]. Big data is currently a fast growing field both from an application and from a research point of view.

Granular computing is a newly computing paradigm in the realm of computational intelligence, which offers a multi-view model to describe and process

D. Ciucci et al. (Eds.): RSKT 2015, LNAI 9436, pp. 15–25, 2015.
DOI: 10.1007/978-3-319-25754-9_2

uncertain, imprecise, and incomplete information. According to the selection of a suitable granule and the granularity conversion in the complex problem space, granular computing provides a granular-based approximate solution for mining big data [4]. There are two key issues in granular computing. One is how to select the right level of granularity in problem solving. The story of "The Blind Men and the Elephant" tells us the importance of how to select the right level of granularity. Like an adage says, "One person standing in the right place at the right time possesses the power to change the world." Here is an example to illustrate the importance of the selection of the right level of granularity. The length granularity could be Kilometer, Meter, Centimeter, Micron, ... If you are asked to use the ruler (30 cm) to measure the length of one book, it surely does the job. If you are asked to measure the distance across a desk, you can simply use the ruler. You may have to pick it up and move it several times, but it still does the job. But what if you are asked to measure the distance from one side of campus to another side or the diameter of a cell? The ruler is NOT the right tool for that job. The other one is how to change granularity efficiently in problem solving. Granularity variation often appears when the data evolves with time in real situations, *e.g.*, the variation of attributes, objects or attributes' values in information systems. If you stay in the wrong granularity, like using the ruler to measure the diameter of a cell, you may not easily obtain the solutions of problems. Then how you can change the granularity efficiently by the previous knowledge to get the solutions of problems becomes an important issue in knowledge discovery.

This paper aims to introduce our solution, PICKT (see Fig. 1), on big data analysis based on the theories of granular computing and rough sets, where P refers to parallel/cloud computing for the Volume, I refers to incremental learning for the Velocity, C refers to composite rough set model for the Variety, K refers to knowledge discovery for the Value and T refers to three-way decisions for the Veracity of Big Data. The PICKT (German) means the peck (English). We compare the process of big data analysis to the activity that a woodpecker searches food in the forest. Thus, we use PICKT as a short name to represent our solution for big data analysis.

2   Parallel Computing for the Volume

Volumes of data have increased exponentially due to ubiquitous information-sensing mobile devices, remote sensing, software logs, cameras and wireless sensor networks as well as the increasing capacity to store information [5]. It is not uncommon for people to deal with petabytes of data, and the analysis is typically performed over the entire data set, not just a sample, which has become a huge challenge not only in science but also in business use [6]. As a powerful mathematical tool that can be used to process inconsistent information, Rough Set Theory (RST) has been used successfully for discovering patterns in a given data set through a pair of concepts, namely, the upper and lower approximations [7]. The efficient computation of these approximations is vital to improve

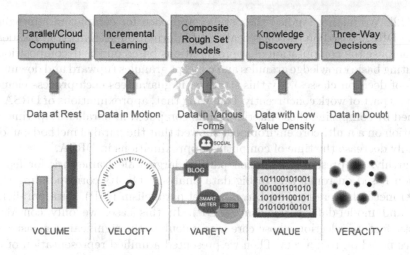

Fig. 1. PICKT—A solution for Big Data Analysis

the performance and obtain the benefits that rough sets can provide for data mining and related tasks.

In the big data environment, we firstly presented a parallel method for computing rough set approximations by the parallel programming model, MapReduce, which is a framework for processing huge data sets using a large number of computers (nodes) [8]. The algorithms corresponding to the parallel method based on MapReduce were designed. Speedup, scaleup and sizeup were used to evaluate the performances of the proposed parallel algorithms. Experiments on the real and synthetic data sets validated that they could deal with large data sets effectively in data mining. Following that, we compared the parallel algorithms of computing approximations of rough sets and knowledge acquisition on three representative MapReduce runtime systems, e.g., Hadoop, Phoenix and Twister [9]. Also, we presented a parallel and incremental algorithm for updating knowledge based on RST in cloud platform [10].

Secondly, we developed three parallel strategies based on MapReduce to compute approximations by introducing the matrix representations of lower and upper approximations to process large-scale incomplete data with RST [11]. The first strategy is that a Sub-Merge operation is used to reduce the space requirement and accelerate the process of merging the relation matrices. The second strategy is that an incremental method is applied to the process of merging the relation matrices and the computational process is efficiently accelerated since the relation matrix is updated in parallel and incrementally. The third strategy is that a sparse matrix method is employed to optimize the proposed matrix-based method and the performance of the algorithm is further improved.

As an extension of classical RST, Dominance-based Rough Set Approach (DRSA) can process information with preference-ordered attribute domain and has been successfully applied in multi-criteria decision analysis and other related

tasks [12]. We further studied the parallel approach for computing approximations in DRSA based on the classification of information systems [13]. Its idea is
to decompose the computation of approximations based on the decomposition of
computing basic knowledge granules and concept granules (upward and downward
unions of decision classes). By this strategy, it guarantees each process element
can do its part of work concurrently. Following that, approximations of DRSA are
obtained by composing those interim results computed in parallel. Experimental
evaluation on a multi-core environment showed that the parallel method can dra-
matically decrease the time of computing approximations in DRSA.

Fourthly, we presented a unified parallel large-scale framework for feature
selection based on rough sets in big data analysis [14]. Its corresponding three
parallel methods were proposed, *e.g.*, model parallelism (MP), data parallelism
(DP), and model-data parallelism (MDP). In this case, we only considered
heuristic feature selection whose core is to calculate the significance measures of
features based on rough sets. Then we presented a unified representation of feature evaluation functions. Furthermore, we showed the divide-and-conquer methods for four representative evaluation functions, and designed MapReduce-based
and Spark-based Parallel Large-scale Attribute Reduction (PLAR) algorithms
[15]. Subsequently, we accelerated the process of feature selection by introducing
granular computing (GrC) theory and presented PLAR-MDP algorithm by combining PLAR with MDP. Finally, we verified the effectiveness of the proposed
algorithms by experimental evaluation on the big data computing platforms,
e.g., Hadoop and Spark, using UCI datasets and astronomical datasets. It was
also shown that PLAR-MDP can maximize the performance of data processing
by combining with MP, DP and GrC methods.

3   Incremental Learning for the Velocity

In real-life applications, data in information systems will evolve over time in
the sense that (1) records (objects) are continuously being added or deleted, (2)
features (attributes) are frequently being added or removed, and (3) criteria values (attribute values) are continually being updated over time. Tracking these
dynamic changes may contribute to improve the ability of decision making. Incremental learning is regarded as the process of using previously acquired learning
results to facilitate knowledge maintenance by dealing with the new added-in
data set without re-implementing the original data mining algorithm, which
leads to considerable reductions in execution time when maintaining knowledge
from the dynamic database [16]. Based on GrC and RST, there have been many
attempts at introducing incremental learning techniques for updating knowledge
in a dynamic data environment.

For the variation of objects, we presented several incremental approaches for
updating approximations and knowledge under the extended models of rough
sets. Firstly, we proposed a novel incremental model and approach to induce
interesting knowledge under both accuracy and coverage from dynamic information systems when the object set varies with time [17]. Its feature is to calculate

both accuracy and coverage by matrixes to obtain the interesting knowledge. Furthermore, we presented an optimization incremental approach for inducing interesting knowledge in business intelligent information systems [18].

Variable Precision Rough Set model (VPRS) has been used to solve noise data problems with great success in many applications [19]. Then, we proposed incremental methods for updating approximations under VPRS when the information system is updated by inserting or deleting an object through investigating the property of information granulation and approximations under the dynamic environment. The variation of an attribute's domain was also considered to perform incremental updating for approximations under VPRS [20].

To handle information with preference-ordered attribute domain, we further proposed an incremental approach for updating the approximations of DRSA under the variation of the object set [21]. The idea is that we only focus on the objects whose P-generalized decision is updated after the addition or deletion of the object. Then, approximations of DRSA can be updated incrementally instead of re-computing from scratch according to the alteration of P-dominated sets and P-dominating sets of those objects. We also extended the proposed method to update approximations based on set-valued ordered information systems in terms of inserting or deleting an object [22].

However, the previous methods can only be used to deal with the categorical data. We presented a new dynamic method for incremental updating knowledge based on the neighborhood rough sets to deal with dynamic numerical data when multiple objects enter into and get out of the information system simultaneously [23]. Following that, we introduced an incremental algorithm for updating approximations of rough fuzzy sets under the variation of the object set in fuzzy decision systems [24]. Moreover, we developed an algorithm to maintain approximations incrementally when adding or removing some objects in composite information systems [25].

For the variation of attributes, based on Chan's work to update approximations by the lower and upper boundary sets [26], we presented a method for updating approximations in incomplete information systems under the characteristic relation-based rough sets when an attribute set varies over time [27]. Then, we proposed an incremental approach for updating approximations in set-valued information systems under the probabilistic neighborhood rough set model [28]. Furthermore, we developed an incremental algorithm for updating approximations under the DRSA by introducing a kind of dominance matrix to update P-dominating sets and P-dominated sets [29]. We also presented incremental approaches for updating approximations in set-valued ordered decision systems by introducing the dominant and dominated matrices with respect to the dominance relation for computing approximations of upward and downward unions of decision classes [30]. We discussed the updating mechanisms for feature selection with the variation of attributes and presented fuzzy rough set approaches for incremental feature selection in hybrid information systems, which consist of different types of data (e.g., boolean, categorical, real-valued and set-valued) [31].

In fact, the ever-changing attribute values may lead some or all of obtained knowledge to failure. For the variation of attribute values, we defined the attribute values' coarsening and refining in information systems as the semantic level changes, and then proposed an incremental algorithm for updating the approximations in case of attribute values' coarsening and refining [32]. Then, we presented a method to dynamically maintain approximations of upward and downward unions when attribute values alter in incomplete ordered decision systems [33]. We also developed incremental algorithms for computing approximations in the set-valued decision systems with the addition and removal of criteria values [34]. Furthermore, we presented matrix-based incremental algorithms for updating decision rules when coarsening or refining attribute values [35]. We have developed "iRoughSet", a toolkit for incremental learning based on RST [36] based on the above research works.

4   Composite Rough Set Model for the Variety

In many real-world applications, attributes in information systems generally have different types, *e.g.*, categorical ones, numerical ones, set-valued ones, interval-valued ones and missing ones. However, most of the previous rough set models fail to deal with information systems with more than two different types of attributes. We firstly defined composite information systems. Then, we introduced an extension of rough set model, composite rough set (CRS), which may deal with multiple different types of attributes and provide a novel approach for complex data fusion. Furthermore, we introduced a matrix-based method to compute the CRS approximations, positive, boundary and negative regions intuitively from composite information systems, which may help to combine different types of data to make the most of the analysis [37]. Finally, we developed parallel algorithms based on Single-GPU and Multi-GPU for computing CRS approximations and carried out experiments on UCI and user-defined datasets to verify the effectiveness of the proposed algorithms [38].

However, CRS cannot be employed directly to mine the useful knowledge hidden in the Incomplete Composite Information System (ICIS). We then introduced a new composite rough set model based on a composite binary relation by the intersection of tolerance and dominance relations. This model can be applied to the ICIS directly when all criteria and attributes are considered at the same time [25]. In addition, considering the preference relations between objects may be different under different criteria in the ordered information system, we proposed a composite dominance relation based on the integration of multiple dominance relations and then presented a composite DRSA model for the composite ordered information system [39].

5   Knowledge Discovery for the Value

Knowledge Discovery in Database (KDD) is regarded as the process of discovering previously unknown and potentially interesting patterns in large databases, which includes the storage and accessing of such data, scaling of algorithms to massive data sets, interpretation and visualization of results, and the modeling and support of the overall human machine interaction [40]. However, the previous KDD process models do not pay attention to the data collection, which is vital to apply the KDD techniques in some real applications, *e.g.*, information security and medical treatment. Therefore, we proposed an extension of the KDD process model (see Fig. 2) of Fayyad et al. which incorporates the step of data collection that employs techniques to collect data for real applications according to the current discovery task and previous mining results [41]. The cost of data collection is generally huge, *e.g.*, making different kinds of investigations for nuclear safeguards information management. By applying the proposed model, we can not only reduce the cost including storage, preprocessing, *etc.*, but also directly enhance the ability and efficiency of knowledge discovery, *e.g.*, reducing the time required to execute a discovery. Then, we illustrated a case study by the reduct method in RST to show in what situation it can be used in practice and how it can support KDD applications better. We have developed "RSHadoop", a rough set toolkit for massive data analysis on Hadoop for large-scale knowledge discovery based on RST [42].

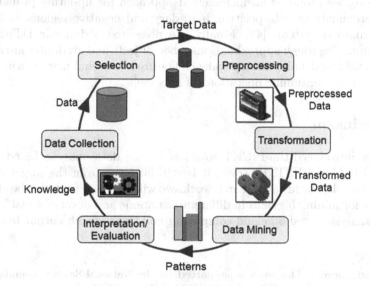

Fig. 2. An updated process model of KDD

6   Three-Way Decisions for the Veracity

In real application situations, there are many biases, noise or abnormality in data. Some people believe that veracity in data analysis is the biggest challenge when compared to things like volume and velocity. The theory of three-way decisions proposed by Yao is an effective mathematical tool to deal with the veracity [43]. Decision-theoretic rough set (DTRS) provides a three-way decision framework for approximating a target concept, with an error-tolerance capability to handle uncertainty problems by using a pair of thresholds with probability estimation. The three-way decision rules of acceptance, rejection and deferment decisions can be derived directly from the three regions implied by rough approximations [44].

In a dynamic data environment, we considered the situation of the variation of objects. With the addition of different objects, different updating patterns of conditional probability will appear, and then the decision regions implied by rough approximations will change dynamically in a non-monotonic way. To address this problem, we presented the incremental estimating methods for the conditional probability in the DTRS [45]. Then, we designed the incremental mechanisms for updating rough approximations with an increase or a decrease of conditional probability. On the other hand, since the rough approximations will be varied inevitably with the data changed continuously, the induced three-way decision rules will be altered accordingly. We developed four different maintenance strategies of the three-way decision rules which can update decision rules by modifying partial original rule sets without forgetting prior knowledge, thereby avoiding re-computing decision rules from the very beginning [46,47]. Furthermore, we proposed an incremental approach for updating probabilistic rough approximations, *i.e.*, positive, boundary and negative regions, in incomplete information systems [48]. Finally, we presented a dynamic DTRS approach for updating rough approximations when objects and attributes are altered simultaneously by defining the equivalence feature vector and matrix, which may avoid some time-consuming operations of sets [49].

7   Conclusion

This paper introduced the PICKT solution for big data analysis based on the theories of GrC and RST. However, it is still far away from the target of fully using big data. In our future research work, we will continue to study the efficient algorithms for mining big data in different situations and develop a platform for big data analysis based on cloud computing and the research output from GrC and RST.

Acknowledgments. This work is supported by the National Science Foundation of China (Nos. 61175047 and 71201133) and NSAF (No. U1230117).

References

1. Li, T.R., Lu, J., Luis, M.L.: Preface: Intelligent techniques for data science. Int. J. Intel. Syst. **30**(8), 851–853 (2015)
2. Dealing with data: Science **331**(6018), 639–806 (2011)
3. Jin, X.L., Wah, B.W., Cheng, X.Q., Wang, Y.Z.: Significance and challenges of big data research. Big Data Res. **2**(2), 59–64 (2015)
4. Yao, Y.Y.: Granular computing: past, present and future. In: Proceedings of GrC 2008, pp. 80–85 (2008)
5. Big data. Nature **455**(7209), 1–136 (2008)

6. Wu, X.D., Zhu, X.Q., Wu, G.Q., Ding, W.: Data mining with big data. IEEE Trans. Knowl. Data Eng. **26**(1), 97–107 (2014)

7. Pawlak, Z., Skowron, A.: Rudiments of rough sets. Inf. Sci. **177**, 3–27 (2007)

8. Zhang, J.B., Li, T.R., Ruan, D., Gao, Z.Z., Zhao, C.B.: A parallel method for computing rough set approximations. Inf. Sci. **194**, 209–223 (2012)

9. Zhang, J.B., Wong, J.S., Li, T.R., Pan, Y.: A comparison of parallel large-scale knowledge acquisition using rough set theory on different MapReduce runtime systems. Int. J. Approx. Reasoning **55**(3), 896–907 (2014)

10. Zhang, J.B., Li, T.R., Pan, Y., Luo, C., Teng, F.: Parallel and incremental algorithm for knowledge update based on rough sets in cloud platform. J. Softw. **26**(5), 1064–1078 (2015). (in Chinese)

11. Zhang, J.B., Wong, J.S., Pan, Y., Li, T.R.: A parallel matrix-based method for computing approximations in incomplete information systems. IEEE Trans. Knowl. Data Eng. **27**(2), 326–339 (2015)

12. Greco, S., Matarazzo, B., Slowinski, R.: Rough set theory for multicriteria decision analysis. Eur. J. Oper. Res. **129**, 1–47 (2001)

13. Li, S.Y., Li, T.R., Zhang, Z.X., Chen, H.M., Zhang, J.B.: Parallel computing of approximations in dominance-based rough sets approach. Knowledge-Based Systems (in Press, 2015)

14. Zhang, J.: Research on efficient feature selection and learning algorithms for big data. Ph.D. dissertation, Southwest Jiaotong University (2015)

15. Zhang, J.B., Li, T.R., Pan, Y.: PLAR: Parallel large-scale attribute reduction on cloud systems. In: Proceedings of PDCAT 2013, pp. 184–191 (2013)

16. Raghavan, V., Hafez, A.: Dynamic data mining. In: Proceedings of the IEA of AI & ES, pp. 220–229 (2000)

17. Liu, D., Li, T.R., Ruan, D.: An incremental approach for inducing knowledge from dynamic information systems. Fundamenta Informaticae **94**(2), 245–260 (2009)

18. Liu, D., Li, T.R., Ruan, D., Zhang, J.B.: Incremental learning optimization on knowledge discovery in dynamic business intelligent systems. J. Glob. Optim. **51**(2), 325–344 (2011)

19. Ziarko, W.: Variable precision rough set model. J. Comput. Syst. Sci. **46**(1), 39–59 (1993)

20. Chen, H.M., Li, T.R., Ruan, D., Lin, J.H., Hu, C.X.: A rough-set based incremental approach for updating approximations under dynamic maintenance environments. IEEE Trans. Knowl. Data Eng. **25**(2), 274–284 (2013)

21. Li, S.Y., Li, T.R., Liu, D.: Dynamic maintenance of approximations in dominance-based rough set approach under the variation of the object set. Int. J. Intel. Syst. **28**, 729–751 (2013)

22. Luo, C., Li, T.R., Chen, H.M., Liu, D.: Incremental approaches for updating approximations in set-valued ordered information systems. Knowl.-Based Syst. **50**, 218–233 (2013)

23. Zhang, J.B., Li, T.R., Ruan, D., Liu, D.: Neighborhood rough sets for dynamic data mining. Int. J. Intel. Syst. **27**, 317–342 (2012)

24. Zeng, A.P., Li, T.R., Zhang, J.B., Chen, H.M.: Incremental maintenance of rough fuzzy set approximations under the variation of object set. Fundamenta Informaticae **132**(3), 401–422 (2014)

25. Li, S.Y., Li, T.R., Hu, J.: Update of approximations in composite information systems. Knowl.-Based Syst. **83**, 138–148 (2015)

26. Chan, C.: A rough set approach to attribute generalization in data mining. Inf. Sci. **107**(1–4), 169–176 (1998)

27. Li, T.R., Ruan, D., Geert, W.: A rough sets based characteristic relation approach for dynamic attribute generalization in data mining. Knowl.-Based Syst. **20**(5), 485–494 (2007)

28. Zhang, J.B., Li, T.R., Ruan, D.: Rough sets based matrix approaches with dynamic attribute variation in set-valued information systems. Int. J. Approx. Reasoning **53**, 620–635 (2012)

29. Li, S.Y., Li, T.R., Liu, D.: Incremental updating approximations in dominance-based roughsets approach under the variation of the attirbute set. Knowl.-Based Syst. **40**, 17–26 (2013)

30. Luo, C., Li, T.R., Chen, H.M.: Dynamic maintenance of approximations in set-valued ordered decision systems under the attribute generalization. Inf. Sci. **257**, 210–228 (2014)

31. Zeng, A.P., Li, T.R., Liu, D., Zhang, J.B., Chen, H.M.: A fuzzy rough set approach for incremental feature selection on hybrid information systems. Fuzzy Sets Syst. **258**, 39–60 (2015)

32. Chen, H.M., Li, T.R., Qiao, S.J., Ruan, D.: A rough set based dynamic maintenance approach for approximations in coarsening and refining attribute values. Int. J. Intel. Syst. **25**(10), 1005–1026 (2010)

33. Chen, H.M., Li, T.R., Ruan, D.: Maintenance of approximations in incomplete ordered decision systems while attribute values coarsening or refining. Knowl.-Based Syst. **31**, 140–161 (2012)

34. Luo, C., Li, T.R., Chen, H.M., Lu, L.X.: Fast algorithms for computing rough approximations in set-valued decision systems while updating criteria values. Inf. Sci. **299**, 221–242 (2015)

35. Chen, H.M., Li, T.R., Luo, C., Horng, S.J., Wang, G.Y.: A rough set-based method for updating decision rules on attribute values' coarsening and refining. IEEE Trans. Knowl. Data Eng. **26**(12), 2886–2899 (2014)

36. http://sist.swjtu.edu.cn:8080/ccit/project/iroughset.html

37. Zhang, J.B., Li, T.R., Chen, H.M.: Composite rough sets for dynamic data mining. Inf. Sci. **257**, 81–100 (2014)

38. Zhang, J., Zhu, Y., Pan, Y., Li, T.: A parallel implementation of computing composite rough set approximations on GPUs. In: Lingras, P., Wolski, M., Cornelis, C., Mitra, S., Wasilewski, P. (eds.) RSKT 2013. LNCS, vol. 8171, pp. 240–250. Springer, Heidelberg (2013)

39. Luo, C., Li, T.R., Chen, H.M., Zhang, J.B.: Composite ordered information systems. J. Chin. Comput. Syst. **35**(11), 2523–2527 (2014). (In Chinese)

40. Fayyad, U.M., Piatetsky-Shapiro, G., Smyth, P.: The KDD process for extracting useful knowledge from volumes of data. Commun. ACM **39**(11), 27–34 (1996)

41. Li, T.R., Ruan, D.: An extended process model of knowledge discovery in database. J. Enterp. Inf. Manage. **20**(2), 169–177 (2007)

42. http://sist.swjtu.edu.cn:8080/ccit/project/rshadoop.html

43. Yao, Y.: An outline of a theory of three-way decisions. In: Yao, J.T., Yang, Y., Słowiński, R., Greco, S., Li, H., Mitra, S., Polkowski, L. (eds.) RSCTC 2012. LNCS, vol. 7413, pp. 1–17. Springer, Heidelberg (2012)

44. Yao, Y.Y.: Three-way decisions with probabilistic rough sets. Inf. Sci. **180**, 341–353 (2010)

45. Luo, C., Li, T.R., Chen, H.M.: Dynamic three-way decisions method based on probabilistic rough sets. In: Yu, H., Wang, G.Y., Li, T.R., Liang, J.Y., Miao, D.Q., Yao, Y.Y. (eds.) Three-Way Decisions: Methods and Practices for Complex Problem Solving. Science Press, Beijing (2015). (in Chinese)

46. Luo, C., Li, T., Chen, H.: Dynamic maintenance of three-way decision rules. In: Miao, D., Pedrycz, W., Slezak, D., Peters, G., Hu, Q., Wang, R. (eds.) RSKT 2014. LNCS, vol. 8818, pp. 801–811. Springer, Heidelberg (2014)
47. Luo C.: Research on approaches for dynamic knowledge acquisition from incomplete data. PhD dissertation, Southwest Jiaotong University (2015) (In Chinese)
48. Luo, C., Li, T.: Incremental three-way decisions with incomplete information. In: Cornelis, C., Kryszkiewicz, M., Ślęzak, D., Ruiz, E.M., Bello, R., Shang, L. (eds.) RSCTC 2014. LNCS, vol. 8536, pp. 128–135. Springer, Heidelberg (2014)
49. Chen, H.M., Li, T.R., Luo, C., Horng, S.J.: Wang G.Y: A decision-theoretic rough setapproach for dynamic data mining. IEEE Trans. Fuzzy Syst. (2015). doi:10.1109/TFUZZ.2014.2387877

A Further Investigation to Relative Reducts of Decision Information Systems

Duoqian Miao[1](✉) and Guangming Lang[1,2]

[1] Department of Computer Science and Technology, Tongji University,
Shanghai 201804, People's Republic of China
{dqmiao,langguangming1984}@tongji.edu.cn
[2] School of Mathematics and Computer Science, Changsha University of Science
and Technology, Changsha 410114, Hunan, People's Republic of China

Abstract. In practical situations, there are many definitions of relative reducts with respect to different criterions, but researchers don't notice their application backgrounds, and less efforts have been done on investigating the relationship between them. In this paper, we first discuss the relationship between these relative reducts and present the generalized relative reduct. Then we investigate the relationship between several discernibility matrixes and present the generalized discernibility matrix and discernibility function. Finally, we employ several examples to illustrate the related results accordingly.

Keywords: Rough sets · Discernibility matrix · Discernibility function · Relative reduct

1 Introduction

Rough set theory [20], the theoretical aspects and applications of which have been intensively investigated, is a useful mathematical tool for knowledge acquisition of information systems. Originally, Pawlak's rough set theory was constructed on the basis of an equivalence relation, and it has been developed by extending the equivalence relation to the tolerance relation, similarity relation and so on. Until now, rough set theory has successfully been applied in many fields such as granular computing, pattern recognition, data mining, and knowledge discovery, with applications increasingly being adopted in the development of rough set theory.

In rough set theory, attribute reduction [1–6,8–17,19,21,22,24,27–37,39–41,43] plays an important role in classification and feature selection, and a reduct is interpreted as a minimal set of attributes that can provide the same descriptive or classification ability as the entire set of attributes. Many attempts have been done on attribute reduction of consistent and inconsistent information systems. On one hand, many concepts of reducts have been proposed for attribute reduction with respect to different criteria. For example, Chen et al. [1] presented the concept of reduct for consistent and inconsistent covering decision

© Springer International Publishing Switzerland 2015
D. Ciucci et al. (Eds.): RSKT 2015, LNAI 9436, pp. 26–38, 2015.
DOI: 10.1007/978-3-319-25754-9_3

information systems with covering rough sets. Jia et al. [6] introduced the generalized attribute reduct of information systems in rough set theory. Kryszkiewicz et al. [7] focused on boundary region reduct of decision information systems. Mi et al. [17] proposed the concepts of β lower and β upper distribution reducts of information systems based on variable precision rough set model. Miao et al. [18] presented relative reducts in consistent and inconsistent decision information systems. Pawlak et al. [20] proposed the concept of positive region reduct of decision information systems. Qian et al. [22] provided the concepts of the lower and upper approximation reducts of decision information systems. Slezak et al. [23] studied decision reduct of decision information systems. Zhao et al. [42] presented a general definition of an attribute reduct. Zhang et al. [40] proposed the concept of assignment reducts and maximum assignment reducts for decision information systems. In practical situations, different reducts are equivalent to each other in consistent decision information systems, but they are not equivalent in inconsistent decision information systems. For example, the boundary region of a consistent decision information system is empty, and the positive region reducts are equivalent to boundary region reducts in consistent decision information systems, but they are not equivalent to each other in inconsistent decision information systems since the boundary regions of inconsistent decision information systems are not empty. Therefore, we only need to consider one of the positive region reducts and boundary region reducts in consistent information systems, and we need to consider the positive region reducts and boundary region reducts simultaneously in inconsistent information systems. But researchers usually utilize the concepts of relative reducts without considering their application backgrounds and many efforts have been paid to useless work. Therefore, it is urgent to discuss the relationship between them.

On the other hand, many methods [1,7,9,11,14,17–19,23,25,26,38,39] have been proposed for constructing relative reducts of decision information systems. For example, Kryszkiewicz et al. [7] provided the notion of discernibility matrix and discernibility function for computing decision reducts. Leung et al. [11] investigated dependence space-based attribute reduction in inconsistent decision information systems. Li et al. [14] studied quick attribute reduction in inconsistent decision information systems. Miao et al. [18] presented the generalized discernibility matrix and discernibility function of three types of relative reducts. Meng et al. [19] studied extended rough set-based attribute reduction in inconsistent incomplete decision systems. Slezak et al. [23] introduced the notion of discernibility matrix and discernibility function for constructing distribution reducts. Skowron et al. [25] proposed the concept of the classical discernibility matrix and discernibility function for constructing relative reducts of decision information systems. Yao et al. [37] investigated discernibility matrix simplification for constructing attribute reducts. Ye et al. [38] also presented the notion of discernibility matrix and discernibility function for calculating positive region reducts. Clearly, the discernibility matrix and discernibility function are effective and feasible to construct attribute reducts of decision information systems. In practice, there are several types of discernibility matrixes and discernibility functions, and we need to propose the generalized discernibility matrix and discernibility function for constructing relative reducts of decision information systems.

The rest of this paper is organized as follows: Sect. 2 briefly reviews the basic concepts of rough set theory. In Sect. 3, we investigate the relationship between relative reducts and present the generalized relative reduct. In Sect. 4, we review several discernibility matrixes and discernibility functions and present the generalized discernibility matrix and discernibility function. We conclude the paper in Sect. 5.

2   Preliminaries

In this section, we review some concepts of rough set theory.

Definition 1. *[20] An information system is a 4-tuple $S = (U, A, V, f)$, where $U = \{x_1 x_2, ..., x_n\}$ is a finite set of objects, A is a finite set of attributes, $V = \{V_a | a \in A\}$, where V_a is the set of values of attribute a, and $card(V_a) > 1$, f is a function from $U \times A$ into V.*

For convenience, we take $S = (U, C \cup D)$ and $f(x, a) = a(x)$ for $x \in U$ and $a \in C \cup D$ in this work, where C and D denote a non-empty finite set of conditional attributes and a non-empty finite set of decision attributes, respectively.

Definition 2. *[20] Let $S = (U, C \cup D)$ be an information system, and $B \subseteq C$. Then an indiscernibility relation $IND(B) \subseteq U \times U$ is defined as $IND(B) = \{(x, y) \in U \times U | \forall a \in B, a(x) = a(y)\}$, where $a(x)$ and $a(y)$ denote the values of objects x and y on a, respectively.*

If $(x, y) \in IND(B)$ for $x, y \in U$, then x and y are indiscernible based on the attribute set B. We can receive a quotient set which is called the family of equivalence classes or the blocks of the universe U by $IND(B)$, denoted as $U/B = \{[x]_B | x \in U\}$. Moreover, if $IND(C) \subseteq IND(D)$, then S is called a consistent information system. Otherwise, it is inconsistent.

Definition 3. *[20] Let $S = (U, C \cup D)$ be an information system, and $B \subseteq C$. For any $X \subseteq U$, the lower and upper approximations of X with respect to B are defined as $\underline{R}_B(X) = \bigcup\{[x]_B | [x]_B \subseteq X\}$ and $\overline{R}_B(X) = \bigcup\{[x]_B | [x]_B \cap X \neq \emptyset\}$.*

By Definition 3, we have the positive region, boundary region and negative region of $X \subseteq U$ as follows: $POS_R(X) = \underline{R}(X), BND_R(X) = \overline{R}(X) - \underline{R}(X)$ and $NEG_R(X) = U - \overline{R}(X)$. Furthermore, we have the positive region and boundary region of D as follows: $\underline{R}(D) = POS_R(D) = \bigcup_{i=1}^{|U/D|} \underline{R}(D_i), \overline{R}(D) = \bigcup_{i=1}^{|U/D|} \overline{R}(D_i)$ and $BND_R(D) = \overline{R}(D) - \underline{R}(D)$, and the Confidence of the rule $[x]_B \rightarrow D_i$ is defined as $Confidence([x]_B \rightarrow D_i) = \frac{|[x]_B \cap D_i|}{|[x]_B|} = P(D_i | [x]_B)$, where $U/D = \{D_i | 1 \leq i \leq r\}$ and $| \bullet |$ denotes the cardinality of $\bullet$.

In a consistent decision information system, the boundary region is empty, but it is not empty in an inconsistent decision information system.

Table 1. An inconsistent decision information system.

	c_1	c_2	c_3	c_4	c_5	c_6	d
x_1	0	0	0	1	1	0	0
x_2	1	0	0	1	1	0	1
x_3	1	1	0	1	1	0	1
x_4	1	1	0	1	1	0	1
x_5	1	1	0	1	1	0	0
x_6	1	1	0	1	0	1	0
x_7	1	1	0	1	0	1	1
x_8	1	1	0	1	0	1	1
x_9	1	1	0	0	0	1	0
x_{10}	1	1	0	0	0	1	1
x_{11}	1	1	0	0	0	1	1
x_{12}	1	1	0	0	0	1	1
x_{13}	1	1	0	0	0	1	1
x_{14}	1	1	1	0	0	1	1
x_{15}	1	1	1	0	0	1	2

Example 1. Let Table 1 be an inconsistent decision information system $S = (U, C \cup \{d\})$, $C = \{c_1, c_2, ..., c_6\}$, $U/d = \{D_1, D_2, D_3\}$, $D_1 = \{x_1, x_5, x_6, x_9\}$, $D_2 = \{x_2, x_3, x_4, x_7, x_8, x_{10}, x_{11}, x_{12}, x_{13}, x_{14}\}$, $D_3 = \{x_{15}\}$. By Definition 3, we have $\overline{R}(D) = \{x_1, x_2, x_3, x_4, x_5, x_6, x_7, x_8, x_9, x_{10}, ..., x_{13}, x_{14}, x_{15}\}$ and $\underline{R}(D) = \{x_1, x_2\}$. Therefore, we obtain $BND_R(D) = \{x_3, x_4, x_5, x_6, x_7, x_8, x_9, x_{10}, ..., x_{13}, x_{14}, x_{15}\}$ and the partition $\{\{x_3, x_4, x_5\}, \{x_6, x_7, x_8\}, \{x_9, x_{10}, ..., x_{13}\}, \{x_{14}, x_{15}\}\}$ of $BND_R(D)$.

Definition 4. *[18] Let $S = (U, C \cup D)$ be an information system, and $B \subseteq C$. Then the relative indiscernibility and discernibility relations defined by B with respect to D are defined as follows:*

$$IND(B|D) = \{(x, y) \in U \times U | \forall c \in B, c(x) = c(y) \vee d(x) = d(y)\};$$
$$DIS(B|D) = \{(x, y) \in U \times U | \exists c \in B, c(x) \neq c(y) \wedge d(x) \neq d(y)\},$$

where $c(x)$ and $c(y)$ denote the values of objects x and y on c, respectively.

By Definition 4, the relative indiscernibility relation is reflexive, symmetric, but not transitive; the relative discernibility relation is irreflexive, symmetric, but not transitive. By Definitions 2 and 4, we observe that if $(x, y) \in IND(C)$ and $(x, y) \in IND(D)$ for $(x, y) \in U \times U$, then S is inconsistent; S is consistent, otherwise. In other words, all objects in an equivalence class $[x]_C$ satisfying one and only one class in a consistent decision information system; while all objects in an equivalence class $[x]_C$ may satisfy different classes in an inconsistent decision information system.

Definition 5. *[18] Let $S = (U, C \cup D)$ be an information system, $A \subseteq C$, and $\tau(x|A) = \{d(y)|y \in [x]_A\}$ for any $x \in U$. Then the set of generalized decisions $\Delta(A)$ of all objects in U is denoted as $\Delta(A) = \{\tau(x_1|A), \tau(x_2|A), ..., \tau(x_n|A)\}$.*

If there exists x such that $\tau(x|C) > 1$, then the decision information system is inconsistent; otherwise, it is inconsistent.

Definition 6. *[32] Let $S = (U, C \cup D)$ be a decision information system, $U/P = \{X_1, X_2, ..., X_n\}$, $U/Q = \{Y_1, Y_2, ..., Y_m\}$, $p(Y_i) = \frac{|Y_i|}{|U|}$, and $p(Y_i|X_i) = \frac{|Y_i \cap X_i|}{|X_i|}$. Then the entropy $H(Q)$ and conditional entropy $H(Q|P)$ are defined as follows;*

$$H(Q) = -\sum_{j=1}^{m} p(Y_i) log(p(Y_i)), H(Q|P) = -\sum_{i=1}^{n} p(X_i) \sum_{j=1}^{m} p(Y_j|X_i) log(p(Y_j|X_i)).$$

The entropy $H(Q)$ can be interpreted as a measure of the information content of the uncertainty about knowledge Q, which reaches a maximum value $log(|U|)$ and the minimum value 0 when the knowledge Q becomes finest and the distribution of the knowledge Q focuses on a particular value, respectively.

Definition 7. *[32] Let $S = (U, C \cup D)$ be a decision information system. Then the mutual information $I(P; Q)$ is defined as $I(P; Q) = H(Q) - H(Q|P)$.*

The mutual information $I(P; Q)$ measures the decrease of uncertainty about Q caused by P. Furthermore, the amount of information contained in P about itself is $H(P)$ obviously.

3 Relative Reducts

In this section, we investigate the relationship between relative reducts.

Definition 8. *Let $S = (U, C \cup D)$ be a decision information system, and $B \subseteq C$. Then*

(1) *If $POS_B(D) = POS_C(D)$, then B is called a positive consistent set of C with respect to D.*
(2) *If $IND(B|D) = IND(C|D)$, then B is called a boundary consistent set of C with respect to D.*
(3) *If $\tau(x|B) = \tau(x|C)$ for any $x \in U$, then B is called a decision consistent set of C with respect to D.*
(4) *If $I(B; D) = I(C; D)$, then B is called a mutual information consistent set of C with respect to D.*
(5) *If $\mu_B(x) = \mu_C(x)$ for any $x \in U$, then B is called a distribution consistent set of C with respect to D, where $\mu_B(x) = (\frac{|[x]_B \cap D_1|}{|[x]_B|}, \frac{|[x]_B \cap D_2|}{|[x]_B|}, ..., \frac{|[x]_B \cap D_{|U/D|}|}{|[x]_B|})$.*
(6) *If $\gamma_B(x) = \gamma_C(x)$ for any $x \in U$, then B is called a maximum distribution consistent set of C with respect to D, where $\gamma_B(x) = max\{\frac{|[x]_B \cap D_i|}{|[x]_B|}|1 \leq i \leq |U/D|\}$.*

Definition 9. *[7, 20, 23, 32, 39] Let $S = (U, C \cup D)$ be a decision information system, and $B \subseteq C$. Then*

(1) *B is called a positive region reduct of C with respect to D if*
(I) $POS_B(D) = POS_C(D)$; (II) $POS_{B'}(D) \neq POS_B(D)$ for any $B' \subset B$.
(2) *B is called a boundary region reduct of C with respect to D if*
(I) $IND(B|D) = IND(C|D)$; (II) $\forall B' \subset B, IND(B'|D) \neq IND(B|D)$.
(3) *B is called a decision reduct of C with respect to D if*
(I) $\forall x \in U, \tau(x|B) = \tau(x|C)$; (II) $\forall B' \subset B, \forall x \in U, \tau(x|B') \neq \tau(x|B)$.
(4) *B is called a mutual information reduct of C with respect to D if*
(I) $I(B; D) = I(C; D)$; (II) $\forall B' \subset B, I(B'; D) \neq I(B; D)$.
(5) *B is called a distribution reduct of C with respect to D if*
(I) $\forall x \in U, \mu_B(x) = \mu_C(x)$; (II) $\forall B' \subset B, \forall x \in U, \mu_{B'}(x) \neq \mu_B(x)$.
(6) *B is called a maximum distribution reduct of C with respect to D if*
(I) $\forall x \in U, \gamma_B(x) = \gamma_C(x)$; (II) $\forall B' \subset B, \forall x \in U, \gamma_{B'}(x) \neq \gamma_B(x)$.

In consistent information systems, the positive region reducts which are the minimum attribute sets remaining the positive regions are equivalent to the boundary region reducts which are the minimum attribute sets keeping the boundary region, but they are not equivalent to each other in inconsistent information systems; the decision reduct is the minimum attribute set remaining the decision sets of equivalence classes; the mutual information reduct is the minimum attribute set keeping the mutual information; the distribution reduct is the minimum attribute set remaining the distributions of all objects with respect to the decision equivalence classes; the maximum distribution reduct is the minimum attribute set keeping the maximum distributions of all objects with respect to the decision equivalence classes. Therefore, we should consider the corresponding concepts of reducts for solving the problems in practical situations.

Theorem 1. *Let $S = (U, C \cup D)$ be a decision information system, and $B \subseteq C$. Then*

(1) *If $IND(B|D) = IND(C|D)$, then $\tau(x|B) = \tau(x|C)$ for any $x \in U$.*
(2) *If $I(B; D) = I(C; D)$, then $\mu_B(x) = \mu_C(x)$ for any $x \in U$.*
(3) *If $\mu_B(x) = \mu_C(x)$ for any $x \in U$, then $I(B; D) = I(C; D)$.*
(4) *If $\mu_B(x) = \mu_C(x)$ for any $x \in U$, then $\tau(x|B) = \tau(x|C)$ for any $x \in U$.*
(5) *If $\tau(x|B) = \tau(x|C)$ for any $x \in U$, then $POS_B(D) = POS_C(D)$.*
(6) *If $IND(B|D) = IND(C|D)$, then $\tau(x|B) = \tau(x|C)$ for any $x \in U$.*
(7) *If $\mu_B(x) = \mu_C(x)$ for any $x \in U$, then $\gamma_B(x) = \gamma_C(x)$ for any $x \in U$.*

Proof: The results can be proved by Definitions 8 and 9. □

By Theorem 1, the boundary consistent set is a distribution consistent set; the mutual information consistent set is a distribution consistent set, and vice versa; the distribution consistent set is a decision consistent set; the boundary region is a positive consistent set; the boundary consistent set is a decision consistent set; the distribution consistent set is a maximum distribution consistent set.

By Definition 9, we can obtain several reducts of the inconsistent decision information system shown in Table 1 as follows.

Example 2 (Continued from Example 1). By Definition 9, we have that

(1) positive region reduct $\{c_1, c_2\}$ and the partition of the boundary region $\{x_3, x_4, ..., x_{15}\}$;
(2) decision reduct $\{c_1, c_2, c_3\}$ and the partition of the boundary region $\{\{x_3, x_4, ..., x_{13}\}, \{x_{14}, x_{15}\}\}$;
(3) distribution reduct $\{c_1, c_2, c_3, c_4\}$ and the partition of the boundary region $\{\{x_3, x_4, ..., x_8\}, \{x_9, x_{10}, ..., x_{13}\}, \{x_{14}, x_{15}\}\}$;
(4) boundary region reduct $\{c_1, c_2, c_3, c_4, c_5\}$ and the partition of the boundary region $\{\{x_3, x_4, x_5\}, \{x_6, x_7, x_8\}, \{x_9, x_{10}\}, \{x_{14}, x_{15}\}\}$;

In Example 2, we obtain the same boundary region by using different reducts, but there are different partitions for the same boundary regions of different reducts. Therefore, it is of interest to investigate the properties of boundary regions with respect to different reducts for inconsistent information systems.

Example 3 (Continued from Examples 1 and 2).

(1) For $B = \{c_1, c_2\}$, we have that

$$Confidence([x_1]_B \rightarrow D_1) = 1; Confidence([x_2]_B \rightarrow D_2) = 1;$$
$$Confidence([x_3]_B \rightarrow D_1) = \frac{3}{13}; Confidence([x_3]_B \rightarrow D_2) = \frac{9}{13};$$
$$Confidence([x_3]_B \rightarrow D_3) = \frac{1}{13}.$$

(2) For $B = \{c_1, c_2, c_3\}$, we have that

$$Confidence([x_1]_B \rightarrow D_1) = 1; Confidence([x_2]_B \rightarrow D_2) = 1;$$
$$Confidence([x_3]_B \rightarrow D_1) = \frac{3}{11}; Confidence([x_3]_B \rightarrow D_2) = \frac{8}{11};$$
$$Confidence([x_{14}]_B \rightarrow D_2) = \frac{1}{2}; Confidence([x_{14}]_B \rightarrow D_3) = \frac{1}{2}.$$

(3) For $B = \{c_1, c_2, c_3, c_4\}$, we have that

$$Confidence([x_1]_B \rightarrow D_1) = 1; Confidence([x_2]_B \rightarrow D_2) = 1;$$
$$Confidence([x_3]_B \rightarrow D_1) = \frac{1}{3}; Confidence([x_3]_B \rightarrow D_2) = \frac{2}{3};$$
$$Confidence([x_9]_B \rightarrow D_1) = \frac{1}{5}; Confidence([x_9]_B \rightarrow D_2) = \frac{4}{5};$$
$$Confidence([x_{14}]_B \rightarrow D_2) = \frac{1}{2}; Confidence([x_{14}]_B \rightarrow D_3) = \frac{1}{2}.$$

(4) For $B = \{c_1, c_2, c_3, c_4, c_5\}$, we have that

$$Confidence([x_1]_B \rightarrow D_1) = 1; Confidence([x_2]_B \rightarrow D_2) = 1;$$
$$Confidence([x_3]_B \rightarrow D_1) = \frac{1}{3}; Confidence([x_3]_B \rightarrow D_2) = \frac{2}{3};$$
$$Confidence([x_6]_B \rightarrow D_1) = \frac{1}{3}; Confidence([x_6]_B \rightarrow D_2) = \frac{2}{3};$$

$$Confidence([x_9]_B \rightarrow D_1) = \frac{1}{5}; Confidence([x_9]_B \rightarrow D_2) = \frac{4}{5};$$

$$Confidence([x_{14}]_B \rightarrow D_2) = \frac{1}{2}; Confidence([x_{14}]_B \rightarrow D_3) = \frac{1}{2}.$$

By Definition 8, we present the concept of generalized consistent set as follows:

Definition 10. *Let $S = (U, C \cup D)$ be a decision information system, and a function $e : 2^C \rightarrow L$, which evaluates the property $\mathcal{P}$, where L is a poset. If $e(B) = e(C)$ for $B \subseteq C$, then B is called a generalized consistent set of C with respect to D.*

By Definition 10, we observe that a positive consistent set, a boundary consistent set, a decision consistent set, a mutual information consistent set, a distribution consistent set and a maximum distribution consistent set are special cases of the generalized consistent sets.

Definition 11 *[18]. Let $S = (U, C \cup D)$ be a decision information system, and a function $e : 2^C \rightarrow L$, which evaluates the property $\mathcal{P}$, where L is a poset. If $e(B) = e(C)$ for $B \subseteq C$ and $e(B') \neq e(B)$ for any $B' \subset B$, then B is called a generalized relative reduct of C with respect to D.*

By Definition 11, we see that a positive region reduct, a boundary region reduct, a decision reduct, a mutual information reduct, a distribution reduct and a maximum distribution reduct are special cases of the generalized relative reducts.

Definition 12 *[18]. Let $S = (U, C \cup D)$ be a decision information system. Then $CORE_{\mathcal{P}} = \{a \in C | g(C - \{a\}) \neq g(C)\}$ is called the generalized relative core of C with respect to D.*

By Definitions 10, 11 and 12, Miao et al. presented the following proposition for the generalized relative core.

Proposition 1 *[18]. Let $S = (U, C \cup D)$ be a decision information system. Then the generalized relative core $CORE_{\mathcal{P}} = \bigcap RED_{\mathcal{P}}(C|D)$.*

4 Discernibility Matrix and Discernibility Function

In this section, we present several discernibility matrixes and discernibility functions.

4.1 Classical Discernibility Matrix and Discernibility Function

To construct relative reducts, we present the typical discernibility matrix and discernibility function.

Definition 13 *[25]. Let $S = (U, C \cup D)$ be an information system. Then the discernibility matrix $M = (M(x, y))$ is defined as follows:*

$$M(x, y) = \begin{cases} \{c \in C | c(x) \neq c(y)\}, & \text{if } d(x) \neq d(y); \\ \emptyset, & \text{otherwise.} \end{cases}$$

By Definition 13, we obtain the discernibility function as follows.

Definition 14 *[25]. Let $S = (U, C \cup D)$ be an information system, and $M = (M(x, y))$ the discernibility matrix. Then the discernibility function is defined as follows:*

$$f(M) = \bigwedge \{\bigvee \{(M(x, y)) | \forall x, y \in U | M(x, y) \neq \emptyset\}\}.$$

The expression $\bigvee \{(M(x, y)) | \forall x, y \in U | M(x, y) \neq \emptyset\}$ is the disjunction of all attributes in $M(x, y)$, which indicates that the object pair (x, y) can be distinguished by any attribute in $M(x, y)$; the expression $\bigwedge \{\bigvee \{(M(x, y)) | \forall x, y \in U | M(x, y) \neq \emptyset\}\}$ is the conjunction of all attributes in $\bigvee M(x, y)$, which indicates that the family of discernible object pairs can be distinguished by a set of attributes satisfying $\bigwedge \{\bigvee \{(M(x, y)) | \forall x, y \in U | M(x, y) \neq \emptyset\}\}$.

4.2 Elements-Based Discernibility Matrixes and Discernibility Functions

In this section, we provide several elements-based discernibility matrixes for constructing attribute reducts of decision information systems.

Definition 15 *[7, 20, 23]. Let $S = (U, C \cup D)$ be an information system. Then*

(1) $M_{positive} = (M_{positive}(x, y))$, where

$$M(x, y) = \begin{cases} \{c \in C | c(x) \neq c(y)\}, & \text{if } [x]_C \vee [y]_C \in POS_C(D); \\ \emptyset, & \text{otherwise.} \end{cases}$$

(2) $M_{boundary} = (M_{boundary}(x, y))$, where

$$M_{boundary}(x, y) = \begin{cases} \{c \in C | c(x) \neq c(y)\}, & \text{if } (x, y) \in IND(C|D); \\ \emptyset, & \text{otherwise.} \end{cases}$$

(3) $M_{decision} = (M_{decision}(x, y))$, where

$$M_{decision}(x, y) = \begin{cases} \{c \in C | c(x) \neq c(y)\}, & \text{if } \tau_C(x) \neq \tau_C(y); \\ \emptyset, & \text{otherwise.} \end{cases}$$

In Definition 15, $M_{positive}, M_{boundary}$ and $M_{decision}$ denote the discernibility matrixes for constructing the positive region reduct, boundary region reduct and decision reduct, respectively.

Subsequently, Miao et al. proposed the elements-based discernibility matrix and discernibility function for constructing relative reducts.

Definition 16 *[18]. Let $S = (U, C \cup D)$ be an information system, and $D = \{d\}$. Then the elements-based discernibility matrix $M_{\mathcal{P}} = (M_{\mathcal{P}}(x, y))$ is defined as follows:*

$$M_{\mathcal{P}}(x, y) = \begin{cases} \{c \in C | c(x) \neq c(y)\}, & \text{if } x, y \text{ are distinguishable w.r.t. } \mathcal{P}; \\ \emptyset, & \text{otherwise.} \end{cases}$$

By Definition 16, we see that $M_{positive}$, $M_{boundary}$ and $M_{decision}$ are special cases of the elements-based discernibility matrix $M_{\mathcal{P}}$. Furthermore, Miao et al. presented the concept of the elements-based discernibility function as follows.

Definition 17 *[18]. Let $S = (U, C \cup D)$ be an information system, and $M_{\mathcal{P}}$ the discernibility matrix. Then the elements-based discernibility function $f(M_{\mathcal{P}})$ is defined as follows:*

$$f(M_{\mathcal{P}}) = \bigwedge \{ \bigvee \{ (M_{\mathcal{P}}(x, y)) | \forall x, y \in U | M_{\mathcal{P}}(x, y) \neq \emptyset \} \}.$$

4.3 Blocks-Based Discernibility Matrixes and Discernibility Functions

In this section, we provide several blocks-based discernibility matrixes for constructing attribute reducts of decision information systems.

Definition 18 *[39]. Let $S = (U, C \cup D)$ be a decision information system. (1) If $D_1^{\bullet} = \{([x]_C, [y]_C) | \mu_A(x) \neq \mu_A(y)\}$. Then the discernibility matrix $M_{DR} = (M_{DR}(X, Y))$ for constructing distribution reducts is defined as follows:*

$$M_{DR}(X, Y) = \begin{cases} \{c \in C | c(X) \neq c(Y)\}, & \text{if } (X, Y) \in D_1^{\bullet}; \\ C, & (X, Y) \notin D_1^{\bullet}. \end{cases}$$

(2) If $D_2^{\bullet} = \{([x]_C, [y]_C) | \gamma_A(x) \neq \gamma_A(y)\}$. Then the discernibility matrix $M_{MD} = (M_{MD}(X, Y))$ for constructing the maximum distribution reducts is defined as follows:

$$M_{MD}(X, Y) = \begin{cases} \{c \in C | c(X) \neq c(Y)\}, & \text{if } (X, Y) \in D_2^{\bullet}; \\ C, & (X, Y) \notin D_2^{\bullet}. \end{cases}$$

In Definition 18, M_{DR} and M_{MD} denote the discernibility matrixes for constructing the distribution reduct and the maximum distribution reduct, respectively. Subsequently, we presented the concept of blocks-based discernibility matrixes as follows:

Definition 19. *Let $S = (U, C \cup D)$ be a decision information system, and $D^{\mathcal{Q}} = \{(X, Y) \in U/C \times U/C | X \text{ and } Y \text{ satisfy the property } \mathcal{Q}\}$. Then the blocks-based discernibility matrix $M_{\mathcal{Q}} = (M_{\mathcal{Q}}(X, Y))$ is defined as follows:*

$$M_{\mathcal{Q}}(X, Y) = \begin{cases} \{c \in C | c(X) \neq c(Y)\}, & \text{if } (X, Y) \in D^{\mathcal{Q}}; \\ C, & (X, Y) \notin D^{\mathcal{Q}}. \end{cases}$$

Definition 20. *Let $S = (U, C \cup D)$ be a decision information system, and $M_Q = (M_Q(X, Y))$ the discernibility matrix. Then the blocks-based discernibility functions $f(M_Q)$ is defined as follows:*

$$f(M_Q) = \bigwedge\{\bigvee\{(M_Q(X,Y))|\forall X, Y \in U/C|M_Q(X,Y) \neq \emptyset\}\}.$$

By Definitions 16 and 19, we present the concept of the generalized discernibility matrix.

Definition 21. *Let $S = (U, C \cup D)$ be a decision information system, $\mathcal{X}, \mathcal{Y} \in F(U, C)$, where $F(U, C)$ is a quotient set of U with respect to C, and $\mathcal{D} = \{(\mathcal{X}, \mathcal{Y})|\mathcal{X}$ and $\mathcal{Y}$ satisfy the property $\mathcal{Q}\}$. Then the generalized discernibility matrix $M_G = (M_G(\mathcal{X}, \mathcal{Y}))$ is defined as follows:*

$$M_G(\mathcal{X}, \mathcal{Y}) = \begin{cases} \{c \in C | c(\mathcal{X}) \neq c(\mathcal{Y})\}, & \text{if } (\mathcal{X}, \mathcal{Y}) \in \mathcal{D}; \\ C, & (\mathcal{X}, \mathcal{Y}) \notin \mathcal{D}. \end{cases}$$

5   Conclusions

In this paper, we have investigated the relationship between relative reducts and presented the generalized relative reduct. We have studied the relationship between discernibility matrixes and presented the generalized discernibility matrix. We have employed several examples to illustrate the related results simultaneously.

In the future, we will further investigate the relationship between different relative reducts. Additionally, we will present the generalized relative reducts for information systems and construct a foundation for attribute reduction of information systems.

Acknowledgments. This work is supported by the National Natural Science Foundation of China (NO. 61273304), the Scientific Research Fund of Hunan Provincial Education Department (No. 14C0049).

References

1. Chen, D.G., Wang, C.Z., Hu, Q.H.: A new approach to attribute reduction of consistent and inconsistent covering decision systems with covering rough sets. Inf. Sci. **177**, 3500–3518 (2007)
2. Chen, Y.M., Miao, D.Q., Wang, R.Z., Wu, K.S.: A rough set approach to feature selection based on power set tree. Knowl. Based Syst. **24**, 275–281 (2011)
3. Du, F.W., Qin, K.Y.: Analysis of several reduction standards and their relationship for inconsistent decision table. Acta Electronica Sinica **39**(6), 1336–1340 (2011)
4. Dai, J.H., Wang, W.T., Tian, H.W., Liu, L.: Attribute selection based on a new conditional entropy for incomplete decision systems. Knowl. Based Syst. **39**, 207–213 (2013)

5. Hu, Q.H., Yu, D.R., Xie, Z.X.: Information-preserving hybrid data reduction based on fuzzy-rough techniques. Pattern Recogn. Lett. **27**(5), 414–423 (2006)
6. Jia, X.Y., Shang, L., Zhou, B., Yao, Y.Y.: Generalized attribute reduct in rough set theory. Knowledge-Based Systems (2015). doi:10.1016/j.knosys.2015.05.017
7. Kryszkiewicz, M.: Comparative study of alternative types of knowledge reduction in inconsistent systems. Int. J. Intell. Syst. **16**, 105–120 (2001)
8. Lang, G.M., Li, Q.G., Cai, M.J., Yang, T.: Characteristic matrices-based knowledge reduction in dynamic covering decision information systems. Knowl. Based Syst. (2015). doi:10.1016/j.knosys.2015.03.021
9. Lang, G.M., Li, Q.G., Cai, M.J., Yang, T., Xiao, Q.M.: Incremental approaches to knowledge reduction based on characteristic matrices. Int. J. Mach. Learn. Cybern. (2014) https://doi.org/10.1007/s13042-014-0315-4
10. Lang, G.M., Li, Q.G., Yang, T.: An incremental approach to attribute reduction of dynamic set-valued information systems. Int. J. Mach. Learn. Cybern. **5**, 775–788 (2014)
11. Leung, Y., Ma, J.M., Zhang, W.X., Li, T.J.: Dependence-space-based attribute reductions in inconsistent decision information systems. Int. J. Approximate Reasoning **49**, 623–630 (2008)
12. Li, J.H., Mei, C.L., Lv, Y.J.: Knowledge reduction in real decision formal contexts. Inf. Sci. **189**, 191–207 (2012)
13. Li, T.R., Ruan, D., Song, J.: Dynamic maintenance of decision rules with rough set under characteristic relation. Wireless Communications, Networking and Mobile Computing, pp. 3713–3716 (2007)
14. Li, M., Shang, C.X., Feng, S.Z., Fan, J.P.: Quick attribute reduction in inconsistent decision tables. Inf. Sci. **254**, 155–180 (2014)
15. Liang, J.Y., Mi, J.R., Wei, W., Wang, F.: An accelerator for attribute reduction based on perspective of objects and attributes. Knowl. Based Syst. **44**, 90–100 (2013)
16. Liu, D., Li, T.R., Zhang, J.B.: A rough set-based incremental approach for learning knowledge in dynamic incomplete information systems. Int. J. Approximate Reasoning **55**(8), 1764–1786 (2014)
17. Mi, J.S., Wu, W.Z., Zhang, W.X.: Approaches to knowledge reduction based on variable precision rough set model. Inf. Sci. **159**(3–4), 255–272 (2004)
18. Miao, D.Q., Zhao, Y., Yao, Y.Y., Li, H.X., Xu, F.F.: Relative reducts in consistent and inconsistent decision tables of the Pawlak rough set model. Inf. Sci. **179**, 4140–4150 (2009)
19. Meng, Z.Q., Shi, Z.Z.: Extended rough set-based attribute reduction in inconsistent incomplete decision systems. Inf. Sci. **204**, 44–69 (2012)
20. Pawlak, Z.: Rough sets. Int. J. Comput. Inf. Sci. **11**(5), 341–356 (1982)
21. Qian, J., Lv, P., Yue, X.D., Liu, C.H., Jing, Z.J.: Hierarchical attribute reduction algorithms for big data using MapReduce. Knowl. Based Syst. **73**, 18–31 (2015)
22. Qian, Y.H., Liang, J.Y., Li, D.Y., Wang, F., Ma, N.N.: Approximation reduction in inconsistent incomplete decision tables. Knowl. Based Syst. **23**, 427–433 (2010)
23. Slezak, D.: Normalized decision functions and measures for inconsistent decision tables analysis. Fundamenta Informaticae **44**, 291–319 (2000)
24. Shao, M.W., Leung, Y.: Relations between granular reduct and dominance reduct in formal contexts. Knowl. Based Syst. **65**, 1–11 (2014)
25. Skowron, A., Rauszer, C.: The discernibility matrices and functions in information systems. In: Slowinski, R. (ed.) Handbook of Applications and Advances of the Rough Sets Theory, pp. 331–362. Kluwer Academic Publishers, Dordrecht (1992)

26. Skowron, A., Rauszer, C.: The discernibility matrices and functions in information systems. In: Słowiński, R. (ed.) Intelligent Decision Support Handbook of Applications and Advances of the Rough Sets Theory, pp. 331–362. Kluwer Academic Publishers, Dordrecht (1992)

27. Tan, A.H., Li, J.J., Lin, Y.J., Lin, G.P.: Matrix-based set approximations and reductions in covering decision information systems. Int. J. Approximate Reasoning **59**, 68–80 (2015)

28. Tan, A.H., Li, J.J.: A kind of approximations of generalized rough set model. Int. J. Mach. Learn. Cybern. **6**(3), 455–463 (2015)

29. Wu, W.Z.: Attribute reduction based on evidence theory in incomplete decision systems. Inf. Sci. **178**(5), 1355–1371 (2008)

30. Wang, F., Liang, J.Y., Dang, C.Y.: Attribute reduction for dynamic data sets. Appl. Soft Comput. **13**, 676–689 (2013)

31. Wang, G.Y., Ma, X.A., Yu, H.: Monotonic uncertainty measures for attribute reduction in probabilistic rough set model. Int. J. Approximate Reasoning **59**, 41–67 (2015)

32. Xu, F.F., Miao, D.Q., Wei, L.: Fuzzy-rough attribute reduction via mutual information with an application to cancer classification. Comput. Math. Appl. **57**, 1010–1017 (2009)

33. Xu, W.H., Zhang, W.X.: Measuring roughness of generalized rough sets induced by a covering. Fuzzy Sets Syst. **158**(22), 2443–2455 (2007)

34. Yang, T., Li, Q.G.: Reduction about approximation spaces of covering generalized rough sets. Int. J. Approximate Reasoning **51**(3), 335–345 (2010)

35. Yang, Y.Y., Chen, D.G., Dong, Z.: Novel algorithms of attribute reduction with variable precision rough set model. Neurocomputing **139**, 336–344 (2014)

36. Yang, X.B., Qi, Y., Yu, H.L., Song, X.N., Yang, J.Y.: Updating multigranulation rough approximations with increasing of granular structures. Knowl. Based Syst. **64**, 59–69 (2014)

37. Yao, Y.Y., Zhao, Y.: Discernibility matrix simplification for constructing attribute reducts. Inf. Sci. **179**(7), 867–882 (2009)

38. Ye, D.Y., Chen, Z.J.: A new discernibility matrix and the computation of a core. Acta Electronica Sinica **30**(7), 1086–1088 (2002)

39. Zhang, W.X., Mi, J.S., Wu, W.Z.: Knowledge reduction in inconsistent information systems. Chin. J. Comput. **26**(1), 12–18 (2003)

40. Zhang, W.X., Mi, J.S., Wu, W.Z.: Approaches to knowledge reductions in inconsistent systems. Int. J. Intell. Syst. **18**(9), 989–1000 (2003)

41. Zhang, J.B., Li, T.R., Ruan, D., Gao, Z.Z., Zhao, C.B.: A parallel method for computing rough set approximations. Inf. Sci. **194**, 209–223 (2012)

42. Zhao, Y., Luo, F., Wong, S.K.M., Yao, Y.: A general definition of an attribute reduct. In: Yao, J.T., Lingras, P., Wu, W.-Z., Szczuka, M.S., Cercone, N.J., Ślęzak, D. (eds.) RSKT 2007. LNCS (LNAI), vol. 4481, pp. 101–108. Springer, Heidelberg (2007)

43. Zhu, W.: Generalized rough sets based on relations. Inf. Sci. **177**(22), 4997–5011 (2007)

Rough Sets - Past, Present and Future: Some Notes

Piero Pagliani[✉]

Rome, Italy
pier.pagliani@gmail.com

Abstract. Some notes about the state-of-the-art of Rough Set Theory
are discussed, and some future research topics are suggested as well.

1 Rough Set Theory: A Brilliant Past Facing the Future

The present paper is a continuation of [24,25], in that it mainly aims at suggesting future research topics to young researchers. I will avoid too much technicality and formalism that can be easily recovered by the quoted works.

Obviously, I will focus on topics which are connected to my researches and my knowledge. Therefore, a large number of areas will not be mentioned, although very important, and I hope I am excused if many cited papers are mine.

Rough Set Theory is now a well-established and widely accepted research field. Zdzisław Pawlak's original intuition was "elementary". This does not mean "simple", much less "trivial". It means that the basic constituents of a problem (possibly very complex) are singled out and displayed. Many great ideas are elementary.

From the application point of view, Rough Set Theory revealed its large potential since inception. Patter recognition and rule discovery are probably the very basic mechanisms that Rough Set Theory offers to Data Mining, Knowledge Discovery in Database, Machine Learning, Image Recognition and so on.

However, Rough Set Theory happened to be connected with a large number of important mathematical and logical fields. They include point and pointless topology, algebra, many-valued logics, and others, often surprising, that will be discussed in the text. Conversely, the information roots of Rough Set Theory made it possible to better understand the features and the meanings of many constructions in the above fields, thus permitting new achievements. In turn, as it always happens, the mathematical more abstract point of view reacts on applications by suggesting new solutions and new applications as well.

The idea of analysing items by means of lower and upper approximations, on the basis of an equivalence relation naturally induced by attribute-values, was indeed both elementary and brilliant. It was the rock-standing starting point over which extensions and generalizations are built which reveal to be more and more promising.

At present, because of the international economic and financial crisis, research funding are more and more decreasing and both private and institutional bodies

© Springer International Publishing Switzerland 2015
D. Ciucci et al. (Eds.): RSKT 2015, LNAI 9436, pp. 39–49, 2015.
DOI: 10.1007/978-3-319-25754-9_4

think it wiser to demand application results instead of theoretical findings. Such a mindset risks to start a negative loop. It is well-known that only the 20 % of all scientific researches is used in applications. But our rulers seem to ignore that without the other 80 % the "good-for-real-problems" 20 % disappears.

Unfortunately, Rough Set community has to deal with this narrow-minded situation. Therefore, in the following brief discussion on the past and the future of Rough Set Theory I will keep in mind the importance of applications.

2   Multi-agent Approximation Spaces

Multi-agent Approximation Spaces is an urgent research field. Indeed, usually applications have to combine data coming from the observations of different agents (or sensors and the like), or data coming from different observations of one agent varying over time. Or a mix of the two cases.

Sometimes the solution is to synthesize different information systems into a single one, on which the usual (in a broader sense) rough set techniques are applied (in the text, "information system" has the usual meaning in Rough Set Theory, that is, a data table $\langle Objects, Attributes, Values \rangle$). Such a synthesis can be obtained in many a way. For instance by computing the average value of the attribute-values. However, other solutions are possible.

If one considers the set of values of the same attributes provided by the different information systems, a *non-deterministic information system* is obtained. This kind of information systems allows for particular multimodal rough set analysis that can be very useful in particular cases. Nonetheless, this topic has been studied by a few researchers (for instance [14]). Thus this subfield of Rough Set Theory needs to be developed further, along with its applications.

A different approach is to combine the approximations obtained by different information systems. Professor M. K. Chakraborty and I used to call it the *dialogical approach* to Rough Set Theory and we maintained it as a fundamental development for the future of the theory itself. It is extremely useful for applications and challenging from a theoretical point of view. However, as far as now, at least to my knowledge, just a few steps have been done: [23] (or [28], ch. 12), [9] and subsequent papers by Khan and Banerjee, and [29]. The topic was pioneered by C. Rauszer from a logical point of view (see [30]).

3   The Mathematics of Rough Sets

Given any relational structure $\langle U, R \rangle$, with $R \subseteq U \times U$, and $X \subseteq U$, one can set:

$$(lR)(X) = \{x : R(\{x\}) \subseteq X\} = \{x : \forall y(\langle x, y \rangle \in R \Rightarrow y \in X)\} \tag{1}$$
$$(uR)(X) = \{x : R(\{x\}) \cap X \neq \emptyset\} = \{x : \exists y(\langle x, y \rangle \in R \wedge y \in X)\} \tag{2}$$

where for any $Y \subseteq U$,

$$R(Y) = \{x : \exists y \in Y (\langle y, x \rangle \in R)\} \tag{3}$$

$(lR)(X)$ is called the *lower approximation* of X (*via R*), while $(uR)(X)$ is called the *upper approximation* of X (*via R*), and $\langle U, (uR), (lR) \rangle$ is called an *approximation space*.

As is well known, if R is an equivalence relation, as it happens in the case of Pawlak's rough sets, then the lower and upper approximations of a set X are the *interior* and, respectively, *closure* of a topological space generated by a basis of *clopens* (closed and open) sets. These clopens are the sets atomically *definable* by means of the information system at hand. That is, they are the equivalence classes modulo R where $\langle a, b \rangle \in R$ if and only if a and b have the same attribute-values for all the attributes of the information system. By extension, also the family of all open (closed) subsets of U will be called an Approximation Space.

If R is a preorder (or a partial order, of course), the two approximations are topological operators as well, but of topological spaces of different kind, namely Alexandrov spaces (see [8,13]).

Similar results are important because they enable a body of well-established results in topology to be applied to rough sets.

Clearly, rough sets based on preorders and partial orders have been the next natural step after rough sets defined by equivalence relations and they enjoy well-established nice properties. But if R is a generic binary relation, things get more complicated. Sometimes one obtains pre-topological operators, sometimes operators which are difficult to classify.

This is, for instance, the situation for rough sets based on *coverings*. Understanding their mathematical behaviour requires some effort and some ad hoc techniques. However, in many cases it can be shown that approaching these operators from the point of view of *formal* (or *pointless*) *topology* simplifies their mathematical analysis a great deal (see [26]). Since formal topology is connected to another key topic, that is, the pure relational approach to rough sets and approximation theory, we shall briefly examine it later on.

By now, we want to underline that coverings are frequent results of real-life observations.

Also, Rough Set Theory revealed to have surprising connections with important and far distant mathematical fields, such as those related to Algebraic Geometry.

Indeed, classical rough set systems, that is, the system of all the rough sets induced by a Pawlak Approximation Spaces, can be described in terms of Lawvere-Tierney operators and, thus, of Grothendiek Topologies (see [22] or [28] ch. 7[1]). In both cases the union B of all the singleton definable sets plays the role of fundamental parameter. B has a similar fundamental role also in the logical analysis of rough set systems, as we shall see later on. Both roles, in Logic and in Mathematics, express a notion of *local validity*.

Lawvere-Tierney operators and connected Grothendieck Topologies can be applied to Pawlak's rough set systems because classical Approximation Spaces are Boolean algebras, hence Heyting algebras. As far as now the same interpretation has not yet been studied for Approximation Spaces induced by preorders

---

[1] In the examples there are some typos. I shall email the corrections upon request.

and partial orders, that is, to Approximation Spaces which are Heyting but not Boolean algebras. Indeed, this is an interesting point to develop, also because of the following facts.

It was noticed that rough set systems can be interpreted in terms of *stalk* (or *etalé*) spaces (see [28], ch. 10.12.5). Moreover, rough set systems form bi-Heyting algebras, and these structures can be defined in terms of topoi (see [31]). Therefore, a unitary and coherent description of (generalised) rough sets from the point of view of etalé spaces and *topos theory* would be an important result (some hints to approach this topic can be found in [24], Issue E).

4 The Relational Approach to Rough Sets

Interpreting rough sets from a pure relational point of view has a practical importance. With "pure relational" I mean an interpretation in which not only relations between elements but also sets of elements are interpreted as binary relations, thus matrices (see [20]). Many programming languages provide a full set of primitive operators which powerfully manipulate matrices (such as generalised matrix multiplications). Since in this setting both lower and upper approximations are particular kinds of matrix multiplications, the practical importance of a pure relational approach to rough sets is crystal clear (also, consider that dependencies between sets of attributes are particular cases of approximations).

Moreover, such an approach makes it possible to explore a very important, but not sufficiently developed research area: the approximation of relations. This topic was introduced in [36] and enhanced in [37] and deals with the approximation of relations by means of other relations. Such a computation has an evident practical importance, because in many applications one has to understand the meaning of a relation with respect to other relations, for instance the classification of items induced by their features and the preference of the items themselves from the point of view of one or more subjects.

Skowron and Stepaniuk took into account an even more general setting where one has a family $\{R_i \subseteq U_i \times U_i\}_{1 \leq i \leq n}$ of n binary relations over n sets and an approximating relation Z defined point-wise on the product of these sets by means of the given relations:

$$\langle \langle x_1, ..., x_n \rangle, \langle y_1, ..., y_n \rangle \rangle \in Z \text{ iff } \langle x_i, y_i \rangle \in R_i, \text{ for } 1 \leq i \leq n.$$

By means of Z one can approximate any other arbitrary relation R defined on the product $U_1 \times ... \times U_n$ of the family of sets. In the case of a family of two relations R_1 and R_2 over two sets U_1 and, respectively, U_2, for any binary relation $R \subseteq U_1 \times U_2$, the lower approximation is:

$$(lZ)(R) = \{\langle x_1, x_2 \rangle : \forall \langle y_1, y_2 \rangle (\langle \langle x_1, x_2 \rangle, \langle y_1, y_2 \rangle \rangle \in Z \Rightarrow \langle y_1, y_2 \rangle \in R\} \quad (4)$$

It is possible to show that $(lZ)(R) = R_1 \longrightarrow R \longleftarrow R_2$, where $\longrightarrow$ and $\longleftarrow$ are the right and, respectively, left residuations between binary relations[2].

---

[2] If $R \subseteq W \times W'$ and $Z \subseteq U \times U'$, then $R \longrightarrow Z = -(R^\smile \otimes -Z)$ (this operation is defined if $|W| = |U|$), and $Z \longleftarrow R = -(-Z \otimes R^\smile)$ (this operation is defined if $|W'| = |U'|$), where "$-Z$" is the Boolean complement of Z and $R^\smile$ is the reverse relation of R.

Since the two residuations are computed just by means of logical matrix multiplication (i.e. $\wedge$ as $\times$ and $\vee$ as $+$), complementation and reverse, the result is of practical importance (see [21,28], ch. 3.5.18).

However, much more work is required to generalise the above result to cases with more then two relations and two sets.

In a obvious sense, this topic is connect to dialogical rough set theory. However, since matrix multiplication is not commutative, the order of application implies that the "dialogue" among different Approximation Spaces is not between peers.

As far as now we have talked of the pure relational interpretation of the usual machinery of Approximation Spaces and Rough Sets induced by an Information System in the sense of Pawlak. Further, one can apply a more integral relational approach by starting with any *property system*, that is, a binary relation R between a set of objects, G, and a set of properties, M, instead of a usual Information System. Consider that any Information System can be made into such a property system. In this case by means of logical definitions one obtains six basic operators:

- $\langle e \rangle : \wp(M) \longmapsto \wp(G); \langle e \rangle(Y) = \{a \in G : \exists b (b \in Y \wedge \langle a, b \rangle \in R)\}$;
- $[e] : \wp(M) \longmapsto \wp(G); [e](Y) = \{a \in G : \forall b(\langle a, b \rangle \in R \Rightarrow b \in Y)\}$;
- $[[e]] : \wp(M) \longmapsto \wp(G); [e](Y) = \{a \in G : \forall b(b \in Y \Rightarrow \langle a, b \rangle \in R)\}$;
- $\langle i \rangle : \wp(G) \longmapsto \wp(M); \langle i \rangle(X) = \{b \in M : \exists a(a \in X \wedge \langle a, b \rangle \in R)\}$;
- $[i] : \wp(G) \longmapsto \wp(M); [i](X) = \{b \in M : \forall a(\langle a, b \rangle \in R \Rightarrow a \in X)\}$;
- $[[i]] : \wp(G) \longmapsto \wp(M); [i](X) = \{b \in M : \forall a(a \in X \Rightarrow \langle a, b \rangle \in R)\}$.

The operators $\langle i \rangle$, $\langle e \rangle$, $[i]$ and $[e]$ are the core of the definitions of a (generalised) *interior* and a (generalized) *closure* in pointless topology. For instance, for any $X \subseteq G$ the interior is $int(X) = \langle e \rangle [i](X)$ and the closure is $cl(X) = [e]\langle i \rangle(X)$. In *Formal Concept Analysis* the *extent* of X is $est(X) = [[e]][[i]](X)$ while the *intent* is $ITS(X) = [[i]][[e]](X)$ ([40]).

All the above operators can be computed by means of relation algebras. For instance, the right cylindrification of $est(X)$ is (see [17] or [28], ch. 3.5.18)[3]:

$$(est(X))^c = R \longleftarrow (X^c \longrightarrow R).$$

Because of the wide applicability of these generalised operators, it is clear that approximation of relations by means of relations is a powerful machinery. Moreover, it is clear that it can be applied for easily dealing with cases in which, for instance, rough set and formal concept approaches must be mixed together, and so on.

Moreover, suppose M is a covering of G, that is, M is a set of subsets of G which covers G. Suppose $R \subseteq G \times M$ is the usual membership relation $\in$. It can be shown that by combining the operators $[e], [i], \langle e \rangle$ and $\langle i \rangle$ a number

---

[3] If $R \subseteq X \times Y$ the right cylindrification of $A \subseteq X$ is the relation $A \times Y$. Notice that the relational formula of est is symmetric to the formula of (lZ). The reason is discussed in [21].

of approximation operators induced by coverings can be defined. More or less, this number includes all such operators introduced in literature so far. Moreover, the two operators decorated by the same letter are dual, while $\langle e \rangle, [i]$ and $\langle i \rangle, [e]$ form adjoint pairs. From these facts, almost all the properties and relations between the approximation operators induced by coverings can be easily deduced (see [26]).

However, all the above results are simply starting points for a long journey to be done.

5   Rough Set Theory and Logic

In [6] classical rough sets were represented as ordered pairs $\langle (lE)(X), (uE)(X) \rangle$ where E is the indiscernibility relation among objects and X is any subsets of objects. We call it the *increasing representation* of rough sets. Three years later, in a workshop at Warsaw organized by professor Pawlak, the author independently presented rough sets as ordered pairs of the form $\langle (lE)(X), -(uE)(X) \rangle$. We call it the *disjoint representation* of rough sets. Such a representation came from a study about a subvariety of Nelson algebras, called $E_0 - lattices$ ([15]). We shall meet these lattices later on because they are connected to rough sets. By now we want to underline that the disjoint representation made it possible to easily show that Pawlak's rough set systems can be made into semi-simple Nelson algebras (see [16]), hence into three-valued Łukasiewicz algebras. Since then, the representation of rough sets by means of ordered pairs of elements from an Approximation Space has been used as a powerful tool to understand the logico-algebraic properties of rough set systems. I am sorry I cannot account here for a more comprehensive story of this interesting topic. However, many details and some historical notes can be found in [28].

I just point out a couple of facts.

First, it must be noticed that as far as one takes into account ordered pairs of elements of Pawlak's Approximation Spaces, hence Boolean algebras, rough set systems in disjoint representation and rough set systems in increasing representation (or in decreasing representation, of course), are isomorphic. The following table shows the corresponding operations (it is understood that $a = \langle a_1, a_2 \rangle, b = \langle b_1, b_2 \rangle$ and the operations between elements of the pairs are those of the Approximation Space *qua* Boolean algebra):

Symbol	*decreasing repr.*	*disjoint repr.*	Name
$0, 1$	$\langle 0, 0 \rangle, \langle 1, 1 \rangle$	$\langle 0, 1 \rangle, \langle 1, 0 \rangle$	Bottom, resp. Top
$\sim a$	$\langle \neg a_2, \neg a_1 \rangle$	$\langle a_2, a_1 \rangle$	Strong negation
$a \longrightarrow b$	$\langle a_2 \Rightarrow b_1, a_2 \Rightarrow b_2 \rangle$	$\langle a_1 \Rightarrow b_1, a_1 \wedge b_2 \rangle$	Weak implication
$a \wedge b$	$\langle a_1 \wedge b_1, a_2 \wedge b_2 \rangle$	$\langle a_1 \wedge b_1, a_2 \vee b_2 \rangle$	Inf
$a \vee b$	$\langle a_1 \vee b_1, a_2 \vee b_2 \rangle$	$\langle a_1 \vee b_1, a_2 \wedge b_2 \rangle$	Sup

But if one has to deal with ordered pairs of elements of a Heyting algebra which is not Boolean, then the set of decreasing pairs is not closed with respect to the above operations, while the set of disjoint pairs is.

Secondly, a rough set system is not given by the set of all pairs of decreasing (increasing) or disjoint elements of an Approximation Space. Indeed, singleton atomic elements $\{x\}$ of the Approximation Space, if any, are such that for any set X, $x \in (lR)(X)$ if and only if $x \in (uR)(X)$. It follows that if we set B as the union of all singleton atomic elements of the Approximation Space, then an ordered pair of decreasing or increasing elements $\langle Y, W \rangle$ represents a rough set only if $Y \cap B = W \cap B$, which means $(lR)(X) \cap B = (uR)(X) \cap B$. Equivalently, an ordered pair $\langle Y, W \rangle$ of disjoint elements represents a rough set only if $Y \cup W \supseteq B$, which means $(lR)(X) \cup -(uR)(X) \supseteq B$ or, in topological terms, singletons are either included in the interior of X or in its exterior (i.e. the complement of the closure): isolated elements cannot be in the boundary of any set.

Indeed in the construction of the carriers of Łukasiewicz three-valued algebras or double Stone algebras from a Boolean algebra, an element α of the Boolean algebra is used to filter out some ordered pairs (see [1]). The same happens for the carriers of Nelson algebras built as sets of ordered pairs of disjoint elements of a Heyting algebra. In this case α must fulfill a particular property (see [33]). Without entering into much details, I would like to stress that the informational character of Rough Set Theory made it possible to provide α with a sound and well-funded informational interpretation: α is the union of all the elements for which we have complete information. That is, information that makes it possible to *isolate* them from the others. Hence, α is the subdomain in which Classical Logic holds, in that the excluded middle holds (see [19,22,28]). The importance of the conceptual approach to Rough Set Theory cannot be underestimated, as is pointed out for instance in [41].

Actually, let $P_B(AS)$ be the the set of ordered pairs of increasing, decreasing or disjoint elements of an Approximation Space AS, filtered by the set B. If AS is a Pawlak Approximation Space, then it is a Boolean algebra, $P_B(AS)$ is the set of all Pawlak's rough sets (classical case), and one has:

- If $B \neq \emptyset, B \neq U$, then $P_B(AS)$ can be made into a three-valued Łukasiewicz algebra.
- If $B = U$, then $P_B(AS)$ is a Boolean algebra isomorphic to AS.
- If $B = \emptyset$, them $P_B(AS)$ is a Post algebra of order three.

But we can go deeper into the logic of these structures. Let δ be $\langle B, \emptyset \rangle$ in disjoint representation, $\langle U, B \rangle$ in decreasing representation and $\langle B, U \rangle$ in increasing representation. Then δ is the least dense element of the lattice $P_B(AS)$. Thus $P_B(AS)$ can be seen also as a $P_2 - lattice$ with chain of values $0 \leq \delta \leq 1$ (notice that if $B = \emptyset$ this $P_2 - lattice$ is a Post algebra of order three). It happens that if τ is any classical tautology (expressed in algebraic terms), then $\delta \leq \tau \leq 1$. Conversely, if κ is a classical contradiction, then $0 \leq \kappa \leq \sim \delta$, where $\sim \delta$ is the strong negation of δ (for instance, in disjoint representation it is $\langle \emptyset, B \rangle$).

This suggests studying Rough Set Theory from the point of view of *paraconsistent logic*. It is no surprise, because for instance Nelson algebras have already been used for paraconsistent logic (see for instance [39]). But there are other facts to be investigated. For instance, semi-simple Nelson algebras (hence classical rough set systems) can be equipped by many interesting operations. One for all is $\lrcorner$ defined as $\lrcorner a = a \longrightarrow 0$. Let us set, moreover, $a \rightharpoonup b = \lrcorner \sim a \longrightarrow \lrcorner \sim b$. It is not difficult to prove that if $1 = \langle 1, 0 \rangle$, $\delta = \langle 0, 0 \rangle$, $0 = \langle 0, 1 \rangle$ and $A = \{1, \delta, 0\}$, then $\langle A, \wedge, \rightharpoonup, \lrcorner, \{1, \delta\} \rangle$ is equivalent to the paraconsistent logic P^1. Simmetrically, using $\neg a = \sim \lrcorner \sim a$ instead of $\lrcorner$ one can recover the paracomplete logic I^1 (see [34, 35]).

Moreover, elaborating on these suggestions, one could link Rough Set Theory to *Society semantics* (see [2] - some hints can be indirectly obtained from [18] and directly from [3]).

The fact that B is a subdomain where locally Classical Logic is valid, surrounded by an environment where other kinds of logic hold (e. i. three-valued or intuitionistic), is highlighted when one considers preorder or partial order relations, R, between objects, instead of equivalence relations. This happens, for instance, when one applies preference relations on an Information System (see [5]). In this case the Approximation Space AS is a Heyting algebra and the derived rough set system is a Nelson algebra.

Since any lower approximation has the form $R(X)$, for some subset of objects X, if x is a maximal element then $R(\{x\}) = \{x\}$. Otherwise stated, x is an isolated element. As in the case of Pawlak's rough sets, the union B of all isolated elements should be a subdomain where Classical Logic is valid. Suppose now that for any x there is a maximal y such that $x \leq y$ (this always happens in the finite case). Then B is the least dense element of AS. Notice that if AS were a Boolean algebra, the least dense element would equal the entire universe U itself and $P_B(AS)$ would be an isomorphic copy of AS. Therefore this case is not trivial only if AS is a generic Heyting algebra. In this case $P_B(AS)$ is a $E_0 - lattice$. These lattices form a subvariety of Nelson algebras and are models for the logic E_0, that is, Classical Logic with Strong Negation plus the following modal axioms: $(\sim A \longrightarrow \perp) \longrightarrow \mathbf{T}(A)$ and $(A \longrightarrow \perp) \longrightarrow \sim \mathbf{T}(A)$. Logic E_0 was introduced in [12] where it is proved, for $\mathcal{CL}$ a classical calculus: $\vdash_{\overline{\mathcal{CL}}} A$ if and only if $\vdash_{\overline{\varepsilon_0}} \mathbf{T}(A)$, also in the predicative case, thus extending the well-known Gödel-Glivenko theorem stating $\vdash_{\overline{\mathcal{CL}}} A$ if and only if $\vdash_{\overline{\mathcal{INT}}} \neg\neg A$, for A any classical *propositional* theorem and $\mathcal{INT}$ an intuitionistic calculus.

If one considers the above discussion about local classical validity and singleton definable sets in Rough Set Theory, then the result just mentioned receives an additional "informational" explanation. Indeed, Rough Set Theory here provides an interpretation based on data analysis of the fact that some pieces of data are "atomic", that is, completely defined, because a set of attributes is given which is able to completely define them among the others. But if a piece of data is completely defined it does not require investigation, hence it does not require a particular effective "detective" logic such as Intuitionism, which, instead, has to be applied outside B.

Since E_0 is just the first step in the investigation of effective logics with a classical subdomain, further connections can be explored. It is an example of the potentiality of Rough Set Theory to help investigating refined and sophisticated logical fields. Another example is given by the *Logic of conjectures and assertions* (see G. Bellin's page profs.sci.univr.it/~bellin/papers.html) which received a seminal rough set semantic.

They are fields to be explored.

6   Additional Topics to Be Investigated

I end this paper with a couple of challenging but very interesting research topics. In view of the above discussion about the local classical subdomain B, one can affirm that neither Constructive Logic with Strong Negation nor three-valued Łukasiewicz logic can be considered the "true" logic of rough sets. Probably the "true" logic of rough sets, if any, requires typed variables and typed deduction steps (as in Labelled Deductive Systems - see [4]). Moreover, Nelson logic with strong negation is a substructural logic over FL_{ew} (the sequent calculus with exchange and weakening) [38]. What does it mean in rough set terms? What is the role of structural rules in the logic of rough set systems?

Secondly, Formal Concept Analysis has received at least two important enhancements. The first is the introduction of *triadic formal contexts*, that are relational data structures in which an object fulfills a property *under certain conditions* ([10]). It is a very useful generalisation, close to real-life problems. The second important improvement is, to me, *multi-adjoint formal concepts* ([11]). I think that Rough Set Theory would profit from similar improvements.

References

1. Balbes, R., Dwinger, P.: Distributive Lattices. University of Missouri Press, Columbia (1974)
2. Carnielli, W.A., Marques, M.L.: Society semantics and multiple-valued logics. In: Carnielli W.A., D'Ottaviano, I. (Eds.): Contemporary Mathematics, vol. 235, pp. 149–163. American Mathematical Society (1999)
3. Ciucci, D., Dubois, D.: Three-valued logics, uncertainty management and rough sets. Trans. Rough Sets **17**, 1–32 (2014)
4. Gabbay, D.M.: Labelled Deductive Systems. Oxford University Press, Oxford (1997)
5. Greco, S., Matarazzo, B., Slowinski, R.: Algebra and topology for dominance-based rough set approach. In: Ras, Z.W., Tsay, L.-S. (eds.) Advanced in Intelligent Information Systems. SCI, vol. 265, pp. 43–78. Springer, Heidelberg (2010)
6. Iwinski, T.B.: Algebraic approach to rough sets. Bull. Pol. Acad. Sci. Math. **35**(3–4), 673–683 (1987)
7. Järvinen, J., Pagliani, P., Radeleczki, S.: Information completeness in Nelson algebras of rough sets induced by quasiorders. Stud. Logica **101**(5), 1073–1092 (2013)
8. Järvinen, J., Radeleczki, S.: Representation of Nelson algebras by rough sets determined by quasiorders. Algebra Univers. **66**, 163–179 (2011)
9. Khan, M.A., Banerjee, M.: A study of multiple-source approximation systems. In: Peters, J.F., Skowron, A., Słowiński, R., Lingras, P., Miao, D., Tsumoto, S. (eds.) Transactions on Rough Sets XII. LNCS, vol. 6190, pp. 46–75. Springer, Heidelberg (2010)

10. Lehmann, F., Wille, R.: A triadic approach to formal concept analysis. In: Ellis, G., Rich, W., Levinson, R., Sowa, J.F. (eds.) ICCS 1995. LNCS, vol. 954, pp. 32–43. Springer, Heidelberg (1995)
11. Medina, J., Ojeda-Aciego, M., Ruiz-Calviño, J.: Formal concept analysis via multiadjoint concept lattices. Fuzzy Sets Syst. **160**(2), 130–144 (2009)
12. Miglioli, P.A., Moscato, U., Ornaghi, M., Usberti, U.: A constructivism based on classical truth. Notre Dame J. Formal logic **30**(1), 67–90 (1989)
13. Nagarajan, E.K.R., Umadevi, D.: A method of representing rough sets system determined by quasi orders. Order **30**(1), 313–337 (2013)
14. Orłowska, E.: Logic for nondeterministic information. Stud. Logica **44**, 93–102 (1985)
15. Pagliani, P.: Remarks on special lattices and related constructive logics with strong negation. Notre Dame J. Formal Logic **31**, 515–528 (1990)
16. Pagliani, P.: A pure logic-algebraic analysis on rough top and rough bottom equalities. In: Ziarko, W.P. (ed.) Rough Sets, Fuzzy Sets and Knowledge Discovery. Workshops in Computing, pp. 225–236. Springer, Heidelberg (1994)
17. Pagliani, P.: A modal relation algebra for generalized approximation spaces. In: Tsumoto, S., Kobayashi, S., Yokomori, T., Tanaka, H., Nakamura, A. (eds.) Proceedings of the 4th International Workshop on Rough Sets, Fuzzy Sets, and Machine Discovery. The University of Tokyo, Japan, Invited Section "Logic and Algebra", pp. 89–96, 6–8 November 1996
18. Pagliani, P.: From information gaps to communication needs: a new semantic foundation for some non-classical logics. J. Logic Lang. Inf. **6**(1), 63–99 (1997)
19. Pagliani, P.: Rough set systems and logic-algebraic structures. In: Orłowska, E. (ed.) Incomplete Information: Rough Set Analysis. STUDFUZZ, vol. 13, pp. 109–190. Physica-Verlag, Heidelberg (1997)
20. Pagliani, P.: A practical introduction to the modal relational approach to approximation spaces. In: Polkowski, L., Skowron, A. (eds.) Rough Sets in Knowledge Discovery 1, pp. 209–232. Physica-Verlag, Heidelberg (1998)
21. Pagliani P.: Modalizing relations by means of relations: a general framework for two basic approaches to knowledge discovery in database. In: Gevers, M. (ed.) Proceedings of the 7th International Conference on Information Processing and Management of Uncertainty in Knowledge-Based Systems. IPMU '98. Paris, France, pp. 1175–1182. Editions E.D.K., 6–10 July 1998
22. Pagliani, P.: Local classical behaviours in three-valued logics and connected systems. An information-oriented analysis. 1 and 2. Journal of Multiple-valued Logics **5**, **6**, pp. 327–347, 369–392 (2000, 2001)
23. Pagliani, P.: Pretopology and dynamic spaces. In: Proceedings of RSFSGRC'03, Chongqing, R.P. China, 2003. Extended version in Fundamenta Informaticae, vol. 59(2–3), pp. 221–239 (2004)
24. Pagliani, P.: Rough sets and other mathematics: ten research programs. In: Chakraborty, M.K., Skovron, A., Maiti, M., Kar, S. (eds.) Facets of Uncertainties and Applications. Springer Proceedings in Mathematics & Statistics, vol. 125, pp. 3–15. Springer, Heidelberg (2015)
25. Pagliani, P.: The relational construction of conceptual patterns - tools, implementation and theory. In: Kryszkiewicz, M., Cornelis, C., Ciucci, D., Medina-Moreno, J., Motoda, H., Raś, Z.W. (eds.) RSEISP 2014. LNCS, vol. 8537, pp. 14–27. Springer, Heidelberg (2014)
26. Pagliani P.: Covering-based rough sets and formal topology. A uniform approach (To appear in Transactions of Rough Sets) (2014)

27. Pagliani P., Chakraborty M.K.: Information quanta and approximation spaces. I: Non-classical approximation operators. In: Proceedings of the IEEE International Conference on Granular Computing, Beijing, R.P. China, Vol. 2, pp. 605–610. IEEE Los Alamitos, 25–27 July 2005

28. Pagliani, P., Chakraborty, M.: A geometry of approximation. Trends in Logic. Springer, Heidelberg (2008)

29. Qian, Y., Liang, J., Yao, Y., Dang, C.: MGRS: a multi-granulation rough set. Inf. Sci. **180**(6), 949–970 (2010)

30. Rauszer, C.M.: Rough logic for multi-agent systems. In: Masuch, M., Polos, L. (eds.) Logic at Work 1992. LNCS, vol. 808, pp. 161–181. Springer, Heidelberg (1994)

31. Reyes, G.E., Zolfaghari, N.: Bi-Heyting algebras, toposes and modalities. J. Philos. Logic **25**, 25–43 (1996)

32. Sambin, G.: Intuitionistic formal spaces and their neighbourhood. In: Ferro, R., Bonotto, C., Valentini, S., Zanardo, A. (eds.) Logic Colloquium '88, pp. 261–285. North-Holland, Elsevier (1989)

33. Sendlewski, A.: Nelson algebras through Heyting ones: I. Stud. Logica **49**(1), 105–126 (1990)

34. Sette, A.M.: On the propositional calculus P^1. Math. Japonicae **18**, 173–180 (1973)

35. Sette, A.M., Carnielli, W.A.: Maximal weakly-intuitionistic logics. Stud. Logica **55**, 181–203 (1995)

36. Skowron, A., Stepaniuk, J.: Approximation of relations. In: Ziarko, W.P. (ed.) Rough Sets, Fuzzy Sets and Knowledge Discovery. Workshops in Computing, pp. 161–166. Springer, Heidelberg (1994)

37. Skowron, A., Stepaniuk, J., Peters, J.F.: Rough sets and infomorphisms: towards approximation of relations in distributed environments. Fundam. Informaticae **54**, 263–277 (2003)

38. Spinks, M., Veroff, R.: Constructive logic with strong negation is a substructural logic - I and II. Studia Logica **88, 89**, 325–348, 401–425 (2008)

39. Wansing, H.: The Logic of Information Structures. Springer, Berlin (1993)

40. Wille, R.: Restructuring lattice theory. In: Rival, I. (ed.) Ordered Sets. NATO ASI Series 83, Reidel, pp. 445–470 (1982)

41. Yao, Y.Y.: The two sides of the theory of rough sets. Knowl.-Based Syst. **80**, 67–77 (2015)

Interactive Granular Computing

Andrzej Skowron[1,2]($\boxtimes$) and Andrzej Jankowski[3]

[1] Institute of Mathematics, Warsaw University, Banacha 2, 02-097 Warsaw, Poland
skowron@mimuw.edu.pl
[2] Systems Research Institute Polish, Academy of Sciences,
Newelska 6, 01-447 Warsaw, Poland
[3] The Dziubanski Foundation of Knowledge Technology,
Nowogrodzka 31, 00-511 Warsaw, Poland
andrzej.adgam@gmail.com

Abstract. Decision support in solving problems related to complex systems requires relevant computation models for the agents as well as methods for incorporating reasoning over computations performed by agents. Agents are performing computations on complex objects (e.g., (behavioral) patterns, classifiers, clusters, structural objects, sets of rules, aggregation operations, (approximate) reasoning schemes etc.). In Granular Computing (GrC), all such constructed and/or induced objects are called granules. To model, crucial for the complex systems, interactive computations performed by agents, we extend the existing GrC approach to Interactive Granular Computing (IGrC) by introducing complex granules (c-granules or granules, for short). Many advanced tasks, concerning complex systems may be classified as control tasks performed by agents aiming at achieving the high quality computational trajectories relative to the considered quality measures over the trajectories. Here, new challenges are to develop strategies to control, predict, and bound the behavior of the system. We propose to investigate these challenges using the IGrC framework. The reasoning, which aims at controlling the computational schemes, in order to achieve the required targets, is called an adaptive judgement. This reasoning deals with granules and computations over them. Adaptive judgement is more than a mixture of reasoning based on deduction, induction and abduction. Due to the uncertainty the agents generally cannot predict exactly the results of actions (or plans). Moreover, the approximations of the complex vague concepts initiating actions (or plans) are drifting with time. Hence, adaptive strategies for evolving approximations of concepts are needed. In particular, the adaptive judgement is very much needed in the efficiency management of granular computations, carried out by agents, for risk assessment, risk treatment, and cost/benefit analysis. In the lecture, we emphasize the role of the rough set based methods in IGrC. The discussed approach

This work was partially supported by the Polish National Science Centre (NCN) grants DEC-2011/01/D/ST6/06981, DEC-2012/05/B/ST6/03215 as well as by the Polish National Centre for Research and Development (NCBiR) under the grant O ROB/0010/03/001.

The original version of this chapter was revised: The acknowledgement was modified. The correction to this chapter is available at https://doi.org/10.1007/978-3-319-25754-9_47

D. Ciucci et al. (Eds.): RSKT 2015, LNAI 9436, pp. 50–61, 2015.
DOI: 10.1007/978-3-319-25754-9_5

is a step towards realization of the Wisdom Technology (WisTech) program, and is developed over years of experiences, based on the work on different real-life projects.

Keywords: Rough set · (Interactive) Granular Computing · Adaptive judgement · Efficiency management · Risk management · Cost/benefit analysis · Cyber-physical system · Wisdom web of things · Ultra-large system

1   Introduction

Ultra-Large-Scale (ULS) systems [1,19] are interdependent webs of software-intensive systems, people, policies, cultures, and economics. ULS are characterized by properties such as: (i) decentralization, (ii) inherently conflicting, unpredictable, and diverse requirements, (iii) continuous evolution and deployment, (iv) heterogeneous, inconsistent, and changing elements, (v) erosion of the people/system boundary, and (vi) routine failures [1,19]. Cyber-Physical Systems (CPSs) [12] and/or systems based on Wisdom Web of Things (W2T) [34] can be treated as special cases of ULS. It is predicted that applications based on the above mentioned systems will have enormous societal impact and economic benefit. However, there are many challenges related to such systems. In this article, we claim that further development of such systems should be based on the relevant computation models.

There are several important issues which should be taken into account in developing such computation models. Among them some are (i) computations are performed on complex objects with very different structures, where the structures themselves are constructed and/or induced from data and domain knowledge, (ii) computations are performed in an open world and they are dependent on interactions of physical objects, (iii) due to uncertainty, the properties and results of interactions can be perceived by agents only partially, (iv) computations are realized in the societies of interacting agents including humans, (v) agents are aiming at achieving their tasks by controlling computations, (vi) agents can control computations by using *adaptive judgement*, in which all of deduction, induction and abduction are used.

We propose to base the relevant computation model on the Interactive Granular Computing (IGrC) framework proposed recently as an extension of the Granular Computing (GrC). The label Granular Computing was suggested by T.Y. Lin in late 1990s.

Granulation of information is inherent in human thinking and reasoning processes. It is often realized that precision is sometimes expensive and not very meaningful in modeling and controlling complex systems. When a problem involves incomplete, uncertain, and vague information, it may be difficult to discern distinct objects, and one may find it convenient to consider granules for tackling the problem of concern. Granules are composed of objects that are drawn together by indiscernibility, similarity, and/or functionality among the objects [30]. Each of the granules according to its structure and size, with a certain level of granularity, may reflect a specific aspect of the problem, or form a portion of the system's domain. GrC is considered to be an effective framework in the design and implementation of intelligent systems for various real life applications. The systems based on GrC, *e.g.*, for pattern recognition, exploit the tolerance for imprecision, uncertainty, approximate reasoning as well as partial truth of soft computing framework and are capable of achieving tractability, robustness, and close resemblance with human-like (natural) decision-making [2,17,18,22].

In GrC, computations are performed on granules of different structures, where granularity of information plays an important role. Information granules (infogranules, for short) in GrC are widely discussed in the literature [17]. In particular, let us mention here the rough granular computing approach based on the rough set approach and its combination with other approaches to soft computing, such as fuzzy sets. However, the issues related to the interactions of infogranules with the physical world, and perception of interactions in the physical world by means of infogranules are not well elaborated yet. On the other hand, the understanding of interactions is the critical issue of complex systems [4]. For example, the ULS are autonomous or semiautonomous systems, and cannot be designed as closed systems that operate in isolation; rather, the interaction and potential interference among smart components, among CPSs, and among CPSs and humans, require to be modeled by coordinated, controlled, and cooperative behavior of agents representing components of the system [1].

We extend the existing GrC approach to IGrC by introducing *complex granules* (*c-granules*, for short) [6] making it possible to model interactive computations carried out by agents and their teams in complex systems working in an open world.

Any agent operates on a local world of c-granules. The agent aims at controlling computations performed on c-granules from this local world for achieving the target goals. In our approach, computations in systems based on IGrC proceed through complex interactions among physical objects. Some results of such interactions are perceived by agents with help of c-granules.

The discussed approach is a step towards one way of realization of the Wisdom Technology (WisTech) program [6,7]. The approach was developed over years of work on different real-life projects.

This article is organized as follows. In Sect. 2, an introduction to Interactive Granular Computing (IGrC) is presented. In particular, we present intuitions concerning the definition of c-granules and we discuss computations over complex granules realized by agents. In Sect. 3, some issues on reasoning based on adaptive judgement are discussed. Section 4 concludes the paper.

The paper summarizes as well as extends the work developed in [9,10,21].

2   Complex Granules

Infogranules are widely discussed in the literature. They can be treated as specifications of compound objects which are defined in a hierarchical manner together with descriptions regarding their implementations. Such granules are obtained as the result of information granulation [33]:

Information granulation can be viewed as a human way of achieving data compression and it plays a key role in implementation of the strategy of divide-and-conquer in human problem-solving.

Infogranules belong to those concepts which play the main role in developing foundations of Artificial Intelligence (AI), data mining, and text mining [17]. They grew up as some generalizations from fuzzy set theory, [28,31,33], rough set theory, and interval analysis [17]. In GrC, rough sets, fuzzy sets, and interval analysis are used to deal with vague concepts. However, the issues related to the interactions of infogranules with the physical world, and their relationship to perception of interactions in the physical world are not well elaborated yet [4]. On the other hand, in [14], it is mentioned that:

[...] *interaction is a critical issue in the understanding of complex systems of any sorts: as such, it has emerged in several well-established scientific areas other than computer science, like biology, physics, social and organizational sciences.*

Interactive computations in IGrC [6,8,21,24,25] are realized by agents on c-granules linking, *e.g.*, infogranules [17] with physical objects from the environment perceived through "windows" (from the agent spatiotemporal space). Physical objects labeled by such windows are called hunks [5,6].

Computations of agents proceed due to interactions in the physical world and they have roots in c-granules [6].

C-granules are defined relative to a given agent. We assume that the agent can perceive physical objects using "windows", *i.e.*, fragments of spatiotemporal space generated by the agent control. C-granules are making it possible to perceive by the agent properties of hunks and their interactions. Any c-granule is synthesized with three physical components, namely soft_suit, link_suit and hard_suit. The soft_suit component of a given c-granule is used to record perceived by the c-granule properties of hunks and their interactions. The link_suit of a given c-granule is used as a kind of c-granule transmission channel for handling interaction between soft_suit and hard_suit. In the soft_suit are encoded procedures for recording (over the selected fragment of the spatiotemporal space) some properties of interactions among hunks in the hard_suit which are transmitted to soft_suit using link_suit. We assume that the relevant pointers to the link_suit, hard_suit, and/or soft_suit are represented in the soft_suit making it possible to localize these components. We also assume that in the soft_suit is represented an information about the expected result of the perceived interactions in the hard_suit.

Interactions of the agent with the environment are realized using configurations of c-granules. The actual agent configuration of c-granules is evolving in the (local agent) time due to interactions of c-granules with the environment, including the agent control. This is leading to changes of the existing configuration of c-granules caused by (i) extending them by new c-granules selected by the agent control for perceiving new interactions (also stimulated by c-granules), (ii) extending the configuration by new c-granules for encoding the results of the

perceived interactions, or (iii) deleting some c-granules from the current configuration, (iv) other kinds of modifications of some parts of the configuration.

Calculi of c-granules are defined beginning from some elementary c-granules.
Then the calculi defined by the control as well as interactions with the environment are making it possible to generate new c-granules from the already defined
ones. The hard_suits, link_suits, and soft_suits of more complex c-granules are
defined using the relevant networks of already defined c-granules. The networks
are satisfying some constraints which can be interpreted as definitions of types of
networks. The link_suits of such more complex granules are responsible for transmission of interactions between the hard_suits and the soft_suits represented by
networks. The results or properties of transmitted interactions are recorded in
their sof_suits.

We assume that for any agent there is a distinguished family of her/his
c-granules creating the *internal language* of the agent [6]. We assume that elements of the internal language can be encoded by information granules (or,
infogranules for short).

One should note that the process of (discovering) distinguishing the relevant
family of c-granules creating the internal language is a very complex process. The
relevant infogranules are discovered in hierarchical aggregation of infogranules
considered in relevant contexts. In general, such infogranules are called semiotic
c-granules [6]. Infogranules are used for constructing the target infogranules. On
the basis of satisfiability (to a degree) of such target infogranules (interpreted
as approximations of complex vague concepts) relevant actions are undertaken
by the agent aiming to satisfy her/his needs. An example of the agent internal
language can be defined by c-granules representing propositional formulas over
descriptors [15, 16].

An illustrative example of c-granule of an agent ag is presented in Fig. 1,
where (i) h_i are hunks corresponding to "space windows" (*i.e.*, windows in spatiotemporal space constant over the agent time in the example) of c-granules in the
network, (ii) s_i denote link_suits for transmitting interactions form h_i in the
environment ENV to soft_suits of c-granules in the network, (iii) S, S' are trees
representing hierarchical aggregations of c-granules leading from some input c-
granules to some output c-granules grounded on hunks h, h'. These hunks are
encoded by infogranules C, C' from the private agent language, where C, C' represent approximations of complex vague concepts used for initiation of actions
ac_i. The states (in context of the given c-granule g) of hunks h, h' at a given
slot (moment) t of the agent time are recorded as some properties of h, h' and
they are the perception results (at the agent time moment t) of h, h', respectively. The states are interpreted as satisfiability degrees of C, C'. In this way, the
perception of the current situations in the environment ENV are represented
and c-granules representing actions ac_i are initiated on the basis of the satisfiability degrees of C, C' representing the currently perceived situation in the
environment ENV. The process of perceiving the current situation is realized
by transmitting interactions from hunks corresponding to "space windows" h_i
through links s_i and link_suits in S, S' to hunks h, h' representing the perceived
current situation. These interactions lead to changes of states of h, h'. These

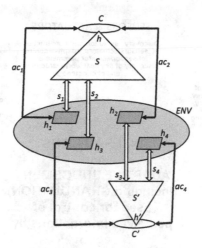

Fig. 1. An example of c-granule g defined as a network (configuration) of other c-granules.

changes are encoded by changes of degrees of satisfiability of C, C'. C-granules for actions ac_i are responsible for initiating interactions in "space windows" h_i corresponding to actions ac_i. The results of the modified interactions caused by actions are transmitted through the network to h, h' leading to modification of their states.

One can consider the following interpretation for c-granule presented in Fig. 1. There are two concepts C and C' representing the agent needs related to, $e.g.$, the security of energy supply from different sources. These sources are influencing the satisfiability of C (C') and are related to h_1, h_2, h_3 (h_2, h_3, h_4). The agent is aiming to keep the satisfiability of these concepts on safe levels, $i.e.$, above given thresholds. For improving satisfiability of C ($i.e.$, to obtain a satisfactory satisfiability of C) first the action ac_1 is performed for compensation of negative influences (on satisfiability of C) of changes in sources related to h_3. If this is not satisfactory the action ac_2 is used to gain more energy sources related to h_2. The results of action ac_2 may influence the satisfiability of C' leading $e.g.$, to decreasing of the sources related to h_2 (which in turn may influence h_1). The actions ac_3, ac_4 are then used to improve the satisfiability of C'. The agent is using the described c-granule in a context of a more compound granule making it possible to preserve the satisfiability of concepts C and C' on safe levels.

The discussed c-granules may represent complex objects. In particular, agents and their societies can be treated as c-granules too. An example of c-granule representing a team of agents is presented in Fig. 2, where some guidelines for implementation of AI projects in the form of a cooperation scheme itself among different agents responsible for relevant cooperation areas is illustrated [6]. This cooperation scheme may be treated as a higher level c-granule. We propose to model a complex system as a society of agents. Moreover, c-granules create the basis for the construction of the agent's language of communication and the language of evolution.

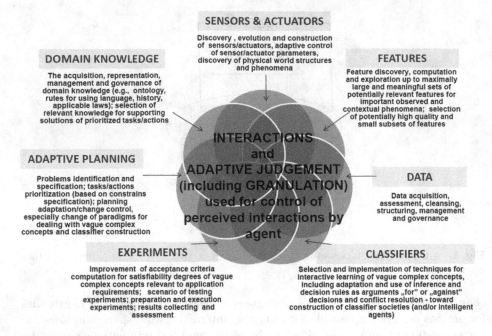

Fig. 2. Cooperation scheme of an agent team responsible for relevant competence area [6].

An agent operates on a local world of c-granules. The control of an agent aims at controlling computations performed on c-granules, from the respective local world of the agent for achieving the target goals. Actions, also represented as c-granules, are used by the agent's control in exploration and/or exploitation of the environment on the way to achieve their targets. C-granules are also used for representation by agents their perception of interactions with the physical world. Due to the limited ability of agent's perception usually only a partial information about the interactions of the physical world may be available to the agents. Hence, in particular the results of performed actions by agents cannot be predicted with certainty. For more details on IGrC based on c-granules the readers are referred to [6].

One of the key issues of the approach related to c-granules presented in [6], is a kind of integration between investigations of physical and mental phenomena. This idea of integration follows from the suggestions presented by many scientists.

3   Adaptive Judgement

The reasoning used to support problem solving, which makes it possible to derive relevant c-granules and control of interactive computations over c-granules for achieving the target goals is called an *adaptive judgement*.

Adaptive judgement over interactive computations is a mixture of reasoning based in on deduction, abduction, induction, which make use of case based or analogy based reasoning, experience, observed changes in the environment, or meta-heuristics from natural computing. However, the meaning of practical judgement goes beyond typical tools for reasoning based on deduction or induction [26]:

> *Practical judgement is not algebraic calculation. Prior to any deductive or inductive reckoning, the judge is involved in selecting objects and relationships for attention and assessing their interactions. Identifying things of importance from a potentially endless pool of candidates, assessing their relative significance, and evaluating their relationships is well beyond the jurisdiction of reason.*

For example, a particular question for the agent's control concerns discovering strategies for models of dynamic changes of the agent's attention. This may be related to discovery of changes in a relevant context necessary for the judgement.

Let us note that the *intuitive judgement* and the *rational judgement* are distinguished as different kinds of judgement in [11].

Among the tasks for adaptive judgement following are the ones supporting reasoning towards,

- inducing relevant classifiers (*e.g.*, searching for relevant approximation spaces, discovery of new features, selection of relevant features (attributes), rule induction, discovery of inclusion measures, strategies for conflict resolution, adaptation of measures based on the minimum description length principle),
- prediction of changes,
- initiation of relevant actions or plans,
- discovery of relevant contexts,
- adaptation of different sorts of strategies (*e.g.*, for existing data models, quality measure over computations realized by agents, objects structures, knowledge representation and interaction with knowledge bases, ontology acquisition and approximation),
- hierarchy of needs, or for identifying problems to be solved according to priority),
- learning the measures of inclusion between granules from sources using different languages (*e.g.*, the formal language of the system and the user natural language) through dialogue,
- strategies for development and evolution of communication language among agents in distributed environments, and
- strategies for efficiency management in distributed computational systems.

Below we discuss some issues related to adaptive judgement in IGrC.

In some cases judgement methods may be based on formal languages, *i.e.*, the expressions from such languages are used as labels (syntax) of granules. However, there are also paradigms such as Computing With Words (CWW), due to Professor Lotfi Zadeh [13,27,29,31–33], where labels of granules are *words* (*i.e.*, words or expressions from a relevant fragment of natural language), and computations are performed on *words* (http://www.cs.berkeley.edu/~zadeh/presentations.html). In IGrC it is necessary to develop new methods extending the approaches for approximating vagues concepts expressed in natural language for *approximating* reasoning on such concepts based on adaptive judgement. It is also important to note that information granulation plays a key role in implementation of the strategy of divide-and-conquer in human problem-solving [28, 33]. Hence, it is important to develop methods which could perform approximate reasoning along such decomposition schemes, delivered by the strategy of divide-and-conquer in human problem-solving, and induce the relevant granules as computational building blocks for constructing the solutions for the considered problems.

In case of systems based on IGrC, the users are often specifying problems in fragments of a natural language with the requirement, that their solutions will satisfy specifications to some satisfactory degrees. Hence, methods for approximation of domain ontology (*i.e.*, ontology on which a fragment is based) as well as approximations of constructions representing solutions based on concepts from the domain ontology should be developed. The rough set approach in combination with other soft computing approaches is used for approximation of the vague concepts [3]. These approximations may help the system to follow, in an approximate sense, the judgement schemes expressed in the relevant (for considered problems) fragment of a natural language. It is worthwhile to emphasize here the importance of dialogues between users and system in the process of obtaining the relevant approximations.

Very often the problems related to systems based on IGrC concern control tasks. Examples of control tasks may be found in different areas, such as the medical therapy support, management of large software projects, algorithmic trading or control on unmanned vehicles, to name a few. Such projects are typical for ULS. Any of such exemplary projects is supported by (large) data and domain knowledge distributed over computer networks and/or Internet. Moreover, interactions of agents with the physical world, which are often unpredictable, are unavoidable. Computations performed by agents are aiming at constructing, learning, or discovering granules, which in turn makes it possible to understand the concerned situation (state) to a satisfactory degree. The relevant controlling of computations based on this situation understanding is realized using approximations of complex vague concepts playing the role of guards, responsible for initiation of actions (or plans) by agents. In particular, for constructing these approximations different kinds of granules, discovered from data, are used. The main processes, namely granulation and degranulation, characterize respectively the synthesis and decomposition of granules in the process of obtaining relevant resultant granules.

The efficiency management in controlling the computations [6] in IGrC are of great importance for the successful behavior of individuals, groups or societies of agents. In particular, such efficiency management is important for constructing systems based on large data for supporting users in problem solving. The efficiency management covers risk assessment, risk treatment, and cost/beneft analysis. The tasks related to this management are related to control tasks aiming at achieving the high quality performance of (societies of) agents. One of the challenges in efficiency management is to develop methods and strategies for adaptive judgement related to adaptive control of computations. The efficiency management in decision systems requires tools to discover, represent, and access approximate reasoning schemes (ARSs) (over domain ontologies) representing the judgement schemes [3,23]. ARSs are approximating, in a sense, judgement expressed in relevant fragments of simplified natural language. Methods for inducing of ARSs are still under development. The systems for problem solving are enriched not only by approximations of concepts and relations from ontologies but also by ARSs.

We would like to stress that still much work should be done to develop approximate reasoning methods about complex vague concepts for the progress of the development of IGrC, in particular for the efficiency management in systems based on IGrC. This idea was very well expressed by Leslie Valiant (see, *e.g.*, http://en.wikipedia.org/wiki/Vagueness, http://people.seas.harvard. edu/~valiant/researchinterests.htm):

> *A fundamental question for artificial intelligence is to characterize the computational building blocks that are necessary for cognition. A specific challenge is to build on the success of machine learning so as to cover broader issues in intelligence. [...] This requires, in particular a reconciliation between two contradictory characteristics – the apparent logical nature of reasoning and the statistical nature of learning.*

4 Conclusions

The approach for modeling interactive computations based on c-granules is presented, and its importance for the efficiency management of controlling interactive computations over c-granules is outlined. It is worthwhile mentioning that in modeling and/or discovering granules, tools from different areas are used. Among these areas some are, machine learning, data mining, multi-agent systems, complex adaptive systems, logic, cognitive science, neuroscience, and soft computing. IGrC is aiming at developing a unified methodology for modeling and controlling computations over complex objects, called c-granules, as well as for reasoning about such objects and computations over them. In particular, such a methodology is of great importance for ULS.

The discussed concepts such as interactive computation and adaptive judgement are among the basic ingredients in the field of WisTech. Let us mention here the WisTech meta-equation:

$$\text{WISDOM} = \tag{1}$$
$$\text{INTERACTIONS} + \text{ADAPTIVE JUDGEMENT} + \text{KNOWLEDGE.}$$

The presented approach has a potential for being used for developing computing models in different areas related to complex systems.

In our research, we plan to further develop the foundations of interactive computations based on c-granules. The approach will be used for development of modeling and analysis of computations in Natural Computing [20], W2Ts [34], CPSs [12], and ULS [1].

References

1. Cyber-physical and ultra-large scale systems (2013). http://resources.sei.cmu.edu/library/asset-view.cfm?assetid=85282
2. Bargiela, A., Pedrycz, W. (eds.): Granular Computing: An Introduction. Kluwer Academic Publishers, Boston (2003)
3. Bazan, J.: Hierarchical classifiers for complex spatio-temporal concepts. In: Peters, J.F., Skowron, A., Rybiński, H. (eds.) Transactions on Rough Sets IX. LNCS, vol. 5390, pp. 474–750. Springer, Heidelberg (2008)
4. Goldin, D., Smolka, S., Wegner, P. (eds.): Interactive Computation: The New Paradigm. Springer, Heidelberg (2006)
5. Heller, M.: The Ontology of Physical Objects. Four Dimensional Hunks of Matter. Cambridge Studies in Philosophy. Cambridge University Press, Cambridge (1990)
6. Gegov, A.: Conclusion. In: Gegov, A. (ed.) Fuzzy Networks for Complex Systems. STUDFUZZ, vol. 259, pp. 275–277. Springer, Heidelberg (2010)
7. Jankowski, A., Skowron, A.: A wistech paradigm for intelligent systems. In: Peters, J.F., Skowron, A., Düntsch, I., Grzymała-Busse, J.W., Orłowska, E., Polkowski, L. (eds.) Transactions on Rough Sets VI. LNCS, vol. 4374, pp. 94–132. Springer, Heidelberg (2007)
8. Jankowski, A., Skowron, A.: Wisdom technology: a rough-granular approach. In: Marciniak, M., Mykowiecka, A. (eds.) Aspects of Natural Language Processing. LNCS, vol. 5070, pp. 3–41. Springer, Heidelberg (2009)
9. Jankowski, A., Skowron, A., Swiniarski, R.W.: Interactive complex granules. Fundamenta Informaticae **133**, 181–196 (2014)
10. Jankowski, A., Skowron, A., Swiniarski, R.W.: Perspectives on uncertainty and risk in rough sets and interactive rough-granular computing. Fundamenta Informaticae **129**, 69–84 (2014)
11. Kahneman, D.: Maps of bounded rationality: psychology for behavioral economics. Am. Econ. Rev. **93**, 1449–1475 (2002)
12. Lamnabhi-Lagarrigue, F., Di Benedetto, M.D., Schoitsch, E.: Introduction to the special theme Cyber-Physical Systems. Ercim News **94**, 6–7 (2014)
13. Mendel, J.M., Zadeh, L.A., Trillas, E., Yager, R., Lawry, J., Hagras, H., Guadarrama, S.: What computing with words means to me. IEEE Comput. Intell. Mag. **5**(1), 20–26 (2010)
14. Omicini, A., Ricci, A., Viroli, M.: The multidisciplinary patterns of interaction from sciences to computer science. In: Goldin, D., et al. (eds.) Interactive Computation: The New Paradigm, pp. 395–414. Springer, Heidelberg (2006)
15. Pawlak, Z., Skowron, A.: Rudiments of rough sets. Inf. Sci. **177**(1), 3–27 (2007)
16. Pawlak, Z.: Rough Sets: Theoretical Aspects of Reasoning about Data, System Theory, Knowledge Engineering and Problem Solving, vol. 9. Kluwer Academic Publishers, Boston (1991)
17. Pedrycz, W., Skowron, S., Kreinovich, V. (eds.): Handbook of Granular Computing. Wiley, New York (2008)

18. Pedrycz, W.: Granular Computing Analysis and Design of Intelligent Systems. Taylor & Francis, CRC Press, Boca Raton (2013)
19. Pollak, B. (ed.): Ultra-Large-Scale Systems. The Software Challenge of the Future. Software Engineering Institute. CMU, Pittsburgh (2006)
20. Rozenberg, G., Bäck, T., Kok, J. (eds.): Handbook of Natural Computing. Springer, Heidelberg (2012)
21. Skowron, A., Jankowski, A., Wasilewski, P.: Risk management and interactive computational systems. J. Adv. Math. Appl. 1, 61–73 (2012)
22. Skowron, A., Pal, S.K., Nguyen, H.S. (eds.): Preface: Special issue on rough sets and fuzzy sets in natural computing. Theor. Comput. Sci. 412(42), 5816–5819 (2011)
23. Skowron, A., Stepaniuk, J.: Information granules and rough-neural computing. In: Pal, S.K., et al. (eds.) Rough-Neural Computing: Techniques for Computing with Words. Cognitive Technologies, pp. 43–84. Springer, Heidelberg (2004)
24. Skowron, A., Stepaniuk, J., Swiniarski, R.: Modeling rough granular computing based on approximation spaces. Inf. Sci. 184, 20–43 (2012)
25. Skowron, A., Wasilewski, P.: Information systems in modeling interactive computations on granules. Theor. Comput. Sci. 412(42), 5939–5959 (2011)
26. Thiele, L.P.: The Heart of Judgment: Practical Wisdom, Neuroscience, and Narrative. Cambridge University Press, Cambridge (2010)
27. Zadeh, A.: Computing with Words: Principal Concepts and Ideas. STUDFUZZ, vol. 277. Springer, Heidelberg (2012)
28. Zadeh, L.A.: Fuzzy sets and information granularity. In: Gupta, M., Ragade, R., Yager, R. (eds.) Advances in Fuzzy Set Theory and Applications, pp. 3–18. North-Holland Publishing Co., Amsterdam (1979)
29. Zadeh, L.A.: Fuzzy Logic = Computing With Words. IEEE Trans. Fuzzy Syst. 4, 103–111 (1996)
30. Zadeh, L.A.: Toward a theory of fuzzy information granulation and its centrality in human reasoning and fuzzy logic. Fuzzy Sets Syst. 90, 111–127 (1997)
31. Zadeh, L.A.: From computing with numbers to computing with words - from manipulation of measurements to manipulation of perceptions. IEEE Trans. Circuits Syst. 45, 105–119 (1999)
32. Zadeh, L.A.: Foreword. In: Pal, S.K., et al. (eds.) Rough-Neural Computing: Techniques for Computing with Words. Cognitive Technologies. Springer, Heidelberg (2004)
33. Zadeh, L.A.: A new direction in AI: toward a computational theory of perceptions. AI Mag. 22(1), 73–84 (2001)
34. Zhong, N., Ma, J.H., Huang, R., Liu, J., Yao, Y., Zhang, Y.X., Chen, J.: Research challenges and perspectives on Wisdom Web of Things (W2T). J. Supercomput. 64, 862–882 (2013)

Rough Sets and Three-Way Decisions

Yiyu Yao[✉]

Department of Computer Science, University of Regina,
Regina, SK S4S 0A2, Canada
yyao@cs.uregina.ca

Abstract. The notion of three-way decisions was originally introduced by the needs to explain the three regions of probabilistic rough sets. Recent studies show that rough set theory is only one of possible ways to construct three regions. A more general theory of three-way decisions has been proposed, embracing ideas from rough sets, interval sets, shadowed sets, three-way approximations of fuzzy sets, orthopairs, square of oppositions, and others. This paper presents a trisecting-and-acting framework of three-way decisions. With respect to trisecting, we divide a universal set into three regions. With respect to acting, we design most effective strategies for processing the three regions. The identification and explicit investigation of different strategies for different regions are a distinguishing feature of three-way decisions.

1 Introduction

In rough set theory [21,23], there exist two representations of approximations of an undefinable set by definable sets [38], namely, a pair of lower and upper approximations or three pair-wise disjoint positive, boundary and negative regions. Although the two representations are mathematically equivalent, they provide different hints when we attempt to generalize rough set theory or to study its relationships to other theories. In fact, the two representations have led to two distinct research directions and two useful generalizations of rough set theory.

The representation with a pair lower and upper approximations was used to establish a close connection between rough set theory and modal logics [20,41]. More specifically, the lower and upper approximation operators in rough set theory are interpreted, respectively, in terms of the necessity and possibility operators in modal logics [1]. This connection immediately suggests generalized rough set models induced by various classes of non-equivalence relations [41]. The representation with three regions motivated the introduction of a theory of three-way decisions (3WD)[1] [33–35]. The three regions are interpreted in terms of three types of classification rules: rules of acceptance, deferment and rejection for the positive, boundary and negative regions, respectively [10,13,15,16,19,30,33,34].

---

[1] TWD was first used as an abbreviation for three-way decisions. Hu [10] suggested 3WD, which seems to be a better abbreviation as we can abbreviate two-way decisions as 2WD.

© Springer International Publishing Switzerland 2015
D. Ciucci et al. (Eds.): RSKT 2015, LNAI 9436, pp. 62–73, 2015.
DOI: 10.1007/978-3-319-25754-9_6

The theory of three-way decision opens new avenues to extend rough set research. In fact, in a recent scientometrics study of rough sets research in the past three decades, JT Yao and Zhang concluded [31], "The theory of three-way decisions, motivated by rough set three-regions but goes beyond rough sets, is a promising research direction that may lead to new breakthrough".

On the one hand, the theory of three-way decisions adopts three pair-wise disjoint regions from rough set theory as one of its basic notions. On the other hand, it moves beyond rough set theory in the sense that the latter is only one of many possible ways to construct and to use three regions. Other theories involving three regions include, for example, interval sets [32], three-way approximations of fuzzy sets [7,46], shadowed sets [24], orthopairs (i.e., a pair of disjoint sets) [2–4], and squares of oppositions [5,8,36]. The theory of three-way decisions embraces ideas from these theories and, at the same time, introduces its own notions and concepts.

There is a fast growing interest in the theory of three-way decisions, resulting in several edited books [11,17,44] and extensive research results (for example, see a sample of papers published in 2015 [12,14,15,18,25–30,45,48,49] and in the session on three-way decisions in this volume). By generalizing results reported in [33,35,43], this paper presents a trisecting-and-acting framework of three-way decisions. The trisecting and acting reflect two important and fundamental components of three-way decisions.

2   Two Representations of Rough Set Approximations

Let U denote a finite non-empty universal set of objects and E denote an equivalence relation on U. The pair $apr = (U, E)$ is called a Pawlak approximation space [21]. The equivalence class containing an object $x \in U$ is given by $[x]_E = [x] = \{y \in U \mid xEy\}$, where the subscript E represents the equivalence relation E and is dropped when no confusion arises. A fundamental notion of rough set theory is the approximation of a subset of U by using the family of equivalence classes $U/E = \{[x] \mid x \in U\}$.

There exist two formulations for defining rough set approximations of a subset of objects $X \subseteq U$. One way to define rough set approximations is by using a pair of lower and upper approximations [21,23]:

$$\underline{apr}(X) = \{x \in U \mid [x] \subseteq X\},$$
$$\overline{apr}(X) = \{x \in U \mid [x] \cap X \neq \emptyset\} = \{x \in U \mid [x] \not\subseteq X^c\}, \qquad (1)$$

where $X^c = U - X$ denotes the complement of X. By definition, it follows that $\underline{apr}(X) \subseteq X \subseteq \overline{apr}(X)$, namely, X lies between its lower and upper approximations. Thus, this definition has a very appealing physical interpretation. Another way is to define rough set approximations by using three pair-wise disjoint positive, negative and boundary regions [40]:

$$\mathrm{POS}(X) = \{x \in U \mid [x] \subseteq X\},$$
$$\mathrm{NEG}(X) = \{x \in U \mid [x] \subseteq X^c\},$$
$$\mathrm{BND}(X) = \{x \in U \mid [x] \not\subseteq X \wedge [x] \not\subseteq X^c\}. \qquad (2)$$

By definition, the three regions are pair-wise disjoint and their union is the universe U. Since some of the regions may be empty, the three regions do not necessarily form a partition of the universe. In this paper, we call the family of three regions a tripartition, or a trisection, of the universe by slightly abusing the notion of a partition. Given any pair of two regions, we can derived the third one through set complement. Thus, the notion of tree regions is related to the notion of orthopairs [2,3], namely, a pair of disjoint sets.

The definition based on a pair of lower and upper approximations has been widely used in the main stream research. Unfortunately, the three regions were treated as a derived notion defined in terms of the lower and upper approximations [21]. By looking at Eqs. (1) and (2), we can conclude that the two formulations can, in fact, be developed independent of each other. In addition, the two formulations are related to each other as follows:

$$\mathrm{POS}(X) = \underline{apr}(X),$$
$$\mathrm{NEG}(X) = (\overline{apr}(X))^c,$$
$$\mathrm{BND}(X) = \overline{apr}(X) - \underline{apr}(X), \tag{3}$$

and

$$\underline{apr}(X) = \mathrm{POS}(X),$$
$$\overline{apr}(X) = \mathrm{POS}(X) \cup \mathrm{BND}(X). \tag{4}$$

Thus, one can formulate rough set theory by using either a pair of approximations or three regions as its primitive notions.

A salient of definitions given by Eqs. (1) and (2) is that approximations are uniformly defined in terms of the set-inclusion relation $\subseteq$. This is obtained at the expenses of using both X and its complement X^c. Eqs. (1) and (2) are qualitative rough set approximations. An advantage of the definitions using only the inclusion relation $\subseteq$ is that we can generalize qualitative approximations into quantitative approximations by considering a certain degree of inclusion [40]. Probabilistic rough sets, in particular, decision-theoretic rough sets [39,42], are examples of quantitative generalizations with three probabilistic positive, boundary and negative regions.

The notion of three-way decisions is introduced to meet the needs to explain three probabilistic regions [33,34]. The definition of approximations in the three-region-based formulation of rough set theory provides hints on building models of three-way classification [6,9,22,34]. The theory of three-way decisions adopts the notion of three regions from rough set theory as one of its basic notions.

3   Refining Formulations of Three-Way Decisions

Research on three-way decisions had been focused mainly on the division of a universal set into three pair-wise disjoint regions. Yao and Yu [43] argued that one must design different strategies for processing the three regions. In this section, we introduce a trisecting-and-acting framework. Within the framework, we examine the three main stages in the evolution of three-way decisions, moving from specific to more general models.

3.1 Probabilistic Three-Way Classifications

The first attempt to formulate three-way decisions is based on a new interpretation of probabilistic rough sets for three-way classification [33]. We assume that the conditional probability $Pr(X|[x])$ denotes the degree to which $[x]$ is a subset of X and $Pr(X^c|[x]) = 1 - Pr(X|[x])$ denotes the degree to which $[x]$ is a subset of X^c. One of the results of the decision-theoretic rough set model [39,42] is three probabilistic positive, boundary and negative regions defined by: for a pair of thresholds (α, β) with $0 \leq \beta < \alpha \leq 1$,

$$
\begin{aligned}
\mathrm{POS}_{(\alpha,\beta)}(X) &= \{x \in U \mid Pr(X|[x]) \geq \alpha\} \\
&= \{x \in U \mid Pr(X^c|[x]) \leq 1 - \alpha\}, \\
\mathrm{BND}_{(\alpha,\beta)}(X) &= \{x \in U \mid \beta < Pr(X|[x]) < \alpha\}, \\
\mathrm{NEG}_{(\alpha,\beta)}(X) &= \{x \in U \mid Pr(X|[x]) \leq \beta\} \\
&= \{x \in U \mid Pr(X^c|[x]) \geq 1 - \beta\}.
\end{aligned}
\tag{5}
$$

The Pawlak rough set model is a special case in which $\alpha = 1$ and $\beta = 0$.

While Pawlak positive and negative regions do not have classification error, probabilistic positive and negative regions have classification errors. This observation calls for a new interpretation of probabilistic regions. Yao [33,34] introduces the notion of three-way decisions for such a purpose. Consider an arbitrary object $y \in [x]$. We make one of the following three decisions:

- If $Pr(X|[x]) \geq \alpha$, y has a high probability at or above α to be in X. We accept y being an object in X, with an understanding that this acceptance is associated with an error rate of $Pr(X^c|[x]) \leq 1 - \alpha$.
- If $Pr(X|[x]) \leq \beta$, or equivalently, $Pr(X^c|[x]) \geq 1-\beta$, y has a low probability at or below β to be in X. We reject y being an object in X, with an understanding that this rejection is associated with an error rate of $Pr(X|[x]) \leq \beta$.
- If $\beta < Pr(X|[x]) < \alpha$, the probability of y in X is neither high nor low, but in the middle. We can neither accept nor reject y being an object in X. In this case, we make a noncomittment decision.

The pair of thresholds is related to the tolerant levels of errors for acceptance and rejection. The three actions are meaningful. In the light of the trisecting-and-acting framework, Fig. 1 illustrates the two basic components of a three-way classification model. Equation (5) achieves the goal of trisecting U. Decisions of acceptance, noncommitment or rejection are strategies for the three regions.

One can observe that three-way classifications directly borrow notions, concepts and terminology from rough set theory. In the context of classification, the physical meaning of positive, boundary and negative regions, and the associated decisions of acceptance, noncommitment and rejection, is meaningful. Unfortunately, these notions may not be appropriate to describe other types of three-way decisions. New formulations of three-way decisions are required.

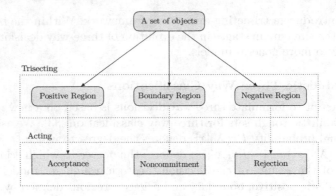

Fig. 1. Three-way classification

3.2 Evaluation-Based Three-Way Decisions

The second attempt to formulate three-way decisions is based on an evaluation function and a description of three regions in generic terms. The result is an evaluation-based model of three-way decisions [35].

Suppose $(\mathbf{L}, \preceq)$ a totally ordered set. For two objects $x, y \in U$, if $x \preceq y$, we also write $y \succeq x$. Suppose $v : U \longrightarrow \mathbf{L}$ is an evaluation function. For an object $x \in U$, $v(x)$ is its evaluation status value (ESV). Based on their evaluation status values, one can arrange objects in U in an increasing order: objects with lower values to the left and objects with larger values to the right. By using a pair of thresholds (α, β) from $\mathbf{L}$ with $\beta \prec \alpha$ (i.e., $\beta \preceq \alpha$ and $\neg(\alpha \preceq \beta)$), one can divide U into three regions [35]:

$$
\begin{aligned}
\mathrm{L}(v) &= \{x \in U \mid v(x) \succeq \alpha\}, \\
\mathrm{M}(v) &= \{x \in U \mid \beta \prec v(x) \prec \alpha\}, \\
\mathrm{R}(v) &= \{x \in U \mid v(x) \preceq \alpha\}.
\end{aligned}
\tag{6}
$$

To be consistent with the increasing ordering of objects from left to right according to their evaluation status values, Yao and Yu [43] call the three regions the left, middle, and right regions, respectively, or simply, L-region, M-region, and R-region. Under the condition $\beta \prec \alpha$, the three regions are pair-wise disjoint and their union is the universe U.

Figure 2 depicts an evaluation-based three-way decision model. We divide the universe U according to Eq. (6). We design different strategies to process the three regions. By using more generic labels and terms in the description, we have a more general model. A three-way classification model may be viewed as a special case of evaluation-based three-way decisions. More specifically, in a three-way classification model, the totally ordered set $(\mathbf{L}, \preceq)$ is $([0, 1], \leq)$, the evaluation function is given by the conditional probability, $v(x) = Pr(X|[x])$, the L-region, M-region, and R-region are, respectively, the positive, boundary and negative regions, and the strategies for L-, M-, and R-regions are the decisions of acceptance, noncommitment and rejection, respectively.

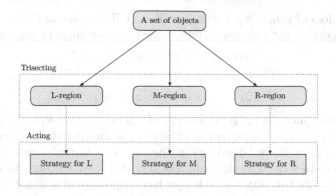

Fig. 2. Evaluation-based three-way decisions

There still exist problems with the second formulation of three-way decisions. By using an totally ordered set $(\mathbf{L}, \preceq)$, we, in fact, impose an ordering on the three regions, as indicated by the names of left, middle and right regions. Such an ordering may unnecessarily limit the generality of three-way decisions. In some situations, it may be difficult to give an analytic formula for computing the evaluation status values of objects. Therefore, we need to further search for new formulations of three-way decisions.

3.3 A Trisecting-and-Acting Framework of Three-Way Decisions

In this paper, we present a third attempt to formulate three-way decisions. The main objective is to avoid limitations of the first two attempts and to describe three-way decisions in even more generic terms.

To avoid an ordering of three regions, we directly consider a division of U into three parts without explicitly referring to an evaluation function. That is, one of the primitive notions is an abstract method to divide U into three pair-wise disjoint regions. This can be defined by a function,

$$\tau : U \longrightarrow \mathbf{T}, \tag{7}$$

where $\mathbf{T}$ is a set with three values, which each value for each of the three regions. We are not concerned with any particular detail procedure, nor are we concerned with the reasons that lead to the three regions. Consider election as an example. By surveying voters about their intended voting decisions, we may divide a set of surveyed voters into three regions: those who support a candidate, those who are undecided or are unwilling to tell their decisions, and those who oppose the candidate. Such a division is obtained without an evaluation function that explains voters' decisions. Given a division, we must work on strategies for processing each of the three regions. Consider again the example of election. Strategies may aim at retaining supporters and transforming those who are undecided or oppose the candidate into supporters.

Without loss of generality, we assume $\mathbf{T} = \{\mathbf{i}, \mathbf{ii}, \mathbf{iii}\}$. According to the values of τ, we can divide the universe U into three pair-wise disjoint regions:

$$\mathbb{I}(\tau) = \{x \in U \mid \tau(x) = \mathbf{i}\},$$
$$\mathbb{II}(\tau) = \{x \in U \mid \tau(x) = \mathbf{ii}\},$$
$$\mathbb{III}(\tau) = \{x \in U \mid \tau(x) = \mathbf{iii}\}. \tag{8}$$

We immediately have $\mathbb{I}(\tau) \cup \mathbb{II}(\tau) \cup \mathbb{III}(\tau) = U$, $\mathbb{I}(\tau) \cap \mathbb{II}(\tau) = \emptyset$, $\mathbb{I}(\tau) \cap \mathbb{III}(\tau) = \emptyset$, and $\mathbb{II}(\tau) \cap \mathbb{III}(\tau) = \emptyset$. To pictorially illustrate an unordered three regions, it is impossible to draw them linearly as in the first two formulations. Instead, we can draw them in a circle so that any region contacts with other two regions, as shown in Fig. 3. In this way, we no longer have a positional naming system used in the evaluation-based model. Instead, we refer to the three regions as regions I, II, and III, respectively. Corresponding to the three regions, we have strategies I, II, and III, respectively. It should be pointed out that we use I, II, and III simply as different labels without considering their numeric values.

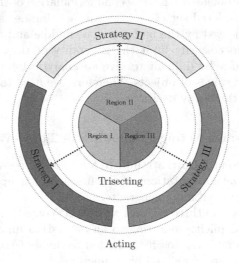

Fig. 3. A conceptual model of three-way decisions

With the new naming system, we avoid the narrow meaning that is implicitly suggested by the two naming systems of a three-way classification model and an evaluation-based model. All three regions are treated on the equal footing, without any preference of one to another. As shown by Fig. 3, any region is connected to the other two regions. No region takes a better position than other two regions. Such an understanding is consistent with a fact that for different applications we may focus on any one, two or all of the three regions.

4   Future Research in Three-Way Decisions

At a first look, the three attempts to formulate three-way decisions are characterized by changes of naming systems and pictorial descriptions. It is important to realize that these changes in fact enable us to gain more insights into three-way decisions. We are moving from concrete levels of specific models to a more abstract level of a general framework of three-way decisions.

The dividing-and-acting framework, as shown by Fig. 3, provides a high-level abstract description of three-way decisions and opens new avenues for further studies. This conceptual model is very useful for us to explain the basic ideas and components of three-way decisions. In order to realize ideas in the conceptual model, we need to build computational models of three-way decisions. By considering different ways to divide a universe and different strategies for processing three regions, we can derive many concrete computational models of three-way decisions.

An explicit separation of the two tasks of three-way decisions, namely, trisecting and acting, is mainly for the purpose of building an easy-to-understand model. In many situations, the two tasks in fact weave together as one and cannot be easily separated. A meaningful trisection of U depends on the strategies and actions used for processing the three regions. Effective strategies and actions of processing depend on an appropriate trisection of U. The two are just different sides of the same coin; we cannot have one without a consideration of the other. It is desirable that we both separate and integrate the tasks of trisecting and acting, depending on particular problems.

We can pursue further studies of three-way decisions by focusing on the two components of the dividing-and-acting framework. With respect to dividing, we suggest the following topics:

- Methods for trisecting a universe: In addition to evaluation-based methods, we may investigate other approaches. One possible is to construct a trisection through a pair of bisections. Another possibility is to consider a ranking of objects, which is a fundamental notion in theories of decision-making. Three regions correspond to the top, middle, and bottom segments of the ranking. A third possibility is to seek for a statistical interpretation of a trisection.

- Different classes of evaluation functions: Probabilistic three-way classifications use probability as an evaluation function. In general, we may consider other types of evaluation functions, including, for example, possibility functions, fuzzy membership functions, Bayesian confirmation measures, similarity measures, and subsethood measures. It is necessary to interpret an evaluation function in operable notions.

- Methods for determining the pair of thresholds: For an evaluation-based model, we need to investigate ways to compute and to interpret a pair of thresholds. An optimization framework can be designed to achieve such a goal. That is, a pair of thresholds should induce a trisection that optimizes a given objective function. By designing different objective functions for different applications, we gain a great flexibility.

A meaningful trisection depends on the strategies and actions for processing three regions. With respect to acting, we suggest the following topics:

- Descriptive rules for three regions: To fully understand three regions and to design effective strategies and actions for processing them, as a prerequisite, we must be able to describe and to represent the three regions. Descriptive rules summary the main features of the three regions, with each rule characterizes a portion of a specific region.

- Predictive rules for three regions: We can also construct predictive rules from the three regions to make decisions for new instances.
- Actionable rules for transferring objects between regions: In some situations, it is desirable to transfer objects from one region to another. We can use actionable rules in order to make such a move possible.
- Effective strategies and actions for processing three regions: For different regions, we must design the most suitable and effective strategies for processing. It may be sufficient to focus on one of the three regions. It may also happen that we must consider two or three regions simultaneously.
- Comparative studies: We can perform comparative studies by considering a pair regions together. In so doing, we can identify the differences between, and similarities of, the two regions. Descriptive rules may play a role in such comparative studies. The results of comparative study may also lead to actionable rules for transferring objects between regions.

In designing strategies for processing one region, we may have to consider the strategies for processing other regions. Interactions of strategies for different regions may play an important role in achieving a best trisection of the universe.

To fully understand and appreciate the value and power of three-way decisions, we must look into fields where the ideas of three-way decisions have been used either explicitly or implicitly. In one direction, one can adopt ideas from these fields to three-way decisions. In the other direction, we can apply new results of three-way decisions to these fields. We can cast a study of three-way decisions in a wider context and extend applications of three-way decisions across many fields. In the next few years, we may investigate three-way decisions in relation to interval sets [32], three-way approximations of fuzzy sets [7,46], shadowed sets [24], orthopairs [2–4], squares of oppositions [5,8,36], granular computing [37], and multilevel decision-making [47].

5　Conclusion

Although three-way decisions were originally motivated by the needs to properly interpret three regions of probabilistic rough sets, extensive studies in the last few years have moved far beyond. By realizing that the notion of three regions has been used in many fields, we can formulate a more general theory of three-way decisions. Ideas from rough sets, interval sets, three-way approximations of fuzzy sets, shadowed sets, orthopairs, squares of oppositions, and granular computing have influenced greatly the development of three-way decisions.

In this paper, we describe three-way decisions as two separated tasks of trisecting and acting. Within this trisecting-and-acting framework, we examine two previous attempts to formulate three-way decisions, i.e., probabilistic three-way classifications and evaluation-based three-way decisions, and present a third attempt. The new formulation uses more generic terms and notations and avoids

limitations of the previous formulations. The new formulation enables us to identify and discuss a number of research topics of three-way decisions.

For a full understanding of three-way decisions, we must cast our study in
the light of results across many fields. In the next few years, we expect to see
a continued growth of research interest in three-way decisions in such a wider
context.

Acknowledgements. This work was supported in part by a Discovery Grant from
NSERC, Canada. The author is grateful to Drs. Davide Ciucci, Jie Lu, Hong Yu
and Guangquan Zhang for constructive discussions on new formulations of three-way
decisions.

References

1. Chellas, B.F.: Modal Logic: An Introduction. Cambridge University Press,
   Cambridge (1980)
2. Ciucci, D.: Orthopairs: a simple and widely used way to model uncertainty. Fundamenta Informaticae **108**, 287–304 (2011)
3. Ciucci, D.: Orthopairs in the 1960s: historical remarks and new ideas. In: Cornelis,
   C., Kryszkiewicz, M., Ślęzak, D., Ruiz, E.M., Bello, R., Shang, L. (eds.) RSCTC
   2014. LNCS, vol. 8536, pp. 1–12. Springer, Heidelberg (2014)
4. Ciucci, D., Dubois, D., Lawry, J.: Borderline vs. unknown: comparing three-valued
   representations of imperfect information. Int. J. Approximate Reasoning **55**, 1866–
   1889 (2014)
5. Ciucci, D., Dubois, D., Prade, H.: Oppositions in rough set theory. In: Li, T.,
   Nguyen, H.S., Wang, G., Grzymala-Busse, J., Janicki, R., Hassanien, A.E., Yu, H.
   (eds.) RSKT 2012. LNCS, vol. 7414, pp. 504–513. Springer, Heidelberg (2012)
6. Deng, X.F.: Three-way classification models. Ph.D. Dissertation, Department of
   Computer Science, University of Regina (2015)
7. Deng, X.F., Yao, Y.Y.: Decision-theoretic three-way approximations of fuzzy sets.
   Inf. Sci. **279**, 702–715 (2014)
8. Dubois, D., Prade, H.: From Blanché hexagonal organization of concepts to formal
   concept analysis and possibility theory. Log. Univers. **6**, 149–169 (2012)
9. Grzymala-Busse, J.W., Yao, Y.Y.: Probabilistic rule induction with the LERS data
   mining system. Int. J. Intell. Syst. **26**, 518–539 (2011)
10. Hu, B.Q.: Three-way decisions space and three-way decisions. Inf. Sci. **281**, 21–52
    (2014)
11. Jia, X.Y., Shang, L., Zhou, X.Z., Liang, J.Y., Miao, D.Q., Wang, G.Y., Li, T.R.,
    Zhang, Y.P. (eds.): Theory of Three-way Decisions and Applications. Nanjing Uni-
    versity Press, Nanjing (2012). (in Chinese)
12. Li, H.X., Zhang, L.B., Huang, B., Zhou, X.Z.: Sequential three-way decision and
    granulation for cost-sensitive face recognition. Knowl. Based Syst. (2015). http://
    dx.doi.org/10.1016/j.knosys.2015.07.040
13. Li, H.X., Zhou, X.Z.: Risk decision making based on decision-theoretic rough set:
    a three-way view decision model. Int. J. Comput. Intell. Syst. **4**, 1–11 (2011)
14. Li, Y., Zhang, Z.H., Chen, W.B., Min, F.: TDUP: an approach to incremental
    mining of frequent itemsets with three-way-decision pattern updating. Int. J. Mach.
    Learn. Cybern. (2015). http://dx.doi.org/10.1007/s13042-015-0337-6

15. Liang, D.C., Liu, D.: Deriving three-way decisions from intuitionistic fuzzy decision-theoretic rough sets. Inf. Sci. **300**, 28–48 (2015)
16. Liu, D., Li, T.R., Liang, D.C.: Three-way government decision analysis with decision-theoretic rough sets. Int. J. Uncertainty Fuzziness Knowl. Based Syst. **20**, 119–132 (2012)
17. Liu, D., Li, T.R., Miao, D.Q., Wang, G.Y., Liang, J.Y. (eds.): Three-way Decisions and Granular Computing. Science Press, Beijing (2013). (in Chinese)
18. Liu, D., Liang, D.C., Wang, C.C.: A novel three-way decision model based on incomplete information system. Knowl. Based Syst. (2015). http://dx.doi.org/10.1016/j.knosys.2015.07.036
19. Luo, C., Li, T., Chen, H.: Dynamic maintenance of three-way decision rules. In: Miao, D., Pedrycz, W., Slezak, D., Peters, G., Hu, Q., Wang, R. (eds.) RSKT 2014. LNCS, vol. 8818, pp. 801–811. Springer, Heidelberg (2014)
20. Orłowska, W., Pawlak, Z.: Representation of nondeterministic information. Theor. Comput. Sci. **29**, 27–39 (1984)
21. Pawlak, Z.: Rough sets. Int. J. Comput. Inf. Sci. **11**, 341–356 (1982)
22. Pawlak, Z.: Rough classification. Int. J. Man Mach. Stud. **20**, 469–483 (1984)
23. Pawlak, Z.: Rough Sets. Theoretical Aspects of Reasoning about Data. Kluwer Academic Publishers, Dordrecht (1991)
24. Pedrycz, W.: Shadowed sets: representing and processing fuzzy sets. IEEE Trans. Syst. Man Cybern. **28**, 103–109 (1998)
25. Peters, J.F., Ramanna, S.: Proximal three-way decisions: theory and applications in social networks. Knowl.-Based Syst. (2015). http://dx.doi.org/10.1016/j.knosys.2015.07.021
26. Sang, Y.L., Liang, J.Y., Qian, Y.H.: Decision-theoretic rough sets under dynamic granulation. Knowl. Based Syst. (2015). http://dx.doi.org/10.1016/j.knosys.2015.08.001
27. Shakiba, A., Hooshmandasl, M.R.: S-approximation spaces: a three-way decision approach. Fundamenta Informaticae **39**, 307–328 (2015)
28. Sun, B.Z., Ma, W.M., Zhao, H.Y.: An approach to emergency decision making based on decision-theoretic rough set over two universes. Soft Comput. (2015). http://dx.doi.org/10.1007/s00500-015-1721-6
29. Xie, Y., Johnsten, T., Raghavan, V.V., Benton, R.G., Bush, W.: A comprehensive granular model for decision making with complex data. In: Pedrycz, W., Chen, S.-M. (eds.) Granular Computing and Decision-Making. SBD, vol. 10, pp. 33–46. Springer, Heidelberg (2015)
30. Yao, J.T., Azam, N.: Web-based medical decision support systems for three-way medical decision making with game-theoretic rough sets. IEEE Trans. Fuzzy Syst. **23**, 3–15 (2014)
31. Yao, J.T., Zhang, Y.: A scientometrics study of rough sets in three decades. In: Lingras, P., Wolski, M., Cornelis, C., Mitra, S., Wasilewski, P. (eds.) RSKT 2013. LNCS, vol. 8171, pp. 28–40. Springer, Heidelberg (2013)
32. Yao, Y.Y.: Interval-set algebra for qualitative knowledge representation. In: Proceedings of the 5th International Conference on Computing and Information, pp. 370–374 (1993)
33. Yao, Y.: Three-way decision: an interpretation of rules in rough set theory. In: Wen, P., Li, Y., Polkowski, L., Yao, Y., Tsumoto, S., Wang, G. (eds.) RSKT 2009. LNCS, vol. 5589, pp. 642–649. Springer, Heidelberg (2009)
34. Yao, Y.Y.: Three-way decisions with probabilistic rough sets. Inf. Sci. **180**, 341–353 (2010)

35. Yao, Y.: An outline of a theory of three-way decisions. In: Yao, J.T., Yang, Y., Słowiński, R., Greco, S., Li, H., Mitra, S., Polkowski, L. (eds.) RSCTC 2012. LNCS, vol. 7413, pp. 1–17. Springer, Heidelberg (2012)

36. Yao, Y.Y.: Duality in rough set theory based on the square of opposition. Fundamenta Informaticae **127**, 49–64 (2013)

37. Yao, Y.: Granular computing and sequential three-way decisions. In: Lingras, P., Wolski, M., Cornelis, C., Mitra, S., Wasilewski, P. (eds.) RSKT 2013. LNCS, vol. 8171, pp. 16–27. Springer, Heidelberg (2013)

38. Yao, Y.Y.: The two sides of the theory of rough sets. Knowl. Based Syst. **80**, 67–77 (2015)

39. Yao, Y.Y.: Probabilistic rough set approximations. Int. J. Approximation Reasoning **49**, 255–271 (2008)

40. Yao, Y.Y., Deng, X.F.: Quantitative rough sets based on subsethood measures. Inf. Sci. **267**, 306–322 (2014)

41. Yao, Y.Y., Lin, T.Y.: Generalization of rough sets using modal logic. Intell. Autom. Soft Comput. Int. J. **2**, 103–120 (1996)

42. Yao, Y.Y., Wong, S.K.M., Lingras, P.: A decision-theoretic rough set model. In: Ras, Z.W., Zemankova, M., Emrich, M.L. (eds.) Methodologies for Intelligent Systems 5, pp. 17–24. North-Holland Publishing, New York (1990)

43. Yao, Y.Y., Yu, H.: An introduction to three-way decisions. In: Yu, H., Wang, G.Y., Li, T.R., Liang, J.Y., Miao, D.Q., Yao, Y.Y. (eds.) Three-Way Decisions: Methods and Practices for Complex Problem Solving, pp. 1–19. Science Press, Beijing (2015). (in Chinese)

44. Yu, H., Wang, G.Y., Li, T.R., Liang, J.Y., Miao, D.Q., Yao, Y.Y. (eds.): Three-Way Decisions: Methods and Practices for Complex Problem Solving. Science Press, Beijing (2015). (in Chinese)

45. Yu, H., Zhang, C., Wang, G.Y.: A tree-based incremental overlapping clustering method using the three-way decision theory. Knowl. Based Syst. (2015). http://dx.doi.org/10.1016/j.knosys.2015.05.028

46. Zadeh, L.A.: Fuzzy sets. Inf. Control **8**, 338–353 (1965)

47. Zhang, G.Q., Lu, J., Gao, Y.: Multi-Level Decision Making. Models, Methods and Application. Springer, Berlin (2014)

48. Zhang, H.R., Min, F.: Three-way recommender systems based on random forests. Knowl. Based Syst. (2015). http://dx.doi.org/10.1016/j.knosys.2015.06.019

49. Zhang, H.Y., Yang, S.Y., Ma, J.M.: Ranking interval sets based on inclusion measures and applications to three-way decisions. Knowl. Based Syst. (2015). http://dx.doi.org/10.1016/j.knosys.2015.07.025

Tutorial

Rough Set Tools for Practical Data Exploration

Andrzej Janusz[1], Sebastian Stawicki[1], Marcin Szczuka[1(✉)],
and Dominik Ślęzak[1,2]

[1] Institute of Mathematics, University of Warsaw,
Banacha 2, 02-097 Warsaw, Poland
{janusza,stawicki,szczuka,slezak}@mimuw.edu.pl
[2] Infobright Inc., Krzywickiego 34, lok. 219, 02-078 Warsaw, Poland

Abstract. We discuss a rough-set-based approach to the data mining process. We present a brief overview of rough-set-based data exploration and software systems for this purpose that were developed over the years. Then, we introduce the *RapidRoughSets* extension for the RapidMiner integrated software platform for machine learning and data mining, along with RoughSets package for R System – the leading software environment for statistical computing. We conclude with discussion of the road ahead for rough set software systems.

Keywords: Rough sets · Data mining software · R system · RapidMiner

1 Introduction

The theory of rough sets (RS) has been around for more than thirty years. From its very beginning this theory was grounded in practical consideration about processing of the data [1]. The basic notions in RS such as reduct, core, approximation, decision rule and so on, are all revolving around a notion of indiscernibility founded on an information system, i.e., a set of experimental data. This orientation towards practical aspect of discovering information and knowledge from the data was one of the main reasons why this emerging direction of research has gained the level of recognition that it enjoys today.

While the development of RS theoretical foundation has been very interesting and brought plethora of new ideas and concepts, the more practical side was also progressing. Over the years more and more advanced and useful RS algorithms have been implemented in various software tools. While mostly research-oriented, some of those software systems managed to make a splash outside of academic circles. The paper presents two of the latest members of the RS software family, the *RapidRoughSets* extension for RapidMiner[1] along with the *RoughSets* package for R system [2].

Partially supported by Polish National Science Centre – grant DEC-2012/05/B/-ST6/03215 and by Polish National Centre for Research and Development – grants: PBS2/B9/20/2013 in frame of Applied Research Programmes and O ROB/0010/-03/001 in frame of Defence and Security Programmes and Projects.

[1] https://rapidminer.com.

D. Ciucci et al. (Eds.): RSKT 2015, LNAI 9436, pp. 77–86, 2015.
DOI: 10.1007/978-3-319-25754-9_7

This paper complements the tutorial given by the authors at the 2015 International Joint Conference on Rough Sets (IJCRS) in Tianjin, China. It is also intended as a basic reference for potential users of the presented software as well as a concise explanation of the RS approach to data exploration. It is intentionally quite informal and rather general.

2   Evolution of Rough Set Software for Data Analysis

From the dawn of the RS research area the practical use of the developed methods and algorithms was in the center of attention of many. From the humble beginnings of a C library of just a few functions, the RS software systems evolved into fully functional, GUI-based data exploration toolboxes. This evolution was never controlled or coordinated in any way. As a result, a wide range of implementations were made, many of which are now all but forgotten.

The work of various researchers in the RS area resulted in creation of several specialized tools, such as LERS (Learning from Examples based on Rough Sets) [3] and PRIMEROSE – an expert system for differential diagnosis with use of inhibitory rules [4]. These tools were limited in functionality or availability, but managed to make a lasting impression, at least in the RS community.

Among more complete, GUI-based RS software environments that left a lasting trace and are still useful to some extent, although no longer developed, are: Rough Set Data Explorer (ROSE) [5], Rough Set Exploration System (RSES) [6], ROSETTA – a general-purpose tool for discernibility-based modelling [7], Rough Set Based Intelligent Data Analysis System (RIDAS) [8].

More recently, there were several attempts at making implementations of the RS methods more accessible and popular by opening the sources (Rseslib project[2]), integrating with popular data mining tools (FRST module for Weka[3]) and creating an open experimental environment (Debellor [9]). Most recently, there were some quite successful attempts on creating RS tools for non-deterministic data (getRNIA [10]) and providing incremental learning capabilities (iRoughSet [11]). Still, all these solutions required the user to possess quite advanced technical skills, sometimes even programming abilities.

A brief comparison of the existing software systems that implement the RS methods reveals a list of algorithms that any such system is expected to feature. When we exclude all non-RS elements, usually associated with data manipulation and visualization, we are left with:

– Approximation methods, i.e., discernibility-based methods for calculation of upper and lower bounds, positive regions, boundary regions and so on.
– Reduction methods, i.e., calculation and management of core and reduct sets, both in the general and in the decision support scenarios.
– Data modification methods, i.e., methods for discretizing and completing the data with use of discernibility and other RS concepts.

---

[2] http://rsproject.mimuw.edu.pl/.
[3] http://users.aber.ac.uk/rkj/Weka.pdf.

– Classification methods, i.e., construction and evaluation of classifiers, usually based on the RS methods, including decision rules and/or trees.

We would like to make the next step in the evolution of RS software. Our intention is to build a set of tools that will bring new qualities in this area. A software system that we want to create, in addition to requirements for inclusion of the RS methods, should provide all capabilities and conveniences that we expect from a *mainstream* environment for data exploration and analysis. This means that such software should answer to the following requirements:

– **Extensibility.** The software should allow for easy inclusion of user-created procedures and algorithms. At the same time integration with external, specialized software should be seamless and straightforward. For example, reaching for the data to an external RDBMS or distribution of workload using Hadoop should be easy and as transparent for the user as possible.
– **Scalability.** The software should scale well with respect to the size (number of objects), complexity (number and type of attributes) and representation (tabular, relational, stream, ...) of the data. For computationally challenging tasks the implemented procedures should offer means for distributing the workload among several processing nodes in a straightforward and efficient manner.
– **Versatility.** The software should be portable, platform-independent and modularized. It has to contain a comprehensive collection of everyday use data manipulation and processing methods, so that only the most specialized operations would require the use of a plug-in or an external tool.
– **Accessibility.** The software should be easy to install, understand and use for people with limited programming abilities. The learning curve for an everyman in the field of data exploration should not be too steep. At the same time, for somebody who is an advanced programmer and data scientist it should be easy to access the internal "machinery". This translates to requirement for the system to be open-source and implemented with use of good practices of software engineering.

In order to meet these requirements we have decided to integrate the algorithms and methods of the RS origin with RapidMiner and R system, popular data exploration environments, leaders in their respective subareas.

3 RapidMiner Predictive Analytics Platform

RapidMiner is a software platform developed to deliver a comprehensive environment for tasks of data mining, machine learning, predictive analytics, etc. It is an integrated environment that provides more than just an invocation of knowledge discovery algorithms. It is intended as a system that supports all steps of a data mining process, including data preparation (load and transformation), data preprocessing, data visualization, statistical modeling, algorithms' parameter tuning, validation and results aggregation, interpretation and visualization. In this regard it fulfills the requirements for versatility stated in Sect. 2.

RapidMiner is developed by a commercial company under the same name. By the company policy, the most recent version of RapidMiner toolbox is licensed (commercially) with limited versions – Trial Version and Starter Edition – available for free. However, the licensing fee is only collected on the extension of the computational core that is regarded company's product. The core itself is an Open Source Software (OSS). With every new major commercial release of RapidMiner its predecessors become free to use. As a result, the "version -1" of all tools is always available under the permissive GNU-AGPL v3 license[4].

RapidMiner is widely utilized in a broad range of applications in business, industry, research, education, prototyping and software development. What makes the platform very attractive is the GUI environment for design and execution of data mining workflows. This means that the user can prepare advanced data processing tasks visually, without a need to write even a line of code, thus reducing possibility of errors during the experiments preparation and speeding up the whole deployment. The workflows in the RapidMiner's nomenclature are called *Processes* and they consist of multiple *Operators* arranged into a graph-like structure. The process may be exported for further usage in a concise XML format. As an alternative to GUI-based approach, one can call the RapidMiner's computational core from other software applications. This makes the potential use of RapidMiner even wider, allowing it to be embedded in applications as an engine for performing predefined (e.g. designed earlier by means of GUI) or generated on the fly (as XML) processes. In this way RapidMiner answers to requirements for versatility, extensibility and accessibility.

Each operator (an elementary building block) in RapidMiner is a node in the process' design (workflow graph) and is performing a single, well-defined task. The connections (directed graph edges) describe the direction of data flow between operators – the outputs of one operator may form the inputs to others. A task may be complex or may employ a meta-algorithm. For example, there is an operator that performs an entire train-and-test procedure on a nested subprocess (a sub-workflow nested inside the operator) as well as an operator that searches through a defined space of algorithm's parameters trying to find the most optimal settings with regards to a specified criterion.

RapidMiner provides a broad range of built-in operators. They range from elementary input/output procedures to meta-level knowledge modeling, including: preprocessing of the data, interactive visual data analysis, generation of artificial data for test purposes, feature generation and selection, classification, regression, clustering, attribute weighting, association and item set mining, hierarchical and ensemble model creation, result evaluation with cross-validation, visual result evaluation, and many more.

4   RapidMiner Extensions

Even the plain RapidMiner distribution constitutes a rich software environment, ready to perform all the steps in the data analysis process. More still is available

---

[4] http://www.gnu.org/licenses/agpl-3.0.en.html.

thanks to additional plug-ins dubbed *extensions* in RapidMiner's nomenclature. A prominent example of such extension is the one providing integration with popular Weka system [12]. Through the *Weka* extension the user may seamlessly access all of the learning schemes, models and algorithms from Weka.

The RapidMiner extension ecosystem facilitates incorporation of various emerging paradigms into the platform. Several components recently added to the system make it possible to integrate with industry standards in big data processing, stream data processing, realtime analytics, and continuous computation frameworks. That includes interfaces to Hadoop, Apache Spark, Apache Storm and even integration with storage and/or computational cloud environments like Amazon AWS or Dropbox. The idea behind introducing these components is to give the user a fully transparent access to all the power of external processing through the same simple GUI, without the need for writing specialized code.

Another prominent extension, used extensively in the development of the *RapidRoughSets* extension (see Sect. 6), is the one that makes it possible to interact through RapidMiner with the powerful R system[5]. The *R* extension (available under the AGPL license) follows the principles of simplicity and transparency of use in the RapidMiner's GUI. The extension provides a limited set of predefined operators that correspond to a small subset of classification and regression methods available in R system. It also contains two very powerful tools – an embedded R console and the *Execute Script (R)* operator. Through these two tools, anybody can incorporate R functionality into the RapidMiner process. R console can be used to interactively write R code, examine the data or even plot data charts inside the tool. Moreover, the results such as data tables and functions, can be stored in a RapidMiner local repository and later used in the GUI. *Execute Script (R)* operator – lets the user create the RapidMiner's operators inside the GUI by placing suitable R statements in their internals. It serves as a form of a wrapper for R code.

The convenience of integration with R encouraged us to design our own extension that provides the user of RapidMiner with RS notions and analytic techniques, as implemented in the *RoughSets* package (see Sect. 5).

5   RoughSets Package for R System

The core of *RapidRoughSets* is founded upon a newly developed library for R System, called *RoughSets*. This open source package is available for download in the CRAN repository[6]. It provides implementations of classical RS methods and their fuzzy-related generalizations for data modeling and analysis [2]. The package not only allows to apply fundamental concepts (e.g., computation of indiscernibility relations, lower/upper approximations, positive regions, etc.) from the both of these methodologies but also facilitates their usage in many

---

[5] http://www.R-project.org/.
[6] http://cran.r-project.org/web/packages/RoughSets/index.html.

practical tasks, such as discretization, feature selection, instance selection, decision rule induction and case-based classification.

The functions available in *RoughSets* are divided into several categories based on their intended functionality. The category of a particular function can be easily identified by a prefix and a suffix added to its name. In general, nine different prefixes are used to distinguish between implemented methods, however, from the point of view of *RapidRoughSets* the most important categories are:

SF: support functions – general purpose methods for transformation of the data.
BC: basic concepts – functions implementing basic building blocks for many of the common RS algorithms and their fuzzy-related generalizations.
D: discretization methods – a few of the most popular data discretization algorithms.
FS: feature selection – implementations of the state-of-the-art reduct computation methods.
RI: rule induction – popular decision rule induction and rule-based classification algorithms.
IS: instance selection – algorithms for selecting the most important instances for a case-based classification.

Additionally, a suffix RST or FRST is used in order to distinguish classical RS methods from the ones that originate in the fuzzy-rough domain.

The basic data structure used in *RoughSets* is a *DecisionTable*. It is implemented as an S3 object class. Apart from the data, these objects are able to store some meta-data information, such as the index of a decision attribute and information regarding intended types of conditional attributes. The architecture of the package allows to conveniently use results of many of the typical RS data analysis tasks to *DecisionTables*. For example, a set of discretization cuts or a decision reduct computed for one data set (i.e., a training *DecisionTable*) can be easily applied to other tables (i.e., test sets) using a single function. In practice, this flexibility enables designing complex experiments with the data and promotes the notion of reusing available knowledge.

The first incarnation of *RapidRoughSets* is based on the version 1.3-0 of *RoughSets*. In comparison to the earlier versions, it allows for much more efficient computation of certain types of decision reducts for large data tables. Several tests conducted on real-life data sets show that the improved implementations of feature selection algorithms are able to reduce the computation times by a factor of several hundred, as compared to the version 1.1-0. This optimization was achieved by rewriting of several key internal functions in C++ and embedding them into R code using the routines provided by the *Rcpp* library. The version of *RoughSets* used in our extension enables efficient processing of high dimensional data sets, such as those from recent AAIA'14 and IJCRS'15 data mining competitions[7].

The *RoughSets* package is being continuously developed and extended by new functionalities. For instance, the most recent version includes implementations

---

[7] https://knowledgepit.fedcsis.org/.

of popular decision rule induction algorithms: LEM2, CN2 and AQ [3,13,14]. It also contains an implementation of a novel heuristic for computation of approximate reducts that we described in details in [15]. Another recent addition to the package is a collection of standard functions for handling missing data values. All of these methods can be easily used for data analytics conducted with *RapidRoughSets*, without any need for R programming.

6 RapidRoughSets Extension for RapidMiner

In order to use *RoughSets* inside RapidMiner it is enough to have an installation of the two – the package in a local R environment and the *R* extension for RapidMiner. However, the major drawback of such a solution is the need for mastering R system, as well as the *RoughSets* package and its functionalities. Naturally, not all users have the time or willingness to learn R system in general and *RoughSets* in particular. For that reason, we decided to facilitate the access to the methods provided by our package. The resulting *RapidRoughSets* extension is an attempt to equip the data explorers with powerful, yet easily approachable tool set for RS-related operations.

It has to be mentioned that the development of *RapidRoughSets* was initiated within the DISESOR project[8] as an answer to a practical need. There, the idea was not only to provide the access to the RS methods to wider community but also to explore possible applications of the methods in solving practical industrial problems. DISESOR is aimed at creation of an integrated decision support system for monitoring processes, equipment and treatments in coal mines in Poland. It addresses the safety of people and equipment, as well as minimization of financial loses, e.g. derived from power cuts when the concentration of methane or seismic energy is predicted to be dangerously high [16]. The decision support system provided as a result of the project is meant to be available to a coal mine worker who monitors coal extraction process, parameters of working devices (e.g. coal harvesters) along with methane concentration, air flow and many other readings, through an array of installed sensors. The access to the developed methods should be maximally simplified and the use of a GUI-like environment supplied by RapidMiner is a lot easier than the alternatives.

Despite the fact that *RapidRoughSets* was conceived for a purpose of a practical application of the RS methods in the specific industrial problem, we believe that the prepared software has wider appeal. It greatly simplifies the process of getting familiar with the data, as well as the preparation of a prototype or even the final solution. It also introduces several customized RapidMiner operators for performing various tasks corresponding to steps in the RS data analysis. There are operators that facilitate the process of data preparation and transformation to a required format. There are also operators that, given a subset of data objects X, allow the user to derive and use the RS concepts, including: upper

---

[8] DISESOR is a Polish national R&D project founded by Polish National Centre for Research and Development grant PBS2/B9/20/2013 in frame of Applied Research Programmes.

and lower approximation of the subset, positive region and approximation quality γ. Finally, *RapidRoughSets* provides heavy duty operators that correspond to more complex and costly computational procedures from the RS inventory, e.g. computation of reducts, supervised discretization, as well as construction and evaluation of rule-based classifiers.

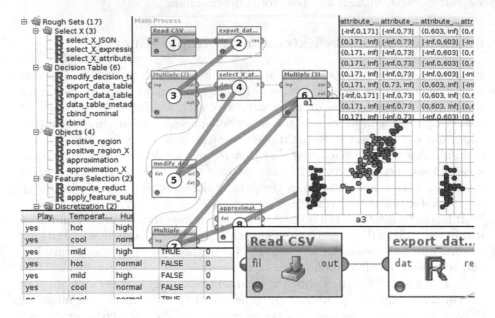

Fig. 1. *RapidRoughSets* at work.

The requirements that we stated in Sect. 2 are satisfied by *RapidRoughSets* to a various degree. The extensibility of our software is a direct consequence of the way that both, R system and RapidMiner, are designed. For those familiar with R, the open architecture of *RoughSets* package makes it possible to introduce changes and additions to the code. Moreover, RapidMiner's user-friendly GUI facilitates additions to the RS functionality in several ways (see Fig. 1). One is by using the embedded R console or the *Execute Script (R)* operator. Another way involves combining available operators (building blocks) using the drag-and-drop capabilities of the platform. The versatility and accessibility are also provided thanks to the choice of the toolchain.

The scalability of *RapidRoughSets* is also a direct consequence of the choice of the underlying software platforms, R and RapidMiner. Both of these software systems are constantly evolving in order to match the growing demand for the ability to process large and complicated data sets. In the narrower context of the software components developed by us, the scalability depends on efficient implementation of communication between RapidMiner and the RS methods from *RoughSets*. Obviously, the scalability of *RapidRoughSets* also depends on the efficiency of algorithms implemented within *RoughSets*.

7 Summary and the Road Ahead

The *RapidRoughSets* extension and the *RoughSets* package represent a new approach to construction of RS software tools. Through integration with professional, comprehensive and highly popular software environments (R system and RapidMiner) they have acquired highly desirable properties. They became easy to use, extensible, flexible and versatile. The tools can greatly increase the accessibility of the RS methods to researchers and practitioners in the field of data exploration. Even advanced users and programmers, who have their own habits and commonly used tools, may appreciate the swiftness of the development of experiments that they can use as a step towards better understanding of available data. It is our strong belief that the implemented software can make the theory of rough sets and its applications even more popular and widely used.

RapidRoughSets and *RoughSets* inherit the strengths and weaknesses of data representation and processing capabilities of respective mother systems. Taken together, they impose additional requirements on the user as the data must flow efficiently back and forth between RapidMiner and R. The challenge we want to take in the future versions of the software is to make this part of data exploration cycle as transparent to the user as possible. This will most likely entail a design and implementation of dedicated data structures and methods tailored specifically to the RS data analysis.

Both of the presented software packages are intended to grow and evolve in the future. By using two of the most popular data analysis tools as a vessel, we intend to launch the RS tools to the open seas. We hope that they will attract enough attention to gather momentum for creation of an open community around them. With such community the software would become sustainable, as the community members could suggest or even implement new functionalities and provide invaluable feedback.

References

1. Pawlak, Z.: Rough Sets - Theoretical Aspects of Reasoning about Data. Theory and Decision Library D. Kluwer, Dordrecht (1991)
2. Riza, L.S., Janusz, A., Bergmeir, C., Cornelis, C., Herrera, F., Ślęzak, D., Benítez, J.M.: Implementing algorithms of rough set theory and fuzzy rough set theory in the R package "RoughSets". Inf. Sci. **287**, 68–89 (2014)
3. Grzymała-Busse, J.W.: LERS - a data mining system. In: Maimon, O., Rokach, L. (eds.) Data Mining and Knowledge Discovery Handbook, pp. 1347–1351. Springer, Heidelberg (2005)
4. Tsumoto, S.: Automated induction of medical expert system rules from clinical databases based on rough set theory. Inf. Sci. **112**, 67–84 (1998)
5. Prędki, B., Wilk, S.: Rough set based data exploration using ROSE system. In: Raś, Z.W., Skowron, A. (eds.) ISMIS 1999. LNCS, vol. 1609, pp. 172–180. Springer, Heidelberg (1999)
6. Bazan, J., Szczuka, M.S.: The rough set exploration system. In: Peters, J.F., Skowron, A. (eds.) Transactions on Rough Sets III. LNCS, vol. 3400, pp. 37–56. Springer, Heidelberg (2005)

7. Komorowski, J., Øhrn, A., Skowron, A.: Case studies: public domain, multiple mining tasks systems: ROSETTA rough sets. In: Klösgen, W., Żytkow, J.M. (eds.) Handbook of Data Mining and Knowledge Discovery, pp. 554–559. Oxford University Press, Oxford (2002)

8. Wang, G., Zheng, Z., Zhang, Y.: RIDAS - a rough set based intelligent data analysis system. In: Proceedings of ICMLC 2002, vol. 2, pp. 646–649. IEEE (2002)

9. Wojnarski, M.: Debellor: a data mining platform with stream architecture. In: Peters, J.F., Skowron, A., Rybiński, H. (eds.) Transactions on Rough Sets IX. LNCS, vol. 5390, pp. 405–427. Springer, Heidelberg (2008)

10. Wu, M., Nakata, M., Sakai, H.: An overview of the getRNIA system for non-deterministic data. In: Watada, J., Jain, L.C., Howlett, R.J., Mukai, N., Asakura, K., (eds.) Proceedings of Procedia Computer Science KES 2013, vol. 22, pp. 615–622. Elsevier (2013)

11. Zhang, J., Li, T., Chen, H.: Composite rough sets for dynamic data mining. Inf. Sci. **257**, 81–100 (2014)

12. Hall, M., Frank, E., Holmes, G., Pfahringer, B., Reutemann, P., Witten, I.H.: The WEKA data mining software: an update. ACM SIGKDD Explor. Newslett. **11**(1), 10–18 (2009)

13. Clark, P., Niblett, T.: The CN2 Induction Algorithm. Mach. Learn. **3**(4), 261–283 (1989)

14. Michalski, R.S., Kaufman, K.A.: Learning patterns in noisy data: The AQ approach. In: Paliouras, G., Karkaletsis, V., Spyropoulos, C.D. (eds.) ACAI 1999. LNCS (LNAI), vol. 2049, pp. 22–38. Springer, Heidelberg (2001)

15. Janusz, A., Ślęzak, D.: Random probes in computation and assessment of approximate reducts. In: Kryszkiewicz, M., Cornelis, C., Ciucci, D., Medina-Moreno, J., Motoda, H., Raś, Z.W. (eds.) RSEISP 2014. LNCS, vol. 8537, pp. 53–64. Springer, Heidelberg (2014)

16. Kabiesz, J., Sikora, B., Sikora, M., Wróbel, Ł.: Application of rule-based models for seismic hazard prediction in coal mines. Acta Montanistica Slovaca **18**(4), 262–277 (2013)

Reducts and Rules

Dominance-Based Neighborhood Rough Sets
and Its Attribute Reduction

Hongmei Chen, Tianrui Li[✉], Chuan Luo, and Jie Hu

School of Information Science and Technology, Southwest Jiaotong University,
Chengdu 610031, China
{hmchen,trli,jiehu}@swjtu.edu.cn, luochuan@my.swjtu.edu.cn

Abstract. In real-life applications, partial order may exist in the
domain of attributes and different data types often coexist in an decision
system. Dominance-based rough set approach has been used widely in
multi-attribute and multi-criteria decision by using the dominating relation. Neighborhood rough set aims to deal with hybrid data types. But
the preference relation in the context of neighborhood rough set has not
been taken into consideration. In this paper, a novel rough set model,
Dominance-based Neighborhood Rough Sets (DNRS), is proposed which
aims to process a decision system with hybrid data types where the partial order between objects is taken into consideration. The properties of
DNRS are studied. Attribute reduction under DNRS is investigated.

Keywords: Rough set theory · Dominance-based neighborhood rough
sets · Neighborhood dominating relation · Attribute reduction

1 Introduction

Rough Set Theory (RST) is one of effective soft computing theories for solving
classification problems [19]. Classical RST deals with complete data and the
relationship between objects is the equivalence relation. It has been extended to
deal with different types of data and different types of relationship in real-life
applications, *e.g.*, tolerance relations in the set-valued information system [22],
similarity relation and limited tolerance relation in the incomplete information
system [21], fuzzy similarity relation in fuzzy rough set [2], etc.

Dominance Rough Set Approach (DRSA) that allows analysing preference-
ordered data is an important extension of RST [6]. The dominating relation has
been extended to Fuzzy Rough Set (FRS) [7], intuitionistic fuzzy information systems [9], incomplete interval-valued information system [23], and incomplete information system [3], etc. By applying probabilistic method to DRSA, Inuiguchi et al.

This work is supported by the National Science Foundation of China (Nos. 61175047,
61100117, 71201133) and NSAF (No. U1230117), the 2013 Doctoral Innovation
Funds of Southwest Jiaotong University, the Fundamental Research Funds for the
Central Universities (No. SWJTU12CX091), and the Beijing Key Laboratory of
Traffic Data Analysis and Mining (BKLTDAM2014001).

© Springer International Publishing Switzerland 2015
D. Ciucci et al. (Eds.): RSKT 2015, LNAI 9436, pp. 89–99, 2015.
DOI: 10.1007/978-3-319-25754-9_8

proposed variable-precision DRSA approach and studied attribute reduction [12]. Kotlowski et al. proposed a stochastic DRSA [17]. Szelazg et al. studied Variable Consistency DRSA (VC-DRSA) and its rule induction in VC-DRSA [20]. Kadzinski et al. investigated minimal sets of rules in DRSA [16].

RST proposed by Pawlak can only deals with nominal data. In order to deal with numerical data, RST was extended to FRS by Dubois and Prade [4]. In [18], Park and Choi applied rough set approach to process categorical data by using information-theoretic dependency measure. But hybrid data coexists in the real-life applications. Then, Hu et al. proposed Neighborhood Rough Set (NRS) to deal with homogeneous feature selection [11]. Neighborhood relation and nearest neighborhood relation have been defined in NRS. Wei et al. compared NRS and FRS method in studying the hybrid data [24]. Zhang et al. proposed composite rough sets (CRS) [25] and Li et al. proposed a new CRS [13] by considering criteria and regular attributes simultaneously. Lin et al. developed a so-called Neighborhood-based Multi-Granulation Rough Set (NMGRS) in the framework of multi-granulation rough sets [15]. Liu et al. presented a quick attribute reduct algorithm by using a series of buckets to shorten the time of computing positive region for NRS [14]. However, the dominance relationship has not been considered in the context of NRS. In this paper, Dominance-based Neighborhood Rough Sets (DNRS) is proposed. The properties of the DNRS are also studied. Furthermore, attribute reduction in DNRS is presented.

The reminder of the paper is structured as follows. Section 2 reviews basic concepts and notions in RST. Section 3 shows DNRS and its properties. Section 4 investigates the method of attribute reduction in DNRS. Section 5 concludes the paper.

2   Preliminaries

In this section, we introduce basic concepts in DRSA and NRS, respectively.

2.1   Dominance-Based Rough Set Approach

In DRSA, the dominating (dominated) relation is defined by considering partial order between attribute values. Then granules induced by the dominating (dominated) relation are defined. Based on these basic element knowledge, the definitions of approximations for upward (downward) union are given.

Definition 1 *[6]. 4-tuple $S = (U, A, V, f)$ is a decision system, where U is a non-empty finite set of objects, called the universe. $A = C \cup D$ is a non-empty finite set of attributes, where C and D are the set of condition attribute and the set of decision attribute, respectively. V is domain of attributes set A. $f : U \times A \to V$ is an information function, $\forall a \in A$, $x \in U$, $f(x, a) \in V_a$. Let $x, y \in U$, $P \subseteq C$. If $\forall q \in P$, $f(y, q) \succeq f(x, q)$, then $yD_P x$ denotes a* **dominating relation**.

$f(y, q) \succeq f(x, q)$ means x is at most as good as y on criteria q. Dominating relation is reflexive, transitive, and antisymmetric, which denotes a partial relation on q between objects. The dominating relation is a weak partial ordering relation. If $\forall q \in P$, $f(x,q) \succ f(y,q)$, then the dominating relation is a strong partial ordering relation.

Definition 2 *[6]. $S = (U, A, V, f)$ is a decision system, $A = C \cup D$, $P \subseteq C$, $x, y \in U$, P-**dominating set** and P-**dominated set** are: $D_P^+(x) = \{y \mid y D_P x\}$, $D_P^-(x) = \{y \mid x D_P y\}$, respectively.*

The set of decision attributes D partitions U into a finite number of classes, denoted as $Cl = \{Cl_t \mid t \in \{1, \dots, n\}\}$, where $Cl_n \succ \dots \succ Cl_s \succ \dots \succ Cl_1$. An upward union and a downward union of classes are defined respectively as follows. $Cl_t^{\geq} = \bigcup_{s \geq t} Cl_s$, $Cl_t^{\leq} = \bigcup_{s \leq t} Cl_s$, where $t, s \in \{1, \dots, n\}$. $x \in Cl_t^{\geq}$ means x at least belongs to class Cl_t. $x \in Cl_t^{\leq}$ means x at most belongs to class Cl_t.

Definition 3 *[6]. $S = (U, A, V, f)$ is a decision system, $P \subseteq C$, $x \in U$, $Cl_t^{\geq}$, $Cl_t^{\leq} \subseteq U$, $t = 1, 2, \dots, n$. The approximations of $Cl_t^{\geq}$ and $Cl_t^{\leq}$ are defined respectively as:*

$$\underline{P}(Cl_t^{\geq}) = \{x \mid D_P^+(x) \subseteq Cl_t^{\geq}\}, \tag{1}$$

$$\overline{P}(Cl_t^{\geq}) = \bigcup_{x \in Cl_t^{\geq}} D_P^+(x); \tag{2}$$

$$\underline{P}(Cl_t^{\leq}) = \{x \mid D_P^-(x) \subseteq Cl_t^{\leq}\}, \tag{3}$$

$$\overline{P}(Cl_t^{\leq}) = \bigcup_{x \in Cl_t^{\leq}} D_P^-(x). \tag{4}$$

If an object is superior to the object x in lower approximation of an upward (downward) union $Cl_t^{\geq}(Cl_t^{\leq})$, then it must belong to the decision class which is superior (inferior) to the decision class that the object x belongs to. If an object is superior to the object x in upper approximation of an upward (downward) union $Cl_t^{\geq}(Cl_t^{\leq})$, then it may belong to the decision class which is superior (inferior) to the decision class that the object x belongs to.

2.2 Neighborhood Rough Set

In NRS, neighborhood granules are used to describe uncertain concepts.

Definition 4 *[11]. Let $B_1 \subseteq C$ and $B_2 \subseteq C$ be the sets of numerical attributes and categorical attributes, respectively. The neighborhood granules of sample x induced by B_1, B_2 and $B_1 \cup B_2$ are defined as follows.*

i. $\delta_{B_1}(x) = \{x_i \mid \triangle_{B_1}(x, x_i) \leqslant \delta, x_i \in U\}$
ii. $\delta_{B_2}(x) = \{x_i \mid \triangle_{B_2}(x, x_i) = 0, x_i \in U\}$
iii. $\delta_{B_1 \cup B_2}(x) = \{x_i \mid \triangle_{B_1}(x, x_i) \leqslant \delta \wedge \triangle_{B_2}(x, x_i) = 0, x_i \in U\}$

The formula of $\triangle$ is given as follows [11].

$$\triangle_P(x,y) = \left(\sum_{i=1}^{N} |f(x, a_i) - f(y, a_i)|^P\right)^{1/P} \tag{5}$$

For Definition 4, the relationship among objects having numerical and categorical data can form via the distance measurement.

Definition 5 *[11]. Given a set of objects U and a neighborhood relation N over U, we call $\langle U, N \rangle$ a neighborhood approximation space. For any $X \subseteq U$, two subsets of objects, called lower and upper approximations of X in $\langle U, N \rangle$, respectively, are defined as*

$$\underline{N}(X) = \{x_i | \delta(x_i) \subseteq X, x_i \in U\},$$
$$\overline{N}(X) = \{x_i | \delta(x_i) \cap X, x_i \in U\}.$$

3 Dominance-Based Neighborhood Rough Set

By considering the partial order, DNRS is introduced in this paper. The definition of δ dominating relation is given, which aims to consider the overall degree of dominating and avoid data error from data collection.

3.1 Hybrid Decision System

The definition of Hybrid Decision System (HDS) considering partial order among attribute values on different attributes is given as follows.

Definition 6. *A quadruple (U, A, V, f) is a Hybrid Decision System (HDS), where U is a non-empty finite set of objects, called the universe. $A = C_S \cup C_N \cup D$ is a non-empty finite set of attributes, where C_S, C_N and D are the set of symbolic condition attribute, the set of numeric condition attribute, and the set of decision attribute, respectively. $C = C_S \cup C_N$ is the condition attribute. V is the domain of attributes set A. $f : U \times A \rightarrow V$ is an information function, $\forall a \in A$, $x \in U$, $f(x,a) \in V_a$.*

Example 31. *Part of data set Postoperative patient down from UC Irvine Machine Learning Repository are listed in Table 1 [8], where $U = \{x_i | i = 1, 2, \cdots, 13\}$, $A = \{a_1, a_2, a_3, a_4, a_5, a_6, a_7, a_8, d\}$, $C_S = \{a_1, a_2, a_3, a_4, a_5, a_6, a_7\}$, $C_N = \{a_8\}$, $D = \{d\}$. Considering partial order related to decision attribute d, the rank of attribute values on different attributes are listed as follows.*

- V_{a_1}: *high $\prec$ low $\prec$ mid*; V_{a_2}: *high $\prec$ low $\prec$ mid*; V_{a_3}: *poor $\prec$ fair $\prec$ good $\prec$ excellent*;
- V_{a_4}: *high $\prec$ low $\prec$ mid*; V_{a_5}: *unstable $\prec$ mod − stable $\prec$ stable*;
- V_{a_6}: *unstable $\prec$ mod−stable $\prec$ stable*; V_{a_7}: *unstable $\prec$ mod−stable $\prec$ stable*;
- V_{a_8}: $0 \prec \cdots \prec 20$; V_d: $I \prec A \prec S$.

3.2 Normalization in HDS

Different normalization policies will be employed for different types of data in a HDS [10].

Table 1. A hybrid decision system

U_1	a_1	a_2	a_3	a_4	a_5	a_6	a_7	a_8	d
x_1	low	low	excellent	mid	stable	stable	stable	10	A
x_2	mid	mid	excellent	mid	stable	stable	mod-stable	10	S
x_3	mid	mid	excellent	high	stable	stable	stable	10	A
x_4	mid	low	excellent	high	stable	stable	mod-stable	10	A
x_5	low	mid	good	mid	stable	stable	unstable	10	A
x_6	mid	mid	excellent	mid	stable	stable	mod-stable	10	A
x_7	mid	mid	excellent	mid	stable	stable	unstable	10	A
x_8	mid	mid	excellent	mid	unstable	unstable	stable	10	S
x_9	mid	mid	good	high	stable	stable	stable	10	A
x_{10}	mid	mid	excellent	mid	stable	stable	stable	15	A
x_{11}	mid	mid	excellent	mid	stable	stable	stable	10	S
x_{12}	mid	low	good	mid	stable	stable	unstable	10	I
x_{13}	high	mid	excellent	mid	unstable	stable	unstable	5	A

Normalization for Numerical Data. Let $\Delta_b\ (x_i, x_j)$ denote the distance between x_i and x_j under attribute b ($b \in C_N$). For numerical data, the data is normalized in $[0, 1]$ before computing the distance between x_i and x_j. The method of normalization is

$$v'_{ij} = \frac{v_{ij} - \min(V_{a_j})}{\max(V_{a_j}) - \min(V_{a_j})}, \qquad (6)$$

where v_{ij} denotes the value $f(x_i, a_j)$, V_{a_j} denotes the domain of a_j, v'_{ij} denotes the normalized value of v_{ij}, $\min(V_{a_j})$ is the minimum value of V_{a_j}, and $\max(V_{a_j})$ is the maximum value of V_{a_j}.

Normalization for Nominal Data. Suppose the domain of attribute a_k which data type is nominal, namely, $V_{a_k} = \{v_{k_1}, v_{k_2}, \cdots, v_{k_s}\}$, where $v_{k_1} \preccurlyeq v_{k_2} \preccurlyeq \cdots \preccurlyeq v_{k_s}$. Because the degree of preference is important, then let v'_{k_i} denote the normalized value of v_{k_i},

$$v'_{k_i} = \frac{i - 1}{|V_{a_k}| - 1}(1 \leqslant i \leqslant |V_{a_k}|) \qquad (7)$$

For a nominal value v'_{k_i}, if $v_{k_i} = v_{k_m}$, then $v'_{k_i} = v'_{k_m}$.

Normalized Distance Between Objects. The normalized distance between objects, denoted as $\Delta'_C(x_i, x_j)$, is defined as

$$\Delta'_C(x_i, x_j) = \left(\sum_{k=1}^{|C|} \left| v'_{k_i} - v'_{k_j} \right|^P \right)^{\frac{1}{P}} |C|^{-\frac{1}{P}} \qquad (8)$$

where v'_{k_i} and v'_{k_j} are normalized values of objects x_i and x_j on attribute C. When $P = 2$, it's a Euclidean distance; when $P = 1$, it's a Manhattan distance. In this paper, Euclidean distance is used.

3.3   Approximations in DNRS

In this section, we present the definitions of neighborhood dominating relation, neighborhood dominating set, dominating relation matrix, and approximations.

Definition 7. $\forall x \in U,\, a_i \in B,\, B \subseteq C,\, 0 \leqslant \delta \leqslant 1$, the neighborhood dominating relation $\delta_B^{\succcurlyeq}$ in B is defined as:

$$\delta_B^{\succcurlyeq} = \{(x, y) \in U \times U \,|\, \Delta'_B(x, y) \geqslant \delta \wedge f(x, a_i) \succcurlyeq f(y, a_i)\}, \tag{9}$$

where $\Delta'_B(x, y)$ is the normalized distance between objects x and y.

Definition 8. Let $S = (U, A, V, f)$ be an HDS, $A = C_S \cup C_N \cup D$, $B \subseteq C_S \cup C_N$, $x, y \in U$. B neighborhood dominating set and B neighborhood dominated set are defined as follows.

$$\delta_B^+(x) = \{y \,|\, y\delta_B^{\succcurlyeq}x\}, \tag{10}$$

$$\delta_B^-(x) = \{y \,|\, x\delta_B^{\succcurlyeq}y\}. \tag{11}$$

Property 1. For $\delta_B^+(x)$ and $\delta_B^-(x)$,

1. $\delta_B^+(x)$ and $\delta_B^-(x)$ are transitive, and unsymmetrical;
2. $\forall \delta_1, \delta_2,\, 0 \leqslant \delta_1 < \delta_2 \leqslant 1,\, \delta_{2B}^+(x) \subset \delta_{1B}^+(x),\, \delta_{2B}^-(x) \subset \delta_{1B}^-(x)$;
3. $\forall E \subseteq F \subseteq C,\, \delta_C^+(x) \subseteq \delta_F^+(x) \subseteq \delta_E^+(x),\, \delta_C^-(x) \subseteq \delta_F^-(x) \subseteq \delta_E^-(x)$.

Proof. It follows from Definition 7 directly.

Then, the dominating relation matrix and the dominated relation matrix are

$$R^{\succcurlyeq} = \begin{bmatrix} r_{11}^{\succcurlyeq} & \cdots & r_{1j}^{\succcurlyeq} & \cdots & r_{1n}^{\succcurlyeq} \\ \vdots & \vdots & \vdots & \vdots & \vdots \\ r_{i1}^{\succcurlyeq} & \cdots & r_{ij}^{\succcurlyeq} & \cdots & r_{in}^{\succcurlyeq} \\ \vdots & \vdots & \vdots & \vdots & \vdots \\ r_{n1}^{\succcurlyeq} & \cdots & r_{nj}^{\succcurlyeq} & \cdots & r_{nn}^{\succcurlyeq} \end{bmatrix},\ R^{\preccurlyeq} = \begin{bmatrix} r_{11}^{\preccurlyeq} & \cdots & r_{1j}^{\preccurlyeq} & \cdots & r_{1n}^{\preccurlyeq} \\ \vdots & \vdots & \vdots & \vdots & \vdots \\ r_{i1}^{\preccurlyeq} & \cdots & r_{ij}^{\preccurlyeq} & \cdots & r_{in}^{\preccurlyeq} \\ \vdots & \vdots & \vdots & \vdots & \vdots \\ r_{n1}^{\preccurlyeq} & \cdots & r_{nj}^{\preccurlyeq} & \cdots & r_{nn}^{\preccurlyeq} \end{bmatrix},\ \text{respectively,}$$

where $r_{ij}^{\succcurlyeq} = \begin{cases} 1, (x_i, x_j) \in \delta_B^+(x) \\ 0, others \end{cases}$, $r_{ij}^{\preccurlyeq} = \begin{cases} 1, (x_i, x_j) \in \delta_B^-(x) \\ 0, others \end{cases}$.

The degree of preference is taken into consideration in neighbourhood dominating relation in an HDS. Dominating sets or dominated sets are then used as elementary elements to describe uncertain decisions in an HDS.

Definition 9. Let $S = (U, A, V, f)$ be an HDS, $A = C_S \cup C_N \cup D$, $B \subseteq C_S \cup C_N$, $x \in U$, $Cl_t^{\geqslant}$, $Cl_t^{\leqslant} \subseteq U$, $t = 1, 2, \ldots, n$. The upper and lower approximations of $Cl_t^{\geqslant}$, $Cl_t^{\leqslant}$ under the neighborhood dominating relation are defined respectively as follows.

$$\underline{\delta_B}(Cl_t^{\geqslant}) = \{x \,\big|\, \delta_B^+(x) \subseteq Cl_t^{\geqslant}\}, \tag{12}$$

$$\overline{\delta_B}(Cl_t^{\geq}) = \{x \,\big|\, \delta_B^+(x) \cap Cl_t^{\geq} \neq \emptyset\}; \tag{13}$$

$$\underline{\delta_B}(Cl_t^{\leq}) = \{x \,\big|\, \delta_B^-(x) \subseteq Cl_t^{\leq} \}, \tag{14}$$

$$\overline{\delta_B}(Cl_t^{\leq}) = \{x \,\big|\, \delta_B^-(x) \cap Cl_t^{\leq} \neq \emptyset\}; \tag{15}$$

The positive region and boundary region are defined respectively as follows.

$$POS_B(Cl_t^{\geq}) = \underline{\delta_B}(Cl_t^{\geq}), \tag{16}$$

$$POS_B(Cl_t^{\leq}) = \underline{\delta_B}(Cl_t^{\leq}); \tag{17}$$

$$BND_B(Cl_t^{\geq}) = \overline{\delta_B}(Cl_t^{\geq}) - \underline{\delta_B}(Cl_t^{\geq}), \tag{18}$$

$$BND_B(Cl_t^{\leq}) = \overline{\delta_B}(Cl_t^{\leq}) - \underline{\delta_B}(Cl_t^{\leq}). \tag{19}$$

4   Attribute Reduction in DNRS

Attribute reduction in RST may remove superfluous attributes while keep the classification ability. It has become an efficiency way in machine learning and data mining. In this section, attribute reduction in DNRS is discussed. The definitions of the discernibility matrix, the minimal discernibility attribute set, and the matrix of the attribute importance in DNRS are presented as follows, which need to be computed in the attribute reduction in DNRS.

4.1   Discernibility Matrix in DNRS

If $D = \{d_1, \cdots, d_s\}$, then let $D_i C = \{f(x_k, d_1), \cdots, f(x_k, d_s)\}(x_k \in D_i, D_i \in {}^U\!/_{R_D})$ denote the characteristic value of decision class D_i. Let $GD(\delta_i^+) = \{D_j C | \delta_B^+(x_i) \cap D_j \neq \emptyset\}(D_j \in {}^U\!/_{R_D})$ denote the generalized decision of a neighborhood dominating set $\delta_B^+(x_i)$, and $GD(\delta_i^-) = \{D_j C | \delta_B^-(x_i) \cap D_j \neq \emptyset\}(D_j \in {}^U\!/_{R_D})$ denote the generalized decision of a neighborhood dominated set $\delta_B^-(x_i)$. Let $GD(x_i) = D_j C(x_i \in D_j)$, which denotes the generalized decision of x_i. For simplicity, let $D = \{d\}$ in this paper.

Definition 10. (U, A, V, f) is a decision system. $M^{\succeq}$ is a $|U| \times |U|$ dominating discernibility matrix in DRNS, $M^{\succeq} = \begin{bmatrix} d_{11}^{\succeq} & \cdots & d_{1j}^{\succeq} & \cdots & d_{1n}^{\succeq} \\ \vdots & \vdots & \vdots & \vdots & \vdots \\ d_{i1}^{\succeq} & \cdots & d_{ij}^{\succeq} & \cdots & d_{in}^{\succeq} \\ \vdots & \vdots & \vdots & \vdots & \vdots \\ d_{n1}^{\succeq} & \cdots & d_{nj}^{\succeq} & \vdots & d_{nn}^{\succeq} \end{bmatrix}$. If $x_j \notin \delta_B^+(x_i), GD$

$(\delta_i^+) \neq GD(\delta_j^+)$, then $d_{ij} = \{a_k \in C : f(x_i, a_k) > f(x_j, a_k) \vee |f(x_i, a_k) - f(x_j, a_k)| < \delta\}$; otherwise $d_{ij} = \emptyset$.

Discernibility formula of the reduction can be calculated according to the values of dominating discernibility matrix in DNRS. The definition of discernibility formula of reduction is given as follows.

Definition 11. *The discernibility formula of reduction is*

$$F = \wedge\{\vee\{a_k : a_k \in d_{ij}^{\succeq}\}\}(i, j \leq n). \tag{20}$$

The minimum disjunction form is

$$M_{min} = \overset{p}{\underset{k=1}{\vee}} (\overset{q_k}{\underset{l=1}{\wedge}} a_{k,l}).$$

Let $B_k = \{a_{k,l} : l = 1, 2, \cdots, q_k\}$. Then $RED(P) = \{B_k : k = 1, 2, \cdots, p\}$ is the set of reductions in DNRS.

4.2 Minimal Discernibility Attribute Set Based Reduction in DNRS

Computing the minimal conjunctive formula directly from the discernibility matrix is an NP problem. We presented the definition of a Minimal Discernibility Attribute Set (MDAS) together with its generated algorithm in [1]. In this paper, the MDAS method for reduction is used in DNRS.

Definition 12. *The MDAS is defined as $Att_{min}^{\succeq} = \{Att_0, \cdots, Att_i, \cdots, Att_t\}$, where $\forall Att_i \in Att_{min}^{\succeq}, \exists d_{jk}^{\succeq}, s.t., Att_i \subseteq d_{ik}^{\succeq}, and \forall d_{jk}^{\succeq}, \exists Att_i \in Att_{min}^{\succeq}, s.t., Att_i \subseteq d_{jk}^{\succeq}. \forall Att_i, Att_j \in Att_{min}^{\succeq}(i \neq j), \neg\exists Att_i \subseteq Att_j or Att_j \subseteq Att_i. Att_i is called a minimal discernibility attribute.*

It means each attribute in the MDAS is different and there doesn't exist an inclusion between them.

Property 2. Given $Att_{min}^{\succeq}, Core \in Att_{min}^{\succeq}$.

Proof: It follows directly from Definition 12 of the MDAS.

In the following, we present an algorithm to generate the MDAS.

Algorithm 1. *Generation of the Minimal Discernibility Attributes Set Gradually (GMDASG)*

GMDASG$(U, R^{\succeq})$
```
 1   Att_min^≽ ← ∅
 2   for i ← 1 to |U|
 3       do
 4           for j ← i + 1 to |U| while j < |U| + 1
 5               do if x_j ∉ δ_B^+(x_i)
 6                   then Compute d_ij^≽ by U
 7                       if Att_min^≽ = ∅
 8                           then Att_min^≽ ← d_ij^≽
 9                           else UPDATEATT(d_ij^≽, Att_min^≽)
10   return Att_min^≽, M^≽
```

UPDATEATT($DisAtt, Att_{min}^{\succeq}$)
1    $sig \leftarrow 0$
2   **for** each Att_k in $Att_{min}^{\succeq}$
3   **do if** $DisAtt \supseteq Att_k$
4   **then** break;
5   **else if** $Att_k \supset DisAtt$
6   **then** $Att_{min}^{\succeq} \leftarrow Att_{min}^{\succeq} - Att_k$,
    $sig \leftarrow 1$
7   **if** $sig = 1$
8   **then** $Att_{min}^{\succeq} \leftarrow Att_{min}^{\succeq} \cup DisAtt$

The MDAS is generated during the course of computing the discernibility attribute set in Algorithm GMDASG.

4.3   Matrix of the Attribute Importance

Definition 13. *Let $Important_i$ denote the importance of attribute a_i in the assignment reduction. Then the matrix of the attribute importance is defined as*

$$M_{Important} = \begin{bmatrix} a_1 & Important_1 \\ \vdots & \vdots \\ a_l & Important_l \end{bmatrix} \quad (l = |C|). \quad AT = \begin{bmatrix} a_1 \\ \vdots \\ a_l \end{bmatrix} \quad and \quad IM =$$

$$\begin{bmatrix} Important_1 \\ \vdots \\ Important_l \end{bmatrix}$$ *denote the attribute vector and the importance vector of the matrix of the attribute importance, respectively. The order matrix of the attribute importance is:*

$$M_{Important}^{\geqslant} = \begin{bmatrix} \vdots & \vdots \\ a_i & Important_i \\ \vdots & \vdots \\ a_j & Important_j \\ \vdots & \vdots \end{bmatrix} \quad (Important_i \geqslant Important_j, i, j \in$$

$\{1, \cdots, l\}, l = |C|$), where $Important_i = \sum\limits_{i=1}^{n} \sum\limits_{j=1}^{n} N_i (N_i = 1, if \ a_i \in d_{ij}^{\succeq}(i, j \in$ $\{1, \cdots, n\})$; otherwise, $N_i = 0$).

4.4   Generation of the Reduct

Generation of the reduct includes the following steps: computing the equivalence feature matrix and the characteristic value matrix, generating the MDAS, computing the matrix of the attribute importance, and generating the reduct. The algorithm is given as follows.

Algorithm 2. *Generation of the Reduct by Minimal Discernibility Attribute Set (GRMDAS)*

GRMDAS(U)

1   Compute $R^\succcurlyeq$
2   GMDASG($U, R^\succcurlyeq$)
3   Compute $M_{Important}$ by $M^\succcurlyeq$
4   Generate $M_{Important}^\geq$ by $M_{Important}$
5   GENERATEREDUCT ($M_{Important}^\geq, Att_{min}^\succcurlyeq$)
6   **return** *Reduct*

5   Conclusion

Different data types coexist in real-life applications. DRSA has been used widely in multi-attribute and multi-criteria decision. In this paper, DNRS was proposed when dealing hybrid data by considering the partial order relation. The definition of the neighborhood dominating relation was given firstly. Then the approximations under neighborhood dominating relation were defined. Finally, attribute reduction in DNRS was investigated. In the future work, we will verify the proposed method in real applications.

References

1. Chen, H.M., Li, T.R., Luo, C., Horng, S.J., Wang, G.Y.: A rough set-based method for updating decision rules on attribute values coarsening and refining. IEEE Trans. Knowl. Data Eng. **26**(12), 2886–2899 (2014)
2. Chen, D.G., Kwong, S., He, Q., Wang, H.: Geometrical interpretation and applications of membership functions with fuzzy rough sets. Fuzzy Sets Syst. **193**, 122–135 (2012)
3. Chen, H.M., Li, T.R., Ruan, D.: Maintenance of approximations in incomplete ordered decision systems. Knowl. Based Syst. **31**, 140–161 (2012)
4. Dubois, D., Prade, H.: Rough fuzzy set sand fuzzy rough sets. Int. J. Gen. Syst. **17**, 191–208 (1990)
5. Guyon, I., Elisseeff, A.: An introduction to variable and feature selection. J. Mach. Learn. Res. **3**, 1157–1182 (2003)
6. Greco, S., Matarazzo, B., Slowinski, R.: Rough approximation of a preference relation by dominance relations. Eur. J. Oper. Res. **117**(1), 63–83 (1999)
7. Greco, S., Matarazzo, B., Slowinski, R.: Fuzzy dominance-based rough set approach. In: Masulli, F., Parenti, R., Pasi, G. (eds.) Advances in Fuzzy Systems and Intelligent Technologies, pp. 56–66. Shaker Publishing, Maastricht (2000)
8. Grzymala-Busse J.W.: UCI Machine Learning Repository. University of California, School of Information and Computer Science, Irvine, CA (1993). http://archive.ics.uci.edu/ml
9. Huang, B., Li, H.X., Wei, D.K.: Dominance-based rough set model in intuitionistic fuzzy information systems. Knowl. Based Syst. **28**, 115–123 (2012)
10. Han, J.W., Kamber, M., Pei, J.: Data Mining, 3rd edn. Morgan Kaufmann Publishers, San Francisco (2012)
11. Hu, Q.H., Yu, D.R., Liu, J.F., Wu, C.X.: Neighborhood rough set based heterogeneous feature subset selection. Inf. Sci. **178**, 3577–3594 (2008)

12. Inuiguchi, M., Yoshioka, Y., Kusunoki, Y.: Variable-precision dominance-based rough set approach and attribute reduction. Int. J. Approximate Reasoning **50**(8), 1199–1214 (2009)
13. Li, S.Y., Li, T.R., Hu, J.: Update of approximations in composite information systems. Knowl. Based Syst. **83**, 138–148 (2015)
14. Liu, Y., Huang, W.L., Jiang, Y.L., Zeng, Z.Y.: Quick attribute reduct algorithm for neighborhood rough set model. Inf. Sci. **271**, 65–81 (2014)
15. Lin, G.P., Qian, Y.H., Li, J.J.: NMGRS: neighborhood-based multigranulation rough sets. Int. J. Approximate Reasoning **53**(7), 1080–1093 (2012)
16. Kadzinski, M., Greco, S., Slowinski, R.: Robust ordinal regression for dominance-based rough set approach to multiple criteria sorting. Inf. Sci. **283**, 211–228 (2014)
17. Kotlowski, W., Dembczynski, K., Greco, S., Slowinski, R.: Stochastic dominance-based rough set model for ordinal classification. Inf. Sci. **178**(21), 4019–4037 (2008)
18. Park, I.-K., Choi, G.-S.: Rough set approach for clustering categorical data using information-theoretic dependency measure. Inf. Syst. **48**, 289–295 (2015)
19. Pawlak, Z., Skowron, A.: Rough sets: some extensions. Inf. Sci. **177**(1), 8–40 (2007)
20. Szelazg, M., Greco, S., Slowinski, R.: Variable consistency dominance-based rough set approach to preference learning in multicriteria ranking. Inf. Sci. **277**, 525–555 (2014)
21. Slowinski, R., Vanderpooten, D.: A generalized definition of rough approximations based on similarity. IEEE Trans. Knowl. Data Eng. **12**, 331–336 (2000)
22. Song, X.X., Zhang, W.X.: Knowledge reduction in inconsistent set-valued decision information system. Comput. Eng. Appl. **45**(1), 33–35 (2009). (In Chinese)
23. Yang, X.B., Yu, D.J., Yang, J.Y., Wei, L.H.: Dominance-based rough set approach to incomplete interval-valued information system. Data Knowl. Eng. **68**(11), 1331–1347 (2009)
24. Wei, W., Liang, J.Y., Qian, Y.H.: A comparative study of rough sets for hybrid data. Inf. Sci. **190**(1), 1–16 (2012)
25. Zhang, J.B., Li, T.R., Chen, H.M.: Composite rough sets for dynamic data mining. Inf. Sci. **257**, 81–100 (2014)

Mining Incomplete Data with Many Lost and Attribute-Concept Values

Patrick G. Clark[1] and Jerzy W. Grzymala-Busse[1,2](✉)

[1] Department of Electrical Engineering and Computer Science, University of Kansas,
Lawrence, KS 66045, USA
patrick.g.clark@gmail.com, jerzy@ku.edu
[2] Department of Expert Systems and Artificial Intelligence,
University of Information Technology and Management, 35-225 Rzeszow, Poland

Abstract. This paper presents experimental results on twelve data sets with many missing attribute values, interpreted as lost values and attribute-concept values. Data mining was accomplished using three kinds of probabilistic approximations: singleton, subset and concept. We compared the best results, using all three kinds of probabilistic approximations, for six data sets with lost values and six data sets with attribute-concept values, where missing attribute values were located in the same places. For five pairs of data sets the error rate, evaluated by ten-fold cross validation, was significantly smaller for lost values than for attribute-concept values (5 % significance level). For the remaining pair of data sets both interpretations of missing attribute values do not differ significantly.

Keywords: Incomplete data · Lost values · Attribute-concept values · Probabilistic approximations · MLEM2 rule induction algorithm

1 Introduction

In mining incomplete data using rough set theory we may take advantage of the interpretation of missing attribute values. In this research, we use two interpretations of missing attribute values: lost values and attribute-concept values. In the former case, a missing attribute value is interpreted as not available in spite of the fact that initially it was accessible. For example, such value could be erased. For mining such data we use only existing, specified attribute values. In the latter case, we are assuming that a missing attribute value may be reconstructed from existing values from the concept in which the case belongs. As an example, for a data set describing patients sick with flu, the temperature of all patients sick with flu is high or very-high. If we know that a patient is sick with flu and the value of patient's temperature is missing, we will assume that the patient's temperature is a value from the set {high, very-high}.

The main objective of this paper is to determine the better interpretation of missing attribute values among: lost values or attribute-concept values.

© Springer International Publishing Switzerland 2015
D. Ciucci et al. (Eds.): RSKT 2015, LNAI 9436, pp. 100–109, 2015.
DOI: 10.1007/978-3-319-25754-9_9

For many data sets with missing attribute values the actual interpretation of the missing attribute values is not known and we may select out own interpretation. If we have a choice, we should select an interpretation that provides for a smaller error rate, evaluated by ten-fold cross validation.

Research on a comparison of lost values and attribute-concept values was presented in [3,4]. In [3] experiments were conducted on eight pairs of data sets with 35 % of missing attribute values, using only one type of probabilistic approximations: concept. As a result, on six pairs of data sets lost values were significantly better than attribute-concept values. In [4] experiments were conducted on data sets with broad range of missing attribute values, but again, using only one kind of probabilistic approximations (concept) but with three values of the parameter α: 0.001, 0.5 and 1. That research distinguished 24 cases because we experimented with eight data sets and with three different values of α. In 10 out of these 24 cases the smaller error rate was associated with lost values, in remaining 14 cases the difference in performance was not statistically significant (5 % significance level).

An idea of the approximation is fundamental for rough set theory. A probabilistic approximation, defined using an additional parameter α, interpreted as a probability, is a generalization of ordinary, lower and upper approximations. If the parameter α is equal to 1, the probabilistic approximation becomes the lower approximation; if α is slightly greater than 0, the probabilistic approximation is reduced to the upper approximation. For incomplete data sets ordinary singleton, subset and concept approximations were generalized to respective probabilistic approximations in [7]. Probabilistic approximations have been investigated in Bayesian rough sets, decision-theoretic rough sets, variable precision rough sets, etc., see, e.g., [9–12,14–18].

Research on probabilistic approximations, until recently, concentrates mostly on theoretical properties of such approximations and is restricted to complete data sets (with no missing attribute values). For incomplete data sets ordinary approximations are extended to probabilistic approximations in [7]. Furthermore, [1,2] are the first papers to report experimental results on these probabilistic approximations.

In this paper we use three kinds of probabilistic approximations (singleton, subset and concept). For a better contrast of the two interpretations of missing attribute values, we use data sets with many missing attribute values. None of the data sets used for our experiments had less than 40 % of missing attribute values. For a small number of missing attribute values the difference between the two interpretations of missing attribute values is not so obvious.

2   Data Sets

We assume that the input data sets are presented in the form of a *decision table*. An example of a decision table is shown in Table 1. Rows of the decision table represent *cases*, while columns are labeled by *variables*. The set of all cases will be denoted by U. In Table 1, $U = \{1, 2, 3, 4, 5, 6, 7\}$. Independent variables are called *attributes* and a dependent variable is called a *decision* and is denoted by d. The set of all attributes will be denoted by A. In Table 1, $A = \{Wind, Humidity, Temperature\}$. The value for a case x and an attribute a will be denoted by $a(x)$.

Table 1. An incomplete data set

	Attributes			Decision
Case	Wind	Humidity	Temperature	Trip
1	low	?	high	yes
2	high	low	−	yes
3	−	high	low	yes
4	low	−	high	yes
5	high	high	?	no
6	high	?	low	no
7	−	low	high	no

In this paper we distinguish between two interpretations of missing attribute values: *lost values*, denoted by "?" and *attribute-concept values*, denoted by "−". We assume that lost values were erased or are unreadable and that for data mining we use only remaining, specified values [8,13]. The attribute-concept value is restricted to attribute values typical for the concept to which the case belongs.

We will assume that for any case at least one attribute value is specified (i.e., is not missing) and that all decision values are specified.

For complete data sets, a *block* of a variable-attribute pair (a, v), denoted by $[(a, v)]$, is the set $\{x \in U \mid a(x) = v\}$ [5]. For incomplete data sets the definition of a block of an attribute-value pair is modified in the following way.

- If for an attribute a there exists a case x such that $a(x) = ?$, i.e., the corresponding value is lost, then the case x should not be included in any blocks $[(a, v)]$ for all values v of attribute a,
- If for an attribute a there exists a case x such that the corresponding value is an attribute-concept value, i.e., $a(x) = -$, then the corresponding case x should be included in blocks $[(a, v)]$ for all specified values $v \in V(x, a)$ of attribute a, where

$$V(x, a) = \{a(y) \mid a(y) \ is \ specified, y \in U, \ d(y) = d(x)\}.$$

For the data set from Table 1, $V(2, Temperature) = \{low, high\}$, $V(3, Wind) = \{low, high\}$, $V(4, Humidity) = \{low, high\}$, and $V(7, Wind) = \{high\}$. For the data set from Table 1, the blocks of attribute-value pairs are:

[(Wind, low)] = {1, 3, 4},
[(Wind, high)] = {2, 3, 5, 6, 7},
[(Humidity, low)] = {2, 4, 7},

[(Humidity, high)] = {3, 4, 5},
[(Temperature, low)] = {2, 3, 6},
[(Temperature, high)] = {1, 2, 4, 7}.

For a case $x \in U$ and $B \subseteq A$, the *characteristic set* $K_B(x)$ is defined as the intersection of the sets $K(x, a)$, for all $a \in B$, where the set $K(x, a)$ is defined in the following way:

- If $a(x)$ is specified, then $K(x, a)$ is the block $[(a, a(x)]$ of attribute a and its value $a(x)$,
- If $a(x) =?$, then the set $K(x, a) = U$, where U is the set of all cases,
- If $a(x) = -$, then the corresponding set $K(x, a)$ is equal to the union of all blocks of attribute-value pairs (a, v), where $v \in V(x, a)$ if $V(x, a)$ is nonempty. If $V(x, a)$ is empty, $K(x, a) = U$.

For Table 1 and $B = A$,

$K_A(1) = \{1, 4\}$,
$K_A(2) = \{2, 7\}$,
$K_A(3) = \{3\}$,
$K_A(4) = \{4\}$,
$K_A(5) = \{3, 5\}$,
$K_A(6) = \{2, 3, 6\}$,
$K_A(7) = \{2, 7\}$.

3   Probabilistic Approximations

The singleton probabilistic approximation of X with the threshold α, $0 < \alpha \le 1$, denoted by $appr_\alpha^{singleton}(X)$, is defined as follows

$$\{x \mid x \in U, \ Pr(X|K_B(x)) \ge \alpha\},$$

where $Pr(X|K_B(x)) = \frac{|X \cap K_B(x)|}{|K_B(x)|}$ is the conditional probability of X given $K_B(x)$.

A subset probabilistic approximation of the set X with the threshold α, $0 < \alpha \le 1$, denoted by $appr_\alpha^{subset}(X)$, is defined as follows

$$\cup\{K_B(x) \mid x \in U, \ Pr(X|K_B(x)) \ge \alpha\}.$$

A concept probabilistic approximation of the set X with the threshold α, $0 < \alpha \le 1$, denoted by $appr_\alpha^{concept}(X)$, is defined as follows

$$\cup\{K_B(x) \mid x \in X, \ Pr(X|K_B(x)) \ge \alpha\}.$$

For Table 1 and the concept $X = [(\mathit{Trip}, yes)] = \{1, 2, 3, 4\}$, for any characteristic set $K_A(x)$, $x \in U$, conditional probabilities $P(X|K_A(x))$ are presented in Table 2.

Table 2. Conditional probabilities

x	1	2	3	4	5	6	7
$K_A(x)$	$\{1, 4\}$	$\{2, 7\}$	$\{3\}$	$\{4\}$	$\{3, 5\}$	$\{2, 3, 6\}$	$\{2, 7\}$
$P(\{1,2,3,4\} \mid K_A(x))$	1.0	0.5	1.0	1.0	0.5	0.667	0.5

For Table 1, all distinct probabilistic approximations (singleton, subset and concept) for [(Trip, yes)] are

$$appr_{0.5}^{singleton}(\{1, 2, 3, 4\}) = U,$$
$$appr_{0.667}^{singleton}(\{1, 2, 3, 4\}) = \{1, 3, 4, 6\},$$
$$appr_{1.0}^{singleton}(\{1, 2, 3, 4\}) = \{1, 3, 4\},$$
$$appr_{0.5}^{subset}(\{1, 2, 3, 4\}) = U,$$
$$appr_{0.667}^{subset}(\{1, 2, 3, 4\}) = \{1, 2, 3, 4, 6\},$$
$$appr_{1.0}^{subset}(\{1, 2, 3, 4\}) = \{1, 3, 4\},$$
$$appr_{0.5}^{concept}(\{1, 2, 3, 4\}) = \{1, 2, 3, 4, 7\},$$
$$appr_{1}^{concept}(\{1, 2, 3, 4\}) = \{1, 3, 4\}.$$

As follows from our example, all three probabilistic approximations, in general, are distinct, even for the same value of the parameter α. Additionally, if for a given set X a probabilistic approximation $appr_\beta(X)$ is not listed, then $appr_\beta(X)$ is equal to the closest probabilistic approximation $appr_\alpha(X)$ of the same type with α larger than or equal to β. For example, $appr_{0.667}^{concept}(\{1, 2, 3, 4\})$ is not listed, so

$$appr_{0.667}^{concept}(\{1, 2, 3, 4\}) = appr_{1.0}^{concept}(\{1, 2, 3, 4\}).$$

If a characteristic relation $R(B)$ is an equivalence relation, all three types of probabilistic approximation: singleton, subset and concept are reduced to the same probabilistic approximation (Figs. 1,2,3,4,5,6).

4   Experiments

For our experiments we used six real-life data sets that are available on the University of California at Irvine *Machine Learning Repository*. For every data set a set of templates was created. Templates were formed by replacing incrementally (with 5 % increment) existing specified attribute values by *lost* values. Thus, we started each series of experiments with no *lost* values, then we added 5 % of *lost* values, then we added additional 5 % of *lost* values, etc., until at least one entire row of the data sets was full of *lost* values. Then three attempts were made to change configuration of new *lost* values and either a new data set with extra 5 % of *lost* values were created or the process was terminated. Additionally, the same formed templates were edited for further experiments by replacing question marks, representing *lost* values by "−"s, representing *attribute-concept* values.

For any data set there was some maximum for the percentage of missing attribute values. For example, for the *Breast cancer* data set, it was 44.81 %.

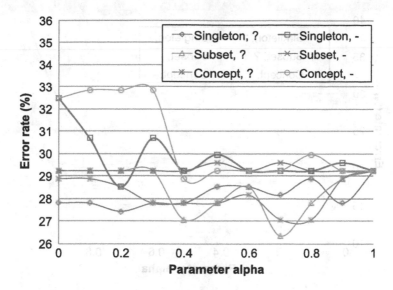

Fig. 1. Error rate for the *breast cancer* data set missing 44.81 % of values

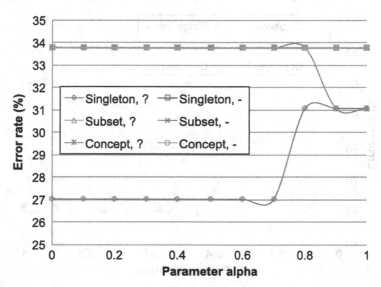

Fig. 2. Error rate for the *echocardiogram* data set missing 40.15 % of values

In our experiments we used only such incomplete data sets, with as many missing attribute values as possible. Note that for some data sets the maximum of the number of missing attribute values was less than 40 %, we have not used such data for our experiments.

For rule induction we used the MLEM2 (Modified Learning from Examples Module version 2) rule induction algorithm, a component of the LERS (Learning from Examples based on Rough Sets) data mining system [5,6].

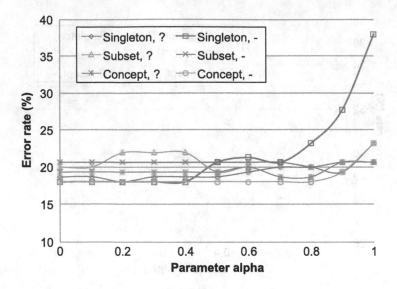

Fig. 3. Error rate for the *hepatitis* data set missing 60.27 % of values

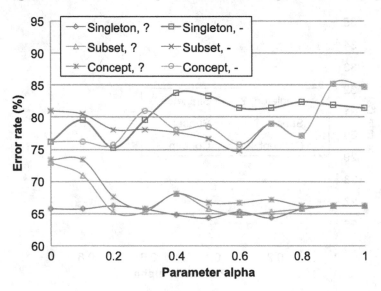

Fig. 4. Error rate for the *image segmentation* data set missing 69.85 % of values

Our objective was to select the better interpretation of missing attribute values among these two: lost values and attribute-concept values, in terms of the error rate measured by ten-fold cross validation. We are assuming that for any data set we may select not only the interpretation of missing attribute values but also one of the three kinds of probabilistic approximations (singleton, subset and concept). Thus, for any value of the parameter α and for any data set, we compared the smallest error rates for all three kinds of probabilistic

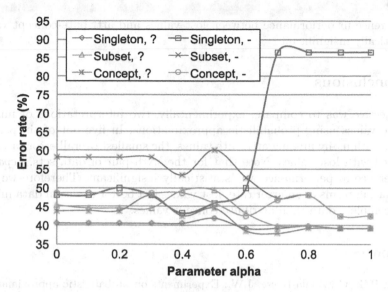

Fig. 5. Error rate for the *lymphography* data set missing 69.89 % of values)

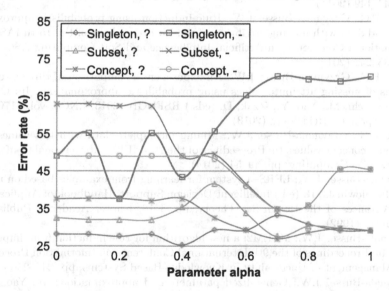

Fig. 6. Error rate for the *wine recognition* data set missing 64.65 % of values

approximations: singleton, subset and concept for lost values with the smallest error rates for the same three probabilistic approximations.

For five pairs of data sets (*breast cancer, echocardiogram, image segmentation, lymphography* and *wine recognition*), the error rate for lost values is always smaller than or equal to the error rate for attribute-concept values. For one pair of data sets (*hepatitis*), results are mixed. Using the Wilcoxon matched-pairs signed rank test (two-tailed test with 5 % significance level), we conclude that the difference in performance between lost values and attribute-concept values is statistically insignificant.

5 Conclusions

Our objective was to compare, experimentally, two interpretations of missing attribute values using probabilistic approximations. In five out of six pairs of data sets with many missing attribute values, the smallest overall error rate, was associated with lost values. Note that for the sixth pair of data sets, *hepatitis*, the difference in performance was statistically insignificant. Therefore we may claim that, in terms of an error rate, lost values is a better choice in data mining from incomplete data than attribute-concept values.

References

1. Clark, P.G., Grzymala-Busse, J.W.: Experiments on probabilistic approximations. In: Proceedings of the 2011 IEEE International Conference on Granular Computing, pp. 144–149 (2011)
2. Clark, P.G., Grzymala-Busse, J.W.: Rule induction using probabilistic approximations and data with missing attribute values. In: Proceedings of the 15-th IASTED International Conference on Artificial Intelligence and Soft Computing ASC 2012, pp. 235–242 (2012)
3. Clark, P.G., Grzymała-Busse, J.W.: An experimental comparison of three interpretations of missing attribute values using probabilistic approximations. In: Ciucci, D., Inuiguchi, M., Yao, Y., śęzak, D. (eds.) RSFDGrC 2013. LNCS, vol. 8170, pp. 77–86. Springer, Heidelberg (2013)
4. Clark, P.G., Grzymala-Busse, J.W.: Mining incomplete data with lost values and attribute-concept values. In: Proceedings of the 2014 IEEE International Conference on Granular Computing, pp. 49–54 (2014)
5. Grzymala-Busse, J.W.: LERS–a system for learning from examples based on rough sets. In: Slowinski, R. (ed.) Intelligent Decision Support. Handbook of Applications and Advances of the Rough Set Theory, pp. 3–18. Kluwer Academic Publishers, Dordrecht (1992)
6. Grzymala-Busse, J.W.: MLEM2: a new algorithm for rule induction from imperfect data. In: Proceedings of the 9th International Conference on Information Processing and Management of Uncertainty in Knowledge-Based Systems, pp. 243–250 (2002)
7. Grzymała-Busse, J.W.: Generalized parameterized approximations. In: Yao, J.T., Ramanna, S., Wang, G., Suraj, Z. (eds.) RSKT 2011. LNCS, vol. 6954, pp. 136–145. Springer, Heidelberg (2011)
8. Grzymala-Busse, J.W., Wang, A.Y.: Modified algorithms LEM1 and LEM2 for rule induction from data with missing attribute values. In: Proceedings of the 5-th International Workshop on Rough Sets and Soft Computing in conjunction with the Third Joint Conference on Information Sciences, pp. 69–72 (1997)
9. Grzymala-Busse, J.W., Ziarko, W.: Data mining based on rough sets. In: Wang, J. (ed.) Data Mining: Opportunities and Challenges, pp. 142–173. Idea Group Publ., Hershey (2003)
10. Pawlak, Z., Skowron, A.: Rough sets: some extensions. Inf. Sci. **177**, 28–40 (2007)

11. Pawlak, Z., Wong, S.K.M., Ziarko, W.: Rough sets: probabilistic versus deterministic approach. Int. J. Man-Mach. Stud. **29**, 81–95 (1988)
12. Ślęzak, D., Ziarko, W.: The investigation of the bayesian rough set model. Int. J. Approximate Reasoning **40**, 81–91 (2005)
13. Stefanowski, J., Tsoukias, A.: Incomplete information tables and rough classification. Comput. Intell. **17**(3), 545–566 (2001)
14. Wang, G.: Extension of rough set under incomplete information systems. In: Proceedings of the IEEE International Conference on Fuzzy Systems, pp. 1098–1103 (2002)
15. Yao, Y.Y.: Probabilistic rough set approximations. Int. J. Approximate Reasoning **49**, 255–271 (2008)
16. Yao, Y.Y., Wong, S.K.M.: A decision theoretic framework for approximate concepts. Int. J. Man Mach. Studies **37**, 793–809 (1992)
17. Ziarko, W.: Variable precision rough set model. J. Comput. Sys. Sci. **46**(1), 39–59 (1993)
18. Ziarko, W.: Probabilistic approach to rough sets. Int. J. Approximate Reasoning **49**, 272–284 (2008)

Inconsistent Dominance Principle Based Attribute Reduction in Ordered Information Systems

Guang-Lei Gou[1,2,3] and Guoyin Wang[2(✉)]

[1] School of Information Science and Technology, Southwest Jiaotong University,
Chengdu 610031, China
[2] Big Data Mining and Applications Center, Chongqing Institute of Green
and Intelligent Technology, CAS, Chongqing 401122, China
`wanggy@cqupt.edu.cn`
[3] School of Computer Science and Engineering, Chongqing University
of Technology, Chongqing 400054, China
`ggl@cqut.edu.cn`

Abstract. Dominance-based rough set is an important model for ordered decision system, in which knowledge reduction is one of the most important problems. The preference ordering of decision between objects is ignored in existed reduction. This paper proposed a knowledge reduction approach based on inconsistent dominance principle, with which two objects are discernable. Furthermore, the judgment theorems and the discernable matrix are investigated, from which we can obtain a new approach to knowledge reduction in ordered decision system.

Keywords: Ordered decision system · Knowledge reduction · Inconsistent dominance principle

1 Introuction

Dominance-based Rough Set Approach (DRSA) [1,2] is always used to deal with Ordered Decision System (ODS) in decision analysis. Different from the Classic Rough Sets (CRS), which is based on equivalence relation (reflectivity, symmetric, transitivity), DRSA is based on dominance relation (reflectivity, anti-symmetric, transitivity).

Knowledge reduction is a core topic in rough set theory. Lots of studies on the topic have been reported and many methods of reduct have been proposed by different concern in classic rough set theory [3–6]. DRSA takes the preference ordering of attribute into account, which is based on dominance relation instead of equivalence relation [7,8]. Thus, the reduction methods of CRS cannot been used in DRSA directly. Greco, Matarazzo, and Slowinski gave the definition of the quality of approximation extended by the CRS and proposed the notion of reduct that preserves the same quality of approximation [9]. Xu and Zhang introduced assignment reduction and approximation reduction [10] and proposed the

© Springer International Publishing Switzerland 2015
D. Ciucci et al. (Eds.): RSKT 2015, LNAI 9436, pp. 110–118, 2015.
DOI: 10.1007/978-3-319-25754-9_10

corresponding discernibility matrix based approaches [11]. Furthermore, Xu and Zhang proposed the distribution reduction and maximum distribution reduction in inconsistent ODS, and discussed properties and relationship between them [12]. The possible distribution reduction (named assignment reduction in [11]) and compatible distribution reduction was discussed in [13]. Chen, Wang and Hu proposed positive domain reduction by defining condition attribute significance [14]. Inuiguchi and Yoshioka investigated several reducts preserving one structure (positive, negative, boundary region) based on upward/downward unions of decision classes and constructed a discernibility matrix for each reduct [15]. Similarly, based on classes btained by the upward/downward unionsclass-based reducts have been introduced in [16]. Susmaga presented a uniform definition framework of reducts and constructs in both CRS and DRSA [17].

The reduction theory of DRSA mentioned above, which regard the decision attribute as a "class" or a "union" from various perspectives, omitted the important characteristic of ODS, the preference order of attribute, called criteria. In ODS, objects under single criteria are comparable because of total ordering, but not all of them are comparable under multi-criteria because of partial ordering. For instance, let objects x, y and z have condition attributes value $(1,2,1)$, $(2,1,2)$, $(2,2,2)$ respectively with increasing preference ordering, so we cannot compare x with y, which means there is no attribute that could discern them. Although the object z dominates x under condition attributes, we still say they are indiscernibility in dominance relation if decision of object z dominates x.

Therefore, we consider it is meaningless to compare two objects without referring to their decision preference ordering under ODS. This paper introduces inconsistent dominance principle, that is, if one object has not worse than another object on all criteria, it has been assigned to a worse decision than the other. The two objects with dominance relation are discernible if they satisfy the inconsistent dominance principle, otherwise they are indiscernible. Then, reduct based on inconsistent dominance principle is proposed. Furthermore, judgment and discernibility matrix are obtained, from which an approach to reduct can be provided in ODS.

The paper is organized as follows. In next section, some preliminary concepts are briefly recalled. In Sect. 3, theories and approaches on inconsistent dominance principle reduction are investigated in ODS. An example is illustrated for the approaches. Finally, we conclude the paper with a summary in Sect. 4.

2   Preliminaries

To facilitate our discussion, some preliminary concepts of ODS and some major reducts in DRSA are briefly recalled in this section.

Let a 4-tuple $DS = (U, A, V, f)$ be a decision system, where U is non-empty finite set of objects, called the universe, $A = C \cup D$ is a set of attributes, with C being the set of conditional attributes and D being the decision attribute, $V = \bigcup_{a \in C \cup D} V_a$ is the set of all attribute values, and $f : U \times A \to V$ assigns every object $x \in U$ with an attribute value for each attribute.

In a decision system, if the domain of an attribute is ordering according to a decreasing or increasing preference, then we say the attribute is a criterion.

Definition 1 *[7–9]. A decision system is called an ordered decision system (ODS) if all attributes are criteria.*

Assume that the domain of a criterion $a \in A$ is completely preference ordering by an outranking relation $\succeq_a$, i.e., $x \succeq_a y$ means that x is at least as good as y with respect to criteria a, in this case, we say that x dominates y. In this paper, we define $x \succeq_a y$ by $f(x, a) \geq f(y, a)$ according to increasing preference ordering, and by $f(x, a) \leq f(y, a)$ for decreasing preference ordering.

Definition 2 *[13]. Let $ODS = (U, A, V, f)$ be an ordered decision system, $A = C \cup D$, $B \subseteq C$, denote*

$$R_B^{\geq} = \{(x, y) \in U \times U : f(y, a) \geq f(x, a), \forall a \in B\}; \tag{1}$$

$$R_D^{\geq} = \{(x, y) \in U \times U : f(y, D) \geq f(x, D)\}. \tag{2}$$

$R_B^{\geq}$ and $R_D^{\geq}$ are called dominance relations of the ordered decision system ODS. The dominance class of the object x with respect to relation $R_B^{\geq}$ and $R_D^{\geq}$ is denoted by

$$[x]_B^{\geq} = \{y \in U : (x, y) \in R_B^{\geq}\}; \tag{3}$$

$$[x]_D^{\geq} = \{y \in U : (x, y) \in R_D^{\geq}\}. \tag{4}$$

For any $X \subseteq U$, and $B \subseteq C$ of ODS, the lower and upper approximation of X with respect to a dominance relation $R_B^{\geq}$ is defined as follows:

$$\underline{R_B^{\geq}}(X) = \{x \in U : [x]_B^{\geq} \subseteq X\}; \tag{5}$$

$$\overline{R_B^{\geq}}(X) = \{x \in U : [x]_B^{\geq} \cap X \neq \emptyset\}. \tag{6}$$

Let $ODS = (U, A, V, f)$ be an ordered decision system, $A = C \cup D$, $B \subseteq C$, denote

$$\sigma_B(x) = \{D_j : D_j \cap [x]_B^{\geq} \neq \emptyset\}; \tag{7}$$

$$\delta_B(x) = \{D_j : [x]_B^{\geq} \subseteq D_j\}. \tag{8}$$

Definition 3 *(possible distribution reduct [13] or assignment reduct [10]). If $\sigma_B(x) = \sigma_C(x)$ for any $x \in U$, we say that B is a possible distribution (or assignment) consistent set of ODS. And if none proper subsets of B is a possible distribution (or assignment) consistent set, then B is a possible distribution (or assignment) reduct.*

Definition 4 *(compatible reduct [13]). If $\delta_B(x) = \delta_C(x)$ for any $x \in U$, we say that B is a possible distribution consistent set of ODS. And if none proper subsets of B is a possible distribution consistent set, then B is a possible distribution reduct.*

Definition 5 *(union-based reduct [15]). If $R_{\overline{B}}^{\geq}(D_i) = R_{\overline{C}}^{\geq}(D_i)$. and $\nexists B' \subset B, R_{\overline{B'}}^{\geq}(D_i) = R_{\overline{C}}^{\geq}(D_i)$ then B is called a reduct preserving lower approximations of upward unions.*

For more details of other union-based reducts and class-based reducts, please refer to [15,16].

3   Inconsistent Dominance Principle Reduction

In CRS theory, two objects are indiscernibility if they are equivalence relation. Analogously, we consider two objects x, y to be indiscernibility if they satisfy the dominance principle [1,18], that is, if there is an object x not worse than another object y on all conditional attributes, then the decision assigned to x is not worse than that assigned to y. Thus, it is meaningless to compare two objects without referring to the preference ordering of their decision in ODS because they are partial ordered under the multi-criteria.

We choose the instance in [13] presented in Table 1, denoted by $ODS = (U, A, V, f)$, where $U = \{x_1, x_2, \cdots, x_6\}$, $A = C \cup D$, $C = \{a_1, a_2, a_3\}$, D is the decision attribute. Assume that $A = C \cup D$ is all the preference ordering with increasing.

Table 1. An ordered decision system ODS

	a_1	a_2	a_3	D
x_1	1	2	1	3
x_2	3	2	2	2
x_3	1	1	2	1
x_4	2	1	3	2
x_5	3	3	2	3
x_6	3	2	3	1

From Table 1, the objects x_1 and x_3 are not comparable on the set of all condition attributes $C = \{a_1, a_2, a_3\}$ by the dominance relation. Although, x_3 dominated x_4, they are still indiscernibility because the decision of x_3 dominated x_4 according to the dominance principle.

However, objects satisfy the dominance principle, such as x_3, x_4, are still discerned in the existed reduction theories in dominance relation. For example, the condition attributes set $\{a_1, a_3\}$ discerns the object x_4 from x_3 in possible distribution reduct [13]. For detailed approaches to the other reducts in DRSA, readers can look back into the reference.

Investigating the objects x_1 and x_2, we can find that x_2 is not worse than x_1 in condition $C = \{a_1, a_2, a_3\}$, however, it has a worse decision than x_1. So, the two objects x_1 and x_2 are discernible in condition $\{a_1, a_3\}$ which cause inconsistent dominance principle.

3.1   Theories of Reduction Based Inconsistent Dominance Principle

Let $ODS = (U, A, V, f)$ be an ordered decision system, $A = C \cup D$, $B \subseteq C$. $R_C^\geq$ and $R_D^\geq$ are dominance relations derived from the condition attributes set C and the decision attribute D respectively. To introduce reduction based inconsistent dominance principle, the formal definitions of dominance principle and inconsistent dominance principle are given as follows.

In [1, 18], dominance principle described as, if there exists an object not worse than another object on all considered criteria, it has not been assigned to a worse decision class than the other. Thus, we use definition of relation between objects as the formal description of the dominance principle.

Definition 6 *(Dominance Principle, DP). For any $x, y \in U$, if they are relationship defined as follows, then we say x,y satisfy the dominance principle.*

$$DP^\geq(C|D) = \{(x,y) : (x,y) \in R_C^\geq \wedge (x,y) \in R_D^\geq\}; \tag{9}$$

$$DP^\leq(C|D) = \{(x,y) : (x,y) \in R_C^\leq \wedge (x,y) \in R_D^\leq\} \tag{10}$$

According to the definition, two objects are the relation of dominance principle if one object dominate the other object both all criteria and the decision.

Similarly, we use definition of relation between objects as the formal description of the inconsistent dominance principle.

Definition 7 *(Inconsistent Dominance Principle, IDP). For any $x, y \in U$, if they are relationship defined as follows, then we say x,y satisfy the inconsistent dominance principle.*

$$IDP^\geq(C|D) = \{(x,y) : (x,y) \in R_C^\geq \wedge (x,y) \notin R_D^\geq\}; \tag{11}$$

$$IDP^\leq(C|D) = \{(x,y) : (x,y) \in R_C^\leq \wedge (x,y) \notin R_D^\leq\} \tag{12}$$

The definition shows two objects x,y are relation of inconsistent dominance principle if object x is not worse than the object y on all criteria, but the decision of x is worse than y.

From the above, we can have the following properties.

Proposition 1. *Let $ODS = (U, A, V, f)$ be an ordered decision system, $A = C \cup D$, $B \subseteq C$.*
   (1)$DP^\geq(C|D) \subseteq DP^\geq(B|D), DP^\leq(C|D) \subseteq DP^\leq(B|D)$
   (2)$IDP^\geq(C|D) \subseteq IDP^\geq(B|D), IDP^\leq(C|D) \subseteq IDP^\leq(B|D)$

Definition 8. *If $IDP^\geq(C|D) = IDP^\geq(B|D)$ for any $x \in U$, we say that B is an $IDP^\geq$ consistent set of ODS. And if none proper subsets of B is an $IDP^\geq$ consistent set, then B is an $IDP^\geq$ reduct.*

Definition 9. *If $IDP^\leq(C|D) = IDP^\leq(B|D)$ for any $x \in U$, we say that B is an $IDP^\leq$ consistent set of ODS. And if none proper subsets of B is an $IDP^\leq$ consistent set, then B is an $IDP^\leq$ reduct.*

Theorem 1. *(Judgment Theorem 1)B is an $IDP^{\geq}$ consistent set if and only if $(x,y) \in IDP^{\geq}(C|D) \Rightarrow (y,x) \notin R_B^{\geq}$ holds for any $(x,y) \in U^2$.*

Proof. "$\Rightarrow$" Assume that $(x,y) \in IDP^{\geq}(C|D) \Rightarrow (y,x) \notin R_B^{\geq}$ does not hold, which implies that there exist $(x,y) \in U^2$ such that $(x,y) \in IDP^{\geq}(C|D)$ but $(y,x) \in R_B^{\geq}$. On the other hand, since B is an $IDP^{\geq}$ consistent set, namely, $IDP^{\geq}(C|D) = IDP^{\geq}(B|D)$, we have $(x,y) \in R_B^{\geq}$, which is in contradiction with $(y,x) \in R_B^{\geq}$.

"$\Leftarrow$" Proposition 3.1, we have $IDP^{\geq}(C|D) \subseteq IDP^{\geq}(B|D)$. So we only prove $IDP^{\geq}(B|D) \subseteq IDP^{\geq}(C|D)$. Assume that there exists $(x,y) \in U^2$ such that $(x,y) \notin IDP^{\geq}(B|D)$ and $(x,y) \in IDP^{\geq}(C|D)$. According to the necessary condition of inverse negative proposition, if any $(y,x) \in R_B^{\geq}$, we have $(x,y) \notin IDP^{\geq}(C|D)$, which is in contradiction with the assumption.

The proof is completed.

Theorem 2. *(Judgment Theorem 2)B is an $IDP^{\leq}$ consistent set if and only if $(x,y) \in IDP^{\leq}(C|D) \Rightarrow (y,x) \notin R_B^{\leq}$ holds for any $(x,y) \in U^2$.*

Proof. It is similar to Theorem 1.

3.2 Approach to Reduction Based Inconsistent Dominance Principle

This section provides an approach to reduction based in IDP. We present the following definition.

Definition 10. *Let $ODS = (U, A, V, f)$ be an ordered decision system, $A = C \cup D$, $B \subseteq C$. For any $(x,y) \in U^2$, we denote*

$$D^{\geq}(x,y) = \begin{cases} \{a \in C : (y,x) \notin R_B^{\geq}\}, (x,y) \in IDP^{\geq}(C|D) \\ C, \quad otherwise \end{cases} \tag{13}$$

$$D^{\leq}(x,y) = \begin{cases} \{a \in C : (y,x) \notin R_B^{\leq}\}, (x,y) \in IDP^{\leq}(C|D) \\ C \qquad\qquad\qquad\qquad otherwise \end{cases} \tag{14}$$

Then $M^{\geq} = (D^{\geq}(x,y), x, y \in U)$ and $M^{\leq} = (D^{\leq}(x,y), x, y \in U)$ are called $IDP^{\geq}$ and $IDP^{\leq}$ discernibility matrix of ODS respectively.

Obviously, the computational complexity of the discernibility matrix is $O(|C| \times |U|^2)$.

Proposition 2. $M^{\geq} = (M^{\leq})'$.

Proof. It follows directly from Definition 10.

Theorem 3. *Let $ODS = (U, A, V, f)$ be an ordered decision system, $A = C \cup D$, $B \subseteq C$. B is an $IDP^{\geq}$ consistent set if and only if $D^{\geq}(x,y) \cap B \neq \emptyset$ holds for any $(x,y) \in IDP^{\geq}(C|D)$.*

Proof. "⇒" B is an $IDP^\geq$ consistent set, we have $IDP^\geq(C|D) = IDP^\geq(B|D)$. According to Theorem 1, for any $(x,y) \in IDP^\geq(C|D)$, have $(y,x) \notin R_B^\geq$. Obviously, $D^\geq(x,y) \cap B \neq \emptyset$ can be obtained.

"⇐" If $(x,y) \in IDP^\geq(C|D)$, then $D^\geq(x,y) \cap B \neq \emptyset$, indicate $(x,y) \notin R_D^\geq$, and $(y,x) \notin R_B^\geq$, that is $(y,x) \in \{f(x,a) \not\geq f(y,a), a \in B\}$, means $(x,y) \in R_B^\geq$. Thus $(x,y) \in IDP^\geq(C|D)$ can be obtained. B is an $IDP^\geq$ consistent set.

The proof is completed.

Theorem 4. *Let $ODS = (U, A, V, f)$ be an ordered decision system, $A = C \cup D$, $B \subseteq C$. B is an $IDP^\leq$ consistent set if and only if $D^\leq(x,y) \cap B \neq \emptyset$ holds for any $(x,y) \in IDP^\leq(C|D)$.*

Proof. It is similar to the proof of Theorem 3.

Definition 11. *Let $ODS = (U, A, V, f)$ be an ordered decision system, $A = C \cup D$, $B \subseteq C$, and $M^\geq$, $M^\leq$ be $IDP^\geq$ and $IDP^\leq$ discernibility matrices of ODS respectively. The corresponding discernibility formulas are defined by*

$$F^\geq = \wedge\{\vee\{a_k : a_k \in D^\geq(x,y)\}, (x,y) \in IDP^\geq(C|D)\};$$

$$F^\leq = \wedge\{\vee\{a_k : a_k \in D^\leq(x,y)\}, (x,y) \in IDP^\leq(C|D)\}.$$

Theorem 5. *The minimal disjunctive normal form of discernibility formula of $IDP^\geq$ is $F^\geq = \vee_{k=1}^p (\wedge_{s=1}^q a_s)$. Denote $B_k = \{a_s, s = 1, 2, \cdots, q_k\}$, then $\{B_k = 1, 2, \cdots, p\}$ is the set of all $IDP^\geq$ reducts of ODS.*

Proof. For any $(x,y) \in IDP^\geq(C|D)$, we have $B_k \cap D^\geq(x,y) \neq \emptyset$ by the definition of minimal disjunctive normal form. According to Theorem 3, B_k is an $IDP^\geq$ consistent set. If we remove an element of B_k, named B_k', then there must exist $(x,y) \in IDP^\geq(C|D)$ make $B_k' \cap D^\geq(x,y) = \emptyset$. Thus, B_k' is not an $IDP^\geq$ consistent set.

Theorem 6. *The minimal disjunctive normal form of discernibility formula of $IDP^\leq$ is $F^\leq = \vee_{k=1}^p (\wedge_{s=1}^q a_s)$. Denote $B_k = \{a_s, s = 1, 2, \cdots, q_k\}$, then $\{B_k = 1, 2, \cdots, p\}$ is the set of all $IDP^\geq$ reducts of ODS.*

Proof. It is similar to the proof of Theorem 5.

Example 1. Reduction based inconsistent dominance principle will be explained by Example 1 (continued by Table 1). According to theories and approach to reduction, we can obtain $M^\geq((M^\geq)')$, which is the $IDP^\geq(IDP^\leq)$ discernibility matrix (Table 2).

From Table 1, the objects x_1 and x_3 are not comparable on the set of attributes $C = \{a_1, a_2, a_3\}$. The objects x_3 and x_4 are the relation of dominance principle so that they are still indiscernibility on attributes set $C = \{a_1, a_2, a_3\}$.

Investigating the objects x_1 and x_2, we have $(x_1, x_2) \in IDP^\geq(C|D)$, and $(x_2, x_1) \notin R_{\{a_1, a_3\}}^\geq$. So, $D^\geq(x_1, x_2) = \{a_1, a_3\}$, namely, the two objects x_1 and x_2 are discernible in condition $\{a_1, a_3\}$ which cause inconsistent dominance principle.

$$F^\geq = (a_1 \vee a_3) \wedge a_3 \wedge (a_1 \vee a_2) = a_3 \wedge (a_1 \vee a_2) = (a_1 \wedge a_3) \vee (a_2 \wedge a_3)$$

Therefore, we acquire that $\{a_1, a_3\}$, $\{a_2, a_3\}$ are all $IDP^\geq$ reducts of the ordered decision system in Table 1. According to Proposition 2, $M^\leq = (M^\geq)'$ can be obtained easily. So, $F^\geq = (a_1 \wedge a_3) \vee (a_2 \wedge a_3)$, $\{a_1, a_3\}$, $\{a_2, a_3\}$ also are all $IDP^\leq$ reducts.

Table 2. $IDP^\geq$ discernibility matrix $M^\geq$

	x_1	x_2	x_3	x_4	x_5	x_6
x_1	C	a_1, a_3	C	C	a_1, a_3	
x_2	C	C	C	C	a_3	
x_3	C	C	C	C	C	
x_4	C	C	C	C	a_1, a_2	
x_5	C	C	C	C	C	
x_6	C	C	C	C	C	

4 Conclusion

Knowledge reduction is a core issue in CRS and its extended model, such as DRSA. Referring to method of reduction in CRS, two objects be indiscenibility if they are equivalence relation, correspondingly, we consider they are indiscenibility with dominance principle in DRSA. In this paper, reduction based inconsistent dominance principle is introduced. Furthermore, the judgment theorems and discernable matrices are investigated, from which we can obtain new approach to knowledge reduction in ordered decision system.

Acknowledgement. This work is supported by the National Science and Technology Major Project (2014ZX07104-006) and the Hundred Talents Program of CAS (NO.Y21Z110A10).

References

1. Dembczyński, K., Greco, S., Kotłowski, W., Słowiński, R.: Quality of Rough Approximation in Multi-criteria Classification Problems. In: Greco, S., Hata, Y., Hirano, S., Inuiguchi, M., Miyamoto, S., Nguyen, H.S., Słowiński, R. (eds.) RSCTC 2006. LNCS (LNAI), vol. 4259, pp. 318–327. Springer, Heidelberg (2006)
2. Greco, S., Matarazzo, B., Słowiński, R.: Rough Approximation by Dominance Relations. Int. J. Intell. Syst. **17**(2), 153–171 (2002)
3. Thangavel, K., Pethalakshmi, A.: Dimensionality reduction based on rough set theory: a review. Appl. Soft Comput. **9**(1), 1–12 (2009)
4. Zhang, W.X., Mi, J.S., Wu, W.Z.: Approaches to knowledge reductions in inconsistent systems. Int. J. Intell. Syst. **18**(9), 989–1000 (2003)

5. Miao, D.Q., Zhao, Y., Yao, Y.Y., Li, H.X., Xu, F.F.: Relative reducts in consistent and inconsistent decision tables of the Pawlak rough set model. Inf. Sci. **179**(24), 4140–4150 (2009)
6. Qian, Y., Liang, J., Pedrycz, W., Dang, C.: Positive approximation: an accelerator for attribute reduction in rough set theory. Artif. Intell. **174**(9), 597–618 (2010)
7. Greco, S., Matarazzo, B., Słowiński, R.: Rough approximation of a preference relation by dominance relations. Eur. J. Oper. Res. **117**(98), 63–83 (1999)
8. Greco, S., Matarazzo, B., Słowiński, R.: Rough sets theory for multicriteria decision analysis. Eur. J. Oper. Res. **129**(1), 1–47 (2001)
9. Greco, S., Matarazzo, B., Słowiński, R.: Rough sets methodology for sorting problems in presence of multiple attributes and criteria. Eur. J. Oper. Res. **138**(2), 247–259 (2002)
10. Xu, W.H., Zhang, W.X.: Knowledge reductions in inconsistent information systems based on dominance relations. Comput. Sci. **33**(2), 182–184 (2006) (in Chinese)
11. Xu, W.H., Zhang W.X.: Matrix computation for assignment reduction in information systems based on dominance relations. Comput. Eng. **33**(14), 4–7 (2007) (in Chinese)
12. Xu, W.H., Zhang, W.X.: Methods for knowledge reduction in inconsistent ordered information systems. J. Appl. Math. Comput. **26**(1), 313–323 (2008)
13. Xu, W.H., Li, Y., Liao, X.: Approaches to attribute reductions based on rough set and matrix computation in inconsistent ordered information systems. Knowl. Based Syst. **27**(3), 78–91 (2012)
14. Chen, J., Wang, G.Y., Hu, J.: Positive domain reduction based on dominance relation in inconsistent system. Comput. Sci. **35**(3), 216–218, 227 (2008) (in Chinese)
15. Inuiguchi, M., Yoshioka, Y.: Several reducts in dominance-based rough set approach. In: Huynh, V.N., Nakamori, Y., Ono, H., Lawry, J., Kreinovich, V., Nguyen, H.T. (eds.) Interval/Probabilistic Uncertainty and Non-Classical Logics. ASC, vol. 46, pp. 163–175. Springer, Heidelberg (2008)
16. Kusunoki, Y., Inuiguchi, M.: A unified approach to reducts in dominance-based rough set approach. Soft Comput. **14**(5), 507–515 (2010)
17. Susmaga, R.: Reducts and constructs in classic and dominance-based rough sets approach. Inf. Sci. **271**(7), 45–64 (2014)
18. Dembczyński, K., Greco, S., Słowiński, R.: Rough set approach to multiple criteria classification with imprecise evaluations and assignments. Eur. J. Oper. Res. **198**(2), 626–636 (2009)

Improving Indiscernibility Matrix Based Approach for Attribute Reduction

Piotr Hońko[✉]

Faculty of Computer Science, Bialystok University of Technology,
Wiejska 45A, 15-351 Białystok, Poland
p.honko@pb.edu.pl

Abstract. The problem of finding all reducts of the attribute set of a data table has often been studied in rough set theory by using the notion of discernibility matrix. An alternative version based on the indiscernibility matrix has been used for this problem to a lesser extent due to its space and time complexity. This paper improves the indiscernibility matrix based approach for computing all reducts. Only indiscernibility matrix cells as well as subsets of the attribute set necessary for computing reducts are processed. The experiments reported in this paper show that the improved version uses less memory to store partial results and can find reducts in a shorter time.

Keywords: Rough sets · Attribute reduction · Indiscernibility matrix

1 Introduction

Attribute reduction is one of the main problems in rough set theory [6]. It attracts many researchers due to not only its importance, but also its complexity. Namely, finding all reducts or even a minimal one is proven to be NP-hard. In fact, one reduct is sufficient to represent a given database in a reduced form that does not lose important information encoded in the data. However, finding all reducts can be useful for performing a more deeper analysis of the data.

Most approaches for computing the reduct set (e.g. [1–3]) are based on the notion of discernibility matrix [7]. Each cell of the matrix shows attributes for which two objects of the data table are different. All reducts correspond to the prime implicants of the Boolean function constructed based on the matrix cells. For more details, see [7].

A structure that is dual to the discernibility matrix and can be used for finding all reducts is an indiscernibility matrix [4]. Each cell of the matrix shows attributes for which two objects of the data table are the same. In spite of the fact that the indiscernibility matrix can easily be obtained from the discernibility one, the process of constructing reducts is fundamentally distinct. It was shown in [4] that the power set of the attribute set decreased by the indiscernibility

---

The project was funded by the National Science Center awarded on the basis of the decision number DEC-2012/07/B/ST6/01504.

D. Ciucci et al. (Eds.): RSKT 2015, LNAI 9436, pp. 119–128, 2015.
DOI: 10.1007/978-3-319-25754-9_11

matrix cells includes all reducts. Namely, elements of the reduced power set that are minimal with respect to the inclusion relation are reducts.

That approach is able to find reducts in a short time for databases with a relatively small number of objects and attributes. In general, the approach is not efficient due to computing and storing the whole indiscernibility matrix and the power set of the attribute set.

The goal of this paper is to improve the indiscernibility matrix based approach for computing all reducts of a data table (an information system or decision table). For this purpose, the indiscernibility matrix is replaced with its reduced version that consists only of the maximal sets with respect to the inclusion relation[1]. The power set, in turn, is replaced with its subset that includes only necessary elements to compute reducts.

Furthermore, a recursive version is proposed that is less memory consuming and thereby enables to computes reducts for databases with a big number of attributes.

The results of experiments conducted under this work show that the proposed approach can substantially speed up the task of finding all reducts.

Section 2 investigates the problem of finding all reducts using the indiscernibility matrix. Section 3 proposes and theoretically verifies two indiscernibility matrix based algorithms improving attribute reduction. Experimental research is described in Sect. 4. Concluding remarks are provided in Sect. 5.

2   Indiscernibility Matrix Based Attribute Reduction

This section proposes ways to improve the indiscernibility matrix based attribute reduction. The problem is defined in rough set theory.

To store data to be processed, an information system is used.

Definition 1 *[6] (information system). An information system is a pair $IS = (U, A)$, where U is a non-empty finite set of objects, called the universe, and A is a non-empty finite set of attributes.*
Each attribute $a \in A$ is treated as a function $a : U \to V_a$, where V_a is the value set of a.

Essential information about data is expressed by an indiscernibility relation.

Definition 2 *[6] (indiscernibility relation). An indiscernibility relation $IND (B)$ generated by $B \subseteq A$ on U is defined by*

$$IND(B) = \{(x, y) \in U \times U : \forall_{a \in B} a(x) = a(y)\} \tag{1}$$

The relation is used to define a reduct of the attribute set.

Definition 3 *[6] (reduct of attribute set). A subset B of A is a reduct of A on U if and only if*

---
[1] Cells equal to the attribute set are not stored in the reduced indiscernibility matrix.

1. $IND(B) = IND(A)$,
2. $\forall_{\emptyset \neq C \subset B} IND(C) \neq IND(B)$.

The set of all reducts of A on U is denoted by $RED(A)$.

Definition 4 *[4] (indiscernibility matrix). The indiscernibility matrix IM of an information system $IS = (U, A)$ is defined by $\forall_{x,y \in U} IM(x, y) = \{a \in A : a(x) = a(y)\}$.*

The reduct set of an information system can be computed using its indiscernibility matrix.

Theorem 1. *A subset B of A is a reduct of A on U if and only if*[2]

1. $\forall_{C \in IM} B \not\subseteq C$,
2. $\forall_{B' \subset B} \exists_{C' \in IM} B' \subseteq C'$.

Proof. We obtain

1. $\forall_{C \in IM} B \not\subseteq C \Leftrightarrow \forall_{\substack{x,y \in U \\ x \neq y}} (x, y) \notin IND(B) \Leftrightarrow IND(B) = IND(A)$.
2. $\forall_{B' \subset B} \exists_{C' \in IM} B' \subseteq C' \Leftrightarrow \forall_{B' \subset B} \exists_{\substack{x,y \in U \\ x \neq y}} (x, y) \in IND(B') \Leftrightarrow$
   $\forall_{\emptyset \neq B' \subset B} IND(B') \neq IND(B)$.

One can show that the whole indiscernibility matrix is not necessary to compute all reducts. It can be replaced with its reduced form.

Let $max(\mathcal{S})$ be the set of maximal elements of a family $\mathcal{S}$ of sets partially ordered by the inclusion relation $\subseteq$. Let also $IM_+ = IM \setminus \{A\}$.

Theorem 2. *Let $B \subseteq A$, where $IS = (U, A)$ is an information system. The following holds $\forall_{C \in IM_+} B \not\subseteq C \Leftrightarrow \forall_{C \in max(IM_+)} B \not\subseteq C$.*

Proof. The theorem is true by the following lemma.

Lemma 1. *Let $X \in \mathcal{S}$, where $\mathcal{S}$ is a family of sets. The following holds $\forall_{Y \in \mathcal{S}} X \not\subseteq Y \Leftrightarrow \forall_{Y \in max(\mathcal{S})} X \not\subseteq Y$.*

Proof (of Lemma). The case "$\Rightarrow$". This holds by $max(\mathcal{S}) \subseteq \mathcal{S}$.
The case "$\Leftarrow$". This holds by $X \not\subseteq Y \Rightarrow \forall_{Z \subset Y} X \not\subseteq Z$.

Computing the whole power set of the attribute set is also unnecessary for attribute reduction. The subsets of the power set that are needed for computing reducts can be defined based on the maximal cardinality of the indiscernibility matrix cells.

Let $maxcard(\mathcal{S}) = max\{|X| : X \in \mathcal{S}\}$, where $\mathcal{S}$ is a family of sets.

Theorem 3. *Let $IS = (U, A)$ be an information system. The following holds $maxcard(RED(A)) \leq maxcard(IM_+) + 1$.*

---

[2] An indiscernibility matrix is treated as a set since the order of its cells is not important.

Proof (by contradiction). Suppose that $maxcard(RED(A)) > l + 1$, where $l = maxcard(IM_+)$. We obtain $\exists_{B \in RED(A)} |B| > l + 1 \Rightarrow \exists_{\substack{B' \subset B \\ |B'|=l+1}} \forall_{C \in IM_+} B' \not\subseteq C \Rightarrow B' \in RED(A)$. This leads to a contradiction with $B \in RED(A)$.

A special case of an information system, called decision table, is used to store data with class distribution.

Definition 5 *[6] (decision table). A decision table is a pair $DT = (U, A \cup \{d\})$, where U is a non-empty finite set of objects, called the universe, A is a non-empty finite set of condition attributes, and $d \notin A$ is the decision attribute. Each attribute $a \in A \cup \{d\}$ is treated as a function $a : U \to V_a$, where V_a is the value set of a.*

For a decision table a relative indiscernibility relation and relative reduct of the attribute set are defined in the following way.

Definition 6 *[5,6] (relative indiscernibility relation). A relative indiscernibility relation $IND(B, d)$ generated by $B \subseteq A$ on U is defined by*

$$IND(B, d) = \{(x, y) \in U \times U : (x, y) \in IND(B) \vee d(x) = d(y)\} \qquad (2)$$

Definition 7 *[5,6] (relative reduct of attribute set). A subset B of A is a relative reduct of A if and only if*

1. $IND(B, d) = IND(A, d)$,
2. $\forall_{\emptyset \neq C \subset B} IND(C, d) \neq IND(B, d)$.

The set of all relative reducts of A on U is denoted by $RED(A, d)$.

Definition 8 *(relative indiscernibility matrix). The relative indiscernibility matrix IM^d of a decision table $DT = (U, A \cup \{d\})$ is defined by*

$$\forall_{x,y \in U} IM^d(x, y) = \begin{cases} \{a \in A : a(x) = a(y)\}, & d(x) \neq d(y); \\ \emptyset, & otherwise. \end{cases}$$

For a decision table one can formulate theorems analogous to Theorems 1–3.

3   Indiscernibility Matrix Based Algorithms for Attribute Reduction

This section proposes two indiscernibility matrix based algorithms that are improved versions of that introduced in [4]. They both take as an input a database that can be considered as an information system or decision table.

The algorithm from [4] can be outlined as follows:

1. Compute the indiscernibility matrix of an information system.
2. Compute the power set of the attribute set.
3. Decrease the power set by the indiscernibility matrix cells.
4. Find elements (i.e. subsets of attribute set) of the decreased power set that are minimal with respect to the inclusion relation.

To evaluate the complexity of the algorithms described in this paper, we use the denotations $n = |U|$ and $m = |A|$, where U and A are sets of all objects and (condition) attributes, respectively.

The time and space complexity of the above algorithm is $O\left(m\frac{n(n-1)}{2}\right) + O(m2^m)$. The first (second) component of the sum is the cost of computing/storing the indiscernibility matrix (the power set).

The first of the proposed algorithms (see Algorithm 1) computes, according to Theorem 2, the reduced indiscernibility matrix of an information system or decision table (the $ComputeRIM$ function). The algorithm uses the $ComputeCurrRED$ function to compute reducts of a given cardinality. The function employs another called $Supersets$ to compute for a given set its supersets with the cardinality higher by 1. The $isSubsetOfAny$ ($isSupersetOfAny$) function checks if a given set is a subset (superset) of any set of a given family.

Function 1. $ComputeCurrRED$

Data: RIM – the reduce IM; T_i – the family of subsets of the attribute set, each of which is of the cardinality i and is included in any RIM cell;

Result: T_{i+1} – the family of subsets of the attribute set, each of which is of the cardinality $i+1$ and is included in any RIM cell; RED – the set of reducts have been found so far;

begin
   $T_{i+1} := \emptyset$;
   foreach $S \in T_i$ **do**
      foreach $S' \in Supersets(S,1)$ **do**
         if $isSubsetOfAny(S', RIM)$ **then** $T_{i+1} := T_{i+1} \cup \{S'\}$;
         else if $!isSupersetOfAny(S', RED)$ **then** $RED := RED \cup \{S'\}$;
      end
   end
end

Algorithm 1. $ComputeRED$

Data: D – a database;

Result: RED – the set of all reducts;

begin
   $RED := \emptyset$; $i := 0$; $T_i := \emptyset$;
   $RIM := ComputeRIM(D)$;
   repeat
      $ComputeCurrRED(RIM, T_i, T_{i+1}, RED)$;
      $i := i + 1$;
   until $T_i = \emptyset$;
end

The time complexity of Algorithm 1 is $O\left(m\frac{n'(n'-1)}{2}\right) + O\left(m\sum_{i=1}^{l+1}\binom{m}{i}\right)$, where $n' = |IM|, l = maxcard(IM_+)$ and $l + 1 \leq m$. The first component of the sum is the cost of computing and reducing the indiscernibility matrix. The second one is the cost of computing the reduced power set.

The space complexity of the algorithm is $min\left\{O\left(m\frac{n(n-1)}{2}\right), O\left(m\binom{m}{[m/2]}\right)\right\}$ $+O\left(m\binom{m}{[m/2]}\right)$. In the pessimistic case, the matrix is irreducible and we need to store $\frac{n(n-1)}{2}$ cells. For databases with a respectively smaller number of attributes, the size of RIM is not bigger than $\binom{m}{[m/2]}$. For the power set it is enough store its subset that consists of elements of the same cardinality. Any such a subset is not bigger than $\binom{m}{[m/2]}$.

For a database with a bigger number of attributes, the size of the subset of the power set to be stored can be too large to fit into memory. Therefore, a recursive version of Algorithm 1 that is able to overcome this problem is proposed.

The main computations of Algorithm 2 are performed by the *Compute CurrREDRec* function. It uses the *Erase* function to erase data (i.e. attribute subsets) from the memory and the *Reduce* function to remove attribute subsets that are not minimal with respect to the inclusion relation.

Function 2. *ComputeCurrREDRec*

Data: S – a subset of the attribute set; RIM – the reduce IM;
Result: RED – the set of reducts have been found so far;
begin
    foreach $S' \in SuperSets(S,1)$ do
        if $isSubsetOfAny(S', RIM)$ then
            $ComputeCurrREDRec(S', RIM, RED)$;
            $Erase(S')$;
        end
        else
            $RED := RED \cup \{S'\}$;
            $Reduce(RED)$;
        end
    end
end

Algorithm 2. *ComputeREDRec*

Data: D – a database;
Result: RED – the set of all reducts;
begin
    $RED := \emptyset$;
    $RIM := ComputeRIM(D)$;
    $ComputeCurrREDRec(\emptyset, RIM, RED)$;
end

The time and complexity of Algorithm 2 is the same as that of Algorithm 1. The space complexity is $min\left\{O\left(m\frac{n(n-1)}{2}\right), O\left(m\binom{m}{[m/2]}\right)\right\} + O\left(\binom{m}{[m/2]}\right)$.

Algorithm 1 needs $m\binom{m}{[m/2]}$ memory to store a subset of the power set, whereas Algorithm 2 needs m memory since only one element of the power set (i.e. attribute subset) is stored at once.

4   Experiments

This section describes experimental research that concerns attribute reduction using the indiscernibility matrix.

Twelve databases (see Table 1) taken from the UCI Repository (http:// archive.ics.uci.edu/ml/) were tested. The indiscernibility matrix based attribute reduction was employed for computing all reducts of an information system and decision table. The approach was implemented in C++ and tested using a laptop (Intel Core i5, 2.3 GHz, 4 GB RAM, Windows 7). Each test was repeated at least once to eliminate the influence of other running processes on the time performance.

Table 1. Characteristics of databases

id	Database's name	Attributes	Objects	Classes	Reducts(IS)	Reducts(DT)
D1	car evolution	7	1728	4	1	1
D2	kingrook vs king (krk)	7	28056	18	1	1
D3	pima ind. diab.	9	768	2	27	28
D4	nursery	9	12960	5	1	1
D5	breast cancer wisconsin	11	699	2	2	27
D6	wine	14	178	3	128	73
D7	australian credit appr.	15	690	2	13	44
D8	adult	15	32561	2	2	2
D9	mushroom	23	8124	7	1	4
D10	parkinson	29	1040	2	316	219
D11	trains	33	10	2	3588	333
D12	sonar, mines vs. rocks	61	208	2	1714	1314

For the comparison purposes, four versions of the indiscernibility matrix based approach were tested: the full indiscernibility matrix and power set, denoted by IM+PS, the reduced indiscernibility matrix and full power set (see Theorem 2), denoted by RIM+PS, the reduced indiscernibility matrix and power set (see Theorem 3 and Algorithm 1), denoted by RIM+RPS, and the recursive version of the previous one (see Algorithm 2), denoted by RIM+RPS(rec).

Tables 2 and 3 show run-time (in seconds) for computing the (reduced) indiscernibility matrix and (reduced) power set – the columns denoted by t1 and t2, respectively.

Table 2. Indiscernibility matrix based attribute reduction of an information system

id[a]	IM+PS		RIM+PS		RIM+RPS		RIM+RPS(rec)	
	t1	t2	t1	t2	t1	t2	t1	t2
D1	0.452	0.53	0.531	0.531	0.546	0.546	0.546	0.546
D2	*	*	131.586	131.602	131.462	131.462	131.696	131.696
D3	0.109	0.156	0.156	0.156	0.141	0.141	0.14	0.14
D4	*	*	35.631	35.631	34.882	34.882	34.897	34.897
D5	0.125	0.78	0.124	0.124	0.124	0.124	0.124	0.124
D6	0.015	0.047	0.015	0.031	0.015	0.031	0.015	0.015
D7	0.141	10.359	0.202	0.249	0.202	0.218	0.203	0.219
D8	*	*	355.197	355.244	353.75	353.765	353.73	353.746
D9	*	*	36.65	1973.35	34.149	37.799	33.852	39.484
D10	0.624	*	0.702	*	0.686	0.702	0.702	0.718
D11	<0.001	*	<0.001	*	<0.001	*	<0.001	1995.61
D12	0.047	*	0.063	*	0.078	0.374	0.078	0.421

[a]The asterisk means that the computations were not finished due to memory overflow.

Due to memory overflow (see Table 2), the IM+PS version is only able to compute reducts for five out of twelve databases. For the databases for which the algorithm returned the reduct sets the time needed to compute the matrix is the same or only slightly shorter compared with that obtained by the RIM+PS algorithm. The final run-time is, in turn, the same for D1 and D2, slightly longer for D5 and D6, and considerable longer for D7, i.e. over 41 times.

The reason for the long-running calculations for D7 lies in the cardinalities of the matrix cells. Namely, the maximal cardinality is 14 (the maximal possible value) which means that IM+PS needs to go through all cardinality levels (from 1 to 14) to reduce the power set. For comparison, for D6, where the number of attributes is close to that of D7, the maximal cardinality of cells is only 5.

Algorithm 1 is able to compute reduct sets for additional six and two databases compared with IM+PS and RIM+PS, respectively. Almost all the results returned by the algorithm are comparable with those produced by RIM+PS. The advantage of Algorithm 1 can be clearly seen for database D9 for which the algorithm is over 52 times faster than RIM+PS. Here we have a situation analogous to that of D7, i.e. the maximal cardinality of the cells is extremely high (22).

Algorithm 2 can compute reduct sets for all databases, but the run-time for D11 is very long. It obtained (almost) the same results for the first eight databases compared with Algorithm 1. The results for D9, D10, and D12 are slightly worse. The reason for that can be found in the way Algorithm 2 verifies if a given attribute subset is a reduct. Namely, unlike Algorithm 1, the recursive version adds to the *RED* set not a reduct but a candidate for it. Whether the current elements of *RED* are reducts is checked any time a new element is added (see the *Reduce* function). The influence of these additional computations on the run-time can be observed for datasets with a bigger number of attributes.

Table 3. Indiscernibility matrix based attribute reduction of a decision table

id	IM+PS		RIM+PS		RIM+RPS		RIM+RPS(rec)	
	t1	t2	t1	t2	t1	t2	t1	t2
D1	0.468	0.531	0.53	0.53	0.546	0.546	0.531	0.531
D2	*	*	246.106	246.122	247.401	247.401	247.136	247.136
D3	0.125	0.125	0.124	0.124	0.125	0.125	0.125	0.125
D4	*	*	49.639	49.639	49.748	49.748	49.936	49.936
D5	0.109	0.202	0.125	0.125	0.125	0.125	0.125	0.125
D6	0.015	0.031	0.015	0.031	0.016	0.016	0.016	0.016
D7	0.141	4.462	0.218	0.249	0.202	0.218	0.219	0.234
D8	*	*	299.333	299.349	299.614	299.629	298.647	298.662
D9	*	*	51.963	406.412	51.37	53.149	51.542	54.397
D10	0.624	*	0.64	*	0.624	0.624	0.624	0.64
D11	<0.001	*	<0.001	*	<0.001	36.8	<0.001	58.999
D12	0.062	*	0.078	*	0.062	0.249	0.062	0.296

The results of computing all reducts of decision tables (see Table 3) show the same dependencies as those observed for databases treated as information systems. Concentrating on more interesting results one can see that for database D7 the IM+PS version is over 17 time slower compared with the remaining versions. For D9 the Algorithm 1 is over seven times faster than RIM+PS. Both the proposed algorithms are able to compute the reduct sets for all the databases; however, for D11 the recursive version is 1.6 times slower. The reasons for the above mentioned run-time discrepancies are the same as for those in Table 2.

5   Conclusions

This paper proposed ways to improve the indiscernibility matrix based approach for computing all reducts of an information system and decision table. It also developed two algorithms that employ the improved attribute reduction app-roach. The algorithms were verified theoretically and experimentally to evaluate their advantages over the basic version that operates on the full indiscernibility matrix and the power set of the attribute set.

The proposed approach can be summarized as follows.

1. The reduced indiscernibility matrix includes all cells necessary for comput-ing reducts. Due to its considerably smaller size the reduced indiscernibility matrix based approach enables to compute reducts for databases with a larger number of objects.

2. The reduced power set of the attribute set includes all elements that are needed for attribute reduction. The reduced version of the power set makes it possible to compute reducts for databases with a bigger number of attributes.
3. The number of computations needed for finding reducts depends to a significant extent on the maximal cardinality of the indiscernibility matrix cells. Algorithms proposed in this paper can considerably shorten the run-time for databases with a bigger number of attributes for which the maximal cardinality of the cells is extremely high.

References

1. Degang, C., Changzhong, W., Qinghua, H.: A new approach to attribute reduction of consistent and inconsistent covering decision systems with covering rough sets. Inf. Sci. **177**(17), 3500–3518 (2007)
2. Hu, X., Cercone, N.: Learning in relational databases: a rough set approach. Comput. Intell. **11**(2), 323–338 (1995)
3. Kryszkiewicz, M.: Rough set approach to incomplete information systems. Inf. Sci. **112**(1–4), 39–49 (1998)
4. Li, H., Zhu, J.: Finding all the absolute reductions based on discernibility matrix. In: ICMLC 2005. vol. 9, pp. 5682–5685. IEEE (2005)
5. Miao, D., Zhao, Y., Yao, Y., Li, H.X., Xu, F.: Relative reducts in consistent and inconsistent decision tables of the Pawlak rough set model. Inf. Sci. **179**(24), 4140–4150 (2009)
6. Pawlak, Z.: Rough Sets - Theoretical Aspects of Reasoning about Data. Kluwer Academic, Dordrecht (1991)
7. Skowron, A., Rauszer, C.: The discernibility matrices and functions in information systems. In: Słowiński, R. (ed.) Intelligent Decision Support. Theory and Decision Library, vol. 11, pp. 331–362. Springer, Amsterdam (1992)

Imprecise Rules for Data Privacy

Masahiro Inuiguchi[1]([✉]), Takuya Hamakawa[1], and Seiki Ubukata[2]

[1] Graduate School of Engineering Science, Osaka University Toyonaka,
Osaka 560-8531, Japan
inuiguti@sys.es.osaka-u.ac.jp
[2] Graduate School of Engineering, Osaka Prefecture University,
Gakuencho 1-1, Sakai, Osaka 599-8531, Japan
subukata@cs.osakafu-u.ac.jp

Abstract. When rules are induced, some rules can be supported only by a very small number of objects. Such rules often correspond to special cases so that supporting objects may be easily estimated. If the rules with small support include some sensitive data, this estimation of objects is not very good in the sense of data privacy. Considering this fact, we investigate utilization of imprecise rules for privacy protection in rule induction. Imprecise rules are rules classifying objects only into a set of possible classes. Utilizing imprecise rules, we propose an algorithm to induce k-anonymous rules, rules with k or more supporting objects. We demonstrate that the accuracy of the classifier with rules induced by the proposed algorithm is not worse than that of the classifier with rules induced by the conventional method. Moreover, the advantage of the proposed method with imprecise rules is examined by comparing other conceivable method with precise rules.

Keywords: Rule induction · Imprecise rules · MLEM2 · Privacy protection · k-anonymity

1 Introduction

There are several approaches to induce rules, association rule mining approach, decision tree learning approach and so on [1]. Rough set approaches [2] provide also rule induction techniques which are characterized by the minimality in the conditions of induced rules and/or the minimality in the number of rules to explain all given training data. Rough set approaches have been applied to various fields including medicine, engineering, management and economy.

In the conventional rough set approaches, rules classifying objects into a single class (simply called "precise rules") have been induced and used to build a classifier system. However, rules classifying objects into a set of classes (simply called "imprecise rules") can be also induced based on the rough set approach. Each of such imprecise rules cannot give a unvocal conclusion. Nevertheless, as demonstrated in our previous papers [3,4], the classifier system with imprecise rules has an advantage in the accuracy of classification over the conventional

© Springer International Publishing Switzerland 2015
D. Ciucci et al. (Eds.): RSKT 2015, LNAI 9436, pp. 129–139, 2015.
DOI: 10.1007/978-3-319-25754-9_12

classifier system with precise rules. The utility of imprecise rules has been studied in Hamakawa and Inuiguchi [4].

On the other hand, as Internet services increase, the capability to collect and store digital information by agencies is enhanced and opportunities to analyze huge amount of data are expanded. Under such circumstance, data mining techniques are making substantial progress. In such huge amount of data, personally identifiable information and other sensitive information can be included. This requires the development of data mining method equipped with sensitive data protection. Techniques for data privacy [5] such as k-anonymity [6,7], differential privacy [8,9], cryptographic approaches [10] and so on are evolving for this purpose. k-anonymity [6,7] protects the privacy by taking only common properties of k or more individual data. Differential privacy [8,9] is a technique used in query systems. It protects the privacy so that the presence or absence of an individual in a database cannot significantly affect the query result". Cryptographic approaches [10] provide methods of storing and transmitting data in an encrypted form so that only those who know the way of decryption can read and process it.

However, in the field of rough sets, the privacy protection methods have not yet been applied considerably. Zhou, Huang and Yun [11] proposed two privacy preserving attribute reduction algorithms using secure multiparty computation. Rokach and Schclar [12] proposed a greedy-based algorithm for attribute reduction with maximal anonymity level. Ye, Wu, Hu and Hu [13] proposed an approach to k-anonymization of decision tables based on hierarchical granulation rough sets and an entropy-based global information gain metric.

In this paper, we propose the utilization of imprecise rules to data privacy [5]. In the conventional approaches to rule induction, precise rules have been induced and some precise rules can be supported only by a very small number of objects. Such precise rules often correspond to special cases so that supporting objects may be identified easily. If rules with small support include some sensitive data, such identification of objects is not very good in the sense of data privacy. Considering this fact, we investigate an utilization of imprecise rules for privacy protection in rule induction. Because imprecise rules classify objects only into a set of classes, they are often supported by many objects. Then we propose an algorithm to induce only k-anonymous rules, precise and imprecise rules supported by at least k objects. We demonstrate that the classifier composed of rules induced by the proposed method performs comparably to the classifier composed of rules induced by the conventional method. Moreover we compare the proposed method to an alternative method for inducing k-anonymous precise rules and show the advantage of the proposed method.

This paper is organized as follows. In next section, rough set approach is briefly reviewed. In Sect. 3, imprecise rules and a classifier using them are described. Moreover, the k-anonymity of a rule is defined. The proposed method is described in Sect. 4. Numerical experiments and their results are shown in Sect. 5. Some concluding remarks are given in Sect. 6.

2   Rough Set Approach to Classification

Rough set theory is useful in the analysis of decision table. A decision table is defined by a four-tuple $DT = \langle U, C \cup \{d\}, V, f \rangle$, where U is a finite set of objects, C is a finite set of condition attributes, d is a decision attribute, $V = \bigcup_{a \in C \cup \{d\}} V_a$ with attribute value set V_a of attribute $a \in C \cup \{d\}$ and $f : U \times C \cup \{d\} \to V$ is called an information function which is a total function. By decision attribute value $v_j^d \in V_d$, class $D_j \subseteq U$ is defined by $D_j = \{u \in U \mid f(u, d) = v_j^d\}$, $j = 1, 2, \ldots, p$. Using condition attributes in $A \subseteq C$, we define equivalence classes $[u]_A = \{x \in U \mid f(x, a) = f(u, a), \forall a \in A\}$.

The lower and upper approximations of an object set $X \subseteq U$ under condition attribute set $A \subseteq C$ are defined by

$$A_*(X) = \{x \in U \mid [x]_A \subseteq X\}, \quad A^*(X) = \{x \in U \mid [x]_A \cap X \neq \emptyset\}. \tag{1}$$

Suppose that members of X can be described by condition attributes in A. If $[x]_A \cap X \neq \emptyset$ and $[x]_A \cap (U - X) \neq \emptyset$ hold, the membership of x to X or to $U - X$ is questionable because objects described in the same way are classified into two different classes. Otherwise, the classification is consistent. From these points of view, each element of $A_*(X)$ can be seen as a consistent member of X while each element of $A^*(X)$ can be seen as a possible member of X. The pair $(A_*(X), A^*(X))$ is called the rough set of X under $A \subseteq C$.

In rough set approaches, the attribute reduction, i.e., the minimal attribute set $A \subseteq C$ satisfying $A_*(D_j) = C_*(D_j)$, $j = 1, 2, \ldots, p$, and the minimal rule induction, i.e., inducing rules inferring the membership to D_j with minimal conditions which can differ members of $C_*(D_j)$ from non-members, are investigated well. In this paper, we use minimal rule induction algorithms proposed in the field of rough sets, i.e., LEM2 and MLEM2 algorithms [14]. By those algorithms, we obtain minimal set of rules with minimal conditions which can explain all objects in lower approximations of X under a given decision table. LEM2 algorithm and MLEM2 algorithm [14] are different in their forms of condition parts of rules: by LEM2 algorithm, we obtain rules of the form of "if $f(u, a_1) = v_1$, $f(u, a_2) = v_2$, ... and $f(u, a_s) = v_s$ then $u \in X$", while by MLEM2 algorithm, we obtain rules of the form of "if $v_1^L \leq f(u, a_1) \leq v_1^R$, $v_2^L \leq f(u, a_2) \leq v_2^R$, ... and $v_s^L \leq f(u, a_s) \leq v_s^R$ then $u \in X$". Namely, MLEM2 algorithm is a generalized version of LEM2 algorithm to cope with numerical/ordinal condition attributes.

For each decision class D_i we induce rules inferring the membership of D_i. Using all those rules, we build a classifier system. To build the classifier system, we apply the idea of LERS [14]. The classification of a new object u is made by the following two steps:

(i) When the condition attribute values of u match to all conditions of at least one of the induced rules, for each D_i, we calculate

$$S(D_i) = \sum_{\text{matching rules } r \text{ for } D_i} Stren(r) \times Spec(r), \tag{2}$$

where r is called a *matching rule* if the condition part of r is satisfied with u. $Stren(r)$ is the total number of objects in given decision table correctly classified by rule r. $Spec(r)$ is the total number of condition attributes in the condition part of rule r. For convenience, when rules from a particular class D_i are not matched by the object, we define $S(D_i) = 0$. If D_j such that $S(D_j) > 0$ exists, class D_i with the largest $S(D_i)$ is selected and terminate the procedure. If class D_i with the largest $S(D_i)$ is not unique, class D_i with smallest index i is selected from them and terminate the procedure.

(ii) When the condition attribute values of u do not match totally to the condition part of any rule in the induced rules, for each D_i, we calculate

$$M(D_i) = \sum_{\text{partially matching rules } r \text{ for } D_i} Mat\_f(r) \times Stren(r) \times Spec(r), \quad (3)$$

where r is called a *partially matching rule* if a part of the premise of r is satisfied with u. $Mat\_f(r)$ is the ratio of the number of matched conditions of rule r to the total number of conditions of rule r. Then class D_i with the largest $M(D_i)$ is selected. If a tie occurs, class D_i with smallest index i is selected from tied classes.

3   Imprecise Rules and k-Anonymity

In the same way as the induction of rules inferring the membership to D_i, we can induce rules inferring the membership to the union of D_i's. Namely, LEM2-based algorithms can be applied because a union of D_i's is a set of objects. Inducing rules inferring the membership to a union $D_i \cup D_j$ for all pairs (D_i, D_j) such that $i \neq j$, we may build a classifier. Moreover, in the same way, we can build a classifier by induced rules inferring the membership to $Z_q = \bigcup_{j \in \{i_1, i_2, \ldots, i_l\}} D_j$ for all combinations of l decision classes. A rule having a union of l classes in its conclusion is called an l-imprecise rule.

To use imprecise rules, we consider a classification method under rules inferring the membership to a union of classes. Although many extensions of LERS classifier described in the previous section are conceivable, in this paper, we use the following classifier performed best in the preliminary experiment:

(i) we calculate $S(Z_q)$ by (2) with substitution of Z_q for D_i for each $Z_q = \bigcup_{j \in \{i_1, i_2, \ldots, i_l\}} D_j$.

(ii) For each Z_q such that $S(Z_q) = 0$, erase all D_j satisfying $D_j \subseteq Z_q$.

(iii) If the remaining class is unique, then we classify the object into that class and terminate the procedure. If the remaining class is empty, we reset all classes as remaining classes.

(iv) For each remaining class D_i, calculate

$$\hat{S}(D_i) = \sum_{\text{matching rule } r \text{ for } Z_i \supseteq D_j} Stren(r) \times Spec(r). \quad (4)$$

(v) Classes D_i with the largest $\hat{S}(D_i)$ are selected. If it is unique, then we classify the object into that class and terminate the procedure. Otherwise, for each remaining class D_i, calculate

$$\hat{M}(D_i) = \sum_{\text{partially matching rule } r \text{ for } Z_i \supseteq D_j} Mat\_f(r) \times Stren(r) \times Spec(r), \quad (5)$$

and class D_i with the largest $\hat{M}(D_i)$ is selected. If a tie occurs, class D_i with smallest index i is selected from tied classes.

Applying a conventional rule induction method such as MLEM2 [14], we obtain a set of precise rules. Some of the induced rules may be supported only a few objects. In the condition part of such a rule, some rare attribute value or some rare combination of attribute values is included. The objects having rare attribute values as well as rare combination of attribute values are easily identified or often memorized by people. Then such a rule can disclose some privacy if the rule includes sensitive data. In such cases, the presentation of rules should be avoided. Accordingly it would be better not to induce rules with a small number of supporting objects from the view of data privacy.

The k-anonymity for a rule is defined as follows.

Definition 1. Let $r=$'if $v_1^L \leq f(x,a_1) \leq v_1^R$, $v_2^L \leq f(x,a_2) \leq v_2^R,\ldots, v_l^L \leq f(x,a_l) \leq v_l^R$ then $f(x,d) \in R$' be an induced rule, where $a_i \in C$, $v_i^L, v_i^R \in V_{a_i}$, $i = 1,2,\ldots,l$ and $\emptyset \neq R \subseteq V_d$. r is a precise rule if $\text{Card}(R) = 1$ and it is imprecise otherwise. r is said to be k-anonymous rule if and only if

$$Stren(r)$$
$$= \text{Card}\left(\{u \in U \mid v_i^L \leq f(u,a_i) \leq v_i^R, \ i=1,2,\ldots,l, \ f(u,d) \in R\}\right) \geq k. \quad (6)$$

Because k-anonymity of rule r is defined by $Stren(r)$, we may use an algorithm enumerating all rules satisfying $Stren(r) \geq k$ such as Apriori developed for knowledge discovery. We investigate a rule induction algorithm for machine learning, i.e., for building a classifier.

4   The Proposed Method for Inducing k-Anonymous Rules

Imprecise rules are supported by more objects than precise rules because the concluding class becomes larger. Then we utilize imprecise rules to obtain k-anonymous rules. Using MLEM2 algorithm [14], as rule induction method, we propose the following procedure for inducing k-anonymous rules from a given decision table $(U, C \cup \{d\}, V, f)$:

(i) Let $\mathcal{R}$ be the set of k-anonymous rules and initialize $\mathcal{R} = \emptyset$. Let $l = 1$.
(ii) Induce a set $\mathcal{S}_1$ of precise rules by MLEM2 algorithm.

(iii) Select rules $r \in S_l$ satisfying $\mathrm{Supp}(r) \geq k$ and put them in $\mathcal{R}$.
(iv) If $S_l - \mathcal{R} \neq \emptyset$ and $l < n$, update $l = l + 1$. Otherwise, terminate this procedure.
 (v) Define object set B by objects match $r \in S_l$ such that $\mathrm{Supp}(r) < k$.
(vi) Induce a set S_l of l-imprecise rules for each possible union Z_q of l sub-classes D_i, $i \in \{1, 2, \ldots, p\}$ by MLEM2 algorithm inputting $B \cap Z_q$ as a set of objects uncovered by presently induced rules. Return to (iii).

This procedure obviously terminates in $(n-1)$ iterations. The set of k-anonymous rules is obtained by the resulting $\mathcal{R}$. Using $\mathcal{R}$, we build the classifier by the way described in Sect. 3. We note that this algorithm neither guarantees that all objects are covered by a rule $r \in \mathcal{R}$ nor guarantees that any $u \in U$ is correctly classified by the classifier using $\mathcal{R}$. Moreover, there is no guarantee that the set of objects covered by a rule $r \in \mathcal{R}$ is maximal set of objects covered by some k-anonymous rule.

5   Numerical Experiments

5.1   Data-Sets and Cross Validation

We examine the performance of the proposed classifier with k-anonymous rules by numerical experiments. For the performance evaluation, we adopt the number of rules used in the classifier and the classification accuracy.

In the numerical experiments, we use eight data-sets obtained from UCI Machine Learning Repository [15]. The eight data-sets are shown in Table 1. In Table 1, $\mathrm{Card}(U)$, $\mathrm{Card}(C)$ and $\mathrm{Card}(V_d)$ mean the number of objects in the given data table, the number of condition attributes and the number of decision classes.

For the evaluation, we apply a 10-fold cross validation method. Namely we divide the data-set into 10 subsets and 9 subsets are used for training data-set and the remaining subset is used for checking data-set. Changing the combination of 9 subsets, we obtain 10 different evaluations. We calculate the averages and the standard deviations in each evaluation measure. We execute this procedure 10 times with different divisions.

Table 1. Eight data-sets

Data-set	$\mathrm{Card}(U)$	$\mathrm{Card}(C)$	$\mathrm{Card}(V_d)$	Attribute type
car	1,728	6	4	ordinal
dermatology	358	34	6	numerical
ecoli	336	7	8	numerical
glass	214	9	6	numerical
hayes-roth	159	4	3	nominal
iris	150	4	3	numerical
wine	178	13	3	numerical
zoo	101	16	7	nominal

5.2   Comparison with the Conventional Classifier with Precise Rules

We first compare the classifier composed of k-anonymous rules induced by the proposed method to the classifier composed of rules induced by the conventional MLEM2 algorithm in terms of the number of induced rules and the classification accuracy. For each data-set listed in Table 1, we conducted the experiment for comparison between the classifier composed of MLEM2 precise rules and the classifiers with k-anonymous imprecise rules for $k = 5$, 10, and 15. The obtained results are shown in Table 2.

In Table 2, the average av and the standard deviation sd is shown in the style of $av \pm sd$ in each cell of table. The underlined numbers are average scores of classification accuracy obtained by the classifiers composed of k-anonymous imprecise rules better than those obtained by the classifiers composed of MLEM2 precise rules. Asterisk $*$ shown in the columns of classification accuracy of k-anonymous imprecise rules stands for the value is significantly different from the case of MLEM2 precise rules by the paired t-test with significance level $\alpha = 0.05$.

As shown in Table 2, the numbers of k-anonymous rules are smaller than those of MLEM2 precise rules for $k = 5$, 10 and 15 in data-sets 'hayes-roth', 'iris' and 'wine'. This implies that in those data-sets, with high probability,

Table 2. MLEM2 precise rules versus k-anonymous imprecise rules

Data-set	MLEM2 precise rules		5-anonymous imprecise rules	
	Number of rules	Accuracy	Number of rules	Accuracy
car	57.22 ± 1.74	98.67 ± 0.97	58.94 ± 2.64	<u>98.76</u> ± 0.88
dermatology	12.09 ± 1.27	92.32 ± 4.42	13.78 ± 5.63	<u>92.38</u> ± 4.28
ecoli	35.89 ± 2.03	77.52 ± 6.21	1343.84 ± 123.08	<u>82.88</u>*±5.64
glass	25.38 ± 1.5	68.34 ± 10.18	84.61 ± 28.87	66.17 ± 10.32
hayes-roth	23.17 ± 1.41	81.38 ± 7.95	18.01 ± 1.40	74.56*± 9.24
iris	7.40 ± 0.72	92.87 ± 5.52	6.53 ± 0.90	<u>93.07</u> ± 5.65
wine	4.65 ± 0.50	93.25 ± 5.87	4.46 ± 0.50	93.25 ± 5.87
zoo	9.67 ± 0.55	95.84 ± 6.63	138.63 ± 29.08	94.55 ± 8.00

Data-set	10-anonymous imprecise rules		15-anonymous imprecise rules	
	Number of rules	Accuracy	Number of rules	Accuracy
car	66.48 ± 4.19	<u>98.94</u>*± 0.85	66.26 ± 8.75	98.21*± 1.34
dermatology	30.69 ± 18.06	91.73 ± 4.80	41.2 ± 31.14	<u>92.77</u> ± 4.74
ecoli	1351.04 ± 58.85	<u>83.68</u>*± 6.31	1094.66 ± 47.43	<u>82.48</u>*± 6.52
glass	254.90 ± 17.07	<u>72.52</u>*± 8.76	200.32 ± 5.95	<u>70.88</u> ± 9.2
hayes-roth	14.25 ± 1.44	59.44*± 16.64	1.90 ± 1.46	39.63*± 12.38
iris	6.87 ± 1.05	<u>93.20</u> ± 5.33	5.11 ± 1.03	92.47 ± 6.16
wine	4.09 ± 0.47	92.92 ± 6.06	4.05 ± 0.46	92.92 ± 6.06
zoo	196.08 ± 14.05	93.98 ± 7.06	198.22 ± 9.66	95.17 ± 6.55

no k-anonymous rules are induced for some of objects in U. Indeed we observed some objects uncovered by induced rules in those data-sets. Especially for data-set 'hayes-roth', no rule is induced in some sets of training data when $k = 15$. We note that any new object is classified into a default class D_1 when no rule is induced. It is very hard to protect the privacy in data-set 'hayes-roth'. We also observe that the three data-sets 'hayes-roth', 'iris' and 'wine' have only three classes. This observation can be understood by the following reason. If we have only a few classes, we obtain a limited number of unions of classes and we cannot make the unions large enough to have many k-anonymous rules. Although the classification accuracy scores of k-anonymous imprecise rules are comparable to those of MLEM2 precise rules in data-sets 'iris' and 'wine', we find that the proposed approach is not always very efficient in data-sets with a few classes. For such data-sets we need a lot of samples to induce a sufficient number of k-anonymous rules.

On the contrary, the proposed approach works well in data-set 'ecoli' having eight classes. We observe that very many k-anonymous imprecise rules are induced in this data-set. In data-sets 'car' and 'glass', 10-anonymous imprecise rules perform best. We observe that the number of rules are most at $k = 10$ in those data-sets. We note that the number of rules does not increase monotonously because, as k increases, the k-anonymity becomes stronger condition while the size of B at (v) of the proposed method increases. The more the size of B, the more imprecise rules are induced at (vi). Indeed, as k increases, the average number of rules increases in data-sets 'dermatology' and 'zoo' while it decreases in data-sets 'hayes-roth', 'iris' and 'wine'. In other data-sets, it attains the largest number at $k = 10$. In general, except 'hayes-roth', the classifier composed of the rules induced by the proposed method preserves the classification accuracy of the conventional classification while the induced rules improve the anonymity.

5.3 Comparison with the Classifier with k-Anonymous Precise Rules

In the previous experiment, it is not very clear to what extent the added k-anonymous imprecise rules improve the classification accuracy. Moreover, we may add k-anonymous precise rules instead of k-anonymous imprecise rules because MLEM2 algorithm does not induce all possible rules from a given decision table $(U, C \cup \{d\}, V, f)$. Therefore, it is not very clear to what extent the k-anonymous imprecise rules are advantageous over the k-anonymous precise rules. In response to those questions, we examine the performances of the following two classifiers: the classifier composed of k-anonymous precise rules induced from (i) to (iii) of the proposed procedure described in Sect. 4, and the classifier composed of k-anonymous precise rules induced by the proposed procedure by replacing (vi) with

(vi') Induce a set $\mathcal{S}_l$ of precise rules for each class $D_i, i \in \{1, 2, \ldots, p\}$ by MLEM2 algorithm inputting $B \cap D_i$ as a set of objects uncovered by presently induced rules. Return to (iii).

Table 3. Comparison of classifiers between k-anonymous rules and precise rules

Data-set	5-anonymous precise rules		5-anonymous precise rules[+]	
	Number of rules	Accuracy	Number of rules	Accuracy
car	53.93 ± 1.87	98.62 ± 1.02	56.37 ± 1.72	98.62 ± 1.82
dermatology	10.79 ± 1.15	92.24 ± 4.40	11.68 ± 2.07	92.24 ± 4.62
ecoli	23.14 ± 2.17	76.60*± 6.06	68.81 ± 6.98	78.29*± 7.4
glass	18.47 ± 1.63	<u>67.66</u> ± 11.03	31.12 ± 5.34	64.14 ± 10.63
hayes-roth	12.11 ± 1.13	<u>84.25</u>*± 8.27	17.91 ± 1.43	72.63 ± 10.71
iris	4.47 ± 0.83	92.47 ± 5.94	6.47 ± 0.91	93.00 ± 5.53
wine	4.46 ± 0.50	93.25 ± 5.87	4.46 ± 0.50	93.25 ± 5.87
zoo	5.01 ± 0.26	86.91*± 10.71	16.65 ± 1.28	84.95*± 10.19
Data-set	10-anonymous precise rules		10-anonymous precise rules[+]	
	Number of rules	Accuracy	Number of rules	Accuracy
car	48.31 ± 1.89	97.73*± 1.32	55.17 ± 3.62	98.39 ± 2.28
dermatology	9.84 ± 1.11	<u>92.07</u> ± 4.56	12.48 ± 3.32	91.65 ± 5.43
ecoli	14.36 ± 2.15	73.74*± 6.21	56.03 ± 5.05	79.66 ± 7.03
glass	9.11 ± 1.43	58.87*± 9.58	28.35 ± 2.50	63.75 ± 10.31
hayes-roth	7.24 ± 0.93	<u>62.13</u>*± 14.33	11.68 ± 1.41	<u>78.19</u> ± 10.40
iris	3.84 ± 0.60	92.47 ± 5.94	5.36 ± 0.88	92.47 ± 5.79
wine	4.09 ± 0.47	92.92 ± 6.06	4.09 ± 0.47	92.92 ± 6.06
zoo	2.99 ± 0.22	72.90*± 12.56	10.00 ± 1.60	72.70*± 12.29
Data-set	15-anonymous precise rules		15-anonymous precise rules[+]	
	Number of rules	Accuracy	Number of rules	Accuracy
car	40.03 ± 2.14	94.75*± 1.95	57.05 ± 5.87	92.21*± 5.67
dermatology	9.14 ± 1.03	91.76*± 5.01	14.81 ± 4.25	91.14*± 5.47
ecoli	10.36 ± 1.69	65.83*± 7.62	44.04 ± 4.27	74.93*± 7.41
glass	4.59 ± 1.04	51.07*± 10.22	17.44 ± 1.79	58.75*± 9.04
hayes-roth	0 ± 0	39.56 ± 11.83	0 ± 0	39.56 ± 11.83
iris	3.65 ± 0.52	92.4 ± 6.18	4.68 ± 0.96	91.87 ± 6.15
wine	4.05 ± 0.46	92.92 ± 6.06	4.05 ± 0.46	92.92 ± 6.06
zoo	2 ± 0	69.52*± 12.63	8.96 ± 1.66	69.52*± 12.60

We conducted the experiment for comparing k-anonymous imprecise rules induced by the proposed method to k-anonymous precise rules induced by the above alternative ways for $k = 5$, 10, and 15. The obtained results are shown in Table 3. In Table 3, 'k-anonymous precise rules' and 'k-anonymous precise rules[+]' ($k = 5$, 10 and 15) means rules induced from (i) to (iii) and rules induced by the proposed method by replacing (vi) with (vi'), respectively. The average av and the standard deviation sd is shown in the style of $av \pm sd$ in each cell of Table 2.

The underlined numbers are average scores of classification accuracy obtained by the classifiers composed of k-anonymous precise rules better than those obtained by the classifiers composed of k-anonymous imprecise rules. Asterisk $*$ shown in the columns of classification accuracy of k-anonymous precise rules stands for the value is significantly different from the case of k-anonymous imprecise rules by the paired t-test with significance level $\alpha = 0.05$.

Comparing numbers of rules between 'MLEM2 precise rules' in Table 2 and 'k-anonymous precise rules' ($k = 5$, 10 and 15) in Table 3, we find that the differences are big in data-sets 'ecoli', 'hayes-roth', 'iris' and 'zoo'. In those data-sets, many of MLEM2 precise rules are rather particular. Moreover, comparing the numbers of rules among 'MLEM2 precise rules' in Table 2, 'k-anonymous precise rules' and 'k-anonymous precise rules$^+$' ($k = 5$, 10 and 15) in Table 3, we find that many k-anonymous precise rules exist in data-sets 'ecoli' and 'zoo' so that the numbers of precise rules are recovered well by the additional induction of k-anonymous precise rules. However, comparing the classification accuracy scores between 'k-anonymous imprecise rules' in Table 2, 'k-anonymous precise rules' and 'k-anonymous precise rules$^+$' in Table 3, the performances of 'k-anonymous precise rules' and 'k-anonymous precise rules$^+$' are significantly worse than that of 'k-anonymous imprecise rules' ($k = 5$, 10 and 15).

Comparing results of 'k-anonymous precise rules$^+$' in Table 3 to results of 'k-anonymous imprecise rules' in Table 2 for data-sets 'glass' and 'zoo', we observe that the compensation by k-anonymous precise rules generated from (iii) to (vi') is weaker than that by k-anonymous imprecise rules generated from (iii) to (vi). In those data-sets, even if all 'k-anonymous precise rules' are enumerated, the compensation by those can be weaker than that by k-anonymous imprecise rules.

Some of MLEM2 precise rules which are not 5-anonymous data-set 'hayes-roth' aggravate the performance of the classification because the classification accuracy sore of 'MLEM2 precise rules' is worse than that of '5-anonymous precise rules'. Together with the fact that it has rather small number of k-anonymous rules, data-set 'hayes-roth' is understood as a data-set composed of particular objects. We note that no 15-anonymous precise rules are induced in data-set 'hayes-roth', in this case all new objects are classified into a default class D_1.

In all data-sets except 'hayes-roth', the classification accuracy scores of 'k-anonymous imprecise rules' are better or not far worse than those of 'k-anonymous precise rules' and 'k-anonymous precise rules$^+$' ($k = 5$, 10 and 15). As a result, the proposed 'k-anonymous imprecise rules' are useful in both classification accuracy and anonymity.

6   Concluding Remarks

In this paper, we proposed to utilize imprecise rules for data privacy. The k-anonymity of a rule is defined by the support. A method for inducing k-anonymous rules without big deterioration of classification accuracy is proposed. The idea of the proposed method is simply the replacement of non-k-anonymous precise rules with k-anonymous imprecise rules in the set of induced precise rules. By numerical experiments, we demonstrated that the k-anonymous imprecise rules induced by the proposed method can classify new objects without big deterioration of classification accuracy of the conventional precise rules. However, the number of induced k-anonymous imprecise rules increases greatly. A method for inducing less number of k-anonymous rules without big deterioration of classification accuracy becomes one of our future research topics.

Acknowledgment. This work was partially supported by JSPS KAKENHI Grant Number 26350423.

References

1. Witten, I.H., Frank, E., Hall, M.A.: Data Mining: Practical Machine Learning Tools and Techniques, 3rd edn. Morgan Kaufmann, Burlington (2011)
2. Pawlak, Z.: Rough Sets. Int. J. Comput. Inf. Sci. **11**(5), 341–356 (1982)
3. Inuiguchi, M., Hamakawa, T.: The utilities of imprecise rules and redundant rules for classifiers. In: Huynh, V.-N., et al. (eds.) Knowledge and Systems Engineering. AISC, vol. 245, pp. 45–56. Springer, Heidelberg (2013)
4. Hamakawa, T, Inuiguchi, M.: On the Utility of Imprecise Rules Induced by MLEM2 in Classification. In: Proceedings of 2014 IEEE International Conference on Granular Computing C, pp. 76–81. IEEE Xplore (2014)
5. Domingo-Ferrer, J., Torra, V.: Disclosure control methods and information loss for microdata, confidentiality, disclosure, and data access. In: Doyle, P., et al. (eds.) Theory and Practical Applications for Statistical Agencies, pp. 91–110. Elsevier, Amsterdam (2001)
6. Samarati, P.: Protecting respondents identities in microdata release. IEEE Trans. Knowl. Data Eng. **13**(6), 1010–1027 (2001)
7. Sweeney, L.: K-anonymity: a model for protecting privacy. Int. J. Uncertainty, Fuzziness Knowl. Based Sys. **10**(5), 557–570 (2002)
8. Dwork, C.: Differential privacy: a survey of results. In: Agrawal, M., Du, D.-Z., Duan, Z., Li, A. (eds.) TAMC 2008. LNCS, vol. 4978, pp. 1–19. Springer, Heidelberg (2008)
9. Dwork, C., Roth, A.: The algorithmic foundations of differential privacy. Found. Trends Theor. Comput. Sci. **9**(3–4), 211–407 (2014)
10. Yakoubov, S., Gadepally, V., Schear, N., Shen, E., Yerukhimovich, A.: A survey of cryptographic approaches to securing big-data analytics in the cloud. In: Proceedings of 2014 IEEE High Performance Extreme Computing Conference, pp. 1–6. IEEE Xplore (2014)
11. Zhou, Z., Huang, L., Yun, Y.: Privacy preserving attribute reduction based on rough set. In: Proceedings of 2nd International Workshop on Knowledge Discovery and Data Mining. WKKD 2009, pp. 202–206. AAAI, Portland (2009)
12. Rokach, L., Schclar, A.: k-anonymized reducts. In: Proceedings of 2010 IEEE International Conference on Granular Computing, pp. 392–395. IEEE Xplore (2010)
13. Ye, M., Wu, X., Hu, X., Hu, D.: Anonymizing classification data using rough set theory. Knowl. Based Sys. **43**, 82–94 (2013)
14. Grzymala-Busse, J.W.: MLEM2 - discretization during rule induction. In: Klopotek, M.A., Wierzchon, S.T., Trojanowski, K. (eds.) Intelligent Information Processing and Web Mining. AISC, vol. 22, pp. 499–508. Springer, Heidelberg (2003)
15. UCI Machine Learning Repository. http://archive.ics.uci.edu/ml/

Proposal for a Statistical Reduct Method for Decision Tables

Yuichi Kato[1]($\boxtimes$), Tetsuro Saeki[2], and Shoutarou Mizuno[1]

[1] Shimane University, 1060 Nishikawatsu-cho, Matsue City, Shimane 690-8504, Japan
ykato@cis.shimane-u.ac.jp
[2] Yamaguchi University, 2-16-1 Tokiwadai, Ube City, Yamaguchi 755-8611, Japan
tsaeki@yamaguchi-u.ac.jp

Abstract. Rough Sets theory is widely used as a method for estimating and/or inducing the knowledge structure of if-then rules from a decision table after a reduct of the table. The concept of the reduct is that of constructing the decision table by necessary and sufficient condition attributes to induce the rules. This paper retests the reduct by the conventional methods by the use of simulation datasets after summarizing the reduct briefly and points out several problems of their methods. Then a new reduct method based on a statistical viewpoint is proposed. The validity and usefulness of the method is confirmed by applying it to the simulation datasets and a UCI dataset. Particularly, this paper shows a statistical local reduct method, very useful for estimating if-then rules hidden behind the decision table of interest.

1 Introduction

Rough Sets theory was introduced by Pawlak [1] and used for inducing if-then rules from a dataset called the decision table. The induced if-then rules simply and clearly express the structure of rating and/or knowledge hiding behind the decision table. Such rule induction methods are needed for disease diagnosis systems, discrimination problems, decision problems and other aspects. Each data in the decision table is a sample data consisting of the tuple of condition attributes' values and the decision attribute value, and the first step for the rule induction is to find the condition attributes which do not have any relationship with the decision attribute, to remove them and finally to reduce the table. The use of those processes to obtain the reduced table is called a reduct. The conventional rough sets theory to induce if-then rules is based on the indiscernibility of the samples of the given table by their attributes, and a reduct by the conventional method also uses the same concept and various types of indiscernibility, their methods and algorithms for the reducts are proposed to date [2–6].

This paper retests the conventional reduct methods through the use of a simulation dataset and points out their problems and how they lack receptivity for datasets containing conflicting and/or indifferent data, such as real-world datasets, after summarizing the conventional rough sets and reduct methods.

© Springer International Publishing Switzerland 2015
D. Ciucci et al. (Eds.): RSKT 2015, LNAI 9436, pp. 140–152, 2015.
DOI: 10.1007/978-3-319-25754-9_13

Table 1. Example of a decision table

U	$(C(1)C(2)C(3)C(4)C(5)C(6))$	D
1	563242	3
2	256124	6
3	116226	1
4	416646	6
...	......	...
$N-1$	151252	2
N	513135	4

Then a new reduct method is proposed to overcome their problems from a statistical point of view. Specifically, the new method recognizes each sample data in the decision table as the outcomes of random variables of the tuple of the condition attributes and the decision attribute since the dataset is obtained from their population of interest. Accordingly, the problem of finding the reduct through the concept of discernibility can be replaced by the problem of finding the condition attributes which are statistically independent of the decision attribute. This paper proposes two reduct methods: one is called a global reduct which examines the statistical independence between each condition attribute and the decision attribute, and the other is called a local reduct which finds each condition attribute statistically independent of a decision attribute value. The validity and usefulness of both reduct methods are confirmed by applying them to the simulation dataset as well as a UCI dataset [7] prepared for machine learning. The local reduct particularly shows that it gives important clues in the search for the if-then rules hidden behind the decision table of interest.

2   Conventional Rough Sets and Reduct Method

Rough Sets theory is used for inducing if-then rules from a decision table S. S is conventionally denoted $S = (U, A = C \cup \{D\}, V, \rho)$. Here, $U = \{u(i)|i = 1, ..., |U| = N\}$ is a sample set, A is an attribute set, $C = \{C(j)|j = 1, ..., |C|\}$ is a condition attribute set, $C(j)$ is a member of C and a condition attribute, and D is a decision attribute. V is a set of attribute values denoted by $V = \bigcup_{a \in A} V_a$ and is characterized by an information function $\rho: U \times A \to V$. Table 1 example where $|C| = 6$, $|V_{a=C(j)}| = M_{C(j)} = 6$, $|V_{a=D}| = M_D = 6$, $\rho(x = u(1), a = C(1)) = 5$, $\rho(x = u(2), a = C(2)) = 5$, and so on.

Rough Sets theory focuses on the following equivalence relation and equivalence set of indiscernibility: $I_C = \{(u(i), u(j)) \in U^2 | \rho(u(i), a) = \rho(u(j), a), \forall a \in C\}$. I_C is an equivalence relation in U and derives the quotient set $U/I_C = \{[u_i]_C | i = 1, 2, ...\}$. Here, $[u_i]_C = \{u(j) \in U | (u(j), u_i) \in I_C, u_i \in U\}$. $[u_i]_C$ is an equivalence set with the representative element u_i and is called an element set of C in Rough Sets theory [2]. Let be $\forall X \subseteq U$ then X can be approximated like $C_*(X) \subseteq X \subseteq C^*(X)$ by use of the element set. Here,

$$C_*(X) = \{u_i \in U | [u_i]_C \subseteq X\}, \tag{1}$$
$$C^*(X) = \{u_i \in U | [u_i]_C \cap X \neq \emptyset\}, \tag{2}$$

$C_*(X)$ and $C^*(X)$ are called the lower and upper approximations of X by C respectively. The pair of $(C_*(X), C^*(X))$ is usually called a rough set of X by C. Specifically, let $X = D_d = \{u(i) | (\rho(u(i), D) = d\}$ then $C_*(X)$ is surely a set satisfying $D = d$ and $C^*(X)$ is possibly, and they derive if-then rules of $D = d$ with necessity and possibility, respectively.

The conventional Rough Sets theory seeks a minimal subset of C denoted with $B(\subseteq C)$ satisfying the following two conditions:

(i) $B_*(D_d) = C_*(D_d)$, $d = 1, 2, ..., M_D$.
(ii) $a(\in B)$ satisfying $(B - \{a\})_*(D_d) = C_*(D_d)$ $(d = 1, 2, ..., M_D)$ does not exist.

$B(\subseteq C)$ is called a relative reduct of $\{D_d | d = 1, ..., M_D\}$ preserving the lower approximation, and is useful for finding if-then rules with necessity, since redundant condition attributes have been already removed from C. In the same way, a relative reduct preserving the upper approximation can be also defined and obtained.

LEM1 algorithm [2] and the discernibility matrix method (DMM) [3] are well known as representative ways to perform reducts. Figure 1 shows an example of LEM1, and A and $\{d\}^*$ at Line 1 of the figure respectively correspond to C and $\{D_d | d = 1, ..., M_D\}$ in this paper. LEM1, from Line 6 to 15 in the figure, in principle, checks and executes (i) and (ii) for all the combinations of the condition attributes.

DMM [3] at first forms a symmetric $N \times N$ matrix having the following (i, j) element δ_{ij}:

$$\delta_{ij} = \{a \in C | \rho(u(i), a) \neq \rho(u(j), a)\};$$

$$\exists d \in D, \rho(u(i), d) \neq \rho(u(j), d) \text{and } \{u(i), u(j)\} \cap Pos(D) \neq \emptyset,$$

$$= *; \quad otherwise(U - Pos(D)).$$

Here, $Pos(D) = \cup_{d=1}^{M_D} C_*(D_d)$ and $*$ denotes "don't care". Then, a relative reduct preserving the lower approximation can be obtained by the following expression:

$$F^{reduct} = \wedge_{i,j:i<j} \vee \delta_{ij}. \tag{3}$$

Here, F^{reduct} is called a discernible function and indicates that reducts are obtained by arranging all of respective discernibility.

3   Retests of the Conventional Reduct Method

Here we retest the ability of the reducts obtained through LEM1 and DMM by use of a simulation dataset. The literatures [8–11] show ways of how to generate simulation datasets. Specifically, let (a) generate the condition attribute values of $u(i)$, that is, $u^C(i) = (v_{C(1)}(i), v_{C(2)}(i), ..., v_{C(|C|)}(i))$ by the use of random numbers with a uniform distribution and (b) determine the decision attribute

Line No.	Algorithm to compute a single global covering
1	(input: the set A of all attributes, partition $\{d\}^*$ on U; output: a single global covering R);
2	Begin
3	compute partition A^*;
4	$P := A$;
5	$R := \emptyset$;
6	if $A^* \leq \{d\}^*$
7	Then
8	Begin
9	for each attribute a in A do
10	Begin
11	$Q := P - \{a\}$;
12	compute partition Q^*;
13	if $Q^* \leq \{d\}^*$ then $P := Q$
14	end {for}
15	end {then}
16	end {algorithm}

Fig. 1. Example of LEM1 algorithm

value of $u(i)$, that is, $u^D(i)$ by use of if-then rules specified in advance and in the hypotheses shown in Table 2, and repeat the (a) and (b) processes by N times. Table 1 shows an example dataset generated by the use of those procedures with the following if-then rule $R(d)$ specified in advance:

$$R(d) : \text{if } Rd \text{ then } D = d \quad (d = 1, ..., M_D = 6), \tag{4}$$

where $Rd = (C(1) = d) \bigwedge (C(2) = d) \bigvee (C(3) = d) \bigwedge (C(4) = d)$. The results of retesting both methods using the $N = 10000$ dataset showed $F_{LEM1}^{reduct} = F_{DMM}^{reduct} = C(1) \wedge C(2) \wedge C(3) \wedge C(4) \wedge C(5) \wedge C(6)$ while the results were expected to be $F^{reduct} = C(1) \wedge C(2) \wedge C(3) \wedge C(4)$ from the rules (4) specified in advance. The retest experiment was repeated three times by changing the generated dataset and obtained the same results, or in other words, the expected reducts were not obtained.

These results are clearly derived from the indiscernibility and/or discernibility caused by the element set in the conventional Rough Sets theory. The element sets couldn't distinguish the differences between samples by the if-then rules (see Hypothesis 1 in Table 2) or those obtained by chance (see Hypothesis 2 and 3 in Table 2). On the other hand, real-world datasets will contain all kinds of data generated by Hypotheses 1, 2 and 3. In other words, the conventional Rough Sets theory does not have any abilities adaptive to sample data caused by chance.

Table 2. Hypotheses with regard to the decision attribute value

Hypothesis 1	$u^C(i)$ coincides with $R(k)$, and $u^D(i)$ is uniquely determined as $D = d(k)$ (uniquely determined data)
Hypothesis 2	$u^C(i)$ does not coincide with any $R(d)$, and $u^D(i)$ can only be determined randomly (indifferent data)
Hypothesis 3	$u^C(i)$ coincides with several $R(d)$ ($d = d1, d2, ...$), and their outputs of $u^C(i)$ conflict with each other. Accordingly, the output of $u^C(i)$ must be randomly determined from the conflicted outputs (conflicted data)

4   Proposal of Statistical Reduct

4.1   Statistical Global Reduct Method

As mentioned in Sect. 3, conventional reduct methods are unable to reproduce the forms of reducts specified in advance from the decision table due to a lack of abilities adaptive to the indifferent and conflicted samples in datasets, despite the fact that real-world datasets will have such samples. This paper studies this problem with reducts from the view of STRIM (Statistical Test Rule Induction Method) [8–11]. STRIM regards the decision table as a sample set obtained from the population of interest. According to a statistical model, $u(i) = (v_{C(1)}(i), v_{C(2)}(i), ..., v_{C(|C|)}(i), u^D(i))$ is an outcome of the random variables of $A = (C(1), C(2), ..., C(|C|), D) = (C, D)$ (hereafter, the names of the attributes are also used as the random variables). Next, the following probability model will be specified: For any j, $P(C(j) = v_{C(j)}(k)) = p(j, k)$, $\sum_{k=1}^{M_{C(j)}} p(j, k) = 1$. For any $j1 \neq j2$, $P(C(j1) = v_{C(j1)}(k1), C(j2) = v_{C(j2)}(k2)) = p(j1, k1)p(j2, k2)$, that is, $C(j1)$ and $C(j2)$ are independent of each other. According to the rules specified in (4), if $C = (1, 1, 2, 3, 4, 5)$ (hereafter (112345) briefly), for example, then $P(D = 1) = 1.0$ by use of Hypothesis 1 in Table 2. If $C = (123456)$ then $P(D = 1) = 1/M_D = 1/6$ by use of Hypothesis 2. If $C = (112256)$ then $P(D = 1) = 1/2$ by use of Hypothesis 3. Generally, the outcome of random variable D is determined by the outcome of C, if-then rules (generally unknown) and the hypothesis shown in Table 2. Consequently, the following expression is obtained:

$$P(D, C) = P(D|C)P(C = (v_{c(1)}(i), v_{c(2)}(i), ..., v_{c(|C|)}(i))). \tag{5}$$

Here, $P(D|C)$ is the condition probability of D by C and dependent on the if-then rules to be induced by the use of the dataset $\{u(i) = (v_{C(1)}(i), v_{C(2)}(i), ..., v_{C(|C|)}(i), u^D(i))|i = 1, ..., N\}$.

In the special case, if $C(j)$ does not exist in the condition part of the if-then rules to be induced, then D is independent of $C(j)$, that is $P(D, C(j)) = P(D|C(j))P(C(j)) = P(D)P(C(j))$. This independency between D and $C(j)$ can be used for a reduct of the decision table. The problem of whether they are independent or not can be easily dealt with using a statistical test of hypotheses by the use of $\{u(i) = (v_{C(1)}(i), v_{C(2)}(i), ..., v_{C(|C|)}(i), u^D(i))|i = 1, ..., N\}$. Specifically, specification and testing of the following null hypothesis $H0(j)$ and its alternative hypothesis $H1(j)$ $(j = 1, ..., |C|)$ were implemented:

H0(j): $C(j)$ and D are independent of each other.
H1(j): $C(j)$ and D are not independent of each other.

This paper adopts a Chi-square test since it is a standard method for testing the independency of two categorical variables by use of the contingency table $M_{C(j)} \times M_D$. Chi-square tests are widely used in a data mining area for feature selection [12]. The test statistic χ^2 of $C(j)$ vs. D is

$$\chi^2 = \sum_{k=1}^{M_{C(j)}} \sum_{l=1}^{M_D} \frac{(f_{kl} - e_{kl})^2}{e_{kl}}, \tag{6}$$

where, $f_{kl} = |U(C(j) = k) \cap U(D = \ell)|$, $U(C(j) = k) = \{u(i)|v_{C(j)}(i) = k\}$, $U(D = l) = \{u(i)|u^D(i) = l\}$, $e_{kl} = n\hat{p}(j,k)\hat{p}(D,l)$, $n = \sum_{k=1}^{M_{C(j)}} \sum_{l=1}^{M_D} f_{kl}$, $\hat{p}(j,k) = f_{k\_}/n$, $\hat{p}(D,l) = f_{\_l}/n$, $f_{k\_} = \sum_{l=1}^{M_D} f_{kl}$, $f_{\_l} = \sum_{k=1}^{M_{C(j)}} f_{kl}$. χ^2 obeys a Chi-square distribution with degrees of freedom $df = (M_{C(j)} - 1) \times (M_D - 1)$ under $H0(j)$ and test condition [13]: $n\hat{p}(j,k)\hat{p}(D,l) \geq 5$. This paper proposes a reduct method to adopt only the $C(j)$s of $H0(j)$ that were rejected and to construct a decision table composed by them since the test of the hypotheses can't control type II errors but only type I errors by a significance level. This paper names the proposed method the statistical global reduct (SGR) to distinguish it from the conventional method.

A simulation experiment was conducted to confirm the validity of the proposed method using the decision table of the samples of $N = 10000$ used in Sect. 2, and the following procedures:

Step1: Randomly select samples by $N_B = 3000$ from the decision table ($N = 10000$), and form a new decision table;
Step2: Apply SGR to the new table, and calculate χ^2 every $C(j)$;
Step3: Repeat Step1 and Step2 $N_r = 100$ times.

Table 3 shows the arrangement of the results of the simulation experiment, that is, the average (AVE), standard deviation (S.D.), maximum (Max) and minimum (Min) of χ^2, and their corresponding p-values for every $C(j)$ $(j = 1, ..., 6)$. The table shows the following:

(1) There are clear differences and big gaps in χ^2 values between those of $C(1) - C(4)$ dependent on D and those of $C(5)$ or $C(6)$ independent of D, and are not overlapped by fluctuating ranges of the χ^2 value.
(2) Accordingly, there are clear differences in p-values between them.
(3) This reduct method of adopting rejected $C(j)$s of $H0(j)$ and to construct a decision table composed by them is valid and useful since the notion reproduces the expected reduct.

Table 3. Results of test for independence by Bootstrap method ($N = 10000$, $N_B = 3000$, $Nr = 100$): $C(j)$ ($j = 1, ..., 6$) vs. D

χ^2		$C(1)$	$C(2)$	$C(3)$	$C(4)$	$C(5)$	$C(6)$
AVE		341.69	360.59	313.27	327.40	24.79	25.23
S.D.		37.35	35.91	35.31	34.55	7.04	6.06
Max		451.19	471.25	412.19	418.39	47.83	40.61
Min		270.7	285.91	232.56	255.42	10.93	8.21
p-value	Max	9.44E-80	6.84E-84	9.87E-72	5.28E-73	3.91E-03	2.52E-02
	Min	4.29E-43	3.98E-46	1.45E-35	4.59E-40	9.93E-01	9.99E-01

4.2 Statistical Local Reduct Method

In this section, a statistical local reduct (SLR) method which is applied for every decision attribute value, that is, every if-then rule, is proposed corresponding to the SGR method. The SLR method is applicable for the decision table which doesn't even have a relative reduct and/or the SGR. Specifically, (5) also holds at $D = l$, that is,

$$P(D = l, C) = P(D = l|C)P(C = (v_{c(1)}(i), v_{c(2)}(i), ..., v_{c(|C|)}(i))). \qquad (7)$$

Accordingly, the discussion in Sect. 4.1 holds even in the case where D is replaced in $H0(j)$ and $H1(j)$ with $D = l$, and the test of statistical independency between $C(j)$ and $D = l$ can be used for the reduct method for every $D = l$ or if-then rule.

Table 4 shows a set of new rules specified in advance in order to confirm the validity and usefulness of the SGL and SLR method. For example, (1100001) at Rule 1 of the table means there is a rule that if $(C(1) = 1) \bigwedge (C(2) = 1)$ then $D = 1$. The set of new rules doesn't include redundant condition attributes, that is, D is dependent on all condition attributes. The same simulation experiment in 4.1 was conducted replacing (4) with the set of rules in Table 4 and the results of the applied SGR method were summarized in Table 5 in the same way as Table 3. The validity of the SGR method is also confirmed since the table suggests the condition attributes seemed to be independent of D never existing, which coincides with the rules specified in advance. However, the structure of the if-then rules specified in advance is left almost unknown.

Then, the SLR method was applied for one of the same datasets of $N_r = 100$ times used in Table 5. As one of examples, Table 6 shows the contingency table of the case of $D = 1$ vs. $C(j)$ ($j = 1, ..., 6$) and the results of a chi-square test of them with $df = (M_{C(j)} - 1)$, and suggests the following knowledge:

(1) The p-values of $C(5)$ and $C(6)$ are quite high compared with those of the other condition attributes and indicate that $C(5)$ and $C(6)$ are independent of $D = 1$, that is, they are redundant and should be removed from the viewpoint of reduct.

Table 4. Example of if-then rules for a simulation experiment

Rule No.	specified rule: $(C(1)C(2), ..., C(6)D)$
1	(1100001)
2	(0011001)
3	(0000222)
4	(2200002)
5	(0033003)
6	(0000333)
7	(4400004)
8	(0044004)
9	(0000555)
10	(5500005)
11	(0066006)
12	(0000666)

Table 5. Results of test for independence by Bootstrap method using sample dataset generated by rules in Table 4 ($N = 10000$, $N_B = 3000$, $Nr = 100$): $C(j)$ ($j = 1, ..., 6$) vs. D

χ^2		$C(1)$	$C(2)$	$C(3)$	$C(4)$	$C(5)$	$C(6)$
AVE		209.56	206.35	244.57	234.94	252.45	219.59
S.D.		29.38	29.08	30.14	30.72	27.41	28.83
Max		293.56	290.93	318.17	317.52	310.48	314.13
Min		140.01	147.43	185.15	162.5	186.58	153.62
p-value	Max	1.17E-47	3.95E-47	1.33E-52	1.80E-52	4.72E-51	8.69E-52
	Min	5.71E-18	2.51E-19	2.14E-26	4.03E-22	1.14E-26	1.81E-20

(2) The frequencies of $C(1) = 1$, $C(2) = 1$, $C(3) = 1$ and $C(4) = 1$ are relatively high compared with those of the rest of the same $C(j)$ ($j = 1, ..., 4$) set. Accordingly, the combinations of $C(j) = 1$ ($j = 1, ..., 4$) will most likely construct the rules of $D = 1$.

The above knowledge of (1) and (2) coincides with the specifications of Rules 1 and 2 in Table 4 and the same results also have been obtained from the cases of $D = l$ ($l = 2, ..., 6$), and thus through them the validity and usefulness of the SLR method have been confirmed.

5   An Example Application on an Open Dataset

The literature [7] provides a lot of datasets for machine learning. This paper applied the SGR and SLR method for the "Car Evaluation" dataset included in

Table 6. Example of contingency table by local reducts ($N = 3000$, $df = 5$): $D = 1$ vs. C

$D = 1$	C					
$|U(D = 1)| = 486$	1	2	3	4	5	6
1	*141*	*122*	*137*	*141*	78	88
2	64	67	89	80	95	74
3	82	86	59	82	74	79
4	58	77	53	52	93	101
5	79	58	83	75	71	64
6	62	76	65	56	75	80
χ^2	52.83	35.01	66.42	53.16	2.99	13.33
p-values	3.64E-10	1.5E-06	5.68E-13	5.00E-10	0.702	2.05E-2

Table 7. Arrangement of Car Evaluation dataset of UCI

unified attribute value	$C(1)$: buying	$C(2)$: maint	$C(3)$: doors	$C(4)$: person	$C(5)$: lug boot	$C(6)$: safety	D: class (freq.)
1	vhigh	vhigh	2	2	small	low	unacc (1210)
2	high	high	3	4	med	med	acc (383)
3	med	med	4	more	big	high	good (69)
4	low	low	5more	–	–	–	vgood (65)

them. Table 7 shows the summaries and specifications of the dataset: $|C| = 6$, $|V_{C(1)}| = |V_{C(2)}| = |V_{C(3)}| = 4$, $|V_{C(4)}| = |V_{C(5)}| = |V_{C(6)}| = 3$, $|V_D| = 4$, $N = |U| = |V_{C(1)}| \times |V_{C(2)}| \times |V_{C(3)}| \times |V_{C(4)}| \times |V_{C(5)}| \times |V_{C(6)}| = 1728$ which consists of every combination of condition attributes' values, and there were not any conflicted or identical samples. The frequencies of the decision attribute values extremely incline toward $D = 1$ as shown in Table 7.

Table 8 shows the results obtained by the SGR method, that is, χ^2 and the corresponding p-values at every $C(j)$ ($j = 1, ..., |C|$) and suggests that $C(3)$ is independent of D and redundant compared to those of the other condition attributes, and should be removed from the viewpoint of the reduct.

Table 8. Results of global reduct for Car Evaluation dataset: D vs. $C(j)$ ($j = 1, ..., 6$)

		$C(1)$	$C(2)$	$C(3)$	$C(4)$	$C(5)$	$C(6)$
D	χ^2	189.24	142.94	10.38	371.34	53.28	479.32
	p-value	5.92E-36	2.54E-26	3.20E-01	4.04E-77	1.03E-09	2.39E-100

Table 9. Results of local reduct for Car Evaluation dataset: $D = \ell$ ($\ell = 1, ..., 4$) vs. $C(j)$ ($j = 1, ..., 6$)

		$C(1)$	$C(2)$	$C(3)$	$C(4)$	$C(5)$	$C(6)$
$D = 1$	χ^2	22.94	19.24	2.58	111.00	8.81	118.81
	p-value	4.16E-05	2.44E-04	4.62E-01	7.88E-25	1.22E-02	1.59E-26
$D = 2$	χ^2	11.77	10.77	3.19	192.56	6.52	194.25
	p-value	8.21E-03	1.30E-02	3.64E-01	1.53E-42	3.85E-02	6.59E-43
$D = 3$	χ^2	84.33	84.33	0.39	34.70	0.26	36.26
	p-value	3.61E-18	3.61E-18	9.42E-01	2.92E-08	8.78E-01	1.34E-08
$D = 4$	χ^2	70.20	28.60	4.23	33.08	37.69	130.00
	p-value	3.87E-15	2.72E-06	2.38E-01	6.57E-08	6.53E-09	5.90E-29

Table 10. Examples of contigency table and χ^2 test by local reducts

(a) $D = 1$ vs. $C(j)$ ($j = 1, ..., 6$)

$V_{C(i)}$	$C(1)$	$C(2)$	$C(3)$	$C(4)$	$C(5)$	$C(6)$
1	**360**	**360**	326	**576**	450	**576**
2	324	314	300	312	392	357
3	268	268	292	322	368	277
4	258	268	292	–	–	–
χ^2	22.94	19.24	2.58	111.00	8.81	118.81
p-value	4.16E-05	2.44E-04	4.62E-01	7.88E-25	1.22E-02	1.59E-26

(b) $D = 4$ vs. $C(j)$ ($j = 1, ..., 6$)

$V_{C(i)}$	$C(1)$	$C(2)$	$C(3)$	$C(4)$	$C(5)$	$C(6)$
1	0	0	10	0	0	0
2	0	13	15	30	25	0
3	26	26	20	**35**	**40**	**65**
4	**39**	**26**	20	–	–	–
χ^2	70.20	28.60	4.23	33.08	37.69	130.00
p-value	3.87E-15	2.72E-06	2.38E-01	6.57E-08	6.53E-09	5.90E-29

Table 9 shows the results obtained by the SLR method, that is, χ^2 and the corresponding p-values at every $D = l$ ($l = 1, ..., 4$) and $C(j)$ ($j = 1, ..., |C|$) and suggests the following:

(1) $C(3)$ is commonly redundant at $D = l$ ($l = 1, ..., 4$), which coincides with the results by the SGR method.
(2) With regard to the if-then rule of $D = 1$, $C(5)$ is redundant. In the same way, $C(2)$ and $C(5)$ at $D = 2$ are, and $C(5)$ at $D = 3$ is.

Table 11. Examples of estimated rules by use of the local reduct results in Table 10

Rule No.	examples of estimated rules $(C(1)\,C(2)\,C(3)$ $C(4)\,C(5)\,C(6)\,D)$	distribution of decision values (n_1, n_2, n_3, n_4)	p-value(χ^2)	accuracy	coverage
1	(0000011)	(576, 0, 0, 0)	3.58E-53(246.59)	1.00	0.476
2	(0001001)	(576, 0, 0, 0)	3.58E-53(246.59)	1.00	0.476
3	(0001011)	(192, 0, 0, 0)	1.03E-17(82.20	1.00	0.157
4	(1000001)	(360, 72, 0, 0)	6.47E-11(50.43)	0.83	0.296
5	(0100001)	(360, 72, 0, 0)	6.47E-11(50.43)	0.83	0.296
6	(1000011)	(144, 0, 0, 0)	2.61E-13(61.64)	1.00	0.119
7	(0100011)	(144, 0, 0, 0)	2.61E-13(61.64)	1.00	0.119
8	(1001001)	(144, 0, 0, 0)	2.61E-13(61.64)	1.00	0.119
9	(0101001)	(144, 0, 0, 0)	2.61E-13(61.64)	1.00	0.119
10	(0000034)	(277, 204, 30, 65)	2.24E-37(173.49)	0.113	1.000
11	(0000304)	(368, 144, 24, 40)	1.24E-4(20.65)	0.069	0.615
12	(0000334)	(88, 64, 0, 40)	1.85E-39(183.14)	0.208	0.615
13	(4000034)	(52, 33, 20, 39)	1.24E-57(267.21)	0.27	0.600
14	(0400034)	(52, 46, 20, 26)	7.32E-31(143.30)	0.18	0.400
15	(4000334)	(16, 8, 0, 24)	–	0.50	0.369
16	(4403334)	(0, 0, 0, 4)	–	1.00	0.062

Table 10 shows examples of the contingency tables of $D = 1$ (a) and $D = 4$ (b), and their χ^2 by the SLR method, and suggests the following knowledge:

(1) With regard to the if-then rules of $D = 1$ (see (a)),the frequencies of $C(1) = 1$, $C(2) = 1$, $C(4) = 1$ and $C(6) = 1$ are distinctively high, which is statistically confirmed by their p-values. Accordingly, the if-then rules of $D = 1$ are supposed to be constructed by the combinations of the set of $\{C(1) = 1, C(2) = 1, C(4) = 1$ and $C(6) = 1\}$ as shown in the knowledge obtained in 4.2.

(2) In the same way, the if-then rules of $D = 4$ are supposed to be constructed by the combinations of the set of $\{C(1) = 4, C(2) = 4, C(4) = 3, C(5) = 3$ and $C(6) = 3\}$ (see (b)).

Table 11 shows the trying rules (Rule 1-9 for $D = 1$, Rule 10-16 for $D = 4$) based on the knowledge obtained above (1) and (2). Here Rule 15 and 16 do not have p-values since they do not satisfy the test condition $n\hat{p}(j, k) \geq 5$ (for $D = 4$, $n \geq \frac{5}{0.04} = 125$). Rule 1 and Rule 2 can be selected as one of the proper rules of $D = 1$ since their p-values are extremely high and their indexes of accuracy and coverage are in good condition compared to the other trying rules. To express those rules by use of the original notation in Table 7, if person=2 $\bigvee$ safety=low then class=unacc, is obtained with the coverage=$(576 + 576 - 192)/1210 \approx 0.793$ and accuracy=1.0.

In the same way, the rule that if buying=low $\bigwedge$ safety=high then class=vgood (Rule 13) is thought to be proper since the p-value is the best and the indexes of accuracy and coverage are moderate among the trying rules for $D = 4$ although the number of sample data for $D = 4$ is small compared to that for $D = 1$. Both rules estimated coincide with our common sense and the validity and usefulness of the SGR and SLR method have been experimentally confirmed through the use of an open dataset.

6 Conclusions

The Rough Sets theory has been used for inducing if-then rules from the decision table and clarifying the structure of rating and/or knowledge hidden behind the dataset. The first step in inducing the rules is to find reducts of the condition attributes. This paper retested the conventional reduct methods LEM1 [2] and DMM [3] by a simulation experiment after summarizing the conventional rough sets theory and pointed out their problems. Then, this paper proposed a new reduct method to overcome the problems of the conventional method from the view of STRIM [8–11]. Specifically, the SGR and SLR methods were proposed and their validity and usefulness were confirmed by a simulation experiment and application to an open dataset of UCI for machine learning. The SLR method should be recognized to be particularly useful for not only reducts of condition attributes but also inducing if-then rules.

References

1. Pawlak, Z.: Rough sets. Int. J. Inf. Comput. Sci. **11**(5), 341–356 (1982)
2. Grzymala-Busse, J.W.: LERS – A system for learning from examples based on rough sets. In: Słowiński, R. (ed.) Handbook of Applications and Advances of the Rough Sets Theory. Theory and Decision Library, vol. 11, pp. 3–18. Kluwer Academic Publishers, Netherlands (1992)
3. Skowron, A., Rauser, C.M.: The Discernibility Matrix and Functions in Information Systems. In: Słowiński, R. (ed.) Handbook of Application and Advances of Rough Sets Theory. Theory and Decision Library, vol. 11, pp. 331–362. Kluwer Academic Publishers, Netherlands (1992)
4. Pawlak, Z.: Rough set fundamentals; KFIS Autumn Coference Tutorial, pp. 1–32 (1996)
5. Ślęzak, D.: Various approaches to reasoning with frequency based decision reducts: a survey. In: Polkowski, L., Tsumoto, S., Lin, T.Y. (eds.) Rough Set Method and Applications, vol. 56, pp. 235–285. Physical-Verlag, Heidelberg (2000)
6. Bao, Y.G., Du, X.Y., Deng, M.G., Ishii, N.: An efficient method for computing all reducts. Trans. Jpn. Soc. Artif. Intell. **19**(3), 166–173 (2004)
7. Asunction, A., Newman, D.J.: UCI Machine Learning Repository, University of California, School of Information and Computer Science, Irvine (2007). http://www.ics.edu/~mlearn/MlRepository.html

8. Matsubayashi, T., Kato, Y., Saeki, T.: A new rule induction method from a decision table using a statistical test. In: Li, T., Nguyen, H.S., Wang, G., Grzymala-Busse, J., Janicki, R., Hassanien, A.E., Yu, H. (eds.) RSKT 2012. LNCS, vol. 7414, pp. 81–90. Springer, Heidelberg (2012)
9. Kato, Y., Saeki, T., Mizuno, S.: Studies on the necessary data size for rule induction by STRIM. In: Lingras, P., Wolski, M., Cornelis, C., Mitra, S., Wasilewski, P. (eds.) RSKT 2013. LNCS, vol. 8171, pp. 213–220. Springer, Heidelberg (2013)
10. Kato, Y., Saeki, T., Mizuno, S.: Considerations on rule induction procedures by STRIM and their relationship to VPRS. In: Kryszkiewicz, M., Cornelis, C., Ciucci, D., Medina-Moreno, J., Motoda, H., Raś, Z.W. (eds.) RSEISP 2014. LNCS, vol. 8537, pp. 198–208. Springer, Heidelberg (2014)
11. Kato, Y., Saeki, T., Mizuno, S.: Proposal of a statistical test rule induction method by use of the decision table. Appl. Soft Comput. **28**, 160–166 (2015). Elsevier
12. Kotu, V., Deshpande, B.: Predictive Analytics and Data Mining, 1st edn. Elsevier, Amsterdam (2014)
13. Walpole, R.E., Myers, R.H., Myers, S.L., Ye, K.: Probability and Statistics for Engineers and Scientists, 8th edn. Pearson Prentice Hal, New Jersey (2007)

Computation of Cores in Big Datasets: An FPGA Approach

Maciej Kopczynski[✉], Tomasz Grzes, and Jaroslaw Stepaniuk

Faculty of Computer Science, Bialystok University of Technology,
Wiejska 45A, 15-351 Bialystok, Poland
{m.kopczynski,t.grzes,j.stepaniuk}@pb.edu.pl
http://www.wi.pb.edu.pl

Abstract. In this paper we propose the FPGA and softcore CPU supported device for performing core calculation for large datasets using rough set methods. Presented architecture has been tested on two real datasets by downloading and running presented solution inside FPGA. Sizes of the datasets were in range 1 000 to 10 000 000 objects. Results show the big acceleration in terms of the computation time using hardware supporting core generation unit.

Keywords: Rough sets · FPGA · Hardware · Core

1 Introduction

The theory of rough sets has been developed in the eighties of the twentieth century by Prof. Z. Pawlak. Rough sets are used as a tool for data analysis and classification including extraction of important characteristics that describe the objects.

Field Programmable Gate Arrays (FPGAs) are a group of integrated circuits, whose functionality is defined by the user. Hardware description language, such as VHDL, allows designers to speed up describing functional properties of the digital system [1].

At the moment some of the hardware implementations of rough set methods were prepared by different authors in the world. Few papers are focused on design or idea of such solutions, while others show the hardware-obtained time results. The idea of sample processor generating decision rules from decision tables was described in [11]. Architecture of rough set processor based on cellular networks was described in [9] and was presented in [7]. Concept of hardware device capable of minimizing the large logic functions created on the basis of discernibility matrix was developed in [3]. More detailed summary of the existing ideas and hardware implementations of rough set methods can be found in [4] and in [16]. Previous authors' research results focused on this subject can be found in [2,5,6,15].

The paper is organized as follows. In Sect. 2 some information about the notion of core and datasets used during research are provided. The Sect. 3 focuses on description of hardware solution, while Sect. 4 is devoted to the experimental results.

© Springer International Publishing Switzerland 2015
D. Ciucci et al. (Eds.): RSKT 2015, LNAI 9436, pp. 153–163, 2015.
DOI: 10.1007/978-3-319-25754-9_14

2   Introductory Information

2.1   Cores in the Rough Set Theory

In decision table some of the condition attributes may be superfluous (redundant in other words). This means that their removal cannot worsen the classification. The set of all indispensable condition attributes is called the core. One can also observe that the core is the intersection of all decision reducts – each element of the core belongs to every reduct. Thus, in a sense, the core is the most important subset of condition attributes. None of its elements can be removed without affecting the classification power of all condition attributes. In order to compute the core we can use discernibility matrix. The core is the set of all single element entries of the discernibility matrix.

A much more detailed description of the concept of the core can be found, for example, in the article [12] or in the book [14].

2.2   Algorithm CORE-DDM for Generating Core Using Discernibility Matrix

Below one can find pseudocode for simple algorithm CORE-DDM (**CORE Direct Discernibility Matrix**) for calculating core using discernibility matrix.

INPUT: discernibility matrix DM
OUTPUT: core $C \subseteq A$

```
1: C ← ∅
2: for x ∈ U do
3:    for y ∈ U do
4:       if |DM(x, y)| = 1 and DM(x, y) ∉ C then
5:          C ← C ∪ DM(x, y)
6:       end if
7:    end for
8: end for
```

The main concept of algorithm CORE-DDM is based on a property of singleton i.e. cell from discernibility matrix consisted of the only one attribute. This property tells that any singleton cannot be removed without affecting the classification power.

Input for the algorithm is the discernibility matrix DM. Output is core C as a subset of condition attributes set denoted as A. Core is initialized as empty set in line 1. Two loops in lines 2 and 3 iterates over all objects (denoted as U) in discernibility matrix. Condition instruction in line 4 checks if matrix cell contains only one attribute. If so, then this attribute is added to the core C.

2.3   Basic Notions for CORE-CT Algorithm

Basing on [10] some background definitions for better understanding pseudocode of algorithm in the following subsection will presented.

Let $\mathbb{A} = (U, A, d)$ be a decision system. Conflict measure $conflict(X)$ of $X \subset U$ is the number of pairs $(x, y) \in X \times X$ of objects from different decision classes.

Counting table CT of a set of attributes B is the two-dimensional array:

$$CT(B) = [n_{v,k}]_{v \in INF(B), k \in V_d} \tag{1}$$

where: $n_{v,k} = card(\{x \in U : INF_B(x) = v \wedge dec(x) = k\})$. $CT(B)$ is a collection created by counting tables for equivalence classes of the indiscernibility relation $IND(B)$.

Discernibility measure relative to a set of attributes B for a given counting table can be calculated by:

$$disc_d(B) = \frac{1}{2} \sum_{v \neq v', k \neq k'} n_{v,k} \cdot n_{v',k'} \tag{2}$$

Discernibility measure can be understood as a number of unresolved conflicts in accordance to the set of attributes B.

Discernibility measure can also be calculated using following formula:

$$disc_d(B) = confict(U) - \sum_{x \in U/IND(B)} conflict([x]_{IND(B)}) \tag{3}$$

It can be shown that attribute $a \in A$ is a core attribute if and only if:

$$disc_d(A \setminus \{a\}) < disc_d(A) \tag{4}$$

For further details and examples please refer to [10].

2.4 Algorithm CORE-CT for Generating Core Using Counting Table

Below one can find pseudocode for an algorithm of calculating core $C \subseteq A$ based on idea of counting tables presented by H. S. Nguyen in [10]. We called this algorithm CORE-CT (**CORE** **C**ounting **T**ables). Let $DT = (U, A \cup \{d\})$ be a decision table, where U is a set of objects, A is a set of condition attributes and d is a decision attribute.

INPUT: decision table $DT = (U, A \cup \{d\})$
OUTPUT: core $C \subseteq A$
1: $C \leftarrow \emptyset$
2: **for** $a \in A$ **do**
3:    $CT_{total} \leftarrow$ set vector values to 0
4:    $CT_{row} \leftarrow$ set vector values to 0
5:    $confs \leftarrow 0$
6:    sort DT over values of $A \setminus \{a\}$
7:    **for** $x \in U$ **do**

```
 8:        for b ∈ A \ {a} do
 9:            if b(x_prev) ≠ b(x) then
10:                difference ← true
11:            end if
12:        end for
13:        if difference = false then
14:            CT_row ← increment CT_row vector cells basing on d(x)
15:            CT_total ← increment CT_total vector cells basing on d(x)
16:        else
17:            confs_row ← calculate conflicts number basing on CT_row
18:            confs ← confs + confs_row
19:            CT_row ← set vector values to 0
20:            difference ← false
21:        end if
22:        x_prev ← x
23:    end for
24:    disc_d(A) ← calculate conflicts basing on CT_total
25:    disc_d(A \ {a}) ← disc_d(A) − confs
26:    if disc_d(A \ {a}) < disc_d(A) then
27:        C ← C ∪ {a}
28:    end if
29: end for
```

The main concept of this algorithm is based on a counting tables mentioned in Sect. 2.3. Input for the algorithm is the decision table DT. Output is core C consisting of a subset of condition attributes set denoted as A. Core is initialized as empty set in line 1. Main loop in line 2 iterates over all condition attributes. In lines 3 and 4 two temporary vectors CT_{total} and CT_{row} are set to 0 in every iteration. Size of the vectors are equal to the number of values on decision attribute d. Main purpose of these vectors is storing number of objects belonging to currently processed equivalence class for each value of decision attribute (decision class). CT_{row} corresponds to the single row of exemplary counting table presented in [10], while CT_{total} stores the values of last row in previously mentioned table. Line 5 sets $confs$ variable to 0. This temporary variable stores total number of conflicts calculated from single CT_{row}. All data in decision table DT is sorted on all attributes in accordance to condition attributes set A except for attribute a. Two loops in lines 7 and 8 are responsible for iterating over all objects and all condition attributes (except for a) in data set. Comparison in line 9 sets variable $difference$ to $true$ if current (denoted x) and previous (denoted x_{prev}) objects belong to different equivalence classes. If current and previous objects are in the same class, then temporary CT_{total} and CT_{row} vectors cells are incremented in lines 13 to 15 depending on value of decision attribute d for current object x. If objects belong to different classes, then number of conflicts are calculated using CT_{total} and CT_{row} vectors in lines 16 to 18. CT_{row} vector and $difference$ variable are set to their initial states (lines 19 to 20). Lines 24 and 25 describe calculation of $disc_d$ values basing on CT_{total} vector (directly) and CT_{row} vector (indirectly by $confs$ variable). Line 26 is the decision if current condition attribute a belongs to the core C.

2.5   Data to Conduct Experimental Research

In this paper, we present the results of the conducted experiments using two datasets: Poker Hand Dataset (created by Robert Cattral and Franz Oppacher) and data about children with insulin-dependent diabetes mellitus (type 1).

First dataset was obtained from UCI Machine Learning Repository [8]. Each of 1 000 000 records is an example of a hand consisting of five playing cards drawn from a standard deck of 52. Each card is described using two attributes (suit and rank), for a total of 10 predictive attributes. There is one decision attribute that describes the "Poker Hand". Decision attribute describes 10 possible combinations of cards in descending probability in the dataset: nothing in hand, one pair, two pairs, three of a kind, straight, flush, full house, four of a kind, straight flush, royal flush.

Diabetes mellitus is a chronic disease of the body's metabolism characterized by an inability to produce enough insulin to process carbohydrates, fat, and protein efficiently. Treatment requires injections of insulin. Twelve condition attributes, which include the results of physical and laboratory examinations and one decision attribute (microalbuminuria) describe the database. The database consisting of 107 objects is shown at the end of the paper [13]. An analysis can be found in Chap. 6 of the book [14].

The Poker Hand database was used for creating smaller datasets consisting of 1 000 to 500 000 of objects by selecting given number of first rows of original dataset. The Diabetes database was base for generating bigger datasets consisting of 1 000 to 10 000 000 of objects. Bigger datasets were created by duplicating original dataset given amount of times. Numerical values were discretized. Each attributes' value was encoded using four bits for both datasets. Every single object was described on 44 bits for Poker Hand and 52 bits for Diabetes. To fit to memory boundaries in both cases, objects descriptions had to be extended to 64 bits words and filling unused attributes with 0.

3   Hardware Implementation

Core calculation process for large data sets based on algorithm CORE-CT described in Sect. 2.4 was implemented using mixed hardware and software approach (Fig. 1). The main goal of hardware block is to calculate the value of discernibility measure $disc_d$. This block utilizes hardware DSP blocks from FPGA circuit.

Hardware module is supported by the software running in a softcore processor which goal is to:

- read and write data to hardware module;
- prepare data from input data set to transformed form as it was briefly described in Sect. 2.5;

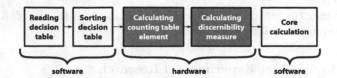

Fig. 1. Workflow of hardware implementation of core calculation algorithm

– perform operations on sets;
– control overall operation;

This type of control software has been implemented using NIOS II softcore processor, that can be instantiated inside FPGA chip. Subsequent parts of decision table are stored in FPGA built-in RAM memories (MLAB, M9k and M144k), while whole dataset is stored in DDR2 RAM module. MLAB blocks are synchronous, dual-port memories with configurable organisation 32 x 20 or 64 x 10. M9k and M144k blocks are synchronous, true dual-port memory blocks with registered inputs and optionally registered outputs with many possible configurable organisations. These blocks give a wide possibity of preparing memories capable of storing almost every type of the objects (words) - from small ones to big ones.

Diagram of the discernibility measure $disc_d$ calculation module is shown on Fig. 2. The performed process is based on algorithm CORE-CT.

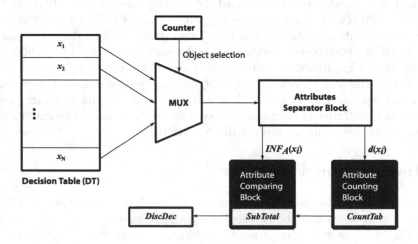

Fig. 2. Diagram of discernibility measure $disc_d$ calculation module

Module consists of the blocks listed below:

– **Decision Table (DT)** – a memory block that stores the dataset objects; every cell (row) consists of one object; structure of every element from decision table must be known by *Attributes Separator Block*;

- **MUX (Multiplexer)** – multiplexer for selecting one row (object) from *Decision Table*; elements are indexed by the value from *Counter*;
- **Counter** – counter for indexing the elements from a decision table.
- **Attributes Separator Block** – block that separates condition and decision attributes values from currently selected object;
- .**Attribute Comparing Block** – block that compares current value of condition attributes values with previous ones and decides whether to add calculated counting table entry value to *discDec* value;
- **Attribute Counting Block** – block that evaluates value of the counting table entry.

Attribute Counting Block is a block for evaluating the counting table entry value using the algorithm CORE-CT described in Sect. 2.4. Diagram of this block is shown on Fig. 3. Elements of this module are:

- **Decoder** – decoder for providing signal that initiates incrementing the value of counter indicated by the decision attribute value;
- **Counter0, Counter1, ..., CounterN** – counters for calculating the number of indiscerned objects for N different values of decision attribute;
- **Count** – adder for calculating the sum of each $Counter_i$ values;
- **Sqrt0, Sqrt1, ..., SqrtN** – registers for storing the values of square root of the values in counters $Counter0, Counter1, ..., CounterN$ respectively;
- **SumSqrt** – register for storing sum of square roots;
- **SqrtSum** – register for storing square root of sum for each counter value;
- **CT** – value of counting table entry.

4 Experimental Results

The results of the software implementation were obtained using a PC equipped with an 8 GB RAM and 4-core Intel Core i7 3632QM with maximum 3.2 GHz

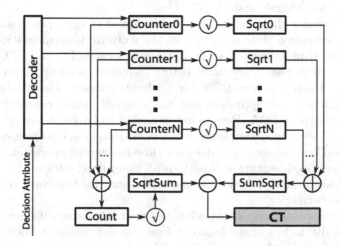

Fig. 3. Diagram of *Attribute Counting Block*

in Turbo mode clock speed running Windows 7 Professional operational system. The source code of application was compiled using the GNU GCC 4.8.1 compiler and was implemented in C language.

Quartus II 13.1 was used for design and implementation of the hardware using VHDL language. Synthesized hardware blocks were tested on TeraSIC DE-3 equipped with Stratix III EP3SL150F1152C2N FPGA chip. FPGA clock running at 50 MHz for the sequential parts of the project was derived from development board oscillator.

Software for NIOS II softcore processor was implemented in C language using NIOS II Software Build Tools for Eclipse IDE.

Timing results were obtained using LeCroy waveSurfer 104MXs-B (1 GHz bandwidth, 10 GS/s) oscilloscope. For longer times, hardware time measurement units instantiated inside FPGA were used.

It should be noticed, that PCs clock is $\frac{clk_{PC}}{clk_{FPGA}} = 64$ times faster than development boards clock source.

Algorithm CORE-CT described in Sect. 2.4 was used both for software and hardware implementation. In current version of hardware implementation, authors used data which was preprocessed on PC in terms of sorting over condition attributes. Presented results show the times for sorting operation (t_S^{sort}) and times and speed-up factors for pure core calculation operations assuming that data was sorted (t_S^{core}, t_H). Table 1 presents the results of the time elapsed for hardware and software solution using Poker Hand and Diabetes datasets. Last two columns in table describe the speed-up factor without (C) and with (C_{clk}) taking clock speed difference between PC and FPGA into consideration. Abbreviations in objects number are: $k = 10^3$, $M = 10^6$.

FPGA resources utilization is independent of the input dataset size and takes 18 622 used of 113 600 Logical Elements (LEs) total available.

Figure 4 presents a graph showing the relationship between the number of objects and execution time for hardware and software solution for both datasets. Axes have the logarithmic scale.

Presented results show big increase in the speed of data processing. Hardware module execution time compared to the software implementation is 3 to 10 times faster. If we take clock speed difference between PC and FPGA under consideration, these results are much better - average speed-up factor is about 500 for Poker Hand and about 200 for Diabetes dataset. The hardware core calculation unit was not optimized and was relatively small comparing to the capacity of available FPGA. Results are expected to be few times better after optimization. If we take sorting times under considerations, the speed-up factors are not so good because sorting is the most time-consuming operation. As it was previously mentioned, sorting was performed using pure software implementation. After transformation of sorting algorithm into its hardware version, the results will be better.

Hardware processing times for both datasets are the same - it doesn't matter what is the width in bits of single object from dataset, unless it fits in assumed memory boundary. Hardware processing unit takes the same time to finish the

Table 1. Comparison of execution time between hardware and software implementation for Poker Hand and Diabetes datasets using CORE-CT algorithm

Objects	Software - t_S^{sort}	Software - t_S^{core}	Hardware - t_H	$C=\dfrac{t_S^{core}}{t_H}$	$C_clk=\dfrac{64t_S^{core}}{t_H}$
—	[ms]	[ms]	[ms]	—	—
Poker Hand dataset					
1 k	10	2	0.249	8.52	545.08
2.5 k	23	5	0.589	7.89	504.97
5 k	43	9	1.12	8.00	512.18
10 k	91	17	2.25	7.49	479.65
25 k	248	49	5.54	8.76	560.35
50 k	536	100	13	7.96	509.15
100 k	1 150	218	27	8.21	525.39
250 k	3 141	589	57	10.26	656.71
500 k	6 697	1 194	112	10.66	682.39
1 M	14 320	2 440	224	10.89	697.14
Diabetes dataset					
1 k	12	1	0.249	4.01	256.82
2.5 k	28	2	0.589	3.39	217.17
5 k	53	3	1.12	3.56	227.72
10 k	113	6	2.25	2.66	170.36
25 k	309	19	5.54	3.43	219.34
50 k	667	36	13	2.86	182.86
100 k	1 430	81	27	3.05	194.89
250 k	3 906	213	57	3.71	237.49
500 k	8 329	386	112	3.45	220.57
1 M	17 810	700	224	3.13	200.00
2.5 M	48 172	2 088	558	3.74	239.23
5 M	101 527	3 549	1 113	3.19	204.08
10 M	213 995	7 703	2 227	3.46	221.33

calculation for every object size, because it always performs the same type of operation.

For software execution time comparing Poker Hand and Diabetes datasets it can be noticed, that number of attributes and characteristic of the dataset have impact on computation time. Diabetes have more attributes, but because of the cloned rows it takes less time to compute core values after the sorting.

Let comparison of attributes' value between two objects be an elementary operation. k denotes number of conditional attributes and n is the number of objects in decision table. Computational complexity of software implementation for the core calculation is $\Theta(kn\log n + k^2 n)$ according to algorithm CORE-CT

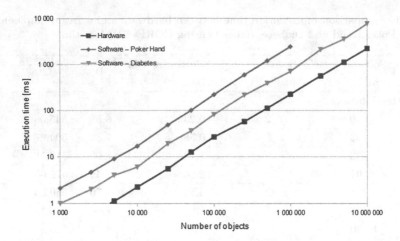

Fig. 4. Relation between number of objects and calculation time for hardware and software implementation using CORE-CT algorithm for both datasets

shown in Sect. 2.4. Algorithm chosen for sorting is merge-sort. Using hardware implementation, complexity of core calculation is $\Theta(knlogn + kn)$. The k from the second part of the formula is missing, because our solution performs comparison between all attributes in $\Theta(1)$ - all attributes values between two objects are compared in single clock cycle. Additionally, core module performs comparisons between many objects at time. In most cases $k << n$, so we can say, that computational complexity for software and hardware implementations are the same.

5   Conclusions and Future Research

The hardware implementation is the main direction of using scalable rough sets methods in real time solutions. As it was presented, performing core calculations using hardware implementations gives us a big acceleration in comparison to software solution.

Core hardware calculation unit was not optimized for performance in this paper. Processing time can be substantially reduced by increasing FPGA clock frequency, modifying control unit and introducing triggering on both edges of clock signal. One of the most important conclusions is the need of preparation hardware implementation for sorting algorithm. It will be developed by our team as the next step of our research plan. Hardware sorting can be used in many different areas of rough sets and data analysis, eg. discretization.

Other research can be focused on optimization of presented solution. Different sizes of hardware core calculation unit should be checked, as well as results related to performing the calculations in parallel by multiplying hardware modules.

Acknowledgements. The research is supported by the Polish National Science Centre under the grant 2012/07/B/ST6/01504 (Jaroslaw Stepaniuk, Maciej Kopczynski) and by the scientific grant S/WI/3/2013 (Tomasz Grzes).

References

1. Bezerra, E., Lettnin, D.V.: Synthesizable VHDL Design for FPGAs. Springer, New York (2014)
2. Grześ, T., Kopczyński, M., Stepaniuk, J.: FPGA in rough set based core and reduct computation. In: Lingras, P., Wolski, M., Cornelis, C., Mitra, S., Wasilewski, P. (eds.) RSKT 2013. LNCS, vol. 8171, pp. 263–270. Springer, Heidelberg (2013)
3. Kanasugi, A., Yokoyama, A.: A basic design for rough set processor. In: The 15th Annual Conference of Japanese Society for Artificial Intelligence (2001)
4. Kopczyński, M., Stepaniuk, J.: Hardware implementations of rough set methods in programmable logic devices. In: Skowron, A., Suraj, Z. (eds.) Rough Sets and Intelligent Systems - Professor Zdzisław Pawlak in Memoriam. ISRL, vol. 43, pp. 309–321. Springer, Heidelberg (2013)
5. Kopczyński, M., Grześ, T., Stepaniuk, J.: FPGA in rough-granular computing : reduct generation. In: WI 2014 : The 2014 IEEE/WCI/ACM International Joint Conferences on Web Intelligence, vol. 2, pp. 364–370. IEEE Computer Society, Warsaw (2014)
6. Kopczynski, M., Grzes, T., Stepaniuk, J.: Generating core in rough set theory: design and implementation on FPGA. In: Kryszkiewicz, M., Cornelis, C., Ciucci, D., Medina-Moreno, J., Motoda, H., Raś, Z.W. (eds.) RSEISP 2014. LNCS, vol. 8537, pp. 209–216. Springer, Heidelberg (2014)
7. Lewis, T., Perkowski, M., Jozwiak, L.: Learning in hardware: architecture and implementation of an FPGA-based rough set machine. In: 25th Euromicro Conference (EUROMICRO 1999), vol. 1, p. 1326 (1999)
8. Lichman, M.: UCI Machine Learning Repository, Irvine, CA: University of California, School of Information and Computer Science (2013). http://archive. ics.uci.edu/ml
9. Muraszkiewicz, M., Rybiński, H.: Towards a parallel rough sets computer. In: Rough Sets, Fuzzy Sets and Knowledge Discovery: Proceedings of the International Workshop on Rough Sets and Knowledge Discovery, RSKD pp. 434–443 (1994)
10. Nguyen, H.S.: Approximate boolean reasoning: foundations and applications in data mining. In: Peters, J.F., Skowron, A. (eds.) Transactions on Rough Sets V. LNCS, vol. 4100, pp. 334–506. Springer, Heidelberg (2006)
11. Pawlak, Z.: Elementary rough set granules: Toward a rough set processor. In: Rough-Neurocomputing: Techniques for Computing with Words. Cognitive Technologies, pp. 5–14. Springer, Berlin (2004)
12. Pawlak, Z., Skowron, A.: Rudiments of rough sets. Inf. Sci. **177**(1), 3–27 (2007)
13. Stepaniuk, J.: Knowledge discovery by application of rough set models. In: Polkowski, L., Tsumoto, S., Lin, T.Y. (eds.) Rough Set Methods and Applications: New Developments in Knowledge Discovery in Information Systems, pp. 137–233. Physica-Verlag, Heidelberg (2000)
14. Stepaniuk, J.: Rough-Granular Computing in Knowledge Discovery and Data Mining. Springer, New York (2008)
15. Stepaniuk, J., Kopczyński, M., Grześ, T.: The first step toward processor for rough set methods. Fundamenta Informaticae **127**, 429–443 (2013)
16. Tiwari, K.S., Kothari, A.G.: Design and implementation of rough set algorithms on FPGA: a survey. (IJARAI) Int. J. Adv. Res. Artif. Intell. **3**(9), 14–23 (2014)

Knowledge Spaces and Reduction of Covering Approximation Spaces

Tong-Jun Li[1,2]($\boxtimes$), Shen-Ming Gu[1,2], and Wei-Zhi Wu[1,2]

[1] School of Mathematics, Physics and Information Science, Zhejiang Ocean
University, Zhoushan 316022, Zhejiang, China
{litj,gsm,wuwz}@zjou.edu.cn
[2] Key Laboratory of Oceanographic Big Data Mining & Application of Zhejiang
Province, Zhoushan 316022, Zhejiang, China

Abstract. Theory of covering rough sets is one kind of effective methods
for knowledge discovery. In Bonikowski covering approximation spaces,
all definable sets on the universe form a knowledge space. This paper
focuses on the theoretic study of knowledge spaces of covering approximation spaces. One kind of dependence relations among covering approximation spaces is introduced, the relationship between the dependence
relation and lower and upper covering approximation operators are discussed in detail, and knowledge spaces of covering approximation spaces
are well characterized by them. By exploring the dependence relation
between a covering approximation space and its sub-spaces, the notion
of the reduction of covering approximation spaces is induced, and the
properties of the reductions are investigated.

Keywords: Covering rough sets · Knowledge spaces · Dependence ·
Reduction

1 Introduction

Rough set theory [10], proposed by Pawlak in the early 1980s, is an important
mathematical tool to deal with inexact, uncertain and insufficient information in
information systems. Equivalence relations are basis of Pawlak rough set models.
By means of lower and upper approximation operators defined by equivalence
relations, the knowledge hidden in information systems can be expressed in the
form of decision rules. The theory of rough sets has found successful applications in attribute reduction, machine learning, pattern recognition and image
processing from large data sets [11,19,21].

The equivalence relations in Pawlak rough set models limit their application. Thus, many generalized rough set models have been proposed in various
ways in recent years, for example, probabilistic rough sets [17], binary relation
based rough sets [12], covering rough sets [1,7,23], etc. As an extension of the
Pawlak rough sets, covering rough sets have obtained much attention in many
domains including machine learning and uncertainty reasoning. Actually, in the

D. Ciucci et al. (Eds.): RSKT 2015, LNAI 9436, pp. 164–174, 2015.
DOI: 10.1007/978-3-319-25754-9_15

view of granular computing [16], a covering of the universe can be considered as a granular space. Zhu and Wang also presented an example application of covering rough sets in [22]. Various types of covering rough set models are proposed [6,18,25]. Relationship between covering rough sets and other types of rough sets is discussed [13,24]. The order relations among six types of covering rough sets are investigated [8]. Some authors have paid attention to reduction of covering rough sets [3,15,23].

In a covering approximation space, lower or upper approximations of all subsets of the universe consist of a knowledge space, and the covering of the universe is the base of the knowledge space. On the same universe, different coverings may generalize different knowledge spaces. The object of this paper is to reveal the mathematical structures of knowledge spaces of covering approximation spaces, by investigating relationship among covering approximation spaces, knowledge spaces and reductions of covering approximation spaces are discussed. The remainder of the paper is organized as follows: Sect. 2 reviews some basic notions and knowledge related to the work. Section 3 presents a dependence relation among covering approximation spaces, by means of the relation knowledge spaces of covering approximation spaces is characterized. With the obtained dependence relations of covering approximation spaces, in Sect. 4, reduction of covering approximation spaces is investigated. Section 5 presents a summary of conclusions.

2 Preliminaries

In this section, we review some basic notions about knowledge spaces and covering rough sets.

2.1 Knowledge Spaces and Interior Operators

Let U be a nonempty set called a universe of discourse. A *set system* S on U is a family of subsets of U [9]. For two set systems, S_1 and S_2, on U, if $S_1 \subseteq S_2$, then S_1 is referred to as a *set sub-system* of S_2. For a set system S, in the following, $\cup S$ represents the union of all members of S.

A set system C on U is called a *covering* of U if its members are nonempty and $\cup\{X|X \in C\} = U$ holds.

A set system K on U is called a *knowledge space* [4] on U, if for any $S \subseteq K$, $\cup S \in K$. For two knowledge spaces K_1 and K_2 on U, if $K_1 \subseteq K_2$, then K_1 is referred to as a *knowledge sub-space* of K_2. A knowledge space is also called a dual closure system [9] or a pro-topology [5] in the literature. In this paper, we use the terminology "knowledge space".

It should be noted that the empty set belongs to any knowledge space K, because for $\emptyset \subseteq K$ we have $\cup\emptyset = \emptyset$.

A knowledge space K on U is a complete lattice under the set inclusion relation $\subseteq$. For any $X_1, X_2 \in K$, their least upper bound or supremum is $X_1 \vee$

$X_2 = X_1 \cup X_2$, and their greatest lower bound or infimum is $X_1 \wedge X_2 = \cup \{X \in \mathcal{K} | X \subseteq X_1, X \subseteq X_2\}$.

An *interior operator* on ϕ [2] on U is a mapping on $\mathcal{P}(U)$ which satisfies the monotonicity ($X \subseteq Y \Rightarrow \phi(X) \subseteq \phi(Y)$), contraction ($\phi(X) \subseteq X$) and idempotence ($\phi(\phi(X)) = \phi(X)$) conditions.

There is a one-to-one correspondence between the set of all knowledge spaces on U and the set of all interior operators on $\mathcal{P}(U)$. For any interior operator ϕ on $\mathcal{P}(U)$, the corresponding knowledge space $\mathcal{K}_\phi$ is the set of all fixed points of ϕ, where a fixed point of ϕ is a subset $X \subseteq U$ with $\phi(X) = X$, which is also called an open set of ϕ. On the other hand, the interior operator $\phi_\mathcal{K}$ corresponding to knowledge space $\mathcal{K}$ is defined by $\phi_\mathcal{K}(X) = \cup \{Y \in \mathcal{K} | Y \subseteq X\}, \forall X \subseteq U$. That is, $\phi_\mathcal{K}(X)$ is the greatest open set in $\mathcal{K}$ contained in X.

For any set system $\mathcal{S} \subseteq \mathcal{P}(U)$, it can be verified that the set system $KS(\mathcal{S}) = \{\cup \mathcal{T} | \mathcal{T} \subseteq \mathcal{S}\}$ is a knowledge space on U, and the minimum knowledge space containing $\mathcal{S}$. Then $KS(\mathcal{S})$ is called the knowledge space generated by $\mathcal{S}$, and $\mathcal{S}$ is called the *basis* of $KS(\mathcal{S})$. It is easy to verify that if S_1 is a set sub-system of S_2 then $KS(\mathcal{S}_1)$ is a knowledge sub-space of $KS(\mathcal{S}_2)$.

The set of all knowledge spaces on U, denoted as $KS(U)$, is a complete lattice under inclusion relation $\subseteq$. For any $\mathcal{K}_1, \mathcal{K}_2 \in KS(U)$, the meet and the join of $\mathcal{K}_1$ and $\mathcal{K}_2$ are $\mathcal{K}_1 \cap \mathcal{K}_2$ and $\cap \{\mathcal{K} \in KS(U) | \mathcal{K}_1 \subseteq \mathcal{K}, \mathcal{K}_2 \subseteq \mathcal{K}\}$, respectively.

With respect to a set system $\mathcal{S}$ on U, it can be checked that $KS(\mathcal{S}) = \cap \{\mathcal{K} \in KS(U) | \mathcal{S} \subseteq \mathcal{K}\}$.

2.2   Covering Rough Approximations

Let U be a nonempty finite universe of discourse, and $\mathcal{C}$ a covering of U. Then the pair $(U, \mathcal{C})$ is called a *covering approximation space*. Based on Pawlak rough set models, various rough sets on covering approximation spaces have been defined [1,18,20,25]. In the following, we review one kind of rough approximations on covering approximation spaces proposed by Z. Bonikowski [1].

In various covering rough set models, almost all lower covering approximations of sets are defined as follows:

Definition 1. *Let $(U, \mathcal{C})$ be a covering approximation space and $X \subseteq U$. The family of sets, $\underline{S_\mathcal{C}}(X) = \{Y \in \mathcal{C} | Y \subseteq X\}$, is called the lower covering approximation set system of X, and the set, $\underline{\mathcal{C}}(X) = \cup \underline{S_\mathcal{C}}(X)$, is called the lower covering approximation of X.*

Seven types of upper covering approximation operators are presented in [15], the first upper covering approximation operator mentioned coincides with that proposed by Z. Bonikowski in [1].

Definition 2. *Let $(U, \mathcal{C})$ be a covering approximation space. For any $x \in U$, the set system*

$$Md_\mathcal{C}(x) = \{X \in \mathcal{C} | x \in X, \forall Y \in \mathcal{C}(x \in Y, Y \subseteq X \Rightarrow Y = X)\}$$

is called the minimal description of x in $(U, \mathcal{C})$.

Obviously, $Md_{\mathcal{C}}(x) \neq \emptyset$ for all $x \in U$.

Definition 3. *Let $(U, \mathcal{C})$ be a covering approximation space and $X \subseteq U$. The family of sets, $BN_{\mathcal{C}}(X) = \cup\{Md_{\mathcal{C}}(x)|x \in X - \underline{\mathcal{C}}(X)\}$, is called the boundary covering approximation set system of X.*
  The family of sets, $\overline{S_{\mathcal{C}}}(X) = S_{\underline{\mathcal{C}}}(X) \cup BN_{\mathcal{C}}(X)$, is called the upper covering approximation set system of X.
  The set $\overline{\mathcal{C}}(X) = \cup\overline{S_{\mathcal{C}}}(X)$ is called the upper covering approximation of X.

The lower and upper covering approximations have the following properties:

Proposition 1. *[14] Let $(U, \mathcal{C})$ be a covering approximation space. The following properties can be satisfied:*

(1) *(Contraction)*: $\underline{\mathcal{C}}(X) \subseteq X$;
(2) *(Extension)*: $X \subseteq \overline{\mathcal{C}}(X)$;
(3) *(Normality)*: $\underline{\mathcal{C}}(\emptyset) = \emptyset$;
(4) *(Normality)*: $\overline{\mathcal{C}}(\emptyset) = \emptyset$;
(5) *(Co-normality)*: $\underline{\mathcal{C}}(U) = U$;
(6) *(Co-normality)*: $\overline{\mathcal{C}}(U) = U$;
(7) *(Monotonicity)*: $X \subseteq Y \Longrightarrow \underline{\mathcal{C}}(X) \subseteq \underline{\mathcal{C}}(Y)$;
(8) *(Idempotency)*: $\underline{\mathcal{C}}(X) = \underline{\mathcal{C}}(\underline{\mathcal{C}}(X))$;
(9) *(Idempotency)*: $\overline{\mathcal{C}}(X) = \overline{\mathcal{C}}(\overline{\mathcal{C}}(X))$.

However, the following properties may not hold for the lower and upper covering approximations:

(10) (Duality): $\underline{\mathcal{C}}(X) =\sim \overline{\mathcal{C}}(\sim X)$,
(11) (Duality): $\overline{\mathcal{C}}(X) =\sim \underline{\mathcal{C}}(\sim X)$;
(12) (Monotonicity): $X \subseteq Y \Longrightarrow \overline{\mathcal{C}}(X) \subseteq \overline{\mathcal{C}}(Y)$;
(13) (Multiplication): $\underline{\mathcal{C}}(X \cap Y) = \underline{\mathcal{C}}(X) \cap \underline{\mathcal{C}}(Y)$;
(14) (Addition): $\overline{\mathcal{C}}(X \cup Y) = \overline{\mathcal{C}}(X) \cup \overline{\mathcal{C}}(Y)$.

In a covering approximation space $(U, \mathcal{C})$, the members of $\mathcal{C}$ and the unions of some members can be considered as understandable or recognizable knowledge. For a unknown concept $X \subseteq U$, it can be described by $\underline{\mathcal{C}}(X)$ and $\overline{\mathcal{C}}(X)$ from upward and downward directions, respectively.
  It can be verified easily that $\underline{\mathcal{C}}(X) = X$ is equivalent to $\overline{\mathcal{C}}(X) = X$ for all $X \subseteq U$. Then a subset A of U is called to be *definable* in $(U, \mathcal{C})$ if and only if $\underline{\mathcal{C}}(A) = A$, or equivalently $\overline{\mathcal{C}}(A) = A$. Thus the knowledge space $KS(\mathcal{C})$ generalized by $\mathcal{C}$ consists of all definable sets of $(U, \mathcal{C})$, that is, $KS(\mathcal{C}) = \{A \subseteq U|\underline{\mathcal{C}}(A) = A\}$. So $KS(\mathcal{C})$ is also called the knowledge space of $(U, \mathcal{C})$.

3   Knowledge Spaces of Covering Approximation Spaces

In this section, we discuss relation among covering approximation spaces on the same universe U. Denote the set of all covering approximation spaces on U as $AS(U)$.

Definition 4. *Let X and $X_1, X_2, \ldots, X_m$ be subsets of U. X is said to be dependent on $X_1, X_2, \ldots, X_m$ if $X = \bigcup\limits_{i=1}^{m} X_i$ holds.*

Definition 5. *Let $(U, \mathcal{C}_1)$ and $(U, \mathcal{C}_2)$ be two covering approximation spaces. $(U, \mathcal{C}_1)$ ia said to be dependent on $(U, \mathcal{C}_2)$, denoted as $(U, \mathcal{C}_1) \leq (U, \mathcal{C}_2)$, if for any $X \in \mathcal{C}_1$, X is dependent on some members of $\mathcal{C}_2$. Then $(U, \mathcal{C}_1)$ and $(U, \mathcal{C}_2)$ are said to be equivalent if $(U, \mathcal{C}_1) \leq (U, \mathcal{C}_2)$ and $(U, \mathcal{C}_2) \leq (U, \mathcal{C}_1)$, denoted as $(U, \mathcal{C}_1) \doteq (U, \mathcal{C}_2)$.*

Clearly, the relation $\doteq$ among covering approximation spaces is an equivalence relation on $AS(U)$, and the relation $\leq$ between two covering approximation spaces is a partial order relation on $AS(U)$ in the sense of the equivalence relation $\doteq$. Therefore, $\langle AS(U), \leq \rangle$ is a partial ordered set or a poset. And it is easy to verify the following conclusions.

Proposition 2. *The poset $\langle AS(U), \leq \rangle$ satisfies the following basic properties: for any $(U, \mathcal{C}), (U, \mathcal{C}_1), (U, \mathcal{C}_2) \in AS(U)$, we have*

(1) *if $\mathcal{C}_1 \subseteq \mathcal{C}_2$ then $(U, \mathcal{C}_1) \leq (U, \mathcal{C}_2)$;*
(2) *if $(U, \mathcal{C}_1) \leq (U, \mathcal{C}_2)$ then $(U, \mathcal{C}_1 \cup \mathcal{C}) \leq (U, \mathcal{C}_2 \cup \mathcal{C})$;*
(3) *if $(U, \mathcal{C}_1) \leq (U, \mathcal{C})$ and $(U, \mathcal{C}_2) \leq (U, \mathcal{C})$, then $(U, \mathcal{C}_1 \cup \mathcal{C}_2) \leq (U, \mathcal{C})$.*

The inclusion relation between two knowledge spaces can be characterized by the dependence relation between two corresponding covering approximation spaces.

Theorem 1. *Let $(U, \mathcal{C}_1), (U, \mathcal{C}_2) \in AS(U)$. Then $KS(\mathcal{C}_1) \subseteq KS(\mathcal{C}_2)$ if and only if $(U, \mathcal{C}_1) \leq (U, \mathcal{C}_2)$.*

Proof. Suppose that $(U, \mathcal{C}_1) \leq (U, \mathcal{C}_2)$ and $X \in KS(\mathcal{C}_1)$. Then there exist $C_i \in \mathcal{C}_1 (1 \leq i \leq m)$ with $X = \bigcup\limits_{i=1}^{m} C_i$. By the supposition and the definition of dependence relation of covering approximation spaces, we have $C_i = \bigcup\limits_{j=1}^{k_i} C_j^i$ for any $C_i (1 \leq i \leq m) \in \mathcal{C}_1$, where $C_j^i \in \mathcal{C}_2 (1 \leq j \leq k_i)$. Hence $X = \bigcup\limits_{i=1}^{m} \bigcup\limits_{j=1}^{k_i} C_j^i$, so $X \in KS(\mathcal{C}_2)$. Therefore, $KS(\mathcal{C}_1) \subseteq KS(\mathcal{C}_2)$.

Conversely, assume that $KS(\mathcal{C}_1) \subseteq KS(\mathcal{C}_2)$. For any $C \in \mathcal{C}_1$, obviously $C \in KS(\mathcal{C}_1)$. By the assumption, we have $C \in KS(\mathcal{C}_2)$, that is, C equals to an union of some members of $\mathcal{C}_2$. As C is an arbitrary member of $\mathcal{C}_1$, it can be concluded that $(U, \mathcal{C}_1) \leq (U, \mathcal{C}_2)$. $\qquad\square$

The below conclusion follows from Theorem 1 immediately.

Corollary 1. *Let $(U, \mathcal{C}_1), (U, \mathcal{C}_2) \in AS(U)$. Then $KS(\mathcal{C}_1) = KS(\mathcal{C}_2)$ if and only if $(U, \mathcal{C}_1) \doteq (U, \mathcal{C}_2)$.*

The poset $\langle AS(U), \leq \rangle$ can be characterized by lower and upper covering approximation operators.

Proposition 3. *Let C_1 and C_2 be two coverings on U, Then $(U, C_1) \leq (U, C_2)$ if and only if $\underline{C_1}(X) \subseteq \underline{C_2}(X)$ for all $X \subseteq U$.*

Proof. Suppose that $(U, C_1) \leq (U, C_2)$ and $X \subseteq U$. For any $x \in \underline{C_1}(X)$, by the definition of $\underline{C_1}(X)$ there is a $Y \in C_1$ such that $x \in Y$ and $Y \subseteq X$. By the supposition there exist $Y_1, \ldots, Y_k \in C_2$ such that $Y = \bigcup_{i=1}^{k} Y_i$. Hence there exists at least one Y_i $(1 \leq i \leq k)$ such that $x \in Y_i$. By the definition of $\underline{C_2}(X)$ we conclude $x \in \underline{C_2}(X)$. Therefore $\underline{C_1}(X) \subseteq \underline{C_2}(X)$.

On the other hand, assume that $\underline{C_1}(X) \subseteq \underline{C_2}(X)$ for all $X \subseteq U$. For any $C \in C_1$, we have $\underline{C_1}(C) = C$. Then $C \subseteq \underline{C_2}(C) \subseteq C$, so $C = \underline{C_2}(C)$. By $\underline{C_2}(C) \in KS(C_2)$ we have $C \in KS(C_2)$, that is, there exist $C_1, \ldots, C_m \in C_2$ such that $C = \bigcup_{i=1}^{m} C_i$. We conclude $(U, C_1) \leq (U, C_2)$. $\qquad\square$

From above proposition it follows that $(U, C_1) \doteq (U, C_2)$ if and only if $\underline{C_1}(X) = \underline{C_2}(X)$ for all $X \subseteq U$.

According to Theorem 1 and Proposition 3, we obtain a kind of characterizations of knowledge spaces of covering approximation spaces by lower covering approximation operators.

Theorem 2. *Let $(U, C_1), (U, C_2) \in AS(U)$. Then $KS(C_1) \subseteq KS(C_2)$ if and only if for all $X \subseteq U$, $\underline{C_1}(X) \subseteq \underline{C_2}(X)$.*

Corollary 2. *Let $(U, C_1), (U, C_2) \in AS(U)$. Then $KS(C_1) = KS(C_2)$ if and only if for any $X \subseteq U$, $\underline{C_1}(X) = \underline{C_2}(X)$.*

About upper covering approximation operators, the following proposition holds.

Proposition 4. *Let C_1 and C_2 be two coverings on U. If for any $X \subseteq U$, $\overline{C_2}(X) \subseteq \overline{C_1}(X)$, then $(U, C_1) \leq (U, C_2)$.*

However, the converse proposition of Proposition 4 does not hold.

Example 1. Let $U = \{1, 2, 3, 4\}$, $C_1 = \{A_1, A_2\}$ and $C_2 = \{B_1, B_2, B_3, B_4\}$, where

$$A_1 = \{1, 2, 3\}, A_2 = \{2, 3, 4\},$$

$$B_1 = \{1, 2\}, B_2 = \{1, 3\}, B_3 = \{1, 4\}, B_4 = \{2, 3, 4\}.$$

Then C_1 and C_2 are two coverings on U, and $(U, C_1) \leq (U, C_2)$ can be checked. Taking $X = \{1\}$, by Definition 3 we figure out that

$$\overline{C_1}(X) = \{1, 2, 3\}, \overline{C_2}(X) = \{1, 2, 3, 4\}.$$

Hence, $\overline{C_1}(X) \subset \overline{C_2}(X)$.

Proposition 5. *Let C_1 and C_2 be two coverings on U. Then the following two statements are equivalent:*

(1) For any $X \subseteq U$, $\overline{C_2}(X) \subseteq \overline{C_1}(X)$.

(2) $(U, C_1) \le (U, C_2)$, and for any $x \in U$, $\cup Md_{C_2}(x) \subseteq \cup Md_{C_1}(x)$.

Proof. (1) $\Rightarrow$ (2) Suppose that $\overline{C_2}(X) \subseteq \overline{C_1}(X)$ for all $X \subseteq U$. For any $C \in C_1$, by $\overline{C_1}(C) = C$ we have $\overline{C_2}(C) \subseteq C$. Then $C = \overline{C_2}(C)$ follows from $C \subseteq \overline{C_2}(C)$ and $\overline{C_2}(C) \subseteq C$, which implies $C \in KS(U)$. Thus $(U, C_1) \le (U, C_2)$.

On the other hand, assume that $\overline{C_2}(X) \subseteq \overline{C_1}(X), \forall X \subseteq U$, and $x \in U$. If $\overline{C_2}(\{x\}) = \{x\}$, then it is easy to prove $Md_{C_2}(x) = \{\{x\}\}$, thus $\cup Md_{C_2}(x) = \{x\}$. Thus we conclude that if $\overline{C_2}(\{x\}) = \{x\}$ then $\cup Md_{C_2}(x) \subseteq \cup Md_{C_1}(x)$. If $\overline{C_1}(\{x\}) = \{x\}$, then $\{x\} \subseteq \overline{C_2}(\{x\}) \subseteq \overline{C_1}(\{x\}) = \{x\}$, it follows that $\{x\} = \overline{C_2}(\{x\})$. Thus $\overline{C_1}(\{x\}) = \{x\}$ means $\overline{C_2}(\{x\}) = \{x\}$. If $\overline{C_2}(\{x\}) \ne \{x\}$ then $\overline{C_1}(\{x\}) \ne \{x\}$, by Definition 3 we have $\overline{C_2}(\{x\}) = \cup Md_{C_2}(x)$ and $\overline{C_1}(\{x\}) = \cup Md_{C_1}(x)$, from the assumption it follows that $\cup Md_{C_2}(x) \subseteq \cup Md_{C_1}(x)$.

(2) $\Rightarrow$ (1) Suppose $(U, C_1) \le (U, C_2)$. For any $X \subseteq U$, by Proposition 3, $\underline{C_1}(X) \subseteq \underline{C_2}(X)$. Thus $X - \underline{C_2}(X) \subseteq X - \underline{C_1}(X)$. Based on the definition of the upper covering approximation operator, we have

$$\overline{C_1}(X) = \underline{C_1}(X) \cup \{\cup \{\cup Md_{C_1}(x) | x \in X - \underline{C_1}(X)\}\}$$

and $\overline{C_2}(X) = \underline{C_2}(X) \cup \{\cup \{\cup Md_{C_2}(x) | x \in X - \underline{C_2}(X)\}\}$. As $\underline{C_1}(X) \subseteq \underline{C_2}(X) \subseteq X$, $\overline{C_1}(X)$ can be rewritten as $\overline{C_1}(X) = \underline{C_2}(X) \cup \{\cup \{\cup Md_{C_1}(x) | x \in X - \underline{C_1}(X)\}\}$. Since $X - \underline{C_2}(X) \subseteq X - \underline{C_1}(X)$, and for any $x \in U$, $\cup Md_{C_2}(x) \subseteq \cup Md_{C_1}(x)$, we have $\cup \{\cup Md_{C_2}(x) | x \in X - \underline{C_2}(X)\} \subseteq \cup \{\cup Md_{C_1}(x) | x \in X - \underline{C_1}(X)\}$. Noticing

$$\overline{C_1}(X) = \underline{C_2}(X) \cup \{\cup \{\cup Md_{C_1}(x) | x \in X - \underline{C_1}(X)\}\},$$

$$\overline{C_2}(X) = \underline{C_2}(X) \cup \{\cup \{\cup Md_{C_2}(x) | x \in X - \underline{C_2}(X)\}\},$$

we concluded that $\overline{C_2}(X) \subseteq \overline{C_1}(X)$.   □

Corresponding to Theorem 1, we easily prove the following characterization of knowledge spaces of covering approximation spaces by upper covering approximation operators as follows:

Theorem 3. *Let $(U, C_1), (U, C_2) \in AS(U)$. Then $KS(C_1) \subseteq KS(C_2)$ if and only if for all $C \in C_1$, $\overline{C_2}(C) \subseteq C$.*

Corollary 3. *Let $(U, C_1), (U, C_2) \in AS(U)$. Then $KS(C_1) = KS(C_2)$ if and only if $\overline{C_2}(C) \subseteq C$ for all $C \in C_1$, and $\overline{C_1}(C) \subseteq C$ for all $C \in C_2$.*

4   Independence and Reduction of Covering Approximation Spaces

For a covering approximation space (U, C), if a covering D of U is a subset of the covering C, then D is said to be a sub-covering of C. In this section, the attention is paid on the dependence among sub-coverings of a covering of the universe.

Definition 6. *Let (U, C) be a covering approximation space. Then (U, C) is said to be dependent if at least one member of C is dependent on some other members of C. Otherwise, (U, C) is said to be independent.*

From Definition 6 we have the following characterizations of dependence of covering approximation spaces.

Proposition 6. *Let (U, C) be a covering approximation space. Then*

(1) *(U, C) is dependent if and only if there exists a proper subset D of C such that (U, D) and (U, C) are equivalent,*

(2) *(U, C) is independent if and only if for any proper covering D of C, (U, D) and (U, C) are not equivalent.*

By Definition 6 we can gain the next proposition easily.

Proposition 7. *Let $(U, C_1), (U, C_2) \in AS(U)$ with $C_1 \subseteq C_2$. Then*

(1) *if (U, C_1) is dependent, then (U, C_2) is also dependent,*

(2) *if (U, C_2) is independent, then (U, C_1) is also independent.*

Proposition 8. *Let (U, C_1) and (U, C_2) be two independent covering approximation spaces. Then (U, C_1) and (U, C_2) are equivalent if and only if $C_1 = C_2$.*

Proof. The sufficiency is obvious, we only need to prove the necessary.

Suppose that (U, C_1) and (U, C_2) are equivalent, that is, $(U, C_1) \leq (U, C_2)$ and $(U, C_2) \leq (U, C_1)$. For any $C \in C_1$, by $(U, C_1) \leq (U, C_2)$ there exist a $D \subseteq C_2$ such that $C = \cup D$. If there is not $D \in D$ with $D = C$, by $(U, C_2) \leq (U, C_1)$ there exist $B_D \subseteq C_1$ for all $D \in D$ such that $D = \cup B_D$, so $C = \bigcup_{D \in D} (\cup B_D)$, it is clear that for any $D \in D$ and $A \in B_D$, $A \neq C$, thus C_1 is not independent, which is in contradiction with the conditions. Therefore, for any $C \in C_1$, there exists a $D \in C_2$ with $D = C$, that is, $C_1 \subseteq C_2$. In the same way, we can prove that $C_2 \subseteq C_1$. It is proved that $C_1 = C_2$. ☐

The below corollary is obvious.

Corollary 4. *Let (U, C_1) and (U, C_2) be two independent covering approximation spaces. If (U, C_1) and (U, C_2) are equivalent, then $card(C_1) = card(C_2)$, where $card(C_1)$ and $card(C_2)$ are the cardinalities of C_1 and C_2, respectively.*

Definition 7. *Let $(U, C), (U, D) \in AS(U)$ with $D \subseteq C$. Then (U, D) is referred to as a reduct of (U, C) if*

(1) *(U, D) is independent, and*

(2) *(U, D) and (U, C) are equivalent.*

Definition 7 means that a reduct of (U, C) is a maximal independent covering approximation sub-space of (U, C) which is equivalent to (U, C), that is also to say, the knowledge space of a reduct equals to $KS(C)$.

Proposition 9. *Let $(U,\mathcal{C})$ be a covering approximation space. Then $(U,\mathcal{C})$ is independent if and only if the reduct of $(U,\mathcal{C})$ just be itself.*

From Propositions 6 and 9 it can be confirmed that there exist reducts for any covering approximation spaces. In terms of Definition 7, it is evident that any two reducts of a covering approximation space are equivalent to each other. At last, by Proposition 8 it can be concluded that there exists single reduct for every covering approximation space. In general, the following conclusion holds.

Proposition 10. *Let $(U,\mathcal{C}_1),(U,\mathcal{C}_2) \in AS(U)$ with $\mathcal{C}_1 \subseteq \mathcal{C}_2$. Then $(U,\mathcal{C}_1)$ is equivalent with $(U,\mathcal{C}_2)$ if and only if $(U,\mathcal{C}_2 - \mathcal{C}_1) \leq (U,\mathcal{C}_1)$.*

By the above Proposition 10, the below conclusion is obtained immediately.

Theorem 4. *Let $(U,\mathcal{C}),(U,\mathcal{D}) \in AS(U)$ with $\mathcal{D} \subseteq \mathcal{C}$. Then $(U,\mathcal{D})$ is the reduct of $(U,\mathcal{C})$ if and only if $(U,\mathcal{D})$ is independent, and $(U,\mathcal{C} - \mathcal{D}) \leq (U,\mathcal{D})$.*

Based on Theorem 4, with respect to lower and upper covering approximation operators, we have the following conclusions.

Theorem 5. *Let $(U,\mathcal{C}),(U,\mathcal{D}) \in AS(U)$ with $\mathcal{D} \subseteq \mathcal{C}$. Then $(U,\mathcal{D})$ is the reduct of $(U,\mathcal{C})$ if and only if one of the following two conditions holds:*

(1) *For any $C \in \mathcal{C} - \mathcal{D}$, $\underline{\mathcal{D}}(C) = C$ holds, and for any $B \in \mathcal{D}$ there exists a $C \in \mathcal{C} - \mathcal{D}$ such that $\underline{(\mathcal{D} - \{B\})}(C) \neq C$.*

(2) *For any $C \in \mathcal{C} - \mathcal{D}$, $\overline{\mathcal{D}}(C) = C$ holds, and for any $B \in \mathcal{D}$ there exists a $C \in \mathcal{C} - \mathcal{D}$ such that $\overline{(\mathcal{D} - \{B\})}(C) \neq C$.*

5   Conclusions

In a Bonikowski covering rough set model, unions of some members of the covering are used to describe any unknown concepts from upward and downward directions, they consist of an algebraic structure called knowledge space. In this paper, we mainly discuss the relation among knowledge spaces of covering approximation spaces with the same universe. By defining a dependence relation of covering approximation spaces, the relation of the knowledge spaces is characterized, which can also be described by covering lower and upper approximation operators completely. By limiting the dependence relation on covering approximation sub-spaces of a covering approximation space, the notion of reduction of a covering approximation space is introduced. As a result, the mathematical properties of the reduction of covering approximation spaces are clarified, and the judgment theorems for the reduction of covering approximation spaces are obtained. The conclusions about the dependence relation among covering approximation spaces are useful for revealing hierarchical structure among members of a covering, and extracting simple decision rules in decision information systems by covering rough set approaches.

Acknowledgements. This work was supported by grants from the National Natural Science Foundation of China (Nos. 11071284, 61075120, 61272021, 61202206) and the Zhejiang Provincial Natural Science Foundation of China (Nos. LY14F030001, LZ12F03002, LY12F02021).

References

1. Bonikowski, Z., Bryniarski, E., Wybraniec, U.: Extensions and intentions in the rough set theory. Inf. Sci. **107**, 149–167 (1998)
2. Caspard, N., Monjardet, B.: The lattices of closure systems, closure operators, and implicational systems on a finite set: a survey. Discrete Appl. Math. **127**, 241–269 (2003)
3. Chen, D., Wang, C., Hu, Q.: A new approach to attributes reduction of consistent and inconsistent covering decision systems with covering rough sets. Inf. Sci. **177**, 3500–3518 (2007)
4. Doignon, J.P., Falmagne, J.C.: Knowledge Spaces. Springer, Heidelberg (1999)
5. Kortelainen, J.: On the relationship between modified sets, topological spaces and rough sets. Fuzzy Sets Syst. **61**, 91–95 (1994)
6. Liu, C., Miao, D.Q., Qian, J.: On multi-granulation covering rough sets. Int. J. Approx. Reason. **44**, 1404–1418 (2015)
7. Li, T.J., Leung, Y., Zhang, W.X.: Generalized fuzzy rough approximation operators based on fuzzy coverings. Int. J. Approx. Reason. **48**, 836–856 (2008)
8. Restrepo, M., Cornelis, C., Gomez, J.: Partial order relation for approximation operators in covering based rough sets. Inf. Sci. **284**, 44–59 (2014)
9. Monjardet, B.: The presence of lattice theory in discrete problems of mathematical social sciences. Why. Math. Soc. Sci. **46**, 103–144 (2003)
10. Pawlak, Z.: Rough sets. Int. J. Comput. Inf. Sci. **11**, 341–356 (1982)
11. Qian, Y.H., Liang, J.Y., Pedrycz, W., Dang, C.Y.: Positive approximation: an accelerator for attribute reduction in rough set theory. Artif. Intell. **174**, 597–618 (2010)
12. Slowinski, R., Vanderpooten, D.: A generalized definition of rough approximations based on similarity. IEEE Trans. Knowl. Data Eng. **12**, 331–336 (2000)
13. Wang, L., Yang, X., Yang, J.: Relationships among generalized rough sets in six coverings and pure reflexive neighborhood system. Inf. Sci. **207**, 66–78 (2012)
14. Xu, W.H., Zhang, W.X.: Measuring roughness of generalized rough sets induced by a covering. Fuzzy Sets Syst. **158**, 2443–2455 (2007)
15. Yang, T., Li, Q.: Reduction about approximation spaces of covering generalized rough sets. Int. J. Approx. Reason. **51**, 335–345 (2010)
16. Yao, Y.Y.: Information granulation and rough set approximation. Int. J. Intell. Syst. **16**, 87–104 (2001)
17. Yao, Y.Y.: The superiority of three-way decisions in probabilistic rough set models. Inf. Sci. **181**, 1080–1096 (2011)
18. Yao, Y., Yao, B.: Covering based rough set approximations. Inf. Sci. **200**, 91–107 (2012)
19. Yue, X.D., Miao, D.Q., Zhang, N., Cao, L.B., Wu, Q.: Multiscale roughness measure for color image segmentation. Inf. Sci. **216**, 93–112 (2012)
20. Zakowski, W.: Approximations in the space $(u, \prod)$. Demonstratio Mathematica **16**, 761–769 (1983)
21. Zhang, J.B., Li, T.R., Chen, H.M.: Composite rough sets for dynamic data mining. Inf. Sci. **257**, 81–100 (2014)

22. Zhu, W., Wang, F.-Y.: Covering based granular computing for conflict analysis. In: Mehrotra, S., Zeng, D.D., Chen, H., Thuraisingham, B., Wang, F.-Y. (eds.) ISI 2006. LNCS, vol. 3975, pp. 566–571. Springer, Heidelberg (2006)

23. Zhu, W., Wang, F.Y.: Reduction and axiomization of covering generalized rough sets. Inf. Sci. **152**, 217–230 (2003)

24. Zhu, W.: Relationship between generalized rough sets based on binary relation and covering. Inf. Sci. **179**, 210–225 (2009)

25. Zhu, W., Wang, F.Y.: The fourth type of covering-based rough sets. Inf. Sci. **201**, 80–92 (2012)

Families of the Granules for Association Rules and Their Properties

Hiroshi Sakai[1]($\boxtimes$), Chenxi Liu[1], and Michinori Nakata[2]

[1] Graduate School of Engineering, Kyushu Institute of Technology,
Tobata, Kitakyushu 804-8550, Japan
sakai@mns.kyutech.ac.jp
[2] Faculty of Management and Information Science,
Josai International University, Gumyo, Togane, Chiba 283, Japan
nakatam@ieee.org

Abstract. We employed the granule (or the equivalence class) defined by a descriptor in tables, and investigated rough set-based rule generation. In this paper, we consider the new granules defined by an implication, and propose a *family of the granules defined by an implication* in a table with exact data. Each family consists of the four granules, and we show that three criterion values, *support*, *accuracy*, and *coverage*, can easily be obtained by using the four granules. Then, we extend this framework to tables with non-deterministic data. In this case, each family consists of the nine granules, and the minimum and the maximum values of three criteria are also obtained by using the nine granules. We prove that there is a table causing *support* and *accuracy* the minimum, and generally there is no table causing *support*, *accuracy*, and *coverage* the minimum. Finally, we consider the application of these properties to *Apriori*-based rule generation from uncertain data. These properties will make *Apriori*-based rule generation more effective.

Keywords: Association rules · Rule generation · Apriori algorithm · Granularity · Uncertainty

1 Introduction

We coped with rule generation and data mining in *Non-deterministic Information Systems* (*NISs*) [8,13,15]. *NIS* was proposed by Pawlak [11], Orłowska [9], and Lipski [6,7] in order to handle information incompleteness in the typical table defined as a *Deterministic Information System* (*DIS*) [10,12,17]. Pawlak's framework in *DIS* is called *Rough Set Theory*, and the equivalence classes take the important role. In [11], we see the definition of the *many valued system*, which is similar to *NIS*. Orłowska and Lipski considered question-answering methods in *NIS* independently.

We tried to extend rule generation in *DIS* to *NIS* by using the modal concepts [3], and proposed the framework *Rough Non-deterministic Information Analysis* (*RNIA*). In *RNIA*, we defined the *certain rules* and the *possible rules*, and proved

© Springer International Publishing Switzerland 2015
D. Ciucci et al. (Eds.): RSKT 2015, LNAI 9436, pp. 175–187, 2015.
DOI: 10.1007/978-3-319-25754-9_16

that the algorithm named *NIS-Apriori* is *sound* and *complete* for the defined rules [15,16]. The *NIS-Apriori* algorithm is an adjusted *Apriori* algorithm [1, 2] to *NIS*. Even though *NIS-Apriori* handles the modal concepts in rules, the computational complexity is about twice the complexity of *Apriori*. Furthermore, we opened a web software tool *getRNIA* [18,19].

In this paper, we propose the *families of the granules* in *NIS*, which are extended from the *division chart* [14]. In rough sets, we usually make use of the granules (or equivalence classes) defined by the descriptors. Some other types of granules are also proposed for handling missing values [5]. Here, we consider the granules defined by the implications. We have already coped with the six granules defined by the implications in [14], and we extend them to the nine granules. By this extension, we can consider the new criterion value *coverage* as well as *support* and *accuracy*.

This paper is organized as follows: Sect. 2 focuses on the case of the tables with exact data. We define the family of the four granules, and show the calculation of the criterion values. Section 3 considers the case of the tables with non-deterministic data. We similarly define the family of the nine granules, and show the extended results from Sect. 2. Section 4 describes the perspective of the *NIS-Apriori* algorithm based on the obtained results. Section 5 concludes this paper.

2   A Family of the Granules in DIS ψ

This section considers a family of the granules and its property in *DIS* ψ.

2.1   Preliminary

A *Deterministic Information System* (*DIS*) ψ is a quadruplet [10–12,17]:

$$\psi = (OB, AT, \{VAL_A \mid A \in AT\}, f), \tag{1}$$

where OB is a finite set whose elements are called *objects*, AT is a finite set whose elements are called *attributes*, VAL_A is a finite set whose elements are called *attribute values* for an attribute $A \in AT$, and f is such a mapping:

$$f : OB \times AT \rightarrow \cup_{A \in AT} VAL_A. \tag{2}$$

We usually consider a table like Table 1 for ψ. A pair $[A, v]$ ($A \in AT$, $v \in VAL_A$) is called a *descriptor*, and we consider a set $CON \subseteq AT$ which we call (a set of) *condition attributes* and an attribute $Dec \in AT$ ($Dec \notin CON$) which we call a *decision attribute*. An *implication* τ for CON and Dec is a formula in the following:

$$\tau : \wedge_{A \in CON} [A, val_A] \Rightarrow [Dec, val] \ (val_A \in VAL_A, val \in VAL_{Dec}). \tag{3}$$

In most of work on rule generation, we try to obtain the appropriate implications, which we call rules. The most famous criterion for defining rules consists of three

Table 1. An exemplary deterministic information system ψ_1.

OB	temperature	headache	nausea	flu
1	very_high	yes	yes	yes
2	high	yes	yes	yes
3	normal	yes	yes	no
4	very_high	yes	no	yes
5	very_high	yes	yes	yes
6	high	no	no	no
7	normal	no	yes	no
8	high	no	no	no

values, i.e., *support*, *accuracy* and *coverage* [10–12,17]. We employ these values, and we define that a *rule* is an implication τ satisfying the constraint

$$support(\tau) \geq \alpha, accuracy(\tau) \geq \beta \text{ and } coverage(\tau) \geq \gamma \quad (4)$$
$$\text{for given } 0 < \alpha, \beta, \gamma \leq 1.$$

The constraint *coverage* may not be employed in some frameworks, for example the *Apriori* algorithm [1,2] does not employ it, and we see $\gamma=0$ in such case. In ψ_1, we have $support(\tau)=3/8$, $accuracy(\tau)=3/4$, and $coverage(\tau)=3/4$ for $\tau : [headache, yes] \wedge [nausea, yes] \Rightarrow [flu, yes]$.

2.2   A Family of the Granules Defined by an Implication

We propose the granules defined by an implication $\tau : \wedge_{A \in CON}[A, val_A] \Rightarrow [Dec, val]$.

Definition 1. *We say an object x supports τ, if $f(x, A)=val_A$ for every $A \in CON$ and $f(x, Dec)=val$. In order to clarify this object x, we may employ the notation τ^x instead of τ.*

In ψ_1, the objects 1, 2 and 5 support $\tau : [headache, yes] \wedge [nausea, yes] \Rightarrow [flu, yes]$, and we have τ^1, τ^2, and τ^5.

For more simplicity of $\tau : \wedge_{A \in CON}[A, val_A] \Rightarrow [Dec, val]$, we employ the following notation:
(1) p denotes the conjunction $\wedge_{A \in CON}[A, val_A]$,
(2) p' denotes any conjunction $\wedge_{A \in CON}[A, val'_A]$ ($val'_A \neq val_A$ for at least one $A \in CON$),
(3) r denotes $[Dec, val]$,
(4) r' denotes any descriptor $[Dec, val']$ ($val' \neq val$).

If we fix an implication $\tau : p \Rightarrow r$, each object defines either $p \Rightarrow r$, $p \Rightarrow r'$, $p' \Rightarrow r$, or $p' \Rightarrow r'$. In ψ_1, let us consider $\tau : [headache, yes] \wedge [nausea, yes] \Rightarrow$

Table 2. Four granules defined by $\tau : p \Rightarrow r$ in ψ.

	r	r'
p	①$=\{x \in OB \mid x$ supports $p \Rightarrow r\}$	②$=\{x \in OB \mid x$ supports $p \Rightarrow r'\}$
p'	③$=\{x \in OB \mid x$ supports $p' \Rightarrow r\}$	④$=\{x \in OB \mid x$ supports $p' \Rightarrow r'\}$

$[flu, yes]$. Then, the objects 1, 2, and 5 define $p \Rightarrow r$, the object 3 defines $p \Rightarrow r'$, the object 4 does $p' \Rightarrow r$, and the objects 6, 7, and 8 do $p' \Rightarrow r'$.

Based on these four types of implications, we define four sets ①, ②, ③, and ④ in Table 2. Since we can show the following,

$$① \cup ② \cup ③ \cup ④ = OB, \quad ① \cap ① = \emptyset \; (i \neq j), \tag{5}$$

the four sets are equivalence classes over OB.

Definition 2. *For DIS ψ, an implication τ and the four equivalence classes in Table 2, we define $FGr(\tau, \psi) = (①, ②, ③, ④)$, and we say $FGr(\tau, \psi)$ is a family of the granules defined by τ in ψ.*

In our previous work, we proposed a set of equivalence classes and named it a *division chart* [14]. In a division chart, we handled two types of implications, namely $p \Rightarrow r$ and $p \Rightarrow r'$, and considered only two granules ① and ②. We calculated *support* and *accuracy* by using ① and ②, however we can also calculate *coverage* by using $FGr(\tau, \psi)$. Thus, we are extending the previous work [14] to the more powerful one. Since $FGr(\tau, \psi)$ takes the role of the contingency table, it is easy to obtain the following proposition.

Proposition 1. *For a family $FGr(\tau, \psi)$ of the granules, the following holds.*

$$(1) \; support(\tau) = \frac{|①|}{|①|+|②|+|③|+|④|} = \frac{|①|}{|OB|},$$

$$(2) \; accuracy(\tau) = \frac{|①|}{|①|+|②|}, \quad (3) \; coverage(\tau) = \frac{|①|}{|①|+|③|}. \tag{6}$$

Remark 1. For $\tau : [temperature, normal] \Rightarrow [flu, no] \; (=p \Rightarrow r)$ in ψ_1, we have $FGr(\tau, \psi_1) = (\{3, 7\}, \emptyset, \{6, 8\}, \{1, 2, 4, 5\})$. Based on Proposition 1,

$$(1) \; support(\tau) = \frac{|\{3,7\}|}{8} = 1/4, \quad (2) \; accuracy(\tau) = \frac{|\{3,7\}|}{|\{3,7\}|+|\emptyset|} = 1.0,$$

$$(3) \; coverage(\tau) = \frac{|\{3,7\}|}{|\{3,7\}|+|\{6,8\}|} = 1/2. \tag{7}$$

Especially, $1, 2 \in ④$ seems quite new, as far as authors know. In the typical equivalence relation, 1 and 2 are not in the same class, however they are equivalent in the aspect that neither object 1 nor 2 is related to the implication τ at all.

Table 3. An exemplary non-deterministic information system Φ_1.

OB	temperature	headache	nausea	flu
1	$\{very\_high\}$	$\{yes, no\}$	$\{yes\}$	$\{yes\}$
2	$\{high, very\_high\}$	$\{yes\}$	$\{yes\}$	$\{yes\}$
3	$\{normal, high\}$	$\{yes\}$	$\{yes\}$	$\{yes, no\}$
4	$\{very\_high\}$	$\{yes\}$	$\{yes, no\}$	$\{yes\}$
5	$\{very\_high\}$	$\{yes, no\}$	$\{yes\}$	$\{yes\}$
6	$\{high\}$	$\{no\}$	$\{yes, no\}$	$\{yes, no\}$
7	$\{normal\}$	$\{no\}$	$\{yes\}$	$\{no\}$
8	$\{normal, high\}$	$\{no\}$	$\{yes, no\}$	$\{no\}$

3　A Family of the Granules in NIS Φ

We have dealt with *NIS* as the case of the tables with uncertainty. In this section, we consider a family of the granules in *NIS* and their property.

3.1　Preliminary

NIS Φ is also a quadruplet $\Phi=(OB, AT, \{VAL_A|A \in AT\}, g)$ [9,11], where g is such a mapping:

$$g : OB \times AT \to P(\cup_{A \in AT} VAL_A) \text{ (a power set of } \cup_{A \in AT} VAL_A). \quad (8)$$

Every set $g(x, A)$ is interpreted as that there is an actual value in this set, but the value is not known. We usually consider tabular representation of Φ like Table 3. Since each VAL_A is a finite set, we can easily define all possible cases from *NIS*. For $\Phi=(OB, AT, \{VAL_A|A \in AT\}, g)$, we name the following *DIS* a *derived DIS* from *NIS* Φ.

$$\psi = (OB, AT, \{VAL_A|A \in AT\}, h) \ (h(x, A) \in g(x, A) \text{ for each } x \text{ and } A). \quad (9)$$

For *NIS* Φ, let $DD(\Phi)$ denote a set below:

$$DD(\Phi) = \{\psi \mid \psi \text{ is a derived } DIS \text{ from } \Phi\}. \quad (10)$$

Actually, ψ_1 in Table 1 is a derived *DIS* from Φ_1. We transfer the problem on information incompleteness to the case analytic problem based on $DD(\Phi)$. We define that τ is a *certain rule*, if τ is a rule in every $\psi \in DD(\Phi)$, and τ is a *possible rule*, if τ is a rule in at least one $\psi \in DD(\Phi)$. However, there are 1024 ($=2^{10}$) derived *DISs* in Φ_1. In Mammographic data set [4], there are more than 10^{100} derived *DISs*. For solving this problem, the family of the granules takes the important role.

Definition 3. *We say an object x supports τ in NIS Φ, if x supports τ in at least one $\psi \in \Phi$.*

Remark 2. For $\tau : [temperature, normal] \Rightarrow [flu, no]$ in Φ_1, we have τ^3, τ^7, and τ^8, and we say each of them is an *instance* of τ. In order to evaluate τ in *NIS* Φ, we consider the instance of τ in Φ. If there is an instance τ^x satisfying the constraint, we say τ is a rule. In *DIS*, we need not to consider such instances, because three criterion values are the same for every τ^x and τ^y $(x \neq y)$. However in *NIS*, they may be different. For example, the instance τ^7 occurs in each of the derived *DIS*, but there is a derived *DIS* where τ^3 nor τ^8 do not occur. They have the different property from τ^7.

3.2   A Family of the Granules in NIS Φ

For Φ, an implication τ, and an object $x \in OB$, we define the following:

$$IMP(x, \Phi, \tau) = \cup_{\psi \in DD(\Phi)} \{ \tau_\psi \mid x \text{ supports } \tau_\psi \text{ in } DIS \ \psi \},$$
$$\text{Here, } \tau_\psi \text{ is either } \tau : p \Rightarrow r, \ p \Rightarrow r', \ p' \Rightarrow r \text{ or } p' \Rightarrow r'. \tag{11}$$

This $IMP(x, \Phi, \tau)$ means a set of the obtainable implications, which are classified by τ, from x. For example in Φ_1, we consider $\tau : [temperature, normal] \Rightarrow [flu, no]$ $(= p \Rightarrow q)$. Then, we have the following:

$$IMP(3, \Phi_1, \tau) = \{ p \Rightarrow r, p \Rightarrow r', p' \Rightarrow r, p' \Rightarrow r' \},$$
$$IMP(7, \Phi_1, \tau) = \{ p \Rightarrow r \}, \tag{12}$$
$$IMP(8, \Phi_1, \tau) = \{ p \Rightarrow r, p' \Rightarrow r \}.$$

Proposition 2. *The relation $\sim_{\tau, \Phi}$ below defines an equivalence relation over OB.*

$$x \sim_{\tau, \Phi} y \Leftrightarrow IMP(x, \Phi, \tau) = IMP(y, \Phi, \tau). \tag{13}$$

Proof. *We can easily show the reflexivity, the symmetry and the transitivity.*

For p $(= \wedge_{A \in CON} [A, val_A])$ in τ, it is necessary to consider three cases, i.e., $\{p\}$, $\{p, p'\}$ and $\{p'\}$. Similarly, we think three cases, $\{r\}$, $\{r, r'\}$ and $\{r'\}$, and we have the nine sets based on the obtainable implications in Table 4.

Table 4. Nine sets based on the obtainable implications for $\tau : p \Rightarrow r$.

	$\{r\}$	$\{r, r'\}$	$\{r'\}$
$\{p\}$	$S_1 : \{p \Rightarrow r\}$	$S_2 : \{p \Rightarrow r, p \Rightarrow r'\}$	$S_3 : \{p \Rightarrow r'\}$
$\{p, p'\}$	$S_4 : \{p \Rightarrow r, \ p' \Rightarrow r\}$	$S_5 : \{p \Rightarrow r, p \Rightarrow r', \ p' \Rightarrow r, p' \Rightarrow r'\}$	$S_6 : \{p \Rightarrow r', \ p' \Rightarrow r'\}$
$\{p'\}$	$S_7 : \{p' \Rightarrow r\}$	$S_8 : \{p' \Rightarrow r, p' \Rightarrow r'\}$	$S_9 : \{p' \Rightarrow r'\}$

Definition 4. *We define the following based on Table 4.*

$$\textcircled{i} = \{x \in OB \mid IMP(x, \Phi, \tau) = S_i\} \ (1 \le i \le 9),$$
$$FGr(\tau, \Phi) = (\textcircled{1}, \textcircled{2}, \textcircled{3}, \textcircled{4}, \textcircled{5}, \textcircled{6}, \textcircled{7}, \textcircled{8}, \textcircled{9}). \tag{14}$$

We say $FGr(\tau, \Phi)$ is a family of the granules defined by τ in Φ.

For example, we have the following for $\tau : [temperature, normal] \Rightarrow [flu, no]$ in Φ_1 and the formulas (12).

$$IMP(3, \Phi_1, \tau) = S_5 \text{ and } 3 \in \textcircled{5},$$
$$IMP(7, \Phi_1, \tau) = S_1 \text{ and } 7 \in \textcircled{1},$$
$$IMP(8, \Phi_1, \tau) = S_4 \text{ and } 8 \in \textcircled{4}, \tag{15}$$
$$FGr(\tau, \Phi_1) = (\{7\}, \emptyset, \emptyset, \{8\}, \{3\}, \emptyset, \emptyset, \{6\}, \{1, 2, 4, 5\}).$$

3.3 Criterion Values of an Implication in NIS Φ

In *NIS Φ*, the criterion values depend upon $\psi \in DD(\Phi)$, so we consider the minimum and the maximum values of $support(\tau^x)$, $accuracy(\tau^x)$, and $coverage(\tau^x)$. We employ the notations $minsupp(\tau^x)$, $maxsupp(\tau^x)$, $minacc(\tau^x)$, $maxacc(\tau^x)$, $mincov(\tau^x)$, $maxcov(\tau^x)$ for them. We say τ^x is *definite* in Φ, if $IMP(x, \Phi, \tau)$ is a singleton set. If τ^x is not definite, there is at least on $\psi \in DD(\Phi)$ where τ^x does not occur. In this ψ, we define $minsupp(\tau^x)=minacc(\tau^x)= mincov(\tau^x)=0$. Now, we sequentially consider three criterion values by generating an actual $\psi \in DD(\Phi)$.

Proposition 3. *For NIS Φ and an implication τ, let us consider $FGr(\tau, \Phi)$ $=(\textcircled{1}, \textcircled{2}, \textcircled{3}, \textcircled{4}, \textcircled{5}, \textcircled{6}, \textcircled{7}, \textcircled{8}, \textcircled{9})$. If $\textcircled{1} \ne \emptyset$, there is an object $x \in \textcircled{1}$ and $\psi_{minSA} \in DD(\Phi)$ satisfying the following:*

(1) $support(\tau^x)$ *in* $\psi_{minSA} = minsupp(\tau^x) = |\textcircled{1}|/|OB|$,

(2) $accuracy(\tau^x)$ *in* $\psi_{minSA} = minacc(\tau^x) = \dfrac{|\textcircled{1}|}{|\textcircled{1}|+|\textcircled{2}|+|\textcircled{3}|+|\textcircled{5}|+|\textcircled{6}|}$, $\qquad$ (16)

(3) $coverage(\tau^x)$ *in* $\psi_{minSA} = \dfrac{|\textcircled{1}|}{|\textcircled{1}|+|\textcircled{4}|+|\textcircled{7}|+|\textcircled{8}|}$.

(Sketch of the proof) By the selection of an implication from the sets S_2, S_4, S_5, S_6, S_8, the criterion values change. For two natural number N and M ($N \le M$), we can easily show the inequality $\frac{N}{M} \le \frac{N+1}{M+1}$. If we select $\tau^x : p \Rightarrow r$ in S_2, this x satisfies the condition on the denominator and the numerator of accuracy. Based on the inequality, this selection causes to increase accuracy. Thus, we select $p \Rightarrow r'$. Namely, we employ the strategy to select the same condition and the different decision. At the same time, we implicitly specify a table by this selection. If we select the underlined part in Table 5, we have ψ_{minSA}, and both support and accuracy are the minimum. If $\textcircled{1}=\emptyset$, we handle τ^x ($x \in \textcircled{2} \cup \textcircled{4} \cup \textcircled{5}$). In this case, we also have the formulas for the calculation, however the formulas are slightly different. We omit this case.

Table 5. The selection (underlined part) of the implications for $\psi_{minSA} \in DD(\Phi)$.

	$\{r\}$	$\{r, r'\}$	$\{r'\}$
$\{p\}$	$S_1 : \{p \Rightarrow r\}$	$S_2 : \{p \Rightarrow r, \underline{p \Rightarrow r'}\}$	$S_3 : \{p \Rightarrow r'\}$
$\{p, p'\}$	$S_4 : \{p \Rightarrow r,$ $\underline{p' \Rightarrow r}\}$	$S_5 : \{p \Rightarrow r, p \Rightarrow r',$ $p' \Rightarrow r, p' \Rightarrow r'\}$	$S_6 : \{p \Rightarrow r',$ $p' \Rightarrow r'\}$
$\{p'\}$	$S_7 : \{\underline{p' \Rightarrow r}\}$	$S_8 : \{\underline{p' \Rightarrow r}, p' \Rightarrow r'\}$	$S_9 : \{p' \Rightarrow r'\}$

Table 6. ψ_{minSA} from $DD(\Phi_1)$.

OB	temperature	flu
1	very_high	yes
2	high	yes
3	normal	yes
4	very_high	yes
5	very_high	yes
6	high	no
7	normal	no
8	high	no

Table 7. ψ_{minSC} from $DD(\Phi_1)$.

OB	temperature	flu
1	very_high	yes
2	high	yes
3	high	no
4	very_high	yes
5	very_high	yes
6	high	no
7	normal	no
8	high	no

Example 1. For $\tau : [temperature, normal] \Rightarrow [flu, no]$, we obtained $FGr(\tau, \Phi_1) = (\{7\}, \emptyset, \emptyset, \{8\}, \{3\}, \emptyset, \emptyset, \{6\}, \{1, 2, 4, 5\})$ in the formulas (15). Since $7 \in ①$, we can apply Proposition 3 to τ^7. Based on Table 5, we select $[temperature, high] \Rightarrow [flu, no]$ from object 8 and $[temperature, normal] \Rightarrow [flu, yes]$ from object 3. Then, we have ψ_{minSA} in Table 6. In ψ_{minSA}, $support(\tau^7) = 1/8 = minsupp(\tau^7)$, $accuracy(\tau^7) = 1/2 = minacc(\tau^7)$, $coverage(\tau^7) = 1/3 > mincov(\tau^7)$.

Proposition 4. *For NIS Φ and an implication τ, let us consider $FGr(\tau, \Phi)$ $= (①, ②, ③, ④, ⑤, ⑥, ⑦, ⑧, ⑨)$. If $① \neq \emptyset$, there is an object $x \in ①$ and $\psi_{minSC} \in DD(\Phi)$ satisfying the following:*

(1) $support(\tau^x)$ in $\psi_{minSC} = minsupp(\tau^x) = |①|/|OB|,$

(2) $accuracy(\tau^x)$ in $\psi_{minSC} = \dfrac{|①|}{|①| + |②| + |③| + |⑥|},$ (17)

(3) $coverage(\tau^x)$ in $\psi_{minSC} = mincov(\tau^x) = \dfrac{|①|}{|①| + |④| + |⑤| + |⑦| + |⑧|}.$

(Sketch of the proof) We similarly have the selection in Table 8, which defines the minimum value of coverage. At the same time, ψ_{minSC} defines $minsupp(\tau^x)$, because $p \Rightarrow r$ occurs only in S_1.

Example 2. In the same condition in Example 1, we apply Proposition 4 to the instance τ^7. Then, we have $support(\tau^7) = 1/8 = minsupp(\tau^7)$, $accuracy(\tau^7) = 1.0 > minacc(\tau^7)$, $coverage(\tau^7) = 1/4 = mincov(\tau^7)$. As the side effect, we obtain a derived ψ_{minSC} in Table 7.

Table 8. The selection (underlined part) of the implications for $\psi_{minSC} \in DD(\Phi)$.

	$\{r\}$	$\{r,r'\}$	$\{r'\}$
$\{p\}$	$S_1 : \{p \Rightarrow r\}$	$S_2 : \{p \Rightarrow r, p \Rightarrow r'\}$	$S_3 : \{p \Rightarrow r'\}$
$\{p,p'\}$	$S_4 : \{p \Rightarrow r,$ $\underline{p' \Rightarrow r}\}$	$S_5 : \{p \Rightarrow r, p \Rightarrow r',$ $\underline{p' \Rightarrow r}, p' \Rightarrow r'\}$	$S_6 : \{\underline{p \Rightarrow r'},$ $p' \Rightarrow r'\}$
$\{p'\}$	$S_7 : \{\underline{p' \Rightarrow r}\}$	$S_8 : \{\underline{p' \Rightarrow r}, p' \Rightarrow r'\}$	$S_9 : \{p' \Rightarrow r'\}$

Proposition 5. *For NIS Φ and an implication τ, let us consider $FGr(\tau, \Phi)$ $=(①,②,③,④,⑤,⑥,⑦,⑧,⑨)$, and let us suppose $① \neq \emptyset$. There is an object $x \in ①$ and $\psi_{minSAC} \in DD(\Phi)$ satisfying the following, if and only if $⑤=\emptyset$.*

(1) $support(\tau^x)$ in $\psi_{minSAC} = minsupp(\tau^x) = |①|/|OB|,$

(2) $accuracy(\tau^x)$ in $\psi_{minSAC} = minacc(\tau^x) = \dfrac{|①|}{|①|+|②|+|③|+|⑥|},$ (18)

(3) $coverage(\tau^x)$ in $\psi_{minSAC} = mincov(\tau^x) = \dfrac{|①|}{|①|+|④|+|⑦|+|⑧|}.$

(*Sketch of the proof*) *Based on Tables 5 and 8, only the selection in $⑤$ is different. If $⑤=\emptyset$, the selections in Tables 5 and 8 are the same.*

We generally know there is no $\psi_{minSAC} \in DD(\Phi)$ defining $minsupp(\tau^x)$, $minacc(\tau^x)$, and $mincov(\tau^x)$ based on Proposition 5. However, there are ψ_{minSA} defining $minsupp(\tau^x)$ and $minacc(\tau^x)$, and ψ_{minSC} defining $minsupp(\tau^x)$ and $mincov(\tau^x)$. Namely, we recognize $minsupp(\tau^x)$, $minacc(\tau^x)$, and $mincov(\tau^x)$ by examining at most two derived *DISs* ψ_{minSA} and ψ_{minSC}. Now, we consider the maximum case. We have the following.

Proposition 6. *For NIS Φ and an implication τ, let us consider $FGr(\tau, \Phi)$ $=(①,②,③,④,⑤,⑥,⑦,⑧,⑨)$. For any object $x \in ① \cup ② \cup ④ \cup ⑤$, there is $\psi_{maxSAC} \in DD(\Phi)$ satisfying the following:*

(1) $support(\tau^x)$ in $\psi_{maxSAC} = maxsupp(\tau^x) = \dfrac{(|①|+|②|+|④|+|⑤|)}{|OB|},$

(2) $accuracy(\tau^x)$ in $\psi_{maxSAC} = maxacc(\tau^x) = \dfrac{(|①|+|②|+|④|+|⑤|)}{(|①|+|②|+|③|+|④|+|⑤|)},$ (19)

(3) $coverage(\tau^x)$ in $\psi_{maxSAC} = maxcov(\tau^x) = \dfrac{(|①|+|②|+|④|+|⑤|)}{(|①|+|②|+|④|+|⑤|+|⑦|)}.$

(*Sketch of the proof*) *Based on Table 9, we can similarly show the equations.*

4   Criterion Values and Apriori-Based Rule Generation

This section applies the obtained results to *Apriori*-based rule generation.

Table 9. The selection (underlined part) of the implications for $\psi_{maxSAC} \in DD(\Phi)$.

	$\{r\}$	$\{r, r'\}$	$\{r'\}$
$\{p\}$	$S_1 : \{p \Rightarrow r\}$	$S_2 : \{\underline{p \Rightarrow r}, p \Rightarrow r'\}$	$S_3 : \{p \Rightarrow r'\}$
$\{p, p'\}$	$S_4 : \{p \Rightarrow r,$ $p' \Rightarrow r\}$	$S_5 : \{\underline{p \Rightarrow r}, p \Rightarrow r',$ $p' \Rightarrow r, p' \Rightarrow r'\}$	$S_6 : \{p \Rightarrow r',$ $\underline{p' \Rightarrow r'}\}$
$\{p'\}$	$S_7 : \{p' \Rightarrow r\}$	$S_8 : \{p' \Rightarrow r, \underline{p' \Rightarrow r'}\}$	$S_9 : \{p' \Rightarrow r'\}$

4.1   Current Rule Generation by Criteria *support* and *accuracy*

By Proposition 3, there is a derived $\psi_{minSA} \in DD(\Phi)$ for τ^x ($x \in ①$), and we can prove (C1) (the definition of a certain rule by *support* and *accuracy*) and (C2) are equivalent [15].

(C1) $support(\tau^x) \geq \alpha$ and $accuracy(\tau^x) \geq \beta$ in each $\psi \in DD(\Phi)$,

(C2) $support(\tau^x) \geq \alpha$ and $accuracy(\tau^x) \geq \beta$ in ψ_{minSA}, namely
$minsupp(\tau^x) \geq \alpha$ and $minacc(\tau^x) \geq \beta$.

Similarly, we can prove (P1) (the definition of a possible rule by *support* and *accuracy*) and (P2) are equivalent by Proposition 6.

(P1) $support(\tau^x) \geq \alpha$ and $accuracy(\tau^x) \geq \beta$ in at least one $\psi \in DD(\Phi)$,

(P2) $support(\tau^x) \geq \alpha$ and $accuracy(\tau^x) \geq \beta$ in ψ_{maxSAC}, namely
$maxsupp(\tau^x) \geq \alpha$ and $maxacc(\tau^x) \geq \beta$.

Even though the conditions (C1) and (P1) depend upon $|DD(\Phi)|$, the conditions (C2) and (P2) do not depend upon $|DD(\Phi)|$. We can calculate the conditions (C2) and (P2) in the polynomial time order, and we escaped from the computational complexity problem on $|DD(\Phi)|$.

We have opened a software *getRNIA* powered by *NIS-Apriori* algorithm [18,19]. In this implementation, we handled *support* and *accuracy*, and did not handle *coverage*. The *getRNIA* actually calculates the conditions (C2) and (P2) instead of the definitions (C1) and (P1), respectively. Moreover, *getRNIA* employs the merging procedure internally. For two families $FGr((p_1 \Rightarrow r), \Phi)$ and $FGr((p_2 \Rightarrow r), \Phi)$, we can obtain $FGr((p_1 \wedge p_2 \Rightarrow r), \Phi)$ [14]. After merging them, we can similarly apply Propositions 3 to 6.

4.2   Rule Generation by Criteria *support*, *accuracy*, and *coverage*

At first, we consider *Apriori*-based possible rule generation. By Proposition 6, there is a derived $\psi_{maxSAC} \in DD(\Phi)$ for any τ^x, so we can easily prove the conditions (P'1) (the definition of a certain rule by *support*, *accuracy*, and *coverage*) and (P'2) are equivalent.

(P'1) $support(\tau^x) \geq \alpha$, $accuracy(\tau^x) \geq \beta$, and $coverage(\tau^x) \geq \gamma$
in at least one $\psi \in DD(\Phi)$,

(P'2) $support(\tau^x) \geq \alpha$, $accuracy(\tau^x) \geq \beta$, and $coverage(\tau^x) \geq \gamma$ in ψ_{maxSAC},
namely $maxsupp(\tau^x) \geq \alpha$, $maxacc(\tau^x) \geq \beta$, and $maxcov(\tau^x) \geq \gamma$.

Therefore, we employ the condition (P'2) for possible rule generation. The following is an overview of *Apriori*-based possible rule generation.

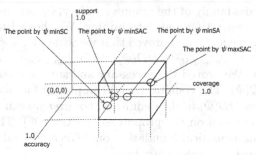

Fig. 1. The relation between ψ_{minSA}, ψ_{minSC}, ψ_{minSAC}, and ψ_{maxSAC}.

An Overview of Apriori-Based Possible Rule Generation in *NIS*

(Base step)

Set $i=1$. Prepare $LIST_i:=\{\tau : [A, val_A] \Rightarrow [Dec, val]\}$, and $ANS:=\{\}$.

(Inductive step)

Generate $FGr(\tau, \Phi)$ for each $\tau \in LIST_i$ by searching the total data set, and examine the following:

(1) $REST:=\{\}$. Set $i:=i+1$.
(2) Apply Proposition 6 for every $FGr(\tau, \Phi)$. If τ^x satisfies the constraint, $ANS :$ $= ANS \cup \{\tau\}$. If $maxsupp(\tau^x) \geq \alpha$, $maxcov(\tau^x) \geq \gamma$, and $maxacc(\tau^x) < \beta$, $REST := REST \cup \{\tau\}$.
(3) $LIST_i:=\{\}$. For $\tau_j : con_j \Rightarrow r, \tau_k : con_k \Rightarrow r \in REST$, generate $\tau :$ $condition \Rightarrow r$ (the condition is a conjunction consisting of i number of descriptors), and $LIST_i:=LIST_i \cup \{\tau\}$.
(4) If $LIST_i$ is an empty set, this program terminates. All certain rules are stored in ANS. Otherwise, repeat the inductive step.

Now, let us consider certain rule generation based on Propositions 3, 4 and 5. For τ^x ($x \in$ ①), the conditions (C'1) and (C'2) are equivalent, if the granule ⑤$=\emptyset$ in $FGr(\tau, \Phi)$.

(C'1) $support(\tau^x) \geq \alpha$, $accuracy(\tau^x) \geq \beta$, and $coverage(\tau^x) \geq \gamma$ in each $\psi \in$ $DD(\Phi)$,
(C'2) $support(\tau^x) \geq \alpha$, $accuracy(\tau^x) \geq \beta$, and $coverage(\tau^x) \geq \gamma$ in ψ_{minSAC}, namely $minsupp(\tau^x) \geq \alpha$, $minacc(\tau^x) \geq \beta$, and $mincov(\tau^x) \geq \gamma$.

If ⑤ $\neq \emptyset$, we need to consider the condition (C'3).

(C'3) $support(\tau^x) \geq \alpha$, $accuracy(\tau^x) \geq \beta$ in ψ_{minSA}, and $coverage(\tau^x) \geq \gamma$ in ψ_{minSC}.

Figure 1 shows the survey. We will employ the condition (C'3) for adding the criterion *coverage* to the *NIS-Apriori* algorithm as well as *support* and *accuracy*.

5   Concluding Remarks

This paper proposed a family of the granules $FGr(\tau, \Phi)$, and examined its property related to rule generation and data mining. We showed the calculation on $mincov(\tau^x)$ and $maxcov(\tau^x)$, and proved that we always have $\psi_{minSA} \in DD(\Phi)$, $\psi_{minSC} \in DD(\Phi)$, and $\psi_{maxSAC} \in DD(\Phi)$. As for $\psi_{minSAC} \in DD(\Phi)$, generally we may not have it. We proved the necessary and sufficient condition for existing $\psi_{minSAC} \in DD(\Phi)$. The computational complexity for calculating criterion values depends upon $DD(\Phi)$ in the definition, however we can calculate them in the polynomial time based on the properties of $FGr(\tau, \Phi)$. The content in this paper will be the mathematical foundation on $FGr(\tau, \Phi)$, and such background will enhance *Apriori*-based rule generation.

Acknowledgment. The authors would be grateful for reviewers' useful comments. This work is supported by JSPS (Japan Society for the Promotion of Science) KAKENHI Grant Number 26330277.

References

1. Agrawal, R., Srikant, R.: Fast algorithms for mining association rules in large databases. In: Proceedings of VLDB'94, pp. 487–499. Morgan Kaufmann (1994)
2. Agrawal, R., Mannila, H., Srikant, R., Toivonen, H., Verkamo, A.I.: Fast discovery of association rules. In: Advances in Knowledge Discovery and Data Mining, pp. 307–328. AAAI/MIT Press (1996)
3. Blackburn, P., et al.: Modal Logic. Cambridge University Press, Cambridge (2001)
4. Frank, A., Asuncion, A.: UCI machine learning repository. Irvine, CA: University of California, School of Information and Computer Science (2010). http://mlearn. ics.uci.edu/MLRepository.html
5. Grzymała-Busse, J.W.: Data with missing attribute values: generalization of indiscernibility relation and rule induction. Trans. Rough Sets **1**, 78–95 (2004)
6. Lipski, W.: On semantic issues connected with incomplete information databases. ACM Trans. Database Syst. **4**(3), 262–296 (1979)
7. Lipski, W.: On databases with incomplete information. J. ACM **28**(1), 41–70 (1981)
8. Nakata, M., Sakai, H.: Twofold rough approximations under incomplete information. Int. J. Gen. Syst. **42**(6), 546–571 (2013)
9. Orłowska, E., Pawlak, Z.: Representation of nondeterministic information. Theor. Comput. Sci. **29**(1–2), 27–39 (1984)
10. Pawlak, Z.: Information systems theoretical foundations. Inf. Syst. **6**(3), 205–218 (1981)
11. Pawlak, Z.: Systemy Informacyjne: Podstawy Teoretyczne (in Polish) WNT (1983)
12. Pawlak, Z.: Rough Sets: Theoretical Aspects of Reasoning about Data. Kluwer Academic Publishers, Dordrecht (1991)
13. Sakai, H., Ishibashi, R., Koba, K., Nakata, M.: Rules and apriori algorithm in non-deterministic information systems. Trans. Rough Sets **9**, 328–350 (2008)
14. Sakai, H., Wu, M., Nakata, M.: Division charts as granules and their merging algorithm for rule generation in nondeterministic data. Int. J. Intell. Syst. **28**(9), 865–882 (2013)

15. Sakai, H., Wu, M., Nakata, M.: Apriori-based rule generation in incomplete information databases and non-deterministic information systems. Fundam. Informaticae **130**(3), 343–376 (2014)
16. Sakai, H., Wu, M., Nakata, M.: The completeness of NIS-Apriori algorithm and a software tool getRNIA. In: Proceedings of International Conference on AAI2014, pp. 115–121 (2014)
17. Skowron, A., Rauszer, C.: The discernibility matrices and functions in information systems. In: Słowiński, R. (ed.) Intelligent Decision Support - Handbook of Advances and Applications of the Rough Set Theory, pp. 331–362. Kluwer Academic Publishers, Dordrecht (1992)
18. Wu, M., Sakai, H.: getRNIA web software (2013). http://getrnia.org
19. Wu, M., Nakata, M., Sakai, H.: An overview of the getRNIA system for non-deterministic data. Procedia Comput. Sci. **22**, 615–622 (2013)

Generalized Rough Sets

New Neighborhood Based Rough Sets

Lynn D'eer[1](✉) and Chris Cornelis[1,2]

[1] Department of Applied Mathematics, Computer Science and Statistics,
Ghent University, Ghent, Belgium
`lynn.deer@ugent.be`
[2] Department of Computer Science and Artificial Intelligence,
University of Granada, Granada, Spain
`chriscornelis@ugr.es`

Abstract. Neighborhood based rough sets are important generalizations of the classical rough sets of Pawlak, as neighborhood operators generalize equivalence classes. In this article, we introduce nine neighborhood based operators and we study the partial order relations between twenty-two different neighborhood operators obtained from one covering. Seven neighborhood operators result in new rough set approximation operators. We study how these operators are related to the other fifteen neighborhood based approximation operators in terms of partial order relations, as well as to seven non-neighborhood-based rough set approximation operators.

Keywords: Neighborhood operator · Rough sets · Approximation operator · Covering

1 Introduction

Pawlak [4] introduced rough set theory in order to deal with uncertainty in data systems due to incompleteness. Originally, an equivalence relation on the universe of discourse was used to describe the indiscernibility between elements. A generalization of rough sets is obtained by replacing the equivalence relation by a neighborhood operator (e.g. [11]). Another generalization results from considering a covering instead of the partition of equivalence classes ([12]).

In [3], twenty-four neighborhood operators obtained from one covering were studied. It was shown that the collection reduces to thirteen different operators. Moreover, it was discussed that two rough set approximation operators which were originally not introduced regarding a neighborhood operator are actually neighborhood based. In particular, they are related to the inverse operators of two of the thirteen neighborhood operators. In this article, we continue the study on inverse neighborhood operators in order to obtain new rough set approximation operators. We discuss nine inverse neighborhood operators, of which seven result in new rough set approximation operators. Furthermore, we study partial order relations between twenty-two neighborhood operators on the one hand, on

D. Ciucci et al. (Eds.): RSKT 2015, LNAI 9436, pp. 191–201, 2015.
DOI: 10.1007/978-3-319-25754-9_17

the other hand partial order relations between twenty-nine approximation operators. Partial order relations between different approximation operators yield information on the accuracy of the approximation operators, i.e., upper approximation operators which yield smaller values will achieve higher accuracy, and therefore, they induce more accurate approximations of a concept.

The outline of this article is as follows. In Sect. 2, we present preliminary concepts. In Sect. 3, we introduce nine neighborhood operators and discuss the partial order relations of twenty-two neighborhood operators. In Sect. 4, the partial order relations of seven new approximation operators with existing rough set approximation operators are discussed. Finally, conclusions and future work are outlined in Sect. 5.

2   Preliminaries

Throughout this paper we assume that the universe U is a non-empty set. Originally, Pawlak [4] used an equivalence relation E to discern elements of the universe of discourse. Two fundamental concepts in rough set theory are the lower and upper approximation of a set $A \subseteq U$. These concepts describe the elements certainly and possibly belonging to A:

$$\underline{\text{apr}}_E(A) = \{x \in U \mid [x]_E \subseteq A\} = \bigcup\{[x]_E \in U/E \mid [x]_E \subseteq A\}, \tag{1}$$
$$\overline{\text{apr}}_E(A) = \{x \in U \mid [x]_E \cap A \neq \emptyset\} = \bigcup\{[x]_E \in U/E \mid [x]_E \cap A \neq \emptyset\}, \tag{2}$$

where U/E represents the set of equivalence classes defined from E. The first equality in Eqs. (1) and (2) is sometimes called the element based definition, while the second one is called the granule based definition [10].

Many generalizations of Pawlak's rough set model can be found in literature. In [10], a survey on dual generalizations is presented. A pair of approximation operators is called dual, if for all $A \subseteq U$, $\overline{\text{apr}}(\text{co}(A)) = \text{co}(\underline{\text{apr}}(A))$.

Equivalence classes can be generalized by neighborhood operators. A neighborhood operator N is a mapping $N: U \to \mathcal{P}(U)$, where $\mathcal{P}(U)$ represents the collection of subsets of U. It is often assumed that the neighborhood operator is reflexive, i.e., $x \in N(x)$ for all $x \in U$. Given N, we define its inverse neighborhood operator N^{-1} by $x \in N^{-1}(y) \Leftrightarrow y \in N(x)$ for $x, y \in U$.

Each neighborhood operator N yields a dual pair of approximation operators $(\underline{\text{apr}}_N, \overline{\text{apr}}_N)$ defined by

$$\underline{\text{apr}}_N(A) = \{x \in U \mid N(x) \subseteq A\}, \tag{3}$$
$$\overline{\text{apr}}_N(A) = \{x \in U \mid N(x) \cap A \neq \emptyset\}, \tag{4}$$

for $A \subseteq U$ ([10]). The approximation operators generalize the element based definition stated in Eqs. (1) and (2). A particular neighborhood of x is the afterset $R(x)$ of x determined by a binary relation R: $R(x) = \{y \in U \mid xRy\}$ ([11]).

Generalization of the granule based approximation operators can be obtained by replacing the partition U/E by a covering of U. Let I be an index set, then

Table 1. Neighborhood operators based on coverings

Group	Operators	Group	Operators
A	$N_1^{\mathbb{C}}, N_1^{\mathbb{C}_1}, N_1^{\mathbb{C}_3}, N_2^{\mathbb{C}_3}, N_1^{\mathbb{C}_\cap}$	H	$N_4^{\mathbb{C}}, N_2^{\mathbb{C}_2}, N_4^{\mathbb{C}_2}, N_4^{\mathbb{C}_\cap}$
B	$N_3^{\mathbb{C}_1}$	I	$N_2^{\mathbb{C}_4}$
C	$N_3^{\mathbb{C}_3}$	J	$N_3^{\mathbb{C}_4}$
D	$N_4^{\mathbb{C}_3}$	K	$N_4^{\mathbb{C}_4}$
E	$N_2^{\mathbb{C}}, N_2^{\mathbb{C}_1}$	L	$N_4^{\mathbb{C}_1}$
F	$N_3^{\mathbb{C}}, N_1^{\mathbb{C}_2}, N_3^{\mathbb{C}_2}, N_3^{\mathbb{C}_\cap}$	M	$N_2^{\mathbb{C}_\cap}$
G	$N_1^{\mathbb{C}_4}$		

a collection $\mathbb{C} = \{K_i \subseteq U \mid i \in I\}$ of non-empty subsets of U is called a covering of U if $\bigcup_{i \in I} K_i = U$.

Given a covering $\mathbb{C}$, we are interested in the sets $K \in \mathbb{C}$ such that $x \in K$. Let $\mathbb{C}$ be a covering of U and $x \in U$, then the neighborhood system $\mathcal{C}(\mathbb{C}, x)$ of x is defined by $\mathcal{C}(\mathbb{C}, x) = \{K \in \mathbb{C} \mid x \in K\}$ ([10]). Moreover, the sets

$$md(\mathbb{C}, x) = \{K \in \mathcal{C}(\mathbb{C}, x) \mid (\forall S \in \mathcal{C}(\mathbb{C}, x))(S \subseteq K \Rightarrow K = S)\},$$
$$MD(\mathbb{C}, x) = \{K \in \mathcal{C}(\mathbb{C}, x) \mid (\forall S \in \mathcal{C}(\mathbb{C}, x))(S \supseteq K \Rightarrow K = S)\}$$

are called the minimal [1] and maximal [14] description of x. They can be seen as the two extreme neighborhoods of x. If we apply the intersection and union to these neighborhood systems, we derive four neighborhood operators based on coverings [10]:

1. $N_1^{\mathbb{C}}(x) = \bigcap\{K \in \mathbb{C} \mid K \in md(\mathbb{C}, x)\} = \bigcap \mathcal{C}(\mathbb{C}, x)$,
2. $N_2^{\mathbb{C}}(x) = \bigcup\{K \in \mathbb{C} \mid K \in md(\mathbb{C}, x)\}$,
3. $N_3^{\mathbb{C}}(x) = \bigcap\{K \in \mathbb{C} \mid K \in MD(\mathbb{C}, x)\}$,
4. $N_4^{\mathbb{C}}(x) = \bigcup\{K \in \mathbb{C} \mid K \in MD(\mathbb{C}, x)\} = \bigcup \mathcal{C}(\mathbb{C}, x)$.

In addition, [10] considered five different coverings derived from a covering $\mathbb{C}$:

1. $\mathbb{C}_1 = \bigcup\{md(\mathbb{C}, x) \mid x \in U\}$,
2. $\mathbb{C}_2 = \bigcup\{MD(\mathbb{C}, x) \mid x \in U\}$,
3. $\mathbb{C}_3 = \{\bigcap md(\mathbb{C}, x) \mid x \in U\} = \{\bigcap \mathcal{C}(\mathbb{C}, x) \mid x \in U\}$,
4. $\mathbb{C}_4 = \{\bigcup MD(\mathbb{C}, x) \mid x \in U\} = \{\bigcup \mathcal{C}(\mathbb{C}, x) \mid x \in U\}$,
5. $\mathbb{C}_\cap = \mathbb{C} \setminus \{K \in \mathbb{C} \mid (\exists \mathbb{C}' \subseteq \mathbb{C} \setminus \{K\})(K = \bigcap \mathbb{C}')\}$.

When combining the four neighborhood operators and the six coverings (one original and five derived ones), then we obtain twenty-four combinations $N_i^{\mathbb{C}_j}$ for $N_i \in \{N_1, N_2, N_3, N_4\}$ and $\mathbb{C}_j \in \{\mathbb{C}, \mathbb{C}_1, \mathbb{C}_2, \mathbb{C}_3, \mathbb{C}_4, \mathbb{C}_\cap\}$. However, in [3], it is stated that there are only thirteen different groups of neighborhood operators. These groups are shown in Table 1. Besides equalities between neighborhood operators, [3] studied partial order relations between the operators. Hence, the

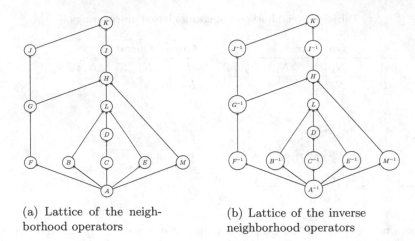

(a) Lattice of the neigh-
borhood operators

(b) Lattice of the inverse
neighborhood operators

Fig. 1. Lattices of neighborhood operators

lattice given in Fig. 1a with respect to the partial order relation $\leq$ was established: let N and N' be two neighborhood operators, then $N \leq N'$ if and only if $\forall x, y \in U \colon x \in N(y) \Rightarrow x \in N'(y)$.

Moreover, it was stated that other covering based approximation operators which are not originally defined according to Eqs. (3) and (4), can be described with neighborhood operators. Let $\mathbb{C}$ be a covering, then the upper approximation operators $H_3^{\mathbb{C}}$ [8] and $H_5^{\mathbb{C}}$ [13] are defined as follows, let $A \subseteq U$, then

$$H_3^{\mathbb{C}}(A) = \bigcup \{N_2^{\mathbb{C}}(x) \colon x \in A\}, \tag{5}$$

$$H_5^{\mathbb{C}}(A) = \bigcup \{N_1^{\mathbb{C}}(x) \colon x \in A\}. \tag{6}$$

We have the following proposition:

Proposition 1. *[3] Let $\mathbb{C}$ be a covering and $A \subseteq U$, then*

$$H_3^{\mathbb{C}}(A) = \{x \in U \mid (N_2^{\mathbb{C}})^{-1}(x) \cap A \neq \emptyset\} = \overline{\mathrm{apr}}_{(N_2^{\mathbb{C}})^{-1}}(A),$$

$$H_5^{\mathbb{C}}(A) = \{x \in U \mid (N_1^{\mathbb{C}})^{-1}(x) \cap A \neq \emptyset\} = \overline{\mathrm{apr}}_{(N_1^{\mathbb{C}})^{-1}}(A).$$

Therefore, $H_3^{\mathbb{C}}$ and $H_5^{\mathbb{C}}$ are element based approximation operators, based on the inverse neighborhood operators of $N_2^{\mathbb{C}}$ and $N_1^{\mathbb{C}}$ respectively.

3   Neighborhood Operators

Proposition 1 suggests that it is interesting to study the inverse neighborhood operators of the operators stated in Table 1. In this section, we discuss which inverse neighborhood operators are new. Moreover, we study the partial order relations and add them to Fig. 1a.

Table 2. Neighborhood operators based on coverings and their inverses

Group	Operators	Group	Operators
A	$N_1^{\mathbb{C}}, N_1^{\mathbb{C}_1}, N_1^{\mathbb{C}_3}, N_2^{\mathbb{C}_3}, N_1^{\mathbb{C}\cap}$	A^{-1}	$(N_1^{\mathbb{C}})^{-1}, (N_1^{\mathbb{C}_1})^{-1}, (N_1^{\mathbb{C}_3})^{-1}, (N_2^{\mathbb{C}_3})^{-1}, (N_1^{\mathbb{C}\cap})^{-1}$
B	$N_3^{\mathbb{C}_1}$	B^{-1}	$(N_3^{\mathbb{C}_1})^{-1}$
C	$N_3^{\mathbb{C}_3}$	C^{-1}	$(N_3^{\mathbb{C}_3})^{-1}$
D	$N_4^{\mathbb{C}_3}$		
E	$N_2^{\mathbb{C}}, N_2^{\mathbb{C}_1}$	E^{-1}	$(N_2^{\mathbb{C}})^{-1}, (N_2^{\mathbb{C}_1})^{-1}$
F	$N_3^{\mathbb{C}}, N_1^{\mathbb{C}_2}, N_3^{\mathbb{C}_2}, N_3^{\mathbb{C}\cap}$	F^{-1}	$(N_3^{\mathbb{C}})^{-1}, (N_1^{\mathbb{C}_2})^{-1}, (N_3^{\mathbb{C}_2})^{-1}, (N_3^{\mathbb{C}\cap})^{-1}$
G	$N_1^{\mathbb{C}_4}$	G^{-1}	$(N_1^{\mathbb{C}_4})^{-1}$
H	$N_4^{\mathbb{C}}, N_2^{\mathbb{C}_2}, N_4^{\mathbb{C}_2}, N_4^{\mathbb{C}\cap}$		
I	$N_2^{\mathbb{C}_4}$	I^{-1}	$(N_2^{\mathbb{C}_4})^{-1}$
J	$N_3^{\mathbb{C}_4}$	J^{-1}	$(N_3^{\mathbb{C}_4})^{-1}$
K	$N_4^{\mathbb{C}_4}$		
L	$N_4^{\mathbb{C}_1}$		
M	$N_2^{\mathbb{C}\cap}$	M^{-1}	$(N_2^{\mathbb{C}\cap})^{-1}$

3.1 Inverse Neighborhood Operators

To study the inverse operators of the thirteen neighborhood operators from Table 1, we first observe that the operators $N_4^{\mathbb{C}_j}$ are symmetric, and therefore, they are equal to their inverses: e.g. let $x, y \in U$, then $x \in N_4^{\mathbb{C}_j}(y)$ if and only if there exists a $K \in \mathbb{C}_j$ such that $x, y \in K$. Hence, $y \in N_4^{\mathbb{C}_j}(x)$, or $x \in (N_4^{\mathbb{C}_j})^{-1}(y)$.

Moreover, the inverses of two different operators are different as well.

Proposition 2. *Let N and N' be two different neighborhood operators, then $N^{-1} \neq N'^{-1}$.*

Proof. If $N \neq N'$, then there exist $x, y \in U$ such that $y \in N(x)$ and $y \notin N'(x)$. Hence, $x \in N^{-1}(y)$ and $x \notin N'^{-1}(y)$. Thus, $N^{-1} \neq N'^{-1}$.

Combining the above two observations, there are nine possible new neighborhood operators. We need to check whether any of the inverse operators coincides with one of the thirteen known groups. However, this is not the case. Here we illustrate that the operator A^{-1} is different from the operators A–D. Other counterexamples can be obtained similarly.

Example 1. Let $U = \{1, 2, 3\}$ and $\mathbb{C} = \{\{1, 2\}, \{1, 2, 3\}\}$. Then $(N_1^{\mathbb{C}})^{-1}(3) = \{3\}$, but $N_1^{\mathbb{C}}(3) = N_3^{\mathbb{C}_1}(3) = N_3^{\mathbb{C}_3}(3) = N_4^{\mathbb{C}_3}(3) = \{1, 2, 3\}$. Therefore, the neighborhood operators of A^{-1} are different than the ones of A, B, C and D.

Hence, we obtain twenty-two neighborhood operators stated in Table 2. For each group X, we denote X^{-1} for the group of inverse operators from X.

3.2   Partial Order Relation for Neighborhood Operators

The objective of this section is to obtain a Hasse diagram containing the twenty-two neighborhood operators from Table 2 with respect to the partial order relation $\leq$. First, we present the following proposition:

Proposition 3. *Let N and N' be two neighborhood operators such that $N \leq N'$, then $N^{-1} \leq N'^{-1}$.*

Proof. Let $N \leq N'$, then for all $x, y \in U$ it holds that $y \in N(x) \Rightarrow y \in N'(x)$. Hence, $x \in N^{-1}(y) \Rightarrow x \in N'^{-1}(y)$ for all $x, y \in U$ and thus, $N^{-1} \leq N'^{-1}$.

Hence, the comparability of the inverse operators follows immediately from the comparability of the thirteen original neighborhood operators. Based on Proposition 3 and Fig. 1a, we obtain the lattice for the inverse neighborhood operators in Fig. 1b. The goal is to combine Figs. 1a and b. To this end, we need to study the comparability between the original neighborhood operators A–M and the nine inverse neighborhood operators $A^{-1} - M^{-1}$. The comparability of operators D, H, K and L is already shown in Fig. 1b, as it is a direct consequence of the symmetry of $N_4^{C_j}$. Moreover, we have the following proposition:

Proposition 4. *It holds that $A^{-1}, B^{-1}, C^{-1}, E^{-1}, F^{-1}, G^{-1}, M^{-1} \leq I$ and A, $B, C, E, F, G, M \leq I^{-1}$.*

Proof. Immediate consequence from $H \leq I$ (Fig. 1a) and $H \leq I^{-1}$ (Fig. 1b).

Other partial order relations do not hold. In the next example we illustrate that the operator A^{-1} is incomparable with A and B. Other counterexamples can be established analogously.

Example 2. Let $U = \{1, 2, 3\}$ and $\mathbb{C} = \{\{3\}, \{1, 2\}, \{1, 3\}, \{1, 2, 3\}\}$, then we compute that $N_1^{\mathbb{C}}(1) = N_3^{\mathbb{C}_1}(1) = \{1\}$ and $N_1^{\mathbb{C}}(2) = N_3^{\mathbb{C}_1}(2) = \{1, 2\}$. However, $(N_1^{\mathbb{C}})^{-1}(1) = \{1, 2\}$ and $(N_1^{\mathbb{C}})^{-1}(2) = \{2\}$. Hence, the neighborhood operators of A^{-1} are incomparable with those of A and B.

Therefore, we obtain the Hasse diagram of the twenty-two neighborhood operators of Table 2 in Fig. 2. Note that there is a clear symmetry. The neighborhood operator K yields the largest neighborhoods, while the operators A and A^{-1} yield the smallest.

4   Hasse Diagram of Approximation Operators

In this section, we discuss the Hasse diagram of different upper approximation operators defined in literature. In Sect. 4.1, we first discuss the element based approximation operators generated by the neighborhood operators stated in Table 2. In Sect. 4.2, we will discuss the partial order relation of other upper approximation operators which are not related with one of the neighborhood operators of Table 2.

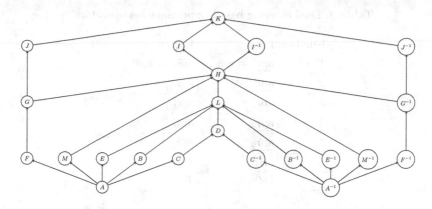

Fig. 2. Hasse diagram of thirteen neighborhood operators and their inverses

The study of the partial order relation of upper approximation operators gives an indication on the accuracy of the dual pair of approximation operators. This accuracy is often given by the ratio of the cardinality of the lower and upper approximation operator $\frac{|\underline{apr}|}{|\overline{apr}|}$ ([4]). By duality it holds that

$$\forall A \subseteq U : \underline{apr}_1(A) \subseteq \underline{apr}_2(A) \Leftrightarrow \overline{apr}_2(A) \subseteq \overline{apr}_1(A).$$

Therefore, smaller upper approximation operators yield higher accuracy.

4.1   Element Based Approximation Operators

The results stated in Sect. 3.1 can be used to define new element based approximations operators, since they are only equal if the generating neighborhood operators are equal: given two neighborhood operators N and N', then $\overline{apr}_N = \overline{apr}_{N'}$ if and only if $N = N'$ ([3]).

As discussed in Proposition 1, the approximation operators based on the neighborhood operators of groups A^{-1} and E^{-1} are H_5^C and H_3^C, respectively. However, the approximation operators related to the other seven inverse neighborhood operators are not yet established in literature.

Moreover, the partial order relation between two neighborhood operators yields immediate results for the order relation between the corresponding approximation operators: if $N \leq N'$, then for all $A \subseteq U$, $\overline{apr}_N(A) \subseteq \overline{apr}_{N'}(A)$ ([7]). Therefore, the Hasse diagram of the upper approximation operators $\overline{apr}_N$ is identical to the Hasse diagram of the neighborhood operators N given in Fig. 2.

From Fig. 2, we derive that the approximation operators based on the neighborhood operators from groups A and A^{-1} have the highest accuracy amongst the element based approximation operators based on neighborhood operators from Table 2. Moreover, the approximation operators based on neighborhood operators from groups B, C, E, F, M and their inverse operators have a high accuracy as well. Approximation operators based on I, I^{-1}, J, J^{-1} and K have a lower accuracy.

Table 3. Dual covering based approximation operators

Number	Lower	Upper
1	$\underline{\mathrm{apr}}'_{\mathbb{C}} = \underline{\mathrm{apr}}'_{\mathbb{C}_1}$	$\overline{\mathrm{apr}}'_{\mathbb{C}} = \overline{\mathrm{apr}}'_{\mathbb{C}_1}$
2	$\underline{\mathrm{apr}}'_{\mathbb{C}_2}$	$\overline{\mathrm{apr}}'_{\mathbb{C}_2}$
3	$\underline{\mathrm{apr}}'_{\mathbb{C}_4}$	$\overline{\mathrm{apr}}'_{\mathbb{C}_4}$
4	$\underline{\mathrm{apr}}'_{\mathbb{C}_\cap}$	$\overline{\mathrm{apr}}'_{\mathbb{C}_\cap}$
5	$\underline{\mathrm{apr}}_{S_\cap}$	$\overline{\mathrm{apr}}_{S_\cap}$
6	$(H_1^{\mathbb{C}})^\partial$	$H_1^{\mathbb{C}}$
7	$(H_4^{\mathbb{C}})^\partial$	$H_4^{\mathbb{C}}$

4.2   Partial Order Relations of Other Approximation Operators

In [3], the partial order relations with respect to other approximation operators than the element based approximation operators are discussed as well. These approximation operators are stated in Table 3. We shortly recall their definition here and study their partial relations with the approximation operators based on B^{-1}, C^{-1}, F^{-1}, G^{-1}, I^{-1}, J^{-1} and M^{-1}. The partial order relations with the other element based approximation operators were already discussed in [3].

The approximation operators 1–4 are called tight covering based rough set approximation operators ([2]). The operator $\underline{\mathrm{apr}}'_{\mathbb{C}}$ is obtained by replacing the partition U/E in Eq. (1) by a covering $\mathbb{C}$ ([5]): let $A \subseteq U$, then

$$\underline{\mathrm{apr}}'_{\mathbb{C}}(A) = \bigcup \{K \in \mathbb{C} \mid K \subseteq A\},$$
$$\overline{\mathrm{apr}}'_{\mathbb{C}}(A) = \mathrm{co}(\underline{\mathrm{apr}}'_{\mathbb{C}}(\mathrm{co}(A))) = \{x \in U \mid (\forall K \in \mathbb{C})(x \in K \Rightarrow K \cap A \neq \emptyset)\}.$$

Note that the tight covering based approximation operator based on $\mathbb{C}_3$ is equal to an element based approximation operator: $\underline{\mathrm{apr}}'_{\mathbb{C}_3} = \underline{\mathrm{apr}}_{N_1^{\mathbb{C}}}$ ([6]).

The pair of approximation operators $(\underline{\mathrm{apr}}_{S_\cap}, \overline{\mathrm{apr}}_{S_\cap})$ is an example of a closure system based pair [10]. Let $S_{\cap,\mathbb{C}}$ be the intersection closure of the covering $\mathbb{C}$, i.e., the minimal subset of $\mathcal{P}(U)$ containing $\mathbb{C}, \emptyset$ and U which is closed under set intersection, and let $S'_{\cap,\mathbb{C}} = \{\mathrm{co}(K) \mid K \in S_{\cap,\mathbb{C}}\}$, then for $A \subseteq U$ it holds that

$$\underline{\mathrm{apr}}_{S_\cap}(A) = \bigcup \{K \in S'_{\cap,\mathbb{C}} \mid K \subseteq A\}, \overline{\mathrm{apr}}_{S_\cap}(A) = \bigcap \{K \in S_{\cap,\mathbb{C}} \mid K \supseteq A\}.$$

Finally, the operators $H_1^{\mathbb{C}}$ ([12]) and $H_4^{\mathbb{C}}$ ([15]) are defined by, for $A \subseteq U$,

$$H_1^{\mathbb{C}}(A) = \underline{\mathrm{apr}}'_{\mathbb{C}}(A) \cup \left(\bigcup \left\{ \bigcup \mathrm{md}(\mathbb{C}, x) : x \in A \setminus \underline{\mathrm{apr}}'_{\mathbb{C}}(A) \right\} \right),$$
$$H_4^{\mathbb{C}}(A) = \underline{\mathrm{apr}}'_{\mathbb{C}}(A) \cup \left(\bigcup \left\{ K \in \mathbb{C} : K \cap (A \setminus \underline{\mathrm{apr}}'_{\mathbb{C}}(A)) \neq \emptyset \right\} \right),$$

and their dual lower approximation operators by $(H_i^{\mathbb{C}})^\partial(A) = \mathrm{co}(H_i^{\mathbb{C}}(\mathrm{co}(A)))$. Both operators were proposed in the so-called non-dual framework of approximation operators [9].

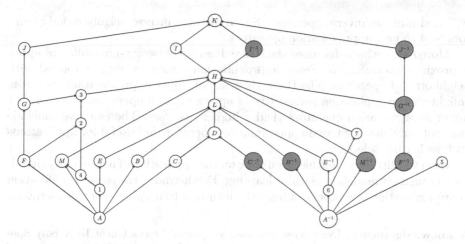

Fig. 3. Hasse diagram of twenty-nine upper approximation operators

In order to extend the Hasse diagram of the element based approximation operators with these seven other covering based approximation operators, we need to study the partial order relations between these seven operators, and the seven element based approximation operators based on the inverse neighborhood operators B^{-1}, C^{-1}, F^{-1}, G^{-1}, I^{-1}, J^{-1} and M^{-1}. However, there are no new partial order relations than the ones established in Fig. 2 and [3] as illustrated in the next example.

Example 3. Let $\mathbb{C} = \{\{1\}, \{3\}, \{1,3\}, \{1,4\}, \{2,4\}, \{3,4\}, \{2,3,4\}\}$ be a covering for $U = \{1,2,3,4\}$. Then $\overline{\text{apr}}_{(N_3^{C_1})^{-1}}(\{1\}) = \{1,4\}$ and $\overline{\text{apr}}_{(N_3^{C_1})^{-1}}(\{4\}) = \{4\}$, but $\overline{\text{apr}}'_{\mathbb{C}}(\{1\}) = \{1\}$ and $\overline{\text{apr}}'_{\mathbb{C}}(\{4\}) = \{2,4\}$. Hence, $\overline{\text{apr}}_{(N_3^{C_1})^{-1}}$ and $\overline{\text{apr}}'_{\mathbb{C}}$ are incomparable.

Therefore, we obtain the Hasse diagram of twenty-nine upper approximation operators in Fig. 3, where we have highlighted the new rough set approximation operators. Observe that we denote the upper approximation operator $\overline{\text{apr}}_N$ with the letter of the associated neighborhood operator N.

Concerning the accuracy of approximation operators, we see that the pairs of approximation operators which are not based on neighborhood operators (Table 3) achieve high accuracy in general, but lower accuracy than the approximation operators based on A and A^{-1}.

5   Conclusion and Future Work

Neighborhood operators are fundamental in the research of covering based rough sets since they generalize the equivalence classes in Pawlak's original rough set model. In this article, we have established the Hasse diagram of twenty-two neighborhood operators, of which thirteen are obtained from the combinations

$N_i^{C_j}$ and nine are inverse operators. Seven of these inverse neighborhood operators lead to new approximation operators.

Moreover, we have discussed the Hasse diagram of twenty-nine different upper approximation operators. Seven approximation operators are not related with neighborhood operators. The Hasse diagram of upper approximation operators enlightens the application perspective of approximation operators, since smaller upper approximation operators yield a higher accuracy. Therefore we conclude that not only element based approximation operators are useful for applications such as feature selection.

However, little research has been done on the application of non-neighborhood-based rough set models in machine learning. Furthermore, the recent proliferation of approximation operators can bury the intuition to usefulness of the operators.

Acknowledgements. Lynn D'eer has been supported by the Ghent University Special Research Fund. Chris Cornelis was partially supported by the Spanish Ministry of Science and Technology under the project TIN2011-28488 and the Andalusian Research Plans P11-TIC-7765 and P10-TIC-6858, and by project PYR-2014-8 of the Genil Program of CEI BioTic GRANADA.

References

1. Bonikowski, Z., Brynarski, E.: Extensions and intensions in rough set theory. Inf. Sci. **107**, 149–167 (1998)
2. Couso, I., Dubois, D.: Rough sets, coverings and incomplete information. Fundam. Informaticae **108**, 223–247 (2011)
3. D'eer, L., Restrepo, M., Cornelis, C., Gómez, J.: Neighborhood operators for covering based rough sets (Submitted)
4. Pawlak, Z.: Rough sets. Int. J. Comput. Inf. Sci. **11**, 341–356 (1982)
5. Pomykala, J.A.: Approximation operations in approximation space. Bull. Acad. Polonaise Sci. **35**(9–10), 653–662 (1987)
6. Restrepo, M., Cornelis, C., Gómez, J.: Duality, conjugacy and adjointness of approximation operators in covering-based rough sets. Int. J. Approximate Reasoning **55**, 469–485 (2014)
7. Restrepo, M., Cornelis, C., Gómez, J.: Partial order relation for approximation operators in covering based rough sets. Inf. Sci. **284**, 44–59 (2014)
8. Tsang, E., Chen, D., Lee, J., Yeung, D.S.: On the upper approximations of covering generalized rough sets. In: Proceedings of the 3rd International Conference on Machine Learning and Cybernetics, pp. 4200–4203 (2004)
9. Yang, T., Li, Q.: Reduction about approximation spaces of covering generalized rough sets. Int. J. Approximate Reasoning **51**, 335–345 (2010)
10. Yao, Y.Y., Yao, B.: Covering based rough sets approximations. Inf. Sci. **200**, 91–107 (2012)
11. Yao, Y.Y.: Relational interpretations of neighborhood operators and rough set approximation operators. Inf. Sci. **101**, 21–47 (1998)
12. Żakowski, W.: Approximations in the space (u, π). Demonstratio Math. **16**, 761–769 (1983)
13. Zhu, W.: Topological approaches to covering rough sets. Inf. Sci. **177**(6), 1499–1508 (2007)

14. Zhu, W., Wang, F.: On three types of covering-based rough sets. IEEE Trans. Knowl. Data Eng. **19**(8), 1131–1144 (2007)
15. Zhu, W., Wang, F.: A new type of covering rough sets. In: 3rd International IEEE Conference Intelligence Systems (2006)

Rough Sets and Textural Neighbourhoods

Murat Diker[1]([✉]), Ayşegül Altay Uğur[2], and Sadık Bayhan[3]

[1] Department of Mathematics, Hacettepe University, Beytepe, 06532 Ankara, Turkey
mdiker@hacettepe.edu.tr
[2] Department of Secondary Science and Mathematics Education,
Hacettepe University, Beytepe, 06532 Ankara, Turkey
altayay@hacettepe.edu.tr
[3] Department of Mathematics, Mehmet Akif Ersoy University,
İstiklal Campus, 15030 Burdur, Turkey
bayhan@mehmetakif.edu.tr

Abstract. In this work, we discuss the neighbourhoods and approximation operators using p-sets and q-sets of a texture. Here, we show that the presections of a direlation correspond to lower and upper approximations in terms of successor neighbourhood operators while the sections of a direlation correspond to lower and upper approximations in terms of predecessor neighbourhood operators. For discrete textures, we observe that the weak forms of definabilities are preserved under the relation preserving bijective functions where the inverses are also relation preserving.

Keywords: Approximation operator · Definability · Direlation · Neighbourhood operators · Rough set · Texture space

1 Introduction

There is a strong connection between theories of rough set and texture [4–9]. For instance, presections of direlations provides natural generalizations of approximation operators in the sense of Pawlak [14]. Recall that difunctions, as morphisms of a ground category **dfTex** in texture space theory, can be characterized using definability [3]. If the textures are discrete, then this characterization is that a relation $r \subseteq U \times V$ is a function from U to V if and only if every subset of V is r-definable in U where U and V are arbitrary universes [7,9]. In rough set theory, the successor and predecessor neighbourhood operators $r_s, r_p : U \to \mathcal{P}(U)$ are defined by

$$r_s(u) = \{v \in U \mid (u,v) \in r\} \text{ and } r_p(u) = \{v \in U \mid (v,u) \in r\}$$

for all $u \in U$. The fundamental properties of neighbourhood operators are extensively discussed by Yao. In particular, he gives a framework for formulation, interpretation of neighbourhood systems and rough set approximations using binary relations [17]. Successor definability and predecessor definability in terms

© Springer International Publishing Switzerland 2015
D. Ciucci et al. (Eds.): RSKT 2015, LNAI 9436, pp. 202–213, 2015.
DOI: 10.1007/978-3-319-25754-9_18

of neighbourhood operators are studied by Grzymala-Busse and Rzasa in [10]. They consider the approximations for reflexive, symmetric or transitive relations. A useful analysis on the relationships between singleton and subset approximations using successor and predecessor neighbourhood operators can be found in the paper of Liu and Zhu [13]. Further, some recent works on neighbourhood operators can be also found in the papers of Liu, and Zhang et al. [12,18].

The aim of this paper is to discuss the neighbourhoods in textures and to give some basic results on approximations using the textural counterparts of neighbourhood operators. Recall that a texturing $\mathcal{U}$ is a family of subsets of a given universe U satisfying certain conditions subjected to the basic properties of the power set $\mathcal{P}(U)$. The pair $(U,\mathcal{U})$ is called a *texture space* or *a texture*, in brief [2]. Actually, the motivation of textures is that providing an alternative point-set based setting for fuzzy lattices, that is, complete, completely distributive lattices with an order reversing involution [1]. A complementation c_U on a texture $(U,\mathcal{U})$ is a mapping $c_U : \mathcal{U} \to \mathcal{U}$ satisfying idempotency and reversing the set-inclusion. Then the triple $(U,\mathcal{U},c_U)$ is called *a complemented texture*. For any universe U, the triple $(U,\mathcal{P}(U),c_U)$ is a complemented texture with ordinary set-complementation $c_U(A) = U \setminus A$ for all $A \subseteq U$ which is called a discrete texture. A direlation from a texture $(U,\mathcal{U})$ to a texture $(V,\mathcal{V})$ is a pair (r,R) where r (relation) and R (co-relation) are the elements of a textural product satisfying certain conditions. If the textures are complemented, then a complement (r',R') of (r,R) can be defined using the complementations of textures, in a natural way [3]. We say that (r,R) is *complemented* if $r' = R$ and $r = R'$. Now let $\mathcal{U} = \mathcal{P}(U)$ where (r,R) is a complemented direlation from $(U,\mathcal{U}))$ to $(U,\mathcal{U})$. Then it is easy to see that $r \subseteq U \times U$, that is, r is an ordinary relation on U and $R = (U \times U) \setminus r$. In this case, for any $A \in \mathcal{U}$, the A-presection of r and R correspond to the sets $\underline{apr}_r(A)$ and $\overline{apr}_r(A)$, respectively where $\underline{apr}_r$ and $\overline{apr}_r$ are the lower and upper approximation operators in the sense of Yao. Moreover, we obtain the point-free formulations $\underline{apr}_r(A) = U \setminus r^{-1}(U \setminus A)$ and $\overline{apr}_r(A) = r^{-1}(A)$ where $r^{-1} = \{(u,v) \mid (v,u) \in r\}$. It is remarkable to note that sections of direlations are also approximations of sets of a texture. Therefore, from the rough set point of view, we compare the section and presection of a set in textures using the counterparts of successor neighbourhoods and predecessor neighbourhoods. Then we present the textural definable sets and discuss them in terms of textural successor and predecessor neighbourhoods, in a natural way. Now let (U,r) be an approximation space where r is an equivalence relation on U. Recall that a subset A of U is definable if it is a union of a family of some equivalence classes with respect to equivalence relation r. Note that A is definable if and only if the upper approximation and lower approximation of A are equal. However, this result is not true anymore if r is not an equivalence relation. In this case, an equivalence class may correspond to a successor or predecessor neighbourhood. Then we have at least two non-equivalent definitions of definability [10]. Suppose that the definability is given in terms of lower and upper approximations, that is, a subset $A \subseteq U$ is r-definable if $\underline{apr}_r(A) = \overline{apr}_r(A)$ where (U,r) is an approximation space. If $f : (U,r) \to (V,h)$ is a relation preserving mapping where r and h are reflexive relations, then $B \subseteq V$ is h-definable implies that $f^{-1}(B)$ is r-definable [9]. For non-equivalence relations, the concepts of successor and predecessor definabilities are weaker than the concept of definability. In this work, we discuss the above arguments in textures.

This paper is organized as follows. The Sects. 2 and 3 are devoted to textures and direlations, respectively. In Sect. 4, we define and discuss the textural counterparts of successor neighbourhood and predecessor neighbourhood operators. In Sects. 5 and 6, we give the textural formulations of singleton and subset approximations. Using duality, we show that subset approximations can be written in terms of singleton approximations. In Sect. 7, we present textural successsor and textural predecessor definable sets. Here, we give the point-free formulations of successor and predecessor definabilities. Finally, in Sect. 8, we show that successor and predecessor definabilities are preserved only under a bijective function f where f and f^{-1} are relation preserving functions.

2   Textures

A texturing on a universe U is a point separating, complete, completely distributive lattice $\mathcal{U}$ of subsets of U with respect to inclusion which contains U and $\emptyset$ and, for which arbitrary meet coincides with intersection and finite joins coincide with union. Then the pair $(U, \mathcal{U})$ is called a texture space, or simply a texture [2]. A mapping $c_U : \mathcal{U} \to \mathcal{U}$ is called a complementation on $(U, \mathcal{U})$ if it satisfies the conditions $c_U(c_U(A)) = A$ for all $A \in \mathcal{U}$ and $A \subseteq B$ in $\mathcal{U}$ implies $c_U(B) \subseteq c_U(A)$. Then the triple $(U, \mathcal{U}, c_U)$ is called a complemented texture space. For $u \in U$, the p-sets and q-sets are defined by $P_u = \bigcap \{A \in \mathcal{U} \mid u \in A\}$ and $Q_u = \bigvee \{A \in \mathcal{U} \mid u \notin A\}$, respectively. A trivial example to textures is the pair $(U, \mathcal{P}(U))$ where $\mathcal{P}(U)$ is the power set of U. It is called a *discrete texture*. For $u \in U$ we have $P_u = \{u\}$ and $Q_u = U \setminus \{u\}$ and $c_U : \mathcal{P}(U) \to \mathcal{P}(U)$ is the ordinary complementation on $(U, \mathcal{P}(U))$ defined by $c_U(A) = U \setminus A$ for all $A \in \mathcal{P}(U)$. A canonical example of textures is the family $\mathcal{M} = \{(0, r] \mid r \in [0, 1]\}$ on $M = (0, 1]$ which is called the Hutton texture. We have $P_r = Q_r = (0, r]$ for all $r \in [0, 1]$. The complementation $c_M : \mathcal{M} \to \mathcal{M}$ is defined by $\forall r \in (0, 1]$, $c_M(0, r] = (0, 1 - r]$. Essentially, Hutton texture corresponds to well-known fuzzy lattice $[0, 1]$. Let us consider the fuzzy lattice $\mathcal{F}(U)$ of all fuzzy subsets $\alpha : U \to [0, 1]$. The product texture $(U \times M, \mathcal{P}(U) \otimes \mathcal{M})$ of $(U, \mathcal{P}(U))$ and $(M, \mathcal{M})$ is isomorphic to $\mathcal{F}(U)$ under the lattice isomorphism [1]. In other words, $(U \times M, \mathcal{P}(U) \otimes \mathcal{M}, c_{U \times M})$ is a point-set based setting for the fuzzy lattice $\mathcal{F}(U)$. We denote the p-sets and q-sets of the texture $\mathcal{P}(U) \otimes \mathcal{M}$ by $\overline{P}_{(u,v)}$ and $\overline{Q}_{(u,v)}$ where $\overline{P}_{(u,v)} = \{u\} \times P_v$ and $\overline{Q}_{(u,v)} = ((U \setminus \{u\}) \times V) \cup (U \times Q_v)$.

3   Direlations

Let $(U, \mathcal{U})$, $(V, \mathcal{V})$ be texture spaces. Then $r \in \mathcal{P}(U) \otimes V$ is called a *relation* from $(U, \mathcal{U})$ to $(V, \mathcal{V})$ if

R1 $r \not\subseteq \overline{Q}_{(u,v)}, P_{u'} \not\subseteq Q_u \implies r \not\subseteq \overline{Q}_{(u',v)}$, and

R2 $r \not\subseteq \overline{Q}_{(u,v)} \implies \exists u' \in U$ such that $P_u \not\subseteq Q_{u'}$ and $r \not\subseteq \overline{Q}_{(u',v)}$.

$R \in \mathcal{P}(U) \otimes \mathcal{V}$ is called a *corelation* from $(U,\mathcal{U})$ to $(V,\mathcal{V})$ if

CR1 $\overline{P}_{(u,v)} \not\subseteq R, P_u \not\subseteq Q_{u'} \implies \overline{P}_{(u',v)} \not\subseteq R$, and
CR2 $\overline{P}_{(u,v)} \not\subseteq R \implies \exists u' \in U$ such that $P_{u'} \not\subseteq Q_u$ and $\overline{P}_{(u',v)} \not\subseteq R$.

A pair (r, R), where r is a relation and R a corelation is called a *direlation* from $(U,\mathcal{U})$ to $(V,\mathcal{V})$ [3]. The identity direlation (i, I) on $(U,\mathcal{U})$ is defined by $i = \bigvee\{\overline{P}_{(u,u)} \mid u \in U\}$ and $I = \bigcap\{\overline{Q}_{(u,u)} \mid u \in U^\flat\}$ where $U^\flat = \{u \mid U \not\subseteq Q_u\}$. Recall that if (r, R) is a direlation on $(U,\mathcal{U})$, then r is *reflexive* if $i \subseteq r$ and R is *reflexive* if $R \subseteq I$. Now let (r, R) be a direlation from $(U,\mathcal{U})$ to $(V,\mathcal{V})$ where $(U,\mathcal{U})$ and $(V,\mathcal{V})$ are any two texture spaces. Then *the inverses* of r and R are defined by $r^\leftarrow = \bigcap\{\overline{Q}_{(v,u)} \mid r \not\subseteq \overline{Q}_{(u,v)}\}$ and $R^\leftarrow = \bigvee\{\overline{P}_{(v,u)} \mid \overline{P}_{(u,v)} \not\subseteq R\}$, respectively where $r^\leftarrow$ is a corelation and $R^\leftarrow$ is a relation. Further, the direlation $(r, R)^\leftarrow = (R^\leftarrow, r^\leftarrow)$ from $(V,\mathcal{V})$ to $(U,\mathcal{U})$ is called the *inverse* of the direlation (r, R). Then (r, R) is called *symmetric* if $r = R^\leftarrow$ and $R = r^\leftarrow$. The A-sections and the B-presections with respect to relation and corelation are given as

$$r^\rightarrow A = \bigcap\{Q_v \mid \forall u, r \not\subseteq \overline{Q}_{(u,v)} \Rightarrow A \subseteq Q_u\},$$
$$R^\rightarrow A = \bigvee\{P_v \mid \forall u, \overline{P}_{(u,v)} \not\subseteq R \Rightarrow P_u \subseteq A\},$$
$$r^\leftarrow B = \bigvee\{P_u \mid \forall v, r \not\subseteq \overline{Q}_{(u,v)} \Rightarrow P_v \subseteq B\}, \text{ and}$$
$$R^\leftarrow B = \bigcap\{Q_u \mid \forall v, \overline{P}_{(u,v)} \not\subseteq R \Rightarrow B \subseteq Q_v\}.$$

for all $A \in \mathcal{U}$ and $B \in \mathcal{V}$, respectively. Now let $(U,\mathcal{U}), (V,\mathcal{V}), (W,\mathcal{W})$ be texture spaces. For any relation p from $(U,\mathcal{U})$ to $(V,\mathcal{V})$ and for any relation q from $(V,\mathcal{V})$ to $(W,\mathcal{W})$ their *composition* $q \circ p$ from $(U,\mathcal{U})$ to $(W,\mathcal{W})$ is defined by $q \circ p = \bigvee\{\overline{P}_{(u,w)} \mid \exists v \in V$ with $p \not\subseteq \overline{Q}_{(u,v)}$ and $q \not\subseteq \overline{Q}_{(v,w)}\}$ and any corelation P from $(U,\mathcal{U})$ to $(V,\mathcal{V})$ and for any corelation Q from $(U,\mathcal{U})$ to $(V,\mathcal{V})$ their *composition* $Q \circ P$ from $(U,\mathcal{U})$ to $(V,\mathcal{V})$ defined by $Q \circ P = \bigcap\{\overline{Q}_{(u,w)} \mid \exists v \in V$ with $\overline{P}_{(u,v)} \not\subseteq P$ and $\overline{P}_{(v,w)} \not\subseteq Q\}$. Further, r is *transitive* if $r \circ r \subseteq r$ and R is *transitive* if $R \subseteq R \circ R$. Then (r, R) is called *transitive* if r and R are transitive. Finally, if (r, R) is reflexive, symmetric and transitive, then it is called *an equivalence direlation*. Now let c_U and c_V be the complementations on $(U,\mathcal{U})$ and $(V,\mathcal{V})$, respectively. The complement r' of the relation r is the corelation $r' = \bigcap\{\overline{Q}_{(u,v)} \mid \exists w, z$ with $r \not\subseteq \overline{Q}_{(w,z)}, c_U(Q_u) \not\subseteq Q_w$ and $P_z \not\subseteq c_V(P_v)\}$. The complement R' of the corelation R is the relation $R' = \bigvee\{\overline{P}_{(u,v)} \mid \exists w, z$ with $\overline{P}_{(w,z)} \not\subseteq R, P_w \not\subseteq c_U(P_u)$ and $c_V(Q_v) \not\subseteq Q_z\}$. We say (r, R) is *complemented* if $r = R'$ and $R = r'$.

4   Textural Neighbourhoods

Recall that if (r, R) is a direlation on $(U,\mathcal{U})$, then the quadruple $(U,\mathcal{U},r,R)$ is called *a textural approximation space* [4].

Definition 1. Let $(U, \mathcal{U}, r, R)$ be a textural approximation space and $w \in U$. Then the set $r^{\rightarrow} P_w = \bigcap \{Q_v \mid \forall u \in U, r \not\sqsubseteq \overline{Q}_{(u,v)} \implies P_w \subseteq Q_u\}$ is called *a textural successor neighbourhood* of w, and the set $R^{\leftarrow} P_w = \bigcap \{Q_u \mid \forall v \in U, \overline{P}_{(u,v)} \not\sqsubseteq R \implies P_w \subseteq Q_v\}$ is called *a textural predecessor neighbourhood* of w.

Example 2. Consider the texture space $(U, \mathcal{U})$ where $U = \{a, b, c\}$ and $\mathcal{U} = \{\emptyset, \{a\}, \{a, b\}, U\}$. For the texturing $\mathcal{U}$, the p-sets are $P_a = \{a\}, P_b = \{a, b\}, P_c = U$ and the q-sets are $Q_a = \emptyset, Q_b = \{a\}, Q_c = \{a, b\}$. For every point of $U \times U$, we may easily determine the p-sets and q-sets of the product texture $\mathcal{P}(U) \otimes \mathcal{U}$. For example, for the points $(c, b), (a, c) \in U \times U$, we have $\overline{P}_{(c,b)} = \{c\} \times P_b = \{c\} \times \{a, b\} = \{(c, a), (c, b)\}$ and $\overline{Q}_{(a,c)} = ((U \setminus \{a\}) \times U) \cup (U \times Q_c) = (\{b, c\} \times U) \cup (U \times \{a, b\}) = \{(b, a), (b, b), (b, c), (c, a), (c, b), (c, c), (a, a), (a, b)\}$. The pair (r, R) is a direlation on $(U, \mathcal{U})$ where $r = \{(a, a), (a, b), (b, a), (b, b), (c, a), (c, b), (c, c)\}$ and $R = \{(b, a), (c, a)\}$. We may easily check the condition CR1. For instance, for $\overline{P}_{(c,b)} \not\sqsubseteq R$ and $P_c \not\sqsubseteq Q_a$, we also have $\{a\} \times P_b = \{a\} \times \{a, b\} = \{(a, a), (a, b)\} = \overline{P}_{(a,b)} \not\sqsubseteq R$. Since $P_a \not\sqsubseteq Q_a, P_b \not\sqsubseteq Q_b$ and $P_c \not\sqsubseteq Q_c$, the condition CR2 is immediate. Similarly, (r, R) satisfies the conditions R1 and R2. Further, the textural successor neighbourhoods and textural predecessor neighbourhoods of a, b and c are $r^{\rightarrow} P_a = r^{\rightarrow} P_b = \{a, b\}, r^{\rightarrow} P_c = U, R^{\leftarrow} P_a = \{a\}, R^{\leftarrow} P_b = R^{\leftarrow} P_c = U$.

Textural successor neighbourhoods and textural predecessor neighbourhoods are the natural counterparts of successor neighbourhoods and predecessor neighbourhoods studied by Slowinski and Vanderpooten [16], Grzymala-Busse and Rzasa [10], Yao [17], Kryszkiewicz [11], Liu [12], and Skowron and Rauszer [15]. Note that if $\mathcal{U} = \mathcal{P}(U)$ and (r, R) is a complemented direlation on $(U, \mathcal{P}(U))$, then $r \subseteq U \times U$ and R is the set theoretical complement of r, that is, $(U \times U) \setminus r = R$. This implies the equivalences $r \not\sqsubseteq \overline{Q}_{(u,v)} \iff (u, v) \in r$, $\overline{P}_{(u,v)} \not\sqsubseteq R \iff (u, v) \in r$. Now we may give the following result:

Proposition 3. *Let $(U, \mathcal{U})$ be a discrete texture and (r, R) be a complemented direlation on $(U, \mathcal{P}(U))$. Then for all $w \in U$, we have the following equalities.*

(i) $r^{\rightarrow} P_w = r_s(w), \quad R^{\rightarrow} Q_w = U \setminus r_s(w)$.
(ii) $R^{\leftarrow} P_w = r_p(w), \quad r^{\leftarrow} Q_w = U \setminus r_p(w)$.
(iii) $r^{\rightarrow} Q_w = r(U \setminus \{w\}), \quad R^{\leftarrow} Q_w = r^{-1}(U \setminus \{w\})$.
(iv) $r^{\leftarrow} P_w = U \setminus r^{-1}(U \setminus w), \quad R^{\rightarrow} P_w = U \setminus r(U \setminus \{w\})$.

Proof. We give the proof using the equalities given in Theorem 6.2 in [4].

(i) For $w \in U$, we have $r^{\rightarrow} P_w = r(\{w\}) = \{v \mid (w, v) \in r\} = r_s(w)$ and $R^{\rightarrow} Q_w = U \setminus r(U \setminus (U \setminus \{w\})) = U \setminus r(\{w\}) = U \setminus r_s(w)$.
(ii) Similarly, we have $R^{\leftarrow} P_w = r^{-1}(\{w\}) = \{u \mid (u, w) \in r\} = r_p(w)$ and $r^{\leftarrow} Q_w = U \setminus r^{-1}(U \setminus (U \setminus \{w\})) = U \setminus r^{-1}(\{w\}) = U \setminus \{v \mid (v, w) \in r\} = U \setminus r_p(w)$.

The proof of (iii) and (iv) are immediate. $\qquad \square$

5   Singleton Approximations

In [17], Yao discussed the lower approximation and upper approximation of a set A using the neighbourhood operators $r_p, r_s : U \to \mathcal{P}(U)$ defined as

$$\underline{apr}_{r_p}(A) = \{u \mid r_p(u) \subseteq A\}, \quad \overline{apr}_{r_p}(A) = \{v \mid r_p(u) \cap A \neq \emptyset\},$$

$$\underline{apr}_{r_s}(A) = \{u \mid r_s(u) \subseteq A\}, \quad \overline{apr}_{r_s}(A) = \{v \mid r_s(u) \cap A \neq \emptyset\}.$$

The above approximations are also studied under the name of singleton approximations by Grzymala-Busse and Rzasa [10]. For discrete textures, presections correspond to the singleton successor lower approximation and singleton successor upper approximation of a set. Further, sections correspond to the singleton predecessor lower approximation and singleton predecessor upper approximation of a set.

Theorem 4. *If (r, R) is a complemented direlation on $(U, \mathcal{P}(U))$, then for all $A \subseteq U$ we have the following equalities:*

(i) $r^{\leftarrow} A = \underline{apr}_{r_s}(A)$ *and* $R^{\leftarrow} A = \overline{apr}_{r_s}(A)$.
(ii) $R^{\rightarrow} A = \underline{apr}_{r_p}(A)$ *and* $r^{\rightarrow} A = \overline{apr}_{r_p}(A)$.

Proof. We give the proof of (ii) leaving the proof of (i) to the interested reader. $r^{\rightarrow} A = \bigcap \{Q_v \mid \forall u, r \not\subseteq \overline{Q}_{(u,v)} \implies A \subseteq Q_u\} = \bigcap \{U \setminus \{v\} \mid \forall u, (u, v) \in r \implies u \notin A\} = U \setminus \{v \mid \forall u, (u, v) \in r \implies u \notin A\} = \{v \mid \exists u, (u, v) \in r \text{ and } u \in A\} = \{v \mid \exists u, u \in r_p(v) \text{ and } u \in A\} = \{v \mid r_p(v) \cap A \neq \emptyset\} = \overline{apr}_{r_p}(A).$
$R^{\rightarrow} A = \bigvee \{P_v \mid \forall u, \overline{P}_{(u,v)} \not\subseteq R \implies P_u \subseteq A\} = \{v \mid \forall u, (u, v) \in r \implies u \in A\} = \{v \mid \forall u, u \in r_p(v) \implies u \in A\} = \{u \mid r_p(u) \subseteq A\} = \underline{apr}_{r_p}(A).$ □

6   Subset Approximations

Subset approximations are also defined in terms of successor neighbourhoods and predecessor neighbourhoods [17] as

$$\underline{apr}'_{r_p}(A) = \bigcup \{r_p(u) \mid r_p(u) \subseteq A\}, \quad \overline{apr}''_{r_p}(A) = \bigcup \{r_p(u) \mid r_p(u) \cap A \neq \emptyset\},$$

$$\underline{apr}'_{r_s}(A) = \bigcup \{r_s(u) \mid r_s(u) \subseteq A\}, \quad \overline{apr}''_{r_s}(A) = \bigcup \{r_s(u) \mid r_s(u) \cap A \neq \emptyset\}.$$

However, the above subset approximations are not dual [17]. In a textural argument, the dual operators can be obtained in a natural way. Now let (r, R) be a direlation on a texture $(U, \mathcal{U})$. By Lemma 2.9 (1) and (2) in [3], for a direlation (r, R), we have $r^{\rightarrow} r^{\leftarrow} A \subseteq A \subseteq R^{\rightarrow} R^{\leftarrow} A$ for all $A \in \mathcal{U}$.

Theorem 5. *If (r, R) is a complemented direlation on a texture $(U, \mathcal{U}, c_U)$, then the approximation operators $r^{\rightarrow} r^{\leftarrow} : \mathcal{U} \to \mathcal{U}$ and $R^{\rightarrow} R^{\leftarrow} : \mathcal{U} \to \mathcal{U}$ are dual.*

For discrete textures, the approximations $r^{\rightarrow} r^{\leftarrow}$ and $R^{\rightarrow} R^{\leftarrow}$ correspond to the approximation operators $\underline{apr}'_{r_s}$ and $\overline{apr}'_{r_s}$ given by Yao in [17] where r_s is the successor neighbourhood operator:

Theorem 6. *Let $(U,\mathcal{U})$ be a discrete texture, that is, $\mathcal{U} = \mathcal{P}(U)$ and (r,R) be a complemented direlation on $(U, \mathcal{P}(U))$. Then*

$$r^{\rightarrow} r^{\leftarrow} A = \underline{apr}'_{r_s}(A) \quad \text{and} \quad R^{\rightarrow} R^{\leftarrow} A = \overline{apr}'_{r_s}(A)$$

for all $A \subseteq U$.

Proof. If (r,R) is a complemented direlation on $(U, \mathcal{P}(U))$, then it is easy to see that r is an ordinary relation on U, that is, $r \subseteq U \times U$ where $R = (U \times U) \setminus r$. Now $R^{\rightarrow} R^{\leftarrow} A = R^{\rightarrow} \bigcap \{Q_u \mid \forall v, \overline{P}_{(u,v)} \not\subseteq R \implies A \subseteq Q_v\} = \bigcap \{R^{\rightarrow} Q_u \mid \forall v, \overline{P}_{(u,v)} \not\subseteq R \implies A \subseteq Q_v\} = \bigcap \{U \setminus r_s(u) \mid \forall v, (u,v) \in r \implies v \notin A\} = U \setminus (\bigcup \{r_s(u) \mid \forall v, (u,v) \in r \implies v \notin A\}) = U \setminus (\{w \mid \exists u, w \in r_s(u)[\forall v, v \in r_s(u) \implies v \notin A]\}) = U \setminus \{w \mid \exists u, w \in r_s(u), r_s(u) \cap A = \emptyset\} = \{w \mid \forall u, w \in r_s(u) \implies r_s(u) \cap A \neq \emptyset\} = \overline{apr}'_{r_s}(A)$. The proof of first equality is similar. $\square$

By Proposition 3.1 in [13], it is already observed that

$$\overline{apr}''_{r_s}(A) = \overline{apr}_{r_p}(\overline{apr}_{r_s}(A)) \quad \text{and} \quad \underline{apr}'_{r_s}(A) = \overline{apr}_{r_p}(\underline{apr}_{r_s}(A)).$$

By Theorem 6, we may also say that the approximations $\underline{apr}'_{r_s}, \overline{apr}'_{r_s}$ can be given in terms of singleton approximations:

Corollary 7. *For the discrete texture $(U, \mathcal{P}(U))$, we have*

$$\underline{apr}'_{r_s}(A) = \overline{apr}_{r_p}(\underline{apr}_{r_s}(A)) \quad \text{and} \quad \overline{apr}'_{r_s}(A) = \underline{apr}_{r_p}(\overline{apr}_{r_s}(A))$$

for all $A \subseteq U$.

On the other hand, if (r,R) is reflexive, then by Lemma 2.9 (1) and (4) in [3], we have $r^{\rightarrow} r^{\leftarrow} A \subseteq A \subseteq r^{\rightarrow} A$ for all $A \in \mathcal{U}$. However, the operators $r^{\rightarrow} r^{\leftarrow}$ and $r^{\rightarrow}$ are not dual. Furthermore, if the texture is discrete, then $r^{\rightarrow} A = r^{\rightarrow} \bigvee \{P_u \mid A \not\subseteq Q_u\} = \bigvee \{r^{\rightarrow} P_u \mid A \not\subseteq Q_u\} = \bigcup \{r_s(u) \mid u \in A\} = \bigcup \{r_s(u) \mid r_s(u) \cap A \neq \emptyset\} = \overline{apr}''_{r_s}(A)$.

Theorem 8. *Let $(U,\mathcal{U})$ be a texture and (r,R) be a reflexive and symmetric direlation. Then we have $R^{\rightarrow} R^{\leftarrow} A = r^{\rightarrow} A$ for all $A \in \mathcal{U}$.*

The above argument can be also considered for predecessor approximations. Indeed, by Lemma 2.9 (1) and (2) in [3], we have $R^{\leftarrow} R^{\rightarrow} A \subseteq A \subseteq r^{\leftarrow} r^{\rightarrow} A$ for all $A \in \mathcal{U}$. Moreover, we have:

Theorem 9. *If (r,R) is a complemented direlation on $(U,\mathcal{U}, c_U)$, then the approximation operators $R^{\leftarrow} R^{\rightarrow} : \mathcal{U} \to \mathcal{U}$ and $r^{\leftarrow} r^{\rightarrow} : \mathcal{U} \to \mathcal{U}$ are dual.*

Theorem 10. *Let (r,R) be a complemented direlation on $(U, \mathcal{P}(U))$. Then*

$$R^{\leftarrow} R^{\rightarrow} A = \underline{apr}'_{r_p}(A) \quad \text{and} \quad r^{\leftarrow} r^{\rightarrow} A = \overline{apr}'_{r_p}(A)$$

for all $A \subseteq U$.

Proof. $r^{\leftarrow}r^{\rightarrow}A = r^{\leftarrow}\bigcap\{Q_v \mid \forall u, r \not\subseteq \overline{Q}_{(u,v)} \implies A \subseteq Q_u\} = \bigcap\{r^{\leftarrow}Q_v \mid \forall u, (u,v) \in r \implies u \notin A\} = \bigcap\{U \setminus r_p(v) \mid r_p(v) \cap A = \emptyset\} = U \setminus (\bigcup\{r_p(v) \mid r_p(v) \cap A = \emptyset\}) = \bigcup\{r_p(v) \mid r_p(v) \cap A \neq \emptyset\} = \overline{apr}'_{r_p}(A)$ and $R^{\leftarrow}R^{\rightarrow}A = R^{\leftarrow}\bigvee\{P_v \mid \forall u, \overline{P}_{(u,v)} \not\subseteq R \implies P_u \subseteq A\}) = \bigvee\{R^{\leftarrow}P_v \mid \forall u, \overline{P}_{(u,v)} \not\subseteq R \implies P_u \subseteq A\}) = \bigcup\{r_p(v) \mid \forall u, (u,v) \in r \implies u \in A\} = \bigcup\{r_p(v) \mid r_p(v) \subseteq A\} = \underline{apr}'_{r_p}(A)$. $\qquad\square$

By duality, we also have:

Theorem 11. *Let $(U,\mathcal{U})$ be a texture and (r, R) be a reflexive and symmetric direlation. Then we have $r^{\rightarrow}r^{\leftarrow}A = R^{\rightarrow}A$ for all $A \in \mathcal{U}$.*

Corollary 12. *For the discrete texture $(U, \mathcal{P}(U))$, we have*

$$\underline{apr}'_{r_p}(A) = \overline{apr}_{r_s}(\underline{apr}_{r_p}(A)) \quad and \quad \overline{apr}'_{r_p}(A) = \underline{apr}_{r_s}(\overline{apr}_{r_p}(A))$$

for all $A \subseteq U$.

7 Definability

Successor and predecessor definabilities are defined in terms of neighbourhoods by Grzymala-Busse and Rzasa in [10].

Definition 13. *Let $(U,\mathcal{U})$ be a texture space and $A \in \mathcal{U}$. Then A is called*

(i) *textural successor definable if $A = \bigvee_{B \not\subseteq Q_u} r^{\rightarrow}P_u$, and*
(ii) *textural predecessor definable if $A = \bigvee_{B \not\subseteq Q_u} R^{\leftarrow}P_u$ for some $B \in \mathcal{U}$.*

If $(U,\mathcal{U})$ is a discrete texture space, then by Proposition 3, a textural predecessor definable set is predecessor definable: $A = \bigvee_{B \not\subseteq Q_u} R^{\leftarrow}P_u = \bigcup_{u \in B} r_p(u)$. Likewise, a textural successor definable set is successor definable: $A = \bigvee_{B \not\subseteq Q_u} r^{\rightarrow}P_u = \bigcup_{u \in B} r_s(u)$.

Proposition 14. *Let $(U,\mathcal{U},r,R)$ be a textural approximation space and $A \in \mathcal{U}$. Then we have the following properties:*

(i)   *A is textural successor definable if and only if $A = r^{\rightarrow}B$ for some $B \in \mathcal{U}$.*
(ii) *A is textural predecessor definable if and only if $A = R^{\leftarrow}B$ for some $B \in \mathcal{U}$.*
(iii) *$r^{\rightarrow}A$ and $R^{\leftarrow}A$ are textural successor definable and textural predecessor definable, respectively.*

Recall that [4] if the texture is discrete, that is, $\mathcal{U} = \mathcal{P}(U)$, then for presections and sections of a direlation (r, R), we have $r^{\rightarrow}A = r(A)$ and $R^{\leftarrow}A = r^{-1}(A)$ where $r(A) = \{v \in U \mid \exists u \in A \ (u,v) \in r\}$ and $r^{-1} = \{(u,v) \mid (v,u) \in r\}$. Therefore, successor definability and predecessor definability can be stated as follows:

Theorem 15. *Let* $\mathcal{U} = \mathcal{P}(U)$. *Then we have the following.*

(i) $A \subseteq U$ *is successor definable if and only if* $A = r(B)$ *for some* $B \subseteq U$.
(ii) $A \subseteq U$ *is predecessor definable if and only if* $A = r^{-1}(B)$ *for some* $B \subseteq U$.

Theorem 16. *Let* (r, R) *be a direlation on* $(U, \mathcal{U})$ *and* $A \in \mathcal{U}$.

(1) If (r, R) *is reflexive, then*
    (i) $r^{\rightarrow}A = R^{\rightarrow}A \implies A$ *is textural successor definable.*
    (ii) $r^{\leftarrow}A = R^{\leftarrow}A \implies A$ *is textural predecessor definable.*
(2) If (r, R) *is a symmetric direlation on* $(U, \mathcal{U})$, *then we have the following equivalences:*
    (i) A *is textural successor definable* $\iff A$ *is textural predecessor definable.*
    (ii) $r^{\rightarrow}A = R^{\rightarrow}A \iff r^{\leftarrow}A = R^{\leftarrow}A$.
(3) If (r, R) *is an equivalence direlation on* $(U, \mathcal{U})$, *then the following conditions are equivalent.*
    (i) A *is textural successor definable.* (ii) A *is textural predecessor definable.*
    (iii) $r^{\leftarrow}A = R^{\leftarrow}A$. (iv) $r^{\rightarrow}A = R^{\rightarrow}A$

Example 17. Consider the texture $(U, \mathcal{U})$ and the direlation (r, R) in Example 2. We have $i_U = \{(a, a), (b, a), (b, b), (c, a), (c, b), (c, c)\}$, $I_U = \{(b, a), (c, a), (c, b)\}$. Since $i_U \subseteq r$ and $R \subseteq I_U$, (r, R) is a reflexive direlation. For $A = \{a, b\} \in \mathcal{U}$, $A = \bigvee_{A \not\subseteq Q_u} r^{\rightarrow}P_u = r^{\rightarrow}P_a \cup r^{\rightarrow}P_b = \{a, b\}$. Hence, A is textural successor definable. Since $r^{\rightarrow}A = A$ and $R^{\rightarrow}A = \{a\}$, we have $r^{\rightarrow}A \neq R^{\rightarrow}A$. This means that the converse of Theorem 16 (1)(i) is not true. On the other hand, we have $R^{\leftarrow} = \{(a, a), (b, a), (c, a), (b, b), (c, b), (b, c), (c, c)\} \neq r$ and hence, (r, R) is not a symmetric direlation. Further, $r^{\leftarrow}A = A$ and $R^{\leftarrow}A = U$ implies that $r^{\leftarrow}A \neq R^{\leftarrow}A$. Therefore, the symmetry condition cannot be removed in Theorem 16 (2)(i).

In view of Theorem 16, we immediately have the following results:

Theorem 18. *Let* (U, r) *be an approximation space.*

(1) If r *is a reflexive relation on* U, *then*
    (i) $\underline{apr}_{r_s}(A) = \overline{apr}_{r_s}(A) \implies A$ *is predecessor definable.*
    (ii) $\underline{apr}_{r_p}(A) = \overline{apr}_{r_p}(A) \implies A$ *is successor definable.*
(2) If r *is a symmetric relation on* U, *then we have the following equivalences:*
    (i) A *is successor definable* $\iff A$ *is predecessor definable.*
    (ii) $\underline{apr}_{r_s}(A) = \overline{apr}_{r_s}(A) \iff \underline{apr}_{r_p}(A) = \overline{apr}_{r_p}(A)$.
(3) If r *is an equivalence relation on* U, *then the following conditions are equivalent.*
    (i) A *is successor definable.* (ii) A *is predecessor definable.*
    (iii) $\underline{apr}_{r_s}(A) = \overline{apr}_{r_s}(A)$. (iv) $\underline{apr}_{r_p}(A) = \overline{apr}_{r_p}(A)$.

8 Relation Preserving Functions

A *difunction* (f, F) from $(U, \mathcal{U})$ to $(V, \mathcal{V})$ is a direlation satisfying the following two conditions:

DF1 For $u, u' \in U$, $P_u \not\subseteq Q_{u'} \implies \exists v \in V$ with $f \not\subseteq \overline{Q}_{(u,v)}$ and $\overline{P}_{(u',v)} \not\subseteq F$.

DF2 For $v, v' \in V$ and $u \in U$, $f \not\subseteq \overline{Q}_{(u,v)}$ and $\overline{P}_{(u,v')} \not\subseteq F \implies P_{v'} \not\subseteq Q_v$.

It is remarkable to note that a direlation (r, R) from $(U, \mathcal{U})$ to $(V, \mathcal{V})$ is a difunction if and only if every set $A \in \mathcal{V}$ is (r, R)-definable [3,9]. Now let $(U, \mathcal{U}, r, R))$, $(V, \mathcal{V}, h, H)$ be textural approximation spaces. Recall that a difunction (f, F) from $(U, \mathcal{U}, r, R)$ to $(V, \mathcal{V}, h, H)$ is called

(i) *relation-preserving* if

$$\forall u, u' \in U,\ r \not\subseteq \overline{Q}_{(u,u')} \implies \exists v, v' \in V, f \not\subseteq \overline{Q}_{(u,v)}, \overline{P}_{(u',v')} \not\subseteq F \text{ and } h \not\subseteq \overline{Q}_{(v,v')},$$

(ii) *corelation-preserving* if

$$\forall u, u' \in U,\ \overline{P}_{(u,u')} \not\subseteq R \implies \exists v, v' \in V, f \not\subseteq \overline{Q}_{(u',v')}, \overline{P}_{(u,v)} \not\subseteq F \text{ and } \overline{P}_{(v,v')} \not\subseteq H.$$

If (f, F) is a relation and corelation-preserving difunction, then we say that (f, F) is *a direlation-preserving difunction* [9]. If (f, F) is a direlation preserving difunction from $(U, \mathcal{P}(U), r, R)$ to $(V, \mathcal{P}(V), h, H)$, then it is easy to see that f is a relation preserving function from (U, r) to (V, h) in ordinary sense.

Theorem 19. *Let (f, F) be a bijective difunction from $(U, \mathcal{U}, r, R)$ to $(V, \mathcal{V}, h, H)$. Then we have the following results:*

(i) *(f, F) and $(f, F)^{\leftarrow}$ are relation preserving if and only if $f \circ r = h \circ f$.*
(ii) *(f, F) and $(f, F)^{\leftarrow}$ are corelation preserving if and only if $H \circ F = F \circ R$.*

Theorem 20. *Let (f, F) and $(f, F)^{\leftarrow}$ be direlation preserving, $A \in \mathcal{U}$ and $B \in \mathcal{V}$. Then we have the following equivalences:*

(i) *A is (r, R)-textural successor definable $\iff f^{\rightarrow} A$ is (h, H)-textural successor definable.*
(ii) *A is (r, R)-textural predecessor definable $\iff f^{\rightarrow} A$ is (h, H)-textural predecessor definable.*

Theorems 18 and 19 can be considered for discrete textures:

Corollary 21. *Let (U, r) and (V, h) be approximation spaces and $f : (U, r) \to (V, h)$ be a bijective function. Then f and f^{-1} are relation preserving functions if and only if $f \circ r = h \circ f$.*

Theorem 22. *Let $f : (U, r) \to (V, h)$ be bijective, f, f^{-1} be relation preserving functions, $A \subseteq U$ and $B \subseteq V$. Then we have the following:*

(i) *A is r-successor definable $\iff f(A)$ is h-successor definable.*
(ii) *A is r-predecessor definable $\iff f(A)$ is h-predecessor definable.*

Let $f : (U, r) \rightarrow (V, h)$ be a relation preserving mapping where r and h are reflexive relations. If definability is given in terms of approximations, that is, for $B \subseteq V$, $\underline{apr}_h(B) = \overline{apr}_h(B)$, then $B \subseteq V$ is h-definable implies that $f^{-1}(B)$ is r-definable [9]. However, the following example shows that this result is not true for predecessor or successor definability:

Example 23. Let $U = \{u, v, w, z\}$, $V = \{a, b, c, d\}$, $\Delta_U = \{(x, x) \mid x \in U\}$ and $\Delta_V = \{(y, y) \mid y \in V\}$. Let us consider the reflexive relations $r = \{(u, v), (w, z), (z, v)\} \cup \Delta_U$ and $h = \{(a, b), (c, d), (d, b), (b, c)\} \cup \Delta_V$ on U and V, respectively. Take the relation preserving function $f : U \rightarrow V$ defined by $f(u) = a$, $f(v) = b$, $f(w) = c$, $f(z) = d$. Note that all successor neighbourhoods of (U, r) are $r_s(u) = \{u, v\}$, $r_s(v) = \{v\}$, $r_s(w) = \{w, z\}$, $r_s(z) = \{z, v\}$. The predecessor neighbourhoods are the sets $r_p(u) = \{u\}$, $r_p(v) = \{u, v, z\}$, $r_p(w) = \{w\}$, $r_p(z) = \{w, z\}$. Moreover, the successor neighbourhoods of (V, h) are $h_s(a) = \{a, b\}$, $h_s(b) = \{b, c\}$, $h_s(c) = \{c, d\}$, $h_s(d) = \{b, d\}$ and the predecessor neighbourhoods of (V, h) are $h_p(a) = \{a\}$, $h_p(b) = \{a, b, d\}$, $h_p(c) = \{b, c\}$, $h_p(d) = \{c, d\}$. Although $A = \{a, b, c\} \subseteq V$ is h-successor definable, $f^{-1}(A) = \{u, v, w\}$ is not r-successor definable. Further, the subset $B = \{b, c, d\} \subseteq U$ is h−predecessor definable. But $f^{-1}(B) = \{v, w, z\}$ is not r-predecessor definable. □

9 Conclusion

Neighbourhood operators provide a useful analysis for approximation operators in rough set theory. Since presections of direlations give natural generalizations for approximaton operators of Pawlak [14], we defined textural neighbourhoods in textures, and using p-sets and q-sets, we compared sections and presections from the rough set point of view. In view of duality in textures, we observed that subset approximations can be stated in terms of singleton approximations. Furthermore, giving the point-free formulations of definabilities, we showed that definabilities with respect to successor neighbourhoods and predecessor neighbourhoods are preserved under a bijective function f where f and f^{-1} are relation preserving.

References

1. Brown, L.M., Ertürk, R.: Fuzzy sets as texture spaces, I. Representation Theorems Fuzzy Sets Sys. **110**(2), 227–236 (2000)
2. Brown, L.M., Diker, M.: Ditopological texture spaces and intuitionistic sets. Fuzzy Sets Sys. **98**, 217–224 (1998)
3. Brown, L.M., Ertürk, R., Dost, Ş.: Ditopological texture spaces and fuzzy topology, I. Basic Concepts Fuzzy Sets Sys. **147**, 171–199 (2000)
4. Diker, M.: Textural approach to rough sets based on relations. Inf. Sci. **180**(8), 1418–1433 (2010)
5. Diker, M.: Textures and fuzzy rough sets. Fundam. Informaticae **108**, 305–336 (2011)

6. Diker, M., Uğur, A.A.: A Textures and covering based rough sets. Inf. Sci. **184**, 44–63 (2012)
7. Diker, M.: Definability and textures. Int. J. Approximate Reasoning **53**(4), 558–572 (2012)
8. Diker, M.: Categories of rough sets and textures. Theor. Comput. Sci. **488**, 46–65 (2013)
9. Diker, M.: A category approach to relation preserving functions in rough set theory. Int. J. Approximate Reasoning **56**, 71–86 (2015)
10. Grzymala-Busse, J.W., Rzasa W.: Definability of approximations for a generalization of the indiscernibility relation, In: Proceedings of the 2007 IEEE Symposium on Foundations of Computational Intelligence pp. 65–72 (2007)
11. Kryszkiewicz, M.: Rough set approach to incomplete information systems. In: Proceedings of the Second Annual Joint Conference on Information Sciences, pp. 194–197 (1995)
12. Liu, G.: Special types of coverings and axiomatization of rough sets based on partial orders. Knowl. Based Sys. **85**, 316–321 (2015)
13. Liu, G., Zhu, K.: The relationship among three types of rough approximation pairs. Knowl. Based Sys. **60**, 28–34 (2014)
14. Pawlak, Z.: Rough Sets. Int. J. Comput. Inf. Sci. **11**(5), 341–356 (1982)
15. Skowron, A., Rauszer, C.: The discernibility matrices and functions in information systems. In: Slowinski, R. (ed.) Handbook of Applications and Advances of the Rough Sets Theory, pp. 331–362. Kluwer Academic Publishers, Dordrecht (1992)
16. Slowinski, R., Vanderpooten, D.: A generalized definition of rough approximations based on similarity. IEEE Trans. Knowl. Data Eng. **12**, 331–336 (2000)
17. Yao, Y.: Relational interpretations of neighborhood operators and rough set approximation operators. Inf. Sci. **111**(1–4), 239–259 (1998)
18. Zhang, Y.-L., Li, C.-Q., Lin, M.-L., Lin, Y.-J.: Relationships between generalized rough sets based on covering and reflexive neighborhood system. Inf. Sci. **319**, 56–67 (2015)

Matrix Approaches for Variable Precision Rough Approximations

Guilong Liu[✉]

School of Information Science, Beijing Language and Culture University,
Beijing 100083, China
liuguilong@blcu.edu.cn

Abstract. Many generalizations of variable precision rough set models(VPRS) have been proposed since Ziarko introduced VPRS. This paper proposes the concept of general VPRS approximations which unifies earlier definitions of variant VPRS and gives an efficient matrix formulae for computing approximations of VPRS. This formulae can simplify the calculation of approximations of VPRS.

Keywords: Approximation · Boolean matrix · Precision degree · Rough set · Variable precision rough set

1 Introduction

Rough set theory, first proposed by Pawlak [13,14], is a new useful tool to dealing with vague concepts and incomplete information. Because the rough sets do not require a priori knowledge, they have enjoyed widespread success in pattern recognition, image processing, feature selection, decision support, data mining and knowledge discovery. Pawlak rough set is based on equivalence relations, however, this requirement is too restrictive for many applications. To address this issue, many generalizations of rough sets [5–7,16] have been introduced, studied and applied to solving problems. Among these generalizations, variable precision rough set (VPRS) [17] is quite interesting.

Much work [1,3,10–12,15,17,18] has been done for VPRS, for example, in order to handle uncertain information and directly derive from the original model without any additional assumptions, Ziarko [17] introduced the concept of VPRS. Mi, Wu and Zhang [10] introduced two types of attribute reduction to study knowledge reduction in VPRS. Yao, Mi and Li [16] propose variable precision (θ, σ)-fuzzy rough set model based on fuzzy granules. Liu, Hu and He [8] propose a new approach to calculate the reduction in VPRS model. Now the concept of VPRS has proven to be particularly useful in analysis of inconsistent decision tables obtained from dynamic control processes [11].

Clearly, one of important problems for VPRS is to calculate approximations of VPRS. We [4] introduced a Boolean matrix approach to study rough set

This work is supported by the National Natural Science Foundation of China (Nos. 60973148 and 61272031).

© Springer International Publishing Switzerland 2015
D. Ciucci et al. (Eds.): RSKT 2015, LNAI 9436, pp. 214–221, 2015.
DOI: 10.1007/978-3-319-25754-9_19

theory. It provides an efficient algorithm for computing upper approximation. Naturally, we hope the algorithm can be used to calculate the approximations of VPRS. The paper aims at presenting a matrix view of VPRS and providing a matrix formula to calculate approximations of VPRS.

In order to unify the earlier definition of VPRS, we propose the concept of β-approximation and strong β-approximation. Via Boolean matrix and characteristic function of sets, this paper gives an efficient algorithm for computing approximations of VPRS. This algorithm also provides a new approach for VPRS from viewpoint of matrices.

This paper is organized as follows. In Sect. 2, we review some basic concepts and properties of approximations of VPRS. In Sect. 3, we propose the concept of β-approximation and strong β-approximation and consider their properties. Section 4 presents Boolean matrix method for computing approximations of VPRS. Finally, Sect. 5 concludes the paper.

2   Preliminaries

This section reviews briefly the fundamental notation and notions based on rough sets and variable precision rough sets.

Let U be a universal set and $P(U)$ be the power set of U. Suppose that R is an arbitrary relation on U, (U, R) is called a generalized approximation space. Recall that the left R- and right R-relative sets for an element x in U are defined as

$$l_R(x) = \{y|y \in U, yRx\} \text{ and } r_R(x) = \{y|y \in U, xRy\}.$$

Recall that the following terminology [9]: (1) R is serial if $r_R(x) \neq \emptyset$ for each $x \in U$; (2) R is reflexive if $x \in r_R(x)$ for each $x \in U$; (3) R is symmetric if $l_R(x) = r_R(x)$ for each $x \in U$; (4) R is transitive if $x \in r_R(y)$ implies that $r_R(x) \subseteq r_R(y)$; and (5) R is an equivalence relation if R is reflexive, symmetric and transitive.

Let R be an equivalence relation on U, $[x]_R$ denotes the equivalent class in R containing an element x in U. For With each subset $X \subseteq U$, Pawlak defined the following sets.

$$\underline{R}(X) = \{x|x \in U, [x]_R \subseteq X\}$$

and

$$\overline{R}(X) = \{x|x \in U, [x]_R \cap X \neq \emptyset\},$$

called the lower and upper approximations of X respectively.

There are many different types of generalization of Pawlak rough sets. We only consider two types of such a generalization. One is generalized rough sets which use a general relation in place of an equivalence relation, and the other is VPRS which is the generalization of the notion of the standard set inclusion relation.

For an arbitrary relation R on U, the following definition is the extension of Pawlak rough sets. For each subset X of U, Yao [16] defined two subsets,

$$\underline{R}(X) = \{x|x \in U, r_R(x) \subseteq X\} \text{ and}$$

$$\overline{R}(X) = \{x | x \in U, r_R(x) \cap X \neq \emptyset\},$$

called the lower and upper approximations of X respectively.

Ziarko [17] first proposed the concept of variable precision rough set based on equivalence relations. In order to overcome the described drawbacks Gong et al. [2] generalized the idea of Ziarko [17] for general binary relations. For a parameter $\alpha(0 \leq \alpha < \frac{1}{2})$ and general binary relation, Gong et al. [2] defined variable precision rough set based on serial relations.

$$\underline{R}_\alpha(X) = \{x | x \in U, \frac{|r_R(x) \cap X|}{|r_R(x)|} \geq 1 - \alpha\} \text{ and}$$

$$\overline{R}_\alpha(X) = \{x | x \in U, \frac{|r_R(x) \cap X|}{|r_R(x)|} > \alpha\},$$

called α-lower and α-upper approximations of X respectively. Clearly, $\underline{R}(X) = \underline{R}_0(X)$ and $\overline{R}(X) = \overline{R}_0(X)$. Similar to Pawlak rough sets, we can define the concept of positive, negative and borderline regions for VPRS, for detail, we refer to see [2].

3 β-approximations and Strong β-approximations

As a generalization of VPRS, this section proposes the concept of β-approximation and strong β-approximation, the concept unifies the early definition of approximations of VPRS. A very useful concept for sets is the characteristic function. Let $U = \{u_1, u_2, \cdots, u_n\}$ be a finite set. If X is a subset of a universal set U, the characteristic function λ_X of X is defined for each $x \in U$ as follows:

$$\lambda_X(x) = \begin{cases} 1, x \in X \\ 0, x \notin X \end{cases}$$

Similarly, for any given binary relation R on U, $\lambda_R(x,y) = \begin{cases} 1, (x,y) \in R \\ 0, (x,y) \notin R \end{cases}$. That is, for $x, y \in U$, $\lambda_R(x,y)$ is equal to 1 if xRy, and it is equal to 0 otherwise.

For a serial relation R on U, suppose that $M_R = (a_{ij})_{n \times n}$ is the relational matrix of R, i.e., $a_{ij} = \lambda_R(u_i, u_j)$. We define a $n \times n$ matrix as follows.

$$N_R = \begin{pmatrix} \frac{1}{|r_R(u_1)|} & & & \\ & \frac{1}{|r_R(u_2)|} & & \\ & & \ddots & \\ & & & \frac{1}{|r_R(u_n)|} \end{pmatrix} \begin{pmatrix} a_{11} & a_{12} & \cdots & a_{1n} \\ a_{21} & a_{22} & \cdots & a_{2n} \\ \vdots & \vdots & \vdots & \vdots \\ a_{n1} & a_{n2} & \cdots & a_{nn} \end{pmatrix}$$

$$= \begin{pmatrix} \frac{a_{11}}{|r_R(u_1)|} & \frac{a_{12}}{|r_R(u_1)|} & \cdots & \frac{a_{1n}}{|r_R(u_1)|} \\ \frac{a_{21}}{|r_R(u_2)|} & \frac{a_{22}}{|r_R(u_2)|} & \cdots & \frac{a_{2n}}{|r_R(u_2)|} \\ \vdots & \vdots & \vdots & \vdots \\ \frac{a_{n1}}{|r_R(u_n)|} & \frac{a_{n2}}{|r_R(u_n)|} & \cdots & \frac{a_{nn}}{|r_R(u_n)|} \end{pmatrix}$$

The following is an example for computing N_R.

Example 3.1. Consider $U = \{1, 2, 3, 4, 5\}$ and $R = \{(1,1), (1,3), (1,5), (2,1),$ $(2,3), (2,5), (3,3), (4,2), (4,3), (4,5), (5,3), (5,4)\}$. Then $M_R = \begin{pmatrix} 1 & 0 & 1 & 0 & 1 \\ 1 & 0 & 1 & 0 & 1 \\ 0 & 0 & 1 & 0 & 0 \\ 0 & 1 & 1 & 0 & 1 \\ 0 & 0 & 1 & 1 & 0 \end{pmatrix}$,

and

$$N_R = \begin{pmatrix} \frac{1}{3} & & & & \\ & \frac{1}{3} & & & \\ & & 1 & & \\ & & & \frac{1}{3} & \\ & & & & \frac{1}{2} \end{pmatrix} M_R = \begin{pmatrix} \frac{1}{3} & & & & \\ & \frac{1}{3} & & & \\ & & 1 & & \\ & & & \frac{1}{3} & \\ & & & & \frac{1}{2} \end{pmatrix} \begin{pmatrix} 1 & 0 & 1 & 0 & 1 \\ 1 & 0 & 1 & 0 & 1 \\ 0 & 0 & 1 & 0 & 0 \\ 0 & 1 & 1 & 0 & 1 \\ 0 & 0 & 1 & 1 & 0 \end{pmatrix} = \begin{pmatrix} \frac{1}{3} & 0 & \frac{1}{3} & 0 & \frac{1}{3} \\ \frac{1}{3} & 0 & \frac{1}{3} & 0 & \frac{1}{3} \\ 0 & 0 & 1 & 0 & 0 \\ 0 & \frac{1}{3} & \frac{1}{3} & 0 & \frac{1}{3} \\ 0 & 0 & \frac{1}{2} & \frac{1}{2} & 0 \end{pmatrix}.$$

Lemma 3.1. Let R be a serial relation on U, for any subset $X \subseteq U$, we have

$$N_R \lambda_X = \left(\frac{|r_R(u_1) \cap X|}{|r_R(u_1)|}, \frac{|r_R(u_2) \cap X|}{|r_R(u_2)|}, \cdots, \frac{|r_R(u_n) \cap X|}{|r_R(u_n)|} \right)^T,$$

where T denotes transpose.

Proof. By direct computation, the ith entry of $N_R \lambda_X$ is

$$\left(\frac{\lambda_R(u_i, u_1)}{|r_R(u_i)|} \lambda_X(u_1) \right) + \left(\frac{\lambda_R(u_i, u_2)}{|r_R(u_i)|} \lambda_X(u_2) \right) + \ldots + \left(\frac{\lambda_R(u_i, u_n)}{|r_R(u_i)|} \lambda_X(u_n) \right) = \frac{|r_R(u_i) \cap X|}{|r_R(u_i)|}.$$

This completes the proof. $\qquad \square$

Gong et al. [2] established variable precision rough set model based on a serial binary relation. In order to unify earlier definition of VPRS, we extend their idea based on a serial relation as follows.

Definition 3.1. Let R be a serial relation on U. For $\beta \in [0, 1]$, and $X \subseteq U$, we define two mappings $R^{(\beta)}, R^{(\beta+)} : P(U) \to P(U)$ as below.

$$R^{(\beta)}(X) = \left\{ x \mid x \in U, \frac{|r_R(x) \cap X|}{|r_R(x)|} \geq \beta \right\}, \text{ and}$$

$$R^{(\beta+)}(X) = \left\{ x \mid x \in U, \frac{|r_R(x) \cap X|}{|r_R(x)|} > \beta \right\}.$$

Then $R^{(\beta)}(X)$ is called β-approximation of X and $R^{(\beta+)}(X)$ is called strong β-approximation of X.

Remark

(1) $(R^{(1)}(X), R^{(0+)}(X))$ is a generalized rough set of X.
(2) If R is an equivalence relation on U, then $(R^{(1)}(X), R^{(0+)}(X))$ is Pawlak rough approximation pair of X.
(3) If R is an equivalence relation on U and $0.5 < \alpha \leq 1$, then $R^{(\alpha)}$ is α-lower approximation of VPRS and $R^{((1-\alpha)+)}$ is α-upper approximation of VPRS [17].

(4) $(R^{(1-\alpha)}(X), R^{(\alpha+)}(X))$ is a VPRS of X defined by Gong et al. [2]. Clearly, Definition 3.1 unifies earlier definitions of variant VPRS.

Example 3.2. Let U and R be as in Example 3.1. If $X = \{1, 4, 5\}$ and $\beta = \frac{1}{2}$, then $R^{(\beta)}(X) = \{1, 2, 4\}$, and $R^{(\beta+)}(X) = \{1, 2\}$.

Recall that if X is a subset of U and $\alpha \in [0, 1]$, the product αX of a scalar α and with the set X is a fuzzy set in U defined by $(\alpha X) = \alpha \wedge \lambda_X(x)$ for $x \in U$. If X is a fuzzy set in U and $\alpha \in [0, 1]$, then α-cut set of X is $X_\alpha = \{x | X(x) \geq \alpha\}$ and strong α-cut set of X is $X_\alpha = \{x | X(x) > \alpha\}$, where $X(x)$ is the membership function for fuzzy set X.

β-approximation and strong β-approximation of X have the following properties.

Theorem 3.1. Let R be a serial relation on U, $\beta \in [0, 1]$, and $X, Y \subseteq U$, then operators $R^{(\beta)}$ and $R^{(\beta+)}$ have the following properties.

(1) $N_R \lambda_X = \cup_{\beta \in [0,1]} \beta(R^{(\beta)}(X))$ and $N_R \lambda_X = \cup_{\beta \in [0,1]} \beta(R^{(\beta+)}(X))$
(2) If $\alpha \leq \beta$, then $R^{(\beta)}(X) \subseteq R^{(\alpha)}(X)$ and $R^{(\beta+)}(X) \subseteq R^{(\alpha+)}(X)$.
(3) $R^{(\beta+)}(X) \subseteq R^{(\beta)}(X)$.
(4) $\overline{R}X = R^{(0+)}(X)$ and $\underline{R}X = R^{(1)}(X)$.
(5) Duality, $R^{(\beta)}(X^C) = (R^{((1-\beta)+)}(X))^C$ and $R^{(\beta+)}(X^C) = (R^{(1-\beta)}(X))^C$, where X^C denotes the complement of X in U.
(6) If $X \subseteq Y$, then $R^{(\beta)}(X) \subseteq R^{(\beta)}(Y)$ and $R^{(\beta+)}(X) \subseteq R^{(\beta+)}(Y)$.
(7) $R^{(\beta)}(X) \cup R^{(\beta)}(Y) \subseteq R^{(\beta)}(X \cup Y)$ and $R^{(\beta+)}(X) \cup R^{(\beta+)}(Y) \subseteq R^{(\beta+)}(X \cup Y)$.
(8) $R^{(\beta)}(X \cap Y) \subseteq R^{(\beta)}(X) \cap R^{(\beta)}(Y)$ and $R^{(\beta+)}(X \cap Y) \subseteq R^{(\beta+)}(X) \cap R^{(\beta+)}(Y)$.

Proof. (1) By decomposition theorem of fuzzy sets, any fuzzy set X in U can be written as $X = \cup_{\beta \in [0,1]} \beta X_\beta$, by applying the decomposition theorem to $N_R \lambda_X$, we have $N_R \lambda_X = \cup_{\beta \in [0,1]} \beta(R^{(\beta)}(X))$, similarly, $N_R \lambda_X = \cup_{\beta \in [0,1]} \beta(R^{(\beta+)}(X))$.

(2) If $\alpha \leq \beta$, suppose $x \in R^{(\beta)}(X)$, then $\frac{|r_R(x) \cap X|}{|r_R(x)|} \geq \beta \geq \alpha$, thus $x \in R^{(\alpha)}(X)$, so $R^{(\beta)}(X) \subseteq R^{(\alpha)}(X)$. Similarly, $R^{(\beta+)}(X) \subseteq R^{(\alpha+)}(X)$.

(3) and (4) are clear.

(5) $x \in R^{(\beta)}(X^C) \Leftrightarrow \frac{|r_R(x) \cap X^C|}{|r_R(x)|} \geq \beta \Leftrightarrow 1 - \frac{|r_R(x) \cap X|}{|r_R(x)|} \geq \beta \Leftrightarrow \frac{|r_R(x) \cap X|}{|r_R(x)|} \leq 1 - \beta$
$\Leftrightarrow x \in (R^{((1-\beta)+)}(X))^C$. Thus $R^{(\beta)}(X^C) = (R^{((1-\beta)+)}(X))^C$. Similarly, $R^{(\beta+)}(X^C) = (R^{(1-\beta)}(X))^C$.

(6) If $X \subseteq Y$, then $\frac{|r_R(x) \cap Y|}{|r_R(x)|} \geq \frac{|r_R(x) \cap X|}{|r_R(x)|}$ for each $x \in U$, so $R^{(\beta)}(X) \subseteq R^{(\beta)}(Y)$ and $R^{(\beta+)}(X) \subseteq R^{(\beta+)}(Y)$.

(7) Since $X \subseteq X \cup Y$, we have $R^{(\beta}(X) \subseteq R^{(\beta}(X \cup Y)$, similarly, $R^{(\beta}(Y) \subseteq R^{(\beta}(X \cup Y)$, thus $R^{(\beta}(X) \cup R^{(\beta}(Y) \subseteq R^{(\beta}(X \cup Y)$. Similarly, $R^{(\beta+)}(X) \cup R^{(\beta+)}(Y) \subseteq R^{(\beta+)}(X \cup Y)$. $\square$

The proof of part (8) is similar to that of part (7) and we omit it.

4   Calculation for Approximations of VPRS

For Pawlak rough sets or generalized rough sets, we [4] gave a Boolean matrix formulae for computing upper approximations, by duality, we can calculate lower approximations. Naturally, we hope to calculate approximations of VPRS with similar formulae. This section will show that β-approximation and strong β-approximation can also be calculated by the following matrix formulae. The following theorem provides a matrix formulae for computing β-approximation and strong β-approximation.

Theorem 4.1. Let R be a serial relation on U and $\beta \in [0,1]$, then $\lambda_{R^{(\beta)}(X)} = (N_R\lambda_X)_\beta$ and $\lambda_{R^{(\beta+)}(X)} = (N_R\lambda_X)_{\beta+}$ for each subset $X \subseteq U$. Particularly, $\lambda_{\underline{R}_\alpha(X)} = \lambda_{R^{(1-\alpha)}(X)} = (N_R\lambda_X)_{(1-\alpha)} =$ and $\lambda_{\overline{R_\alpha}(X)} = \lambda_{R^{(\alpha+)}(X)} = (N_R\lambda_X)_{\alpha+}$.

Proof. By using Lemma 3.1 and definition of β- and strong β-approximations of VPRS, we obtain $\lambda_{R^{(\beta)}(X)} = (N_R\lambda_X)_\beta$ and $\lambda_{R^{(\beta+)}(X)} = (N_R\lambda_X)_{\beta+}$.   □

Theorem 4.1, in fact, provides a simple and efficient Boolean matrix calculate formulae for β-and strong β-approximations. The matrix formulae is not only used to calculation for approximations of VPRS but also used to study structure of VPRS. We use the following example to illustrate our calculate formulae.

Example 4.1. Let $U = \{1,2,3,4,5\}$ and R be as in Example 3.1. Then

$$
N_R = \begin{pmatrix} \frac{1}{3} & & & & \\ & \frac{1}{3} & & & \\ & & 1 & & \\ & & & \frac{1}{3} & \\ & & & & \frac{1}{2} \end{pmatrix} \quad
M_R = \begin{pmatrix} \frac{1}{3} & 0 & \frac{1}{3} & 0 & \frac{1}{3} \\ \frac{1}{3} & 0 & \frac{1}{3} & 0 & \frac{1}{3} \\ 0 & 0 & 1 & 0 & 0 \\ 0 & \frac{1}{3} & \frac{1}{3} & 0 & \frac{1}{3} \\ 0 & 0 & \frac{1}{2} & \frac{1}{2} & 0 \end{pmatrix}.
$$

If $X = \{1,4,5\}$ and $\alpha = 0.5$, then $\lambda_X = (1,0,0,1,1)^T$, and

$$
N_R\lambda_X = \begin{pmatrix} \frac{1}{3} & 0 & \frac{1}{3} & 0 & \frac{1}{3} \\ \frac{1}{3} & 0 & \frac{1}{3} & 0 & \frac{1}{3} \\ 0 & 0 & 1 & 0 & 0 \\ 0 & \frac{1}{3} & \frac{1}{3} & 0 & \frac{1}{3} \\ 0 & 0 & \frac{1}{2} & \frac{1}{2} & 0 \end{pmatrix} \begin{pmatrix} 1 \\ 0 \\ 0 \\ 1 \\ 1 \end{pmatrix} = \begin{pmatrix} \frac{2}{3} \\ \frac{2}{3} \\ 0 \\ \frac{1}{3} \\ \frac{1}{2} \end{pmatrix},
$$

Thus $\lambda_{R^{(\alpha)}(X)} = \begin{pmatrix} \frac{2}{3} \\ \frac{2}{3} \\ 0 \\ \frac{1}{3} \\ \frac{1}{2} \end{pmatrix}_{0.5} = \begin{pmatrix} 1 \\ 1 \\ 0 \\ 0 \\ 1 \end{pmatrix}$ and $\lambda_{R^{(\alpha+)}(X)} = \begin{pmatrix} \frac{2}{3} \\ \frac{2}{3} \\ 0 \\ \frac{1}{3} \\ \frac{1}{2} \end{pmatrix}_{0.5+} = \begin{pmatrix} 1 \\ 1 \\ 0 \\ 0 \\ 0 \end{pmatrix}$.

Therefore, $R^{(\alpha)}(X) = \{1,2,5\}$, $R^{(\alpha+)}(X) = \{1,2\}$. Similarly, $\underline{R}X = \emptyset$, $\overline{R}X = \{1,2,4,5\}$, $\underline{R}_\alpha X = \{1,2,5\}$ and $\overline{R}_\alpha X = \{1,2\}$.

5   Conclusions

This paper has proposed the concept of β-approximation and strong β-approximation based on a serial binary relation and studied their properties, this concept unifies early definition of approximations of VPRS. The paper presented firstly a matrix view of VPRS and providing a matrix formula to calculate approximations of VPRS. The formula not only greatly simplify the calculation of approximations of VPRS but also provides us a new approach for VPRS. Hopefully, the formula can be used in attribute reduction in incomplete information systems.

References

1. Cheng, Y., Zhan, W., Wu, X., Zhang, Y.: Automatic determination about precision parameter value based on inclusion degree with variable precision rough set model. Inf. Sci. **290**, 72–85 (2015)
2. Gong, Z.T., Sun, B.Z., Shao, Y.B., Chen, D.G., He, Q.: Variable precision rough set model based on general relation. In: Proceeding of the Third International Conference On Machine Learning And Cybernetics, Shanghai, pp. 2490–2494 (2014)
3. Inuiguchi, M., Yoshioka, Y., Kusunoki, Y.: Variable-precision dominance-based rough set approach and attribute reduction. Int. J. Approximate Reasoning **50**, 1199–1214 (2009)
4. Liu, G.L.: The axiomatization of the rough set upper approximation operations. Fundamenta Informaticae **69**, 331–342 (2006)
5. Liu, G.L.: Using one axiom to characterize rough set and fuzzy rough set approximations. Inf. Sci. **223**, 285–296 (2013)
6. Liu, G.L.: The relationship among different covering approximations. Inf. Sci. **250**, 178–183 (2013)
7. Liu, G.L., Sai, Y.: Invertible approximation operators of generalized rough sets and fuzzy rough sets. Inf. Sci. **180**, 2221–2229 (2010)
8. Liu, J.N.K., Hua, Y., He, Y.: A set covering based approach to find the reduct of variable precision rough set. Inf. Sci. **275**, 83–100 (2014)
9. Grassmann, W.K., Tremblay, J.P.: Logic and Discrete Mathematics. A computer Science Perspective. Prentice-Hall, Upper Saddle River (1996)
10. Mi, J.S., Wu, W.Z., Zhang, W.X.: Approaches to knowledge reduction based on variable precision rough set model. Inf. Sci. **159**, 255–272 (2004)
11. Mieszkowicz-Rolka, A., Rolka, L.: Variable precision rough rets in analysis of inconsistent decision tables. In: Rutkowski, L., Kacprzyk, J. (eds.) Adv. Soft Comput. Physica-Verlag, Heidelberg (2003)
12. Mieszkowicz-Rolka, A., Rolka, L.: Variable precision fuzzy rough sets. In: Peters, J.F., Skowron, A., Grzymała-Busse, J.W., Kostek, B. (eds.) Transactions on Rough Sets I. LNCS, vol. 3100, pp. 144–160. Springer, Heidelberg (2004)
13. Pawlak, Z.: Rough sets. Int. J. Comput. Inf. Sci. **11**, 341–356 (1982)
14. Pawlak, Z.: Rough Sets: Theoretical Aspects Of Reasoning About Data. Kluwer Academic Publishers, Boston (1991)
15. Yao, Y., Mi, J., Li, Z.: A novel variable precision (θ, σ)-fuzzy rough set model based on fuzzy granules. Fuzzy Sets Syst. **236**, 58–72 (2014)

16. Yao, Y.Y.: Constructive and algebraic methods of theory of rough Sets. Inf. Sci. **109**, 21–47 (1998)
17. Ziarko, W.: Variable precision rough set model. J. Comput. Syst. Sci. **46**, 39–59 (1993)
18. Zhang, H.Y., Leung, Y., Zhou, L.: Variable-precision-dominance-based rough set approach to interval-valued information systems. Inf. Sci. **244**, 75–91 (2013)

The Lower Approximation Number in Covering-Based Rough Set

Hui Liu[✉] and William Zhu[✉]

Lab of Granular Computing, Minnan Normal University, Zhangzhou, China
liuhui313728@163.com, williamfengzhu@gmail.com

Abstract. Covering-based rough set has attracted much research interest with significant achievements. However, there are few analysis that have been conducted to quantify covering-based rough set. The approximation number is viewed as a quantitative tool for analyzing the covering-based rough set. In this paper, we focus on the lower approximation number. Firstly, we investigate some key properties of the lower approximation number. Secondly, we establish a lattice and two semilattice structures in covering-based rough set with the lower approximation number. Finally, based on the lower approximation number, a pair of matroid approximation operators is constructed. Moreover, we investigate the relationship between the pair of matroid approximation operators and a pair of lattice approximation operators.

Keywords: Covering · Rough set · The lower approximation number · Granular computing

1 Introduction

Rough sets [1–4] have been used as a tool to analyze the uncertain and incomplete information systems in data mining, and granular computing models have been well established based on rough sets to process the uncertainly of objects. In data mining research, coverings are a useful form to describe the characteristics of attributes in information systems [5–7]. Covering-based rough sets serve as an efficient technique to deal with covering data [8,9]. In recent years, covering-based rough sets play an important role in data representation and have been attracting more and more researcher interest [10–12]. There are many significant achievements in both theory and application. For example, covering approximation models have been constructed [13–15], covering axiomatic systems have been established [16–18], covering reduction problems have been defined [19,20], and covering decision systems have been proposed [6,21].

Most existing research on covering-based rough sets has been conducted qualitatively. However, few quantitative analysis for covering-based rough sets have been conducted. Recently, Zhu and Wang define [19,22] the upper approximation number function in covering-based approximation space. The upper approximation number of a covering approximation space is equivalent to the rank of

© Springer International Publishing Switzerland 2015
D. Ciucci et al. (Eds.): RSKT 2015, LNAI 9436, pp. 222–230, 2015.
DOI: 10.1007/978-3-319-25754-9_20

matrix. In order to have a better understanding of covering-based rough set quantitatively, it is necessary to investigate the approximation number function.

In this paper, we propose a measurement to study covering-based rough sets quantitatively. Firstly, we define the lower approximation number, and study some properties of the lower approximation number. Secondly, with the low approximation number, we establish a lattice and two semilattice structures in covering-based rough set. Finally, a pair of matroid approximation operators is constructed. What's more, we compare the matroid approximation operators with a pair of lattice approximation operators, and the pair of matroid operators exhibit some quantitative characteristic.

The rest of the paper is organized as follows. Section 2 reviews some fundamental concepts to be used in this paper. Section 3 presents some quantitative analysis for covering-based rough set. Section 4 gives the conclusion.

2 Preliminaries

This section, we present some fundamental concepts to be used in this paper. First, the concept of the poset is given.

Definition 1 (Poset [23]). *A relation $\leq$ on a set P is called a partial order if it is reflexive, antisymmetric and transitive. A set P together with a partial order $\leq$ is called a poset, denoted simply by $(P, \leq)$.*

In the following, we introduce the semilattice, which is widely applied to many areas.

Definition 2 (Semilattice [23]). *An upper-semilattice is a poset (P, R) in which every subset $\{a, b\}$ has a least upper bound $a \vee b$. A lower-semilattice is a poset (P, R) in which every $\{a, b\}$ has a greatest lower bound $a \wedge b$. The upper-semilattice and the lower-semilattice are also called semilattices.*

Based on the semilattice, the lattice is presented. In fact, a lattice is a poset (P, R) which is an upper-semilattice and a lower-semilattice.

Definition 3 (Lattice [23]). *A lattice is a poset $(L, \leq)$ in which every subset $\{x, y\}$ has a least upper bound $x \vee y$ and a greatest lower bound $x \wedge y$.*

The operations $\vee$ and $\wedge$ are disjunction and conjunction respectively, and $(L, \wedge, \vee)$ is an algebraic system induced by the lattice $(L, \leq)$. In a lattice, for any set $\{x, y\}$, and for any $z \in L$, if $x \leq z$ and $y \leq z$, the z is called an upper bound of $\{x, y\}$. The set $\{x, y\}$ have more than one upper bound, and the least upper bound is denoted by $x \vee y$. Similarly, the greatest lower bound of $\{x, y\}$ is denoted by $x \wedge y$. Generally, we also call $(L, \wedge, \vee)$ a lattice. When there is no confusion, we call L a lattice.

Several special types of lattice are introduced in the following.

Definition 4 (Bounded, distributive, modular, and complemented lattices [23]). *(i) A lattice $(L, \wedge, \vee)$ is bounded if there exist $\perp, \top \in L$ such* that $\perp \leq x \leq \top$ for all $x \in L$. Generally $(L, \wedge, \vee, \perp, \top)$ is used to denote a bounded lattice $(L, \wedge, \vee)$, where $\top$ is the greatest element and $\perp$ its least element.

(ii) A lattice $(L, \wedge, \vee)$ is distributive if $x \vee (y \wedge z) = (x \vee y) \wedge (x \vee z)$, and $x \wedge (y \vee z) = (x \wedge y) \vee (x \wedge z)$ for all $x, y, z \in L$.

(iii) A lattice $(L, \wedge, \vee)$ is modular if $x \geq z$, we have $x \wedge (y \vee z) = (x \wedge y) \vee z$ for all $x, y, z \in L$.

(iv) A lattice $(L, \wedge, \vee)$ is complemented if, for any $x \in L$, there exists $x' \in L$, such that $x \vee x' = \top$ and $x \wedge x' = \perp$, we call x' a complement of x in L.

This paper we focus on bounded lattice. In order to better understand bounded lattice, the following we give an example (Fig. 1).

Example 1. Let $L = \{a, b, c, d, e\}$. A bounded lattice $(L, \wedge, \vee, \perp, \top)$ is shown in Fig. 1. In fact, $a = \perp$ and $e = \top$.

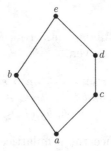

Fig. 1.

The following we will give the definition of the boolean algebra. In fact, a boolean algebra is a special lattice that forms a mathematical structure with high abstraction and broad application.

Definition 5 (Boolean algebra [23]). *Let $(L, \wedge, \vee, \perp, \top)$ be a bounded lattice. Suppose $(L, \wedge, \vee, \perp, \top)$ is a distributive and complemented lattice, then $(L, \wedge, \vee, \perp, \top)$ is called a Boolean algebra.*

Note that in a boolean algebra $(L, \wedge, \vee, \perp, \top)$, any element x has a unique complement which we denote by x'.

The following definition introduce a covering on a boolean algebra.

Definition 6 (Covering [24]). *Let $(L, \wedge, \vee, \perp, \top)$ be a bounded lattice and let $C \subseteq L - \{\perp\}$. C is called a covering of L if $\vee_{x \in C} x = \top$. Further, if C is a covering of L, and $x \wedge y = \perp$ for all $x, y \in C$, $x \neq y$, then C is called a partition of L.*

An example is given to illustrate the definition of covering in the following.

Example 2 (Continued from Example 1). A bounded lattice $(L, \wedge, \vee, \perp, \top)$ is shown in Fig. 1. Let $L = \{a, b, c, d, e\}$ and $C = \{b, c, d\}$. Since $\vee_{x \in C} x = b \vee c \vee d = \top$, we know C is a covering of L. However, $c \wedge d = c \neq \perp$, then C is not a partition of L.

In the following definition, a pair of approximation operators is introduced.

Definition 7. *[24] Let $(L, \wedge, \vee, \bot, \top)$ be a Boolean algebra and let $C \subseteq L - \{\bot\}$. For all $x \in L$,*

$$\underline{apr}_C^{\Theta}(x) = \vee\{c \in C | c \leq x\},$$

$$\overline{apr}_C^{\Theta}(x) = (\underline{apr}_C(x^{'}))^{'}.$$

are called the lower, upper lattice approximation of x with respect to C, respectively. When there is no confusion, we omit the subscript C.

3   The Lower Approximation Number

In this section, we first present some properties of the lower approximation number. Based on the lower approximation number, we establish a lattice and two semilattice structures in covering-based rough set. Finally, a pair of matroid approximation operators is established with the lower approximation number.

3.1   Properties of the Lower Approximation Number

Inspired by the approximation operators in rough sets, the low approximation number based on bounded lattices is defined as follow.

Definition 8 (The lower approximation number). *Let $(L, \wedge, \vee, \bot, \top)$ be a bounded lattice, and let C be a covering of L. For all $x \in L$,*

$$g_C(x) = |\{c \in C | c \leq x\}|$$

is called the lower approximation number of x, and g_C is called the lower approximation number function with respect to C. When there is no confusion, we omit the subscript C.

Note that the low approximation number is defined for a bounded lattice, so we can study covering-based rough set based on bounded lattices within this generalized framework. In the rest of this paper, we restrict ourselves to bounded lattice. To better understand the lower approximation number, an example is given.

Example 3 (Continued from Example 1). A bounded lattice $(L, \wedge, \vee, \bot, \top)$ is shown in Fig. 1. Let $L = \{a, b, c, d, e\}$ and $C = \{a, b, d\}$. Since $\vee_{x \in C} x = a \vee b \vee d = \top$, we know C is a covering of L. We will compute the lower approximation numbers of c and e. As $a \leq c$, then $g_C(c) = |\{a\}| = 1$, and $a \leq e, b \leq e, d \leq e$, so $g_C(e) = |\{a, b, d\}| = 3$.

The following proposition will give some key properties of the lower approximation number, which is very important to analyze covering-based rough set quantitatively.

Proposition 1. *Let* $(L, \wedge, \vee, \perp, \top)$ *be a Boolean algebra, and let* C *be a covering of* L. *The following properties of the lower approximation number function* g_C *hold:*

(1) $g_C(\perp) = 0$;

(2) *for all* $x, y \in L$, *if* $x \leq y$, $g_C(x) \leq g_C(y)$;

(3) *for all* $x, y \in L$, $g_C(x) + g_C(y) \leq g_C(x \vee y) + g_C(x \wedge y)$.

Proof. (1) : For all $c \in C$, we have $\perp \leq c$, so $g_C(\perp) = 0$.

(2) : For all $a \in \{c \in C | c \leq x\}$, we have $a \leq x$. If $x \leq y$, then $a \leq y$. Therefore, $a \in \{c \in C | c \leq y\}$. This prove $\{c \in C | c \leq x\} \subseteq \{c \in C | c \leq y\}$. Thus $|\{c \in C | c \leq x\}| = g(x) \leq g(y) = |\{c \in C | c \leq y\}|$.

(3) : For all $a \in \{c \in C | c \leq x\}$ or $a \in \{c \in C | c \leq y\}$, we have $a \leq x$ or $a \leq y$, so $a \leq x \vee y$. Then $a \in \{c \in C | c \leq x \vee y\}$. For all $a \in \{c \in C | c \leq x\}$ and $a \in \{c \in C | c \leq y\}$, we have $a \leq x$ and $a \leq y$, so $a \leq x \wedge y$. Then $a \in \{c \in C | c \leq x \wedge y\}$. Therefore $g_C(x) + g_C(y) \leq g_C(x \vee y) + g_C(x \wedge y)$.

Based on Definition 8, we present a necessary and sufficient condition for $g_C(x) = n$.

Proposition 2. *Let* $(L, \wedge, \vee, \perp, \top)$ *be a Boolean algebra. Let* C *be a covering of* L *and* $|C| = n$. *For any* $x \in L$, $g_C(x) = n$ *if and only if* $x = \top$.

Proof. $\Leftarrow$: It is straightforward.

$\Rightarrow$: If $g_C(x) = n$, then $\forall c_i \in C$, we have $c_i \leq x$. Hence $\vee_{c_i \in C} c_i \leq x$. Since C is a covering of L, according to Definition 6, we know that $\vee_{c_i \in C} c_i = \top$. Therefore, $x = \top$.

3.2   Lattice Establish with the Lower Approximation Number

It is known to us all that lattices are important algebraical structures, and have a variety of applications in the real world. This section we will establish two semilattice structures with lower approximation number. Moreover, we establish a boolean algebra.

Definition 9. *Let* $(L, \wedge, \vee, \perp, \top)$ *be a Boolean algebra, and let* C *be a covering of* L. *We define two sets as follow:*

$$T_1 = \{y \in L | (x \leq y) \wedge (g_C(x) = g_C(y))\},$$

$$T_2 = \{y \in L | (y \leq x) \wedge (g_C(x) = g_C(y))\}.$$

Based on Definition 9, we will get two semilattice structures. The following will give the proof.

Proposition 3. $(T_1, \wedge)$ *and* $(T_2, \vee)$ *are semilattice.*

Proof. First for all $y_1, y_2 \in T_1$, we need to prove $y_1 \wedge y_2 \in T_1$. Since $y_1, y_2 \in T_1$, then $x \leq y_1$ and $x \leq y_2$, i.e., $x \leq y_1 \wedge y_2$. From $x \leq y_1 \wedge y_2 \leq y_1$, we know $g_C(x) \leq g_C(y_1 \wedge y_2) \leq g_C(y_1)$. For all $y_1 \in T_1$, $g_C(y_1) = g_C(x)$, that is, $g_C(y_1 \wedge y_2) = g_C(x)$. Therefore, $y_1 \wedge y_2 \in T_1$. Similarly, we can prove $y_1 \vee y_2 \in T_2$ for all $y_1, y_2 \in T_2$.

In order to better establish lattice structure, we give the following definition.

Definition 10. *Let* $(L, \wedge, \vee, \perp, \top)$ *be a Boolean algebra. Let* C *be a covering of* L *and* $|C| = n$. T_3 *is defined as follow:*

$$T_3 = \{x \in L | g_C(x) + g_C(x') = n\}.$$

In fact T_3 is not only a lattice, but also a boolean algebra. The following proposition will give the proof.

Proposition 4. $(T_3, \wedge, \vee, \perp, \top)$ *is a Boolean algebra.*

Proof. (*i*) We need to prove $(T_3, \wedge, \vee)$ is a lattice. For all $x_1, x_2 \in T_3$, we have $g_C(x_1) + g_C(x_1') = n$ and $g_C(x_2) + g_C(x_2') = n$. According to Proposition 1, we know $g_C(x_1) + g_C(x_1') + g_C(x_2) + g_C(x_2') \leq g_C(x_1 \wedge x_2) + g_C(x_1 \vee x_2) + g_C(x_1' \wedge x_2') + g_C(x_1' \vee x_2')$. So $g_C(x_1 \wedge x_2) + g_C(x_1 \vee x_2) + g_C(x_1' \wedge x_2') + g_C(x_1' \vee x_2') = 2n$. In fact, $g_C(x_1' \wedge x_2') = g_C((x_1 \vee x_2)')$, $g_C(x_1' \vee x_2') = g_C((x_1 \wedge x_2)')$. It means that $g_C(x_1 \wedge x_2) + g_C((x_1 \wedge x_2)') = n$ and $g_C(x_1 \vee x_2) + g_C((x_1 \vee x_2)') = n$. Thus $x_1 \vee x_2 \in T_3, x_1 \wedge x_2 \in T_3$. This proves $(T_3, \wedge, \vee)$ is a lattice.
(*ii*) It is easy to know that for all $x \in T_3$, $\perp \leq x \leq \top$. Hence $(T_3, \wedge, \vee)$ is bounded lattice.
(*iii*) For all $x, y, z \in T_3$, we know $x, y, z \in L$. Since $(L, \wedge, \vee, \perp, \top)$ is a Boolean algebra, then for all $x, y, z \in L$, $x \vee (y \wedge z) = (x \vee y) \wedge (x \vee z)$, and $x \wedge (y \vee z) = (x \wedge y) \vee (x \wedge z)$. Thus $(T_3, \wedge, \vee)$ is a distributive lattice.
(*iv*) For all $x \in T_3$, we know $g_C(x) + g_C(x') = n$. Then $g_C(x') + g_C(x'') = g_C(x') + g_C(x) = n$, that is, $x' \in T_3$. Thus $(T_3, \wedge, \vee)$ is a complemented lattice. From $(i) - (iv)$, we have the conclusion that $(T_3, \wedge, \vee, \perp, \top)$ is a Boolean algebra.

3.3 A Pair of Matroid Approximation Operators

In this section, a pair of matroid approximation operators is defined using the lower approximation number. Then, we investigate some properties of the pair of matroid operators. What's more, compared with a pair of lattice approximation operators, the pair of matroid operators exhibit some quantitative characteristic. First of all, we give the the the definition of matroid approximation operators as follow.

Definition 11. *Let* $(L, \wedge, \vee, \perp, \top)$ *be a Boolean algebra. Let* C *be a covering of* L *and* $|C| = n$. *For all* $x \in L$,

$$\underline{apr}_C(x) = \wedge\{l \in L | g_C(x) = g_C(x \wedge l)\},$$

$$\overline{apr}_C(x) = (\underline{apr}_C(x'))'.$$

are called the lower, upper matroid approximation of x *with respect to* C, *respectively. When there is no confusion, we omit the subscript* C.

Based on Definition 11, we have some new properties of the matroid approximation operators. Therefore, a quantitative viewpoint to study covering-based rough sets is given.

Proposition 5. *Let* $(L, \wedge, \vee, \bot, \top)$ *be a Boolean algebra. Let* C *be a covering of* L *and* $|C| = n$. *The following properties hold:*
(1) *for all* $x \in L$, $g(x) = g(\underline{apr}(x))$;
(2) *for all* $x, y \in L$, *if* $x \leq y$ *and* $g(x) = g(y)$, *then* $\underline{apr}(y) \leq x$;
(3) *for all* $x, y \in L$, *if* $x \leq y$ *and* $g(x) < g(y)$, *then* $\underline{apr}(x) \neq \underline{apr}(y)$.

Proof. (1): For all $x \in L$, suppose that $a, b \in \{l \in L | g(x) = g(x \wedge l)\}$ and $a \neq b$, so $g(x) = g(x \wedge a)$, $g(x) = g(x \wedge b)$. According to proposition 1, we know $g(x \wedge a) + g(x \wedge b) \leq g((x \wedge a) \vee (x \wedge b)) + g((x \wedge a) \wedge (x \wedge b))$. Since $g((x \wedge a) \wedge (x \wedge b)) = g(x \wedge a \wedge b)$, and $g(x) = g(x \wedge a) \leq g((x \wedge a) \vee (x \wedge b)) \leq g(x)$, then $g((x \wedge a) \vee (x \wedge b)) = g(x)$. Thus $g(x) + g(x) = g(x \wedge a) + g(x \wedge b) \leq g(x) + g(x \wedge a \wedge b)$, that is $g(x \wedge a \wedge b) = g(x)$. Therefore, $g(\underline{apr}(x)) = g(x \wedge a \wedge b \wedge \cdots) = g(x)$.
(2): Since $x \leq y$, then $x \wedge y = x$, that is, $g(x \wedge y) = g(x)$. As $g(x \wedge y) = g(x) = g(y)$, we know $x \in \{l \in L | g(y) = g(y \wedge l)\}$. Hence $\underline{apr}(y) = \wedge\{l \in L | g(y) = g(y \wedge l)\} \leq x$.
(3): If $\underline{apr}(x) = \underline{apr}(y)$, then $g(x) = g(\underline{apr}(x)) = g(\underline{apr}(y)) = g(y)$, which contradicts $g(x) < g(y)$. Therefore $\underline{apr}(x) \neq \underline{apr}(y)$.

The following proposition shows that the matroid approximation operators inherit some traditional properties of approximation operators such as monotony. However, the idempotence does not hold.

Proposition 6. *Let* $(L, \wedge, \vee, \bot, \top)$ *be a Boolean algebra. Let* C *be a covering of* L *and* $|C| = n$. *The following properties of the matroid approximation operators* $\underline{apr}, \overline{apr}$ *hold: for all* $x, y \in L$,
(1) $\underline{apr}(\bot) = \bot, \overline{apr}(\bot) = \bot$;
(2) $\underline{apr}(\top) = \top, \overline{apr}(\top) = \top$;
(3) $\underline{apr}(x) \leq x, x \leq \overline{apr}(x)$;
(4) $\underline{apr}(\underline{apr}(x)) \leq \underline{apr}(x), \overline{apr}(x) \leq \overline{apr}(\overline{apr}(x))$.

Proof. Because of duality, we only need to prove that these properties hold for the matroid lower approximation operator.
(1) and (2): They are both straightforward.
(3): For all $x \in L$, since $g(x) = g(x \wedge x)$, we have $x \in \{l \in L | g(x) = g(x \wedge l)\}$. Therefore, $\wedge\{l \in L | g(x) = g(x \wedge l)\} \leq x$, that is, $\underline{apr}(x) \leq x$.
(4): For all $a \in \{l \in L | g(\underline{apr}(x)) = g(\underline{apr}(x) \wedge l)\}$, we have $g(\underline{apr}(x)) = g(\underline{apr}(x) \wedge a)$. Since $g(x) = g(\underline{apr}(x) \wedge a) \leq g(x \wedge a) \leq g(x)$, then we know $g(x) = g(x \wedge a)$, which implies that $a \in \{l \in L | g(x) = g(x \wedge l)\}$. Thus $\{l \in L | g(\underline{apr}(x)) = g(\underline{apr}(x) \wedge l)\} \subseteq \{l \in L | g(x) = g(x \wedge l)\}$. As $\underline{apr}(\underline{apr}(x)) = \wedge\{l \in L | g(\underline{apr}(x)) = g(\underline{apr}(x) \wedge l)\}$ and $\underline{apr}(x) = \wedge\{l \in L | g(x) = g(x \wedge l)\}$. Therefore $\underline{apr}(\underline{apr}(x)) \leq \underline{apr}(x)$.

Based on the lower approximation operators, we investigate the properties of the lattice approximation operators.

Proposition 7. *Let* $(L, \wedge, \vee, \bot, \top)$ *be a Boolean algebra. Let* C *be a covering of* L. *For all* $x, y \in L$, *if* $x \leq y$ *and* $g(x) = g(y)$, *then*
(1) $\underline{apr}^{\ominus}(x) = \underline{apr}^{\ominus}(y)$;
(2) $\underline{apr}^{\ominus}(x \wedge y) = \underline{apr}^{\ominus}(x \vee y)$.

Proof. (1): For all $x, y \in L$, if $x \leq y$, then $\{c \in C | c \leq x\} \subseteq \{c \in C | c \leq y\}$. Since $g(x) = g(y)$, we know $|\{c \in C | c \leq x\}| = |\{c \in C | c \leq y\}|$, that is $\{c \in C | c \leq x\} = \{c \in C | c \leq y\}$. Therefore, $\vee\{c \in C | c \leq x\} = \underline{apr}^{\Theta}(x) = \underline{apr}^{\Theta}(y) = \vee\{c \in C | c \leq y\}$.

(2): It is straightforward.

In the following, the relationship between the lower matroid approximation operator and the lower lattice approximation operator is studied.

Proposition 8. *Let* $(L, \wedge, \vee, \perp, \top)$ *be a Boolean algebra. Let* C *be a covering of* L. *For all* $x, y \in L$, *if* $x \leq \underline{apr}^{\Theta}(y)$, *then* $\underline{apr}(x) \leq y$.

Proof. For all $z \in \underline{apr}^{\Theta}(y)$, we know $z \leq y$. Since $x \leq \underline{apr}^{\Theta}(y)$, then $x \leq y$, that is, $x \wedge y = x$. Hence $g(x \wedge y) = g(x)$, that is $y \in \{l \in L | g(x \wedge l) = g(x)\}$. As $\underline{apr}(x) = \wedge\{l \in L | g(x \wedge l)\}$. Therefore, $\underline{apr}(x) \leq y$.

Similarly, by duality, the relation between the upper matroid approximation operator and the upper lattice approximation operator is also investigated.

Proposition 9. *Let* $(L, \wedge, \vee, \perp, \top)$ *be a Boolean algebra. Let* C *be a covering of* L. *For all* $x, y \in L$, *if* $\overline{apr}^{\Theta}(x) \leq y$, *then* $x \leq \overline{apr}(y)$.

4 Conclusion

This paper further studies the concept of the lower approximation number function, and shows that the covering-based rough set can be characterized by the lower approximation number. On one hand, we investigate some key properties of the lower approximation number. On the other hand, we establish some lattice structures based on the lower approximation number. Moreover, a pair of matroid approximation operators is constructed with the lower approximation number. Finally, we investigate the relation between the pair of matroid approximation operators and a pair of lattice approximation operators. These studies illustrate that the lower approximation number can be viewed as a quantitative tool for analyzing the covering-based rough set.

Acknowledgments. This work is in part supported by The National Nature Science Foundation of Chi- na under Grant Nos. 61170128, 61379049 and 61379089, the Key Project of Education Department of Fujian Province under Grant No. JA13192, the Project of Education De- partment of Fujian Province under Grant No. JA14194, the Zhangzhou Municipal Nat- ural Science Foundation under Grant No. ZZ2013J03, and the Science and Technology Key Project of Fujian Province, China Grant No. 2012H0043.

References

1. Pawlak, Z.: Rough sets. Int. J. Comput. Inf Sci. **11**, 341–356 (1982)
2. Pawlak, Z.: Rough classification. Int. J. Man-Mach. Stud. **20**, 469–483 (1984)

3. Yao, Y., Chen, Y.: Rough set approximations in formal concept analysis. In: Peters, J.F., Skowron, A. (eds.) Transactions on Rough Sets V. LNCS, vol. 4100, pp. 285–305. Springer, Heidelberg (2006)
4. Drwal, G.: Rough, and fuzzy-rough classification methods implemented in RClass system. In: Ziarko, W.P., Yao, Y. (eds.) RSCTC 2000. LNCS (LNAI), vol. 2005, pp. 152–159. Springer, Heidelberg (2001)
5. Bianucci, D., Cattaneo, G., Ciucci, D.: Entropies and co-entropies of coverings with application to incomplete information systems. Fundamenta Informaticae 75, 77–105 (2007)
6. Chen, D., Wang, C., Hu, Q.: A new approach to attribute reduction of consistent and inconsistent covering decision systems with covering rough sets. Inf. Sci. 177, 3500–3518 (2007)
7. Li, F., Yin, Y.: Approaches to knowledge reduction of covering decision systems based on information theory. Inf. Sci. 179, 1694–1704 (2009)
8. Zhu, F., He, H.: Logical properties of rough sets. In: The Fourth International Conference on High Performance Computing in the Asia-Pacific Region, pp. 670–671. IEEE Press (2000)
9. Zhu, F., He, H.: The axiomization of the rough set. Chin. J. Comput. 23, 330–333 (2000)
10. Zhu, W.: Relationship among basic concepts in covering-based rough sets. Inf. Sci. 179, 2478–2486 (2009)
11. Ma, L.: On some types of neighborhood-related covering rough sets. Int. J. Approx. Reasoning 53, 901–911 (2012)
12. Yao, Y., Yao, B.: Covering based rough set approximations. Inf. Sci. 200, 91–107 (2012)
13. Bonikowski, Z., Bryniarski, E., Wybraniec-Skardowska, U.: Extensions and intentions in the rough set theory. Inf. Sci. 107, 149–167 (1998)
14. Bartol, W., Miró, J., Pióro, K., Rosselló, F.: On the coverings by tolerance classes. Inf. Sci. 166, 193–211 (2004)
15. Liu, G., Sai, Y.: A comparison of two types of rough sets induced by coverings. Int. J. Approx. Reasoning 50, 521–528 (2009)
16. Zhu, W., Wang, F.: The fourth type of covering-based rough sets. Inf. Sci. 201, 80–92 (2012)
17. Liu, G., Sai, Y.: Invertible approximation operators of generalized rough sets and fuzzy rough sets. Inf. Sci. 180, 2221–2229 (2010)
18. Wang, S., Zhu, W., Zhu, Q., Min, F.: Covering base. J. Inf. Comput. Sci. 9, 1343–1355 (2012)
19. Wang, S., Zhu, Q., Zhu, W., Min, F.: Matroidal structure of rough sets and its characterization to attribute reduction. Knowl.-Based Syst. 36, 155–161 (2012)
20. Min, F., Zhu, W.: Attribute reduction of data with error ranges and test costs. Inf. Sci. 211, 48–67 (2012)
21. Qian, Y., Liang, J., Li, D., Wang, F., Ma, N.: Approximation reduction in inconsistent incomplete decision tables. Knowl.-Based Syst. 23, 427–433 (2010)
22. Wang, S., Zhu, W.: Matroidal structure of covering-based rough sets through the upper approximation number. Int. J. Granular Comput., Rough Sets Intel. Syst. 2, 141–148 (2011)
23. Birhoff, G.: Lattice Theory. American Mathematical Society, Rhode Island (1995)
24. Zhang, W., Yao, Y., Liang, Y.: Rough set and concept lattice. Xi'an Jiaotong University Press (2006)

The Matroidal Structures of the Second Type of Covering-Based Rough Set

Yanfang Liu[1] and William Zhu[2($\boxtimes$)]

[1] Institute of Information Engineering, Longyan University, Longyan 364000, China
liuyanfang003@163.com
[2] Lab of Granular Computing, Minnan Normal University, Zhangzhou 363000, China
williamfengzhu@gmail.com

Abstract. Rough set theory is a useful tool for data mining. In recent yeas, ones have combined it with matroid theory to construct an excellent set-theoretical framework for empirical machine learning methods. Hence, the study of its matroidal structure is an interesting research topic, and the structure is part of the foundation of rough set theory. Few people study the combinations the second type of covering-based rough sets with matroids. In this paper, we mainly build the matroidal structures of the second type of covering-based rough sets from the perspective of closure operators. On the one hand, we establish a closure system through the fixed point family of the second type of covering lower approximation operator, and then construct a corresponding closure operator. For a covering of a universe, this closure operator is a matroidal closure operator if and only if the reduct of the covering forms a partition of the universe. On the other hand, we present two sufficient and necessary conditions for the second type of covering upper approximation operator to form a matroidal closure operator through the indiscernible neighborhood and the covering upper approximation operator.

Keywords: Matroid · Covering-based rough set · Closure operator · Covering lower and upper approximation · Reduct

1 Introduction

To deal with the vagueness and granularity in information systems, researchers have proposed several methods such as rough set theory [14] and fuzzy set theory [15]. In recent decades, rough sets have developed significantly due to their independent theoretical background and wide applications, such as medical diagnosis, machine learning, data mining and other fields [2,3,7,12,13,23]. Various generalized rough set models have been established and their properties or structures have been studied, such as relation-based rough sets [4,8,19,25,26] and covering-based rough sets [1,16,17,31].

As a generalization of linear algebra and graph theory, matroid theory [5,9] is an important branch of mathematics, which has the independent theoretical framework and broad applications. Tsumoto combined matroid theory and

© Springer International Publishing Switzerland 2015
D. Ciucci et al. (Eds.): RSKT 2015, LNAI 9436, pp. 231–242, 2015.
DOI: 10.1007/978-3-319-25754-9_21

rough set theory to construct an excellent set-theoretical framework for empirical machine learning methods [21,22]. Therefore, the study of combining matroid theory and rough set theory has deep theoretical and practical significance beyond doubt. Based on this, the relationships among matroids and classical rough sets [6,11,20], relation-based rough sets [10,27] and covering-based rough sets [24] have been investigated. In [18], Restrepo et al. have revealed that the second type of covering lower and upper approximation operators are not dual and also do not form a Galois connection. Then, do they have connections with matroids?

In this paper, we mainly build the matroidal structures of the second type of covering-based rough sets from the viewpoint of closure operators. On the one hand, for a covering of a universe, we prove that the fixed point family of the second type of covering lower approximation operator is a closure system if and only if the covering is unary. We induce a corresponding closure operator by the closure system, and call the closure operator the second type of rough set closure operator. When the family of neighborhoods in the universe forms a partition, the second type of rough set closure operator is a matroidal closure operator. Moreover, we prove that the reduct of a covering is a partition if and only if the covering is unary and the family of all the neighborhoods forms a partition. That is to say, the reduct of a covering is a partition if and only if the second type of rough set closure operator forms a matroidal closure operator.

On the other hand, we study the relationship between the second type of covering upper approximation operator and a matroidal closure operator. In [28], Zhu has studied the properties of the second type of covering-based rough sets and given a sufficient and necessary condition when the second type of covering upper approximation operator satisfies the idempotency. However, this condition is just a necessary one. We give a counterexample to illustrate this issue and present a sufficient and necessary condition. Particularly, for a covering of a universe, the second type of covering upper approximation operator is a matroidal closure operator if and only if the family of all indiscernible neighborhoods forms a partition of the universe.

The rest of this paper is organized as follows: In Sect. 2, we recall some basic definitions and related results of closure systems, the second type of covering-based rough sets and matroids. Section 3 establishes a closure system through the second type of covering lower approximation operator, constructs a corresponding closure operator and investigates a sufficient and necessary condition for the closure operator to form a matroidal closure operator. In Sect. 4, we present two sufficient and necessary conditions when the second type of covering upper approximation operator is a matroidal closure operator. Finally, this paper concludes in Sect. 5.

2   Basic Definitions

In this section, we recall some basic definitions and results of closure systems, the second type of covering-based rough sets and matroids.

In the following discussion, unless it is mentioned specially, the universe U is considered non-empty finite and 2^U denotes the family of all subsets of U.

2.1 Closure Systems and Closure Operators

In recent years, some authors have obtained many important results through the combination of matroids and lattices. Since a closure system is a meet semi-lattice with its greatest element, we introduce the concept of closure systems as follows.

Definition 1 *(Closure system [6]). Let $\mathbf{F}$ be a family of subsets of U. $\mathbf{F}$ is called a closure system if it satisfies the following conditions:*
(F1) If $F_1, F_2 \in \mathbf{F}$, then $F_1 \cap F_2 \in \mathbf{F}$;
(F2) $U \in \mathbf{F}$.

Example 1. Let $U = \{1, 2, 3\}$ and $\mathbf{F} = \{F_1, F_2, F_3, F_4\}$ where $F_1 = \{2\}, F_2 = \{1, 2\}, F_3 = \{2, 3\}, F_4 = \{1, 2, 3\}$. Since $F_1 \cap F_2 = F_1 \cap F_3 = F_1 \cap F_4 = F_2 \cap F_3 = \{2\} \in \mathbf{F}, F_2 \cap F_4 = \{1, 2\} \in \mathbf{F}, F_3 \cap F_4 = \{2, 3\} \in \mathbf{F}$ and $U = \{1, 2, 3\} \in \mathbf{F}$, we have $\mathbf{F}$ is a closure system of U.

Definition 2. *Let φ be an operator on 2^U. φ is a closure operator if it satisfies the following conditions: for all $X, Y \subseteq U$,*
(1) $X \subseteq \varphi(X)$;
(2) If $X \subseteq Y$, then $\varphi(X) \subseteq \varphi(Y)$;
(3) $\varphi(\varphi(X)) = \varphi(X)$.

The following proposition presents the relationship between a closure system and a closure operator. And we see that any closure system can induce a closure operator.

Proposition 1. *Let $\mathbf{F}$ be a closure system of U. For any $X \subseteq U$,*

$$cl_\mathbf{F}(X) = \cap\{F \in \mathbf{F} : X \subseteq F\}.$$

$cl_\mathbf{F}$ is the closure operator induced by $\mathbf{F}$, and $cl_\mathbf{F}(X)$ is the closure of X with respect to $\mathbf{F}$.

In order to provide a clearer picture of the relationship between a closure system and a closure operator, the following example is presented.

Example 2 (Continued from Example 1). According to Proposition 1, we have:

X	$cl_\mathbf{F}(X)$	$cl_\mathbf{F}(cl_\mathbf{F}(X))$
$\emptyset$	$\{2\}$	$\{2\}$
$\{1\}$	$\{1, 2\}$	$\{1, 2\}$
$\{2\}$	$\{2\}$	$\{2\}$
$\{3\}$	$\{2, 3\}$	$\{2, 3\}$
$\{1, 2\}$	$\{1, 2\}$	$\{1, 2\}$
$\{1, 3\}$	$\{1, 2, 3\}$	$\{1, 2, 3\}$
$\{2, 3\}$	$\{2, 3\}$	$\{2, 3\}$
$\{1, 2, 3\}$	$\{1, 2, 3\}$	$\{1, 2, 3\}$

We can easily obtain cl_F holds the properties (1), (2) and (3) of Definition 2. Therefore, cl_F is a closure operator.

2.2   The Second Type of Covering-Based Rough Sets

As a generalization of classical rough sets, covering-based rough sets are obtained through extending partitions to coverings.

Definition 3 *(Covering [31]). Let C be a family of subsets of U. If none of subsets in C is empty and $\cup C = U$, then C is called a covering of U.*

In the description of objects, we do not use all the features to depict those objects. We limit ourself only to the most essential ones. The essential features of an object are established by the following definition.

Definition 4 *(Minimal description [1]). Let C be a covering of U and $x \in U$. The following family:*
$$Md_C(x) = \{K \in C : x \in K \wedge (\forall S \in C)(x \in S \wedge S \subseteq K \to K = S)\}$$
is called the minimal description of x. When the covering is clear, we omit the lowercase C for the minimal description.

In covering-based rough sets unary coverings are important coverings in discussing relationships among basic concepts.

Definition 5 *(Unary covering [32]). Let C be a covering of U. C is called unary if $|Md(x)| = 1$ for all $x \in U$.*

Through different forms of the lower and upper approximations based on coverings, many types of covering-based rough sets are put forward. In this paper, we investigate only the second type of covering-based rough sets.

Definition 6 *(The second type of covering lower and upper approximation operators [32]). Let C be a covering of U. For any $X \subseteq U$,*

$$SL_C(X) = \cup\{K \in C : K \subseteq X\};$$
$$SH_C(X) = \cup\{K \in C : K \cap X \neq \emptyset\}$$

We call SL_C, SH_C the second type of covering lower, upper approximation operators, respectively. When the covering is clear, we omit the lowercase C for the two operators.

Proposition 2 *[1]. Let C be a covering of U. $SL(X) = X$ if and only if X is a union of some elements of C.*

Proposition 3 *[31]. Let C be a covering of U. For any $X, Y \subseteq U$,*
(1L) $SL(U) = U$ *(1H) $SH(U) = U$*
(2L) $SL(\emptyset) = \emptyset$ *(2H) $SH(\emptyset) = \emptyset$*
(3L) $SL(X) \subseteq X$ *(3H) $X \subseteq SH(X)$*
(4H) $SH(X \cup Y) = SH(X) \cup SH(Y)$ (5L) $SL(SL(X)) = SL(X)$
(6L) $X \subseteq Y \Rightarrow SL(X) \subseteq SL(Y)$ *(6H) $X \subseteq Y \Rightarrow SH(X) \subseteq SH(Y)$*
(7LH) $SL(X) \subseteq SH(X)$

2.3   Matroids

A matroid is a structure that captures and generalizes the notion of linear independence in vector spaces. In the following definition, we will introduce a matroid from the viewpoint of independent sets.

Definition 7 *(Matroid [5]). A matroid is a pair $M = (U, \mathbf{I})$ consisting of a universe U and a collection $\mathbf{I}$ of subsets of U called independent sets satisfying the following three properties:*
(I1) $\emptyset \in \mathbf{I}$;
(I2) If $I \in \mathbf{I}$ and $I' \subseteq I$, then $I' \in \mathbf{I}$;
(I3) If $I_1, I_2 \in \mathbf{I}$ and $|I_1| < |I_2|$, then there exists $u \in I_2 - I_1$ such that $I_1 \cup \{u\} \in \mathbf{I}$, where $|I|$ denotes the cardinality of I.

In order to illustrate that linear algebra is an original source of matroid theory, an example is presented as follows.

Example 3. Let A be an 3×4 matrix. Suppose $U = \{a_1, a_2, a_3, a_4\}$. Denote $\mathbf{I} = \{X \subseteq U : X \text{ are linearly independent }\}$, i.e., $\mathbf{I} = \{\emptyset, \{a_2\}, \{a_3\}, \{a_4\}, \{a_2, a_3\}, \{a_2, a_4\}, \{a_3, a_4\}\}$. Then $M = (U, \mathbf{I})$ is a matroid.

$$A = \begin{array}{cccc} a_1 & a_2 & a_3 & a_4 \\ \left[\begin{array}{cccc} 0 & 1 & 0 & 1 \\ 0 & 0 & -1 & 1 \\ 0 & 0 & -1 & 1 \end{array}\right] \end{array}$$

As a generalization of the dimension of a vector space and the rank of a matrix, the rank function of a matroid is introduced in the following definition.

Definition 8 *(Rank function [5]). Let $M = (U, \mathbf{I})$ be a matroid. The rank function r_M of M is defined as $r_M(X) = max\{|I| : I \subseteq X, I \in \mathbf{I}\}$ for all $X \subseteq U$. $r_M(X)$ is called the rank of X in M.*

Example 4 (Continued from Example 3). Suppose $X = \{a_1, a_2\}, X_1 = \{a_1, a_2, a_3\}, X_2 = \{a_1, a_2, a_4\}$ and $X_3 = U$, we have $r_M(X) = 1, r_M(X_1) = 2, r_M(X_2) = 2$ and $r_M(X_3) = 2$.

A notion of the closure operator, which reflects the dependency between a set and elements, is presented based on the rank function of a matroid.

Definition 9 *(Closure operator [5]). Let $M = (U, \mathbf{I})$ be a matroid. The closure operator cl_M of M is defined as $cl_M(X) = \{u \in U : r_M(X) = r_M(X \cup \{u\})\}$ for all $X \subseteq U$. $cl_M(X)$ is called the closure of X in M.*

Example 5 (Continued from Example 4). We have $cl_M(\{a_1, a_2\}) = \{a_1, a_2\}$ and $cl_M(\{a_1, a_2, a_3\}) = cl_M(\{a_1, a_2, a_4\}) = cl_M(U) = U$.

We present some properties of the closure operator of a matroid which will be used in this paper.

Proposition 4 *[5].* *Let* $M = (U, \mathbf{I})$ *be a matroid and* cl_M *the closure operator of* M. *Then we have*
(CL1) For all $X \subseteq U$, $X \subseteq cl_M(X)$;
(CL2) For all $X, Y \subseteq U$, *if* $X \subseteq Y$, *then* $cl_M(X) \subseteq cl_M(Y)$;
(CL3) For all $X \subseteq U$, $cl_M(cl_M(X)) = cl_M(X)$;
(CL4) For all $X \subseteq U, x \in U$, *if* $y \in cl_M(X \cup \{x\}) - cl_M(X)$, *then* $x \in cl_M(X \cup \{y\})$.

There are many different but equivalent ways to define a matroid. In the following proposition, one generate a matroid in terms of the closure operator.

Theorem 1 *(Closure axiom [5]). Let* $cl : 2^U \to 2^U$ *be an operator. Then there exists a matroid* M *such that* $cl = cl_M$ *iff* cl *satisfies the conditions (CL1), (CL2), (CL3) and (CL4) of Proposition 4.*

3 Matroidal Structure of the Second Type of Covering Lower Approximation Operator

From the above section, compared with the closure operator induced by a closure system, a matroidal closure operator has more than one property (CL4) of Proposition 4. Through the second type of covering lower approximation operator, whether we can construct a closure system or not? In order to solve this issue, we first define the fixed point family of the second type of covering lower approximation operator.

Definition 10. *Let* $\mathbf{C}$ *be a covering of* U. *We define the fixed point family of the second type of covering lower approximation operator with respect to* $\mathbf{C}$ *as follows:*
$$\mathbf{S_C} = \{X \subseteq U : SL(X) = X\}.$$

In the following discussion, we call $\mathbf{S_C}$ the fixed point family for short unless otherwise stated.

Lemma 1 *[29]. Let* $\mathbf{C}$ *be a covering of* U. $\forall X, Y \subseteq U, SL(X \cap Y) = SL(X) \cap SL(Y)$ *if and only if* $\mathbf{C}$ *is unary.*

A sufficient and necessary condition for the fixed point family to form a closure system is obtained in the following proposition.

Proposition 5. *Let* $\mathbf{C}$ *be a covering of* U. $\mathbf{S_C}$ *is a closure system if and only if* $\mathbf{C}$ *is unary.*

Proof. According to Definition 10, $\mathbf{S_C} = \{X \subseteq U : SL(X) = X\}$. According to Definition 1, we need to prove $\mathbf{S_C}$ satisfies (F1) and (F2).
(F1) For all $X_1, X_2 \in \mathbf{S_C}$, $SL(X_1) = X_1, SL(X_2) = X_2$. According to Lemma 1, $\mathbf{C}$ is unary $\Leftrightarrow SL(X_1 \cap X_2) = SL(X_1) \cap SH(X_2) \Leftrightarrow SL(X_1 \cap X_2) = X_1 \cap X_2 \Leftrightarrow X_1 \cap X_2 \in \mathbf{S_C}$.
(F2) According to (1L) of Proposition 3, $SL(U) = U$, i.e., $U \in \mathbf{S_C}$.

Any closure system can induce a closure operator. The second type of rough set closure operator is obtained through the fixed point family.

Definition 11. *Let* **C** *be a unary covering of* U *and* $X \subseteq U$. *We call* $\cap\{S \in$ **S**$_\mathbf{C}$ $: X \subseteq S\}$ *the closure of* X *with respect to* **S**$_\mathbf{C}$ *and denote it as* $cl_\mathbf{C}(X)$, *where* $cl_\mathbf{C}$ *is called the second type of rough set closure operator induced by* **S**$_\mathbf{C}$.

For a unary covering of a universe, can the second type of rough set closure operator form a matroidal closure operator? If the answer is yes, what is the condition of the covering? In the following, we solve these issues. We first introduce the notion of neighborhood which is one of important concepts in covering-based rough sets.

Definition 12 *(Neighborhood [30]). Let* **C** *be a covering of* U *and* $x \in U$. $N_\mathbf{C}(x) = \cap\{K \in \mathbf{C} : x \in K\}$ *is called the neighborhood of* x *with respect to* **C**. *When there is no confusion, we omit the subscript* **C**.

For a unary covering of a universe, the relationships between the closure of any subset and neighborhoods are investigated in the following proposition.

Proposition 6. *Let* **C** *be a unary covering of* U *and* $X \subseteq U$. *Then* $cl_\mathbf{C}(X) = \bigcup_{x \in X} N(x)$.

Proof. According to Definitions 10 and 11, $\mathbf{S}_\mathbf{C} = \{Y \subseteq U : SL(Y) = Y\}$, $cl_\mathbf{C}(X) = \cap\{S \in \mathbf{S}_\mathbf{C} : X \subseteq S\}$.
Since **C** is unary, according to Definition 5, we have $|Md(x)| = 1$ for any $x \in U$. According to Definitions 4 and 12, $\cap Md(x) = N(x)$. Hence $N(x) \in \mathbf{C}$. According to Proposition 2, we see $SL(N(x)) = N(x)$, i.e., $N(x) \in \mathbf{S}_\mathbf{C}$. Therefore $cl_\mathbf{C}(\{x\}) = \cap\{S \in \mathbf{S}_\mathbf{C} : x \in S\} = N(x)$.
According to Proposition 2, we see $SL(\bigcup_{x \in X} N(x)) = \bigcup_{x \in X} N(x)$, i.e., $\bigcup_{x \in X} N(x) \in$ $\mathbf{S}_\mathbf{C}$. Since $X \subseteq \bigcup_{x \in X} N(x)$, we have $cl_\mathbf{C}(X) \subseteq \bigcup_{x \in X} N(x)$, i.e., $cl_\mathbf{C}(X) \subseteq$ $\bigcup_{x \in X} cl_\mathbf{C}(\{x\})$. According to (2) of Proposition 1, $\bigcup_{x \in X} cl_\mathbf{C}(\{x\}) \subseteq cl_\mathbf{C}(X)$. Therefore, $cl_\mathbf{C}(X) = \bigcup_{x \in X} cl_\mathbf{C}(\{x\})$, i.e., $cl_\mathbf{C}(X) = \bigcup_{x \in X} N(x)$.

A sufficient and necessary condition under which the second type of rough set closure operator forms a matroidal closure operator is given in the following proposition.

Proposition 7. *Let* **C** *be a unary covering of* U *and* $cl_\mathbf{C}$ *the second type of rough set closure operator.* $cl_\mathbf{C}$ *satisfies* (CL4) *of Proposition 4 if and only if* $\{N(x) : x \in U\}$ *is a partition of* U.

Proof. According to (CL4) of Proposition 4, we need to prove for all $X \subseteq$ $U, x, y \in U, y \in cl_\mathbf{C}(X \cup \{x\}) - cl_\mathbf{C}(X) \Rightarrow x \in cl_\mathbf{C}(X \cup \{y\})$ if and only if $\{N(x) : x \in U\}$ is a partition of U.
($\Rightarrow$): According to (2L) of Proposition 3, $SL(\emptyset) = \emptyset$, i.e., $\emptyset \in \mathbf{S}_\mathbf{C}$. Therefore $cl_\mathbf{C}(\emptyset) = \emptyset$. Suppose $X = \emptyset$. We have if $y \in cl_\mathbf{C}(\{x\})$, then $x \in cl_\mathbf{C}(\{y\})$.

According to Proposition 6, we can obtain for any $x, y \in U$, if $y \in N(x)$, then $x \in N(y)$. Therefore $\{N(x) : x \in U\}$ is a partition of U.

($\Leftarrow$): According to Proposition 6, $cl_{\mathbf{C}}(X) = \bigcup_{x \in X} N(x)$ for all $X \subseteq U$. If $y \in cl_{\mathbf{C}}(X \cup \{x\}) - cl_{\mathbf{C}}(X)$, i.e., $y \in \bigcup_{z \in X \cup \{x\}} N(z) - \bigcup_{z \in X} N(z)$, then $y \in N(x)$. Since $\{N(x) : x \in U\}$ is a partition of U, then $x \in N(y)$, i.e., $x \in cl_{\mathbf{C}}(\{y\}) \subseteq cl_{\mathbf{C}}(X \cup \{y\})$.

For a unary covering of a universe, we see the second type of rough set closure operator is a matroidal closure operator when all neighborhoods of the covering form a partition. Moreover, we will present an equivalent depiction of a covering when this covering is unary and its all neighborhoods form a partition. First, we introduce the definition of reducible elements and related results.

Definition 13 *(Reducible element [31]). Let $\mathbf{C}$ be a covering of U and $K \in \mathbf{C}$. If K is a union of some elements in $\mathbf{C} - \{K\}$, we say K is reducible in $\mathbf{C}$, otherwise K is irreducible.*

As shown in [31], for a covering $\mathbf{C}$ of a universe, if all reducible elements are deleted from $\mathbf{C}$, the remainder is still a covering and has no reducible elements. We call the new covering the reduct of the original covering and denote it as $reduct(\mathbf{C})$.

Lemma 2 *[29]. If $\mathbf{C}$ is unary, then $reduct(\mathbf{C}) = \{K \in Md(x) : x \in U\}$.*

For a covering of a universe U, according to the notions of minimal description and neighborhood, we can obtain $N(x) = \cap Md(x)$ for any $x \in U$. The relationship between a unary covering and its neighborhood is studied in the following proposition.

Proposition 8. *If $\mathbf{C}$ is unary, then $reduct(\mathbf{C}) = \{N(x) : x \in U\}$.*

Proof. According to Definitions 4 and 12, we can obtain $N(x) = \cap Md(x)$. Since $\mathbf{C}$ is unary, we have $|Md(x)| = 1$ for all $x \in U$. Therefore, $Md(x) = \{N(x)\}$ for any $x \in U$. According to Lemma 2, we can obtain if $\mathbf{C}$ is unary, then $reduct(\mathbf{C}) = \{N(x) : x \in U\}$.

In the following proposition, for a covering of a universe, the reduct of the covering is a partition if and only if the covering is unary and its all neighborhoods form a partition.

Proposition 9. *Let $\mathbf{C}$ be a covering of U. $\mathbf{C}$ is unary and $\{N(x) : x \in U\}$ is a partition if and only if $reduct(\mathbf{C})$ is a partition.*

Proof. ($\Rightarrow$): According to Proposition 8, it is easy to prove if $\mathbf{C}$ is unary and $\{N(x) : x \in U\}$ is a partition, then $reduct(\mathbf{C})$ is a partition.

($\Leftarrow$): Suppose $\mathbf{C}$ is not unary, we see there exists $x \in U$ and $x \in K_1, K_2$ such that $K_1, K_2 \in Md(x)$. According to Definitions 4 and 13, K_1 and K_2 are not reducible elements. Therefore $K_1 \in reduct(\mathbf{C})$ and $K_2 \in reduct(\mathbf{C})$, which is contradictory with the condition that $reduct(\mathbf{C})$ is a partition. Hence $\mathbf{C}$ is unary. According to Proposition 8, if $\mathbf{C}$ is unary, then $reduct(\mathbf{C}) = \{N(x) : x \in U\}$. Therefore, we have $\{N(x) : x \in U\}$ is a partition.

A sufficient and necessary condition for the second type of rough set closure operator to be a matroidal closure operator can be briefly described in the following theorem.

Theorem 2. *Let $\mathbf{C}$ be a covering of U. There exists M such that $cl_M = cl_\mathbf{C}$ if and only if $reduct(\mathbf{C})$ is a partition.*

4   Matroidal Structure of the Second Type of Covering Upper Approximation Operator

Generally, properties of upper approximation operator in covering-based rough sets and ones of the closure operator in topology have a lot of similarity. In this section, we will study the relationship between the second type of covering upper approximation operator and a matroidal closure operator.

In [28], Zhu has presented a sufficient and necessary condition of the idempotency of the second type of covering upper approximation operator.

Theorem 3 *[28]. Let $\mathbf{C}$ be a covering of U. SH satisfies $SH(SH(X)) = SH(X)$ if and only if $\mathbf{C}$ satisfies the following property: $\forall K, K_1, \cdots, K_m \in \mathbf{C}$, if $K_1 \cap \cdots \cap K_m \neq \emptyset$ and $K \cap (K_1 \cup \cdots \cup K_m) \neq \emptyset$, then $K \subseteq (K_1 \cup \cdots \cup K_m)$.*

However, the above theorem satisfies only the necessity, that is, Theorem 3 is wrong. A counterexample is presented to illustrate the sufficiency of the above theorem.

Example 6. Let $U = \{a, b, c\}$ and $\mathbf{C} = \{K_1, K_2, K_3, K_4\}$ where $K_1 = \{a, c\}$, $K_2 = \{b, c\}, K_3 = \{b, c, d\}, K_4 = \{a, c, d\}$. We can obtain $\cap \mathbf{C} = \{c\}$, then $\forall K_i \in \mathbf{C}, \cap(\mathbf{C}-\{K_i\}) \neq \emptyset (i = 1, 2, 3, 4)$. Meanwhile, $\forall K_i \in \mathbf{C}, \cup(\mathbf{C}-\{K_i\}) = U$, then $K_i \cap (\cup(\mathbf{C} - \{K_i\})) \neq \emptyset$ and $K_i \subseteq \cup(\mathbf{C} - \{K_i\})$ (i=1, 2, 3, 4). However, according to Definition 6, $SH(\{a\}) = \{a, c, d\}, SH(SH(\{a\})) = SH(\{a, c, d\}) = \{a, b, c, d\}$. That is, $SH(SH(\{a\})) \neq SH(\{a\})$. Therefore, for all $X \subseteq U$, $SH(SH(X)) = SH(X)$ does not always hold.

In the following theorem, we will present a sufficient and necessary condition for the second type of covering upper approximation operator to satisfy the idempotency.

Lemma 3 *[28]. Let $\mathbf{C}$ be a covering of U. If $\{SH(\{x\}) : x \in U\}$ is a partition, then $SH(SH(X)) = SH(X)$ for all $X \subseteq U$.*

Theorem 4. *Let $\mathbf{C}$ be a covering of U. For all $X \subseteq U$, $SH(SH(X)) = SH(X)$ if and only if $\{SH(\{x\}) : x \in U\}$ is a partition.*

Proof. ($\Rightarrow$): Suppose $\{SH(\{x\}) : x \in U\}$ is not a partition, then there exists $x \in U$ such that $x \in SH(\{x_1\}), x \in SH(\{x_2\})$ and $SH(\{x_1\}) \neq SH(\{x_2\})$. According to Definition 6, we see there exist $K_1, K_2 \in \mathbf{C}$ such that $\{x, x_1\} \subseteq K_1, \{x, x_2\} \subseteq K_2$. Therefore, $x_1 \in SH(\{x\}), x_2 \in SH(\{x\})$. According to (6H) of Proposition 3, we obtain $SH(\{x\}) \subseteq SH(SH(\{x_1\})), SH(\{x\}) \subseteq SH(SH(\{x_2\})), SH(\{x_1\}) \subseteq SH(SH(\{x\}))$ and $SH(\{x_1\}) \subseteq SH(SH(\{x\}))$. Since $X \subseteq U$, $SH(SH(X)) = SH(X)$, then $SH(\{x\}) = SH(\{x_1\})$ and $SH(\{x\}) = SH(\{x_2\})$, i.e., $SH(\{x_1\}) = SH(\{x_2\})$ which is contradictory with that $SH(\{x_1\}) \neq SH(\{x_2\})$. Hence, If for all $X \subseteq U$, $SH(SH(X)) = SH(X)$, then $\{SH(\{x\}) : x \in U\}$ is a partition.
($\Leftarrow$): According to Lemma 3, it is straightforward.

When the second type of covering upper approximation operator is a matroidal closure operator, a sufficient and necessary condition is presented in the following theorem.

Theorem 5. *Let $\mathbf{C}$ be a covering of U. SH is a matroidal closure operator if and only if $\{SH(\{x\}) : x \in U\}$ is a partition.*

Proof. We need to prove that SH satisfies (CL1), (CL2), (CL3) and (CL4) of Proposition 4.
(CL1): According to (3H) of Proposition 3, for all $X \subseteq U$, $X \subseteq SH(X)$;
(CL2): According to (6H) of Proposition 3, if $X \subseteq Y \subseteq U$, then $SH(X) \subseteq SH(Y)$;
(CL3): According to Theorem 4, for all $X \subseteq U$, $SH(SH(X)) = SH(X)$ if and only if $\{SH(\{x\}) : x \in U\}$ is a partition.
(CL4): For all $X \subseteq U, x, y \in U$, suppose $y \in SH(X \cup \{x\}) - SH(X)$. According to (4H) of Proposition 3, for all $X, Y \subseteq U, SH(X \cup Y) = SH(X) \cup SH(Y)$. Therefore, $y \in SH(X \cup \{x\}) - SH(X) = SH(\{x\}) - SH(X) \subseteq SH(\{x\})$. So there exists $K \in \mathbf{C}$ such that $\{x, y\} \subseteq K$. According to Definition 6, $x \in SH(\{y\})$. Since $SH(\{y\}) \subseteq SH(X \cup \{y\})$, then $x \in SH(X \cup \{y\})$.

In order to further depict the second type of covering upper approximation operator, we introduce indiscernible neighborhoods in the following definition. Through the indiscernible neighborhoods, another sufficient and necessary condition for the second type of covering upper approximation operator to be a matroidal closure operator is obtained as follows.

Definition 14 *(Indiscernible neighborhood [30]). Let $\mathbf{C}$ be a covering of U and $x \in U$. $I_{\mathbf{C}}(x) = \cup\{K \in \mathbf{C} : x \in K\}$ is called the indiscernible neighborhood of x with respect to $\mathbf{C}$. When there is no confusion, we omit the subscript $\mathbf{C}$.*

Proposition 10. *Let $\mathbf{C}$ be a covering of U. SH is a matroidal closure operator if and only if $\{I(x) : x \in U\}$ is a partition of U.*

5   Conclusions

In this paper, we investigated the relationships between the second type of covering-based rough sets and matroids via closure operators. First, for a covering of a universe, we constructed a closure system through the second type of covering lower approximation operator. Moreover, we proved the closure operator induced by the system was a matroidal closure operator if and only if the reduct of the covering was a partition. Second, the second type of covering upper approximation operator was a matroidal closure operator if and only if the family of all indiscernible neighborhoods formed a partition.

Acknowledgments. This work is in part supported by the National Science Foundation of China under Grant Nos. 61170128, 61379049, 61379089 and 61440047.

References

1. Bonikowski, Z., Bryniarski, E., Skardowska, W.U.: Extensions and intentions in the rough set theory. Inf. Sci. **107**, 149–167 (1998)
2. Chen, D., Wang, C., Hu, Q.: A new approach to attribute reduction of consistent and inconsistent covering decision systems with covering rough sets. Inf. Sci. **177**, 3500–3518 (2007)
3. Dai, J., Xu, Q.: Approximations and uncertainty measures in incomplete information systems. Inf. Sci. **198**, 62–80 (2012)
4. Kryszkiewicz, M.: Rough set approach to incomplete information systems. Inf. Sci. **112**, 39–49 (1998)
5. Lai, H.: Matroid Theory. Higher Education Press, Beijing (2001)
6. Li, X., Liu, S.: Matroidal approaches to rough set theory via closure operators. Int. J. Approximate Reasoning **53**, 513–527 (2012)
7. Liang, J., Li, R., Qian, Y.: Distance: a more comprehensible perspective for measures in rough set theory. Knowl. Based Syst. **27**, 126–136 (2012)
8. Lin, T.Y.: Neighborhood systems and relational databases. In: Proceedings of the 1988 ACM Sixteenth Annual Conference On Computer science, p. 725. ACM (1988)
9. Liu, G., Chen, Q.: Matroid. National University of Defence Technology Press, Changsha (1994)
10. Liu, Y., Zhu, W.: Matroidal structure of rough sets based on serial and transitive relations. J. Appl. Math. **2012**, 16 pages (2012). Article ID 429737
11. Liu, Y., Zhu, W., Zhang, Y.: Relationship between partition matroid and rough set through k-rank matroid. J. Inf. Comput. Sci. **8**, 2151–2163 (2012)
12. Miao, D., Duan, Q., Zhang, H., Jiao, N.: Rough set based hybrid algorithm for text classification. Expert Syst. Appl. **36**, 9168–9174 (2009)
13. Min, F., He, H., Qian, Y., Zhu, W.: Test-cost-sensitive attribute reduction. Inf. Sci. **22**, 4928–4942 (2011)
14. Pawlak, Z.: Rough sets. Int. J. Comput. Inf. Sci. **11**, 341–356 (1982)
15. Pawlak, Z.: Fuzzy sets and rough sets. Fuzzy Sets Syst. **17**, 99–102 (1985)
16. Qian, Y., Liang, J., Yao, Y., Dang, C.: Mgrs: A multi-granulation rough set. Inf. Sci. **180**, 949–970 (2010)
17. Qin, K., Gao, Y., Pei, Z.: On covering rough sets. In: Yao, J.T., Lingras, P., Wu, W.-Z., Szczuka, M.S., Cercone, N.J., Ślęzak, D. (eds.) RSKT 2007. LNCS, vol. 4481, pp. 34–41. Springer, Heidelberg (2007)

18. Restrepo, M., Cornelis, C., Gmez, J.: Duality, conjugacy and adjointness of approximation operators in covering-based rough sets. Int. J. Approximate Reasoning **1**, 469–485 (2014)
19. Skowron, A., Stepaniuk, J.: Tolerance approximation spaces. Fundamenta Informaticae **27**, 245–253 (1996)
20. Tang, J., She, K., Zhu, W.: Matroidal structure of rough sets from the viewpoint of graph theory. J. Appl. Math. **2012**, 27 pages (2012). Article ID 973920
21. Tsumoto, S., Tanaka, H.: Algebraic specification of empirical inductive learning methods. In: Calmet, J., Campbell, J. (eds.) AISMC 1994. LNCS, vol. 958, pp. 224–243. Springer, Heidelberg (1995)
22. Tsumoto, S., Tanaka, H.: A common algebraic framework of empirical learning methods based on rough sets and matroid theory. Fundamenta Informaticae **27**, 273–288 (1996)
23. Wang, G., Hu, J.: Attribute reduction using extension of covering approximation space. Fundamenta Informaticae **115**, 219–232 (2012)
24. Wang, S., Zhu, Q., Zhu, W., Min, F.: Quantitative analysis for covering-based rough sets using the upper approximation number. Inf. Sci. **220**, 483–491 (2013)
25. Yao, Y.: Constructive and algebraic methods of theory of rough sets. Inf. Sci. **109**, 21–47 (1998)
26. Yao, Y.: Relational interpretations of neighborhood operators and rough set approximation operators. Inf. Sci. **111**, 239–259 (1998)
27. Zhang, S., Wang, X., Feng, T., Feng, L.: Reduction of rough approximation space based on matroid. Int. Conf. Mach. Learn. Cybern. **2**, 267–272 (2011)
28. Zhu, W.: Properties of the second type of covering-based rough sets. In: Workshop Proceedings of GrC&BI 2006, pp. 494–497. IEEE WI 06, Hong Kong, China, 18 December (2006)
29. Zhu, W.: Relationship among basic concepts in covering-based rough sets. Inf. Sci. **179**, 2478–2486 (2009)
30. Zhu, W.: Relationship between generalized rough sets based on binary relation and covering. Inf. Sci. **179**, 210–225 (2009)
31. Zhu, W., Wang, F.: Reduction and axiomization of covering generalized rough sets. Inf. Sci. **152**, 217–230 (2003)
32. Zhu, W., Wang, F.: Relationships among three types of covering rough sets. In: 2006 IEEE International Conference on Granular Computing (GrC 2006), pp. 43–48 (2006)

Incremental Updating Rough Approximations in Interval-valued Information Systems

Yingying Zhang[1,2](✉), Tianrui Li[1,2], Chuan Luo[1,2], and Hongmei Chen[1,2]

[1] School of Information Science and Technology, Southwest Jiaotong University, Chengdu 611756, China
yingzhang86@126.com, trli@swjtu.edu.cn, {cluo_swjtu,hm_chen_swjtu}@163.com
[2] Key Lab of Cloud Computing and Intelligent Technique, Sichuan Province, Chengdu 611756, China

Abstract. Interval-valued Information System (IvIS) is a generalized model of single-valued information system, in which the attribute values of objects are all interval values instead of single values. The attribute set in IvIS is not static but rather dynamically changing over time with the collection of new information. The rough approximations may evolve accordingly, which should be updated continuously for data analysis based on rough set theory. In this paper, on the basis of the similarity-based rough set theory in IvIS, we first analyze the relationships between the original approximation sets and the updated ones. And then we propose the incremental methods for updating rough approximations when adding and removing attributes, respectively. Finally, a comparative example is used to validate the effectiveness of the proposed incremental methods.

Keywords: Interval-valued Information System · Similarity · Rough set · Incremental updating · Approximations

1 Introduction

Rough set theory is an efficient method for concept approximation and feature extraction without any external information, which differs from other approximation reasoning methods, such as fuzzy set theory and probability theory. The uncertain concept and knowledge hidden in the information system can be described according to the lower and upper approximations in rough set theory. Nowadays, many rough set based data analysis approaches have been successfully used in medical diagnosis [1], financial analysis [2,3], industrial control [4] and knowledge discovery [5,6].

Knowledge is represented by an information table in rough set theory, which includes the objects, the features of the objects (the attributes) and the attribute values [7]. The classical rough set theory can only deal with single-valued information system with certain category attribute values. In real applications, the attribute values in an information system are often uncertain and real values,

© Springer International Publishing Switzerland 2015
D. Ciucci et al. (Eds.): RSKT 2015, LNAI 9436, pp. 243–252, 2015.
DOI: 10.1007/978-3-319-25754-9_22

which can be expressed by interval numbers. The information system whose attribute values are all interval numbers is called Interval-valued Information System (IvIS), which is a generalized model of single-valued information system.

In recent years, many extensions of classical rough set theory have been applied in IvIS. Leung *et al.* proposed a tolerance-based rough set approach for discovering classification rules from IvIS [9]. Qian *et al.* introduced a dominance-based rough set model in interval ordered information systems for attribute reduction and ordering rules extraction [10]. Miao *et al.* established a new rough set model in IvIS based on the concept of maximal consistent blocks [11]. Yang *et al.* further investigated the dominance-based rough set theory in incomplete IvIS by considering three types of unknown values, and proposed six types of relative reductions to generate the optimal decision rules [12]. Dai *et al.* addressed the uncertainty measurement problem in IvIS by introducing the concepts of possible-degree-based conditional entropy and similarity relation between two interval values, respectively [13,14]. Zhang *et al.* proposed a general framework for the study of IvIS by integrating the variable-precision-dominance-based rough set theory with inclusion measure theory [15]. Du *et al.* introduced a new dominance relation in the interval-valued ordered information system, and established a dominance-based rough set approach to extract the minimal decision rules [16].

In the real-world applications, the attribute set of an information system always varies over time, which leads to the dynamic change of the rough approximations in rough set theory. The incremental approaches by using the original results to update rough approximations are often employed in many dynamic information systems. Li *et al.* proposed a new approach for incrementally updating approximations under the characteristic relation-based rough sets for dynamic attribute generalization [17]. Luo *et al.* investigated the incremental strategies for maintaining rough approximations in dynamic set-valued decision systems [18–20]. Li *et al.* proposed incremental approaches for updating approximations based on dominance-based rough set model when some attributes are added into or deleted from an ordered information system [21]. Zeng *et al.* proposed the fuzzy information granulation methods to incrementally update approximations in hybrid information systems under the variation of the attribute set [22]. Chen *et al.* proposed incremental approaches for dynamic maintenance of rough approximations *w.r.t.* the addition of objects and attributes simultaneously based on decision-theoretic rough sets [23]. However, the incremental maintaining approaches of rough approximations in dynamic IvIS have not been reported. In this paper, we mainly focus on the incremental mechanisms for updating approximations when an attribute set is added into or removed from the IvIS.

The remainder of this paper is organized as follows. Firstly, some basic concepts of IvIS and rough set theory are introduced in Sect. 2. In Sect. 3, based on the similarity-based rough set approach, we first discuss the relationships between the original approximations and the updated ones, and then the incremental principles for updating rough approximations when an attribute set is added into and deleted from the IvIS are proposed, respectively. In Sect. 4, a comparative example is presented to validate the effectiveness of proposed incremental methods. Finally, we conclude this paper with a brief summary and future work in Sect. 5.

2 Preliminaries

In this section, some basic concepts of IvIS and rough set theory are reviewed [8]. First, we recall some basic concepts about interval [11].

A closed interval on the real number set $\mathbb{R}$, denoted by $I = [b^-, b^+]$, is the set of real numbers given by $[b^-, b^+] = \{x \in \mathbb{R} | b^- \leqslant x \leqslant b^+\}$, where b^- and b^+ are called the lower and the upper bounds of I, respectively. If $b^- = b^+$, the interval I will degenerate into a single real number.

Suppose $I_1 = [b_1^-, b_1^+]$ and $I_2 = [b_2^-, b_2^+]$ are two intervals. The intersection $I_1 \cap I_2$ is defined as an interval

$$I_1 \cap I_2 = \begin{cases} [max\{b_1^-, b_2^-\}, min\{b_1^+, b_2^+\}], & max\{b_1^-, b_2^-\} \leq min\{b_1^+, b_2^+\} \\ \emptyset, & otherwise \end{cases} \tag{1}$$

The union $I_1 \cup I_2$ is also defined as an interval

$$I_1 \cup I_2 = [min\{b_1^-, b_2^-\}, max\{b_1^+, b_2^+\}] \tag{2}$$

So we have $I_1 \cap I_2 \subseteq I_1 \cup I_2$.

Definition 1 *[8]. An Interval-valued Information System (IvIS) is a quadruple (U, AT, V, f), where $U = \{x_1, x_2, \ldots, x_n\}$ is a non-empty finite set of objects called the universe, $A = \{a_1, a_2, \ldots, a_l\}$ is a non-empty finite set of attributes, $V = \bigcup_{a \in AT} V_a$ and V_a is called the domain of attribute a, and $f : U \times A \to V$ is called a total function, such that $f(x, a) = [a(x)^-, a(x)^+] \in V_a$ is an interval for every $a \in AT$, $x \in U$, where V_a is a set of intervals.*

Example 1. *An example of IvIS is listed in Table 1, where $U = \{x_1, x_2, \ldots, x_9\}$, $AT = \{a_1, a_2, \ldots, a_5\}$.*

Definition 2 *[8]. Let (U, AT, V, f) be an IvIS, $x_i, x_j \in U$, and $a_k \in AT$. The similarity degree between x_i and x_j under the attribute a_k is defined as follows:*

$$S_{ij}^k = \frac{|f(x_i, a_k) \cap f(x_j, a_k)|}{|f(x_i, a_k) \cup f(x_j, a_k)|} \tag{3}$$

where $i, j = 1, 2, \ldots, n$, $k = 1, 2, \ldots, m$, $|\cdot|$ denotes the length of closed interval, and the length of an empty interval or a single point is zero.

Obviously, the similarity degree S_{ij}^k satisfies the following properties [8]:

1. $0 \leq S_{ij}^k \leq 1$;
2. $S_{ij}^k = 0$ if and only if $f(x_i, a_k) \cap f(x_j, a_k)$ is empty or single value;
3. $S_{ij}^k = 1$ if and only if $f(x_i, a_k) = f(x_j, a_k)$;
4. $S_{ij}^k = S_{ji}^k$.

Table 1. An interval-valued information system.

	a_1	a_2	a_3	a_4	a_5
x_1	$[0.1, 0.6]$	$[0.4, 0.6]$	$[0.2, 0.4]$	$[0.0, 0.7]$	$[0.2, 0.4]$
x_2	$[0.6, 0.8]$	$[0.4, 0.6]$	$[0.3, 0.5]$	$[0.8, 1.0]$	$[0.5, 0.8]$
x_3	$[0.2, 0.4]$	$[0.0, 0.4]$	$[0.2, 0.8]$	$[0.6, 0.7]$	$[0.1, 0.5]$
x_4	$[0.7, 0.8]$	$[0.2, 0.6]$	$[0.3, 0.6]$	$[0.2, 0.5]$	$[0.4, 0.9]$
x_5	$[0.3, 0.4]$	$[0.2, 0.3]$	$[0.4, 0.8]$	$[0.4, 0.6]$	$[0.4, 1.0]$
x_6	$[0.4, 0.9]$	$[0.1, 0.8]$	$[0.2, 0.8]$	$[0.1, 0.6]$	$[0.5, 0.7]$
x_7	$[0.6, 0.7]$	$[0.3, 0.7]$	$[0.1, 0.2]$	$[0.8, 0.8]$	$[0.2, 0.6]$
x_8	$[0.6, 1.0]$	$[0.7, 0.9]$	$[0.2, 0.4]$	$[0.4, 0.8]$	$[0.4, 0.8]$
x_9	$[0.1, 0.2]$	$[0.1, 0.1]$	$[0.3, 0.6]$	$[0.7, 0.9]$	$[0.3, 0.5]$

Definition 3 *[8]. Let (U, AT, V, f) be an IvIS, $x_i, x_j \in U$, $A \subseteq AT$ and $\lambda \in [0.5, 1]$. The λ-similarity relation with respect to A is defined as follows:*

$$S_A^\lambda = \{(x_i, x_j) \in U \times U | S_{ij}^k \geq \lambda, \forall a_k \in A\} \tag{4}$$

where S_{ij}^k is the similarity degree between x_i and x_j, λ is called the similarity degree threshold.

Obviously, S_A^λ is reflexive and symmetric, but not transitive. It is also called the tolerance relation which satisfies $S_A^\lambda = \bigcap_{a \in A} S_a^\lambda$. And the λ-similarity class containing x_i is defined as $S_A^\lambda(x_i) = \{x_j \in U | (x_i, x_j) \in S_A^\lambda\}$.

Lemma 1. *Let (U, AT, V, f) be an IvIS. A_1, A_2 are any two subsets of the attribute set AT. $\forall x \in U$, we have:*
(1)$S_{A_1 \cup A_2}^\lambda(x) \subseteq S_{A_1}^\lambda(x)$;   (2)$S_{A_1 \cup A_2}^\lambda(x) \subseteq S_{A_2}^\lambda(x)$.

Example 2. *In the IvIS given by Table 1, if $\lambda = 0.5$, $A = \{a_1, a_2, a_5\}$, we can obtain the 0.5-similarity classes as follows:*

$$S_A^{0.5}(x_1) = \{x_1\}, \qquad S_A^{0.5}(x_2) = \{x_2, x_4\}, \quad S_A^{0.5}(x_3) = \{x_3\},$$
$$S_A^{0.5}(x_4) = \{x_4, x_2\}, \quad S_A^{0.5}(x_5) = \{x_5\}, \qquad S_A^{0.5}(x_6) = \{x_6\},$$
$$S_A^{0.5}(x_7) = \{x_7\}, \qquad S_A^{0.5}(x_8) = \{x_8\}, \qquad S_A^{0.5}(x_9) = \{x_9\}.$$

Definition 4 *[8]. Let (U, AT, V, f) be an IvIS. $A \subseteq AT$, $\lambda \in [0.5, 1]$, and $\forall X \in U$, the lower and upper approximations of X are defined respectively by:*

$$\underline{A}^\lambda(X) = \{x_i \in U | S_A^\lambda(x_i) \subseteq X\}, \overline{A}^\lambda(X) = \{x_i \in U | (S_A^\lambda(x_i) \cap X) \neq \emptyset\}. \tag{5}$$

Example 3 *(Continued from Example 2). In the IvIS given by Table 1, let $X = \{x_1, x_4, x_6\}$. We can obtain the lower and upper approximations of X according to Definition 4 as follows: $\underline{A}^{0.5}(X) = \{x_1, x_6\}$, $\overline{A}^{0.5}(X) = \{x_1, x_2, x_4, x_6\}$.*

3 Incremental Approaches for Updating Rough Approximations on the Variation of Attributes

In this section, we will discuss the incremental principles for updating approximations through the λ-similarity-based rough set approach when an attribute set is added into or removed from the IvIS, respectively.

For this purpose, we introduce the following propositions about the lower and upper approximations in IvIS:

Proposition 1. Let (U, AT, V, f) be an IvIS, $\lambda \in [0.5, 1]$, $A_1, A_2 \subseteq AT$. $\forall X \subseteq U$, we have:

(1) $\underline{A_1}^{\lambda}(X) \cup \underline{A_2}^{\lambda}(X) \subseteq \underline{A_1 \bigcup A_2}^{\lambda}(X)$;

(2) $\overline{A_1}^{\lambda}(X) \cap \overline{A_2}^{\lambda}(X) \supseteq \overline{A_1 \bigcup A_2}^{\lambda}(X)$.

Proof. (1)$\forall x \in \underline{A_1}^{\lambda}(X) \cup \underline{A_2}^{\lambda}(X) \Rightarrow x \in \underline{A_1}^{\lambda}(X) \vee x \in \underline{A_2}^{\lambda}(X) \Rightarrow S_{A_1}^{\lambda}(x) \subseteq X \vee S_{A_2}^{\lambda}(x) \subseteq X$. According to Lemma 1, we can get $S_{A_1 \cup A_2}^{\lambda}(x) \subseteq X$, so $x \in \underline{A_1 \bigcup A_2}^{\lambda}(X)$. Finally, we have $\underline{A_1}^{\lambda}(X) \cup \underline{A_2}^{\lambda}(X) \subseteq \underline{A_1 \bigcup A_2}^{\lambda}(X)$.

(2)$\forall x \in \overline{A_1 \bigcup A_2}^{\lambda}(X) \Rightarrow S_{A_1 \cup A_2}^{\lambda}(x) \cap X \neq \emptyset$, according to Lemma 1, we can have $S_{A_1}^{\lambda}(x) \cap X \neq \emptyset \wedge S_{A_2}^{\lambda}(x) \cap X \neq \emptyset \Rightarrow x \in \overline{A_1}^{\lambda}(X) \wedge x \in \overline{A_2}^{\lambda}(X)$, that is, $x \in \overline{A_1}^{\lambda}(X) \cap \overline{A_2}^{\lambda}(X)$. Finally, $\overline{A_1}^{\lambda}(X) \cap \overline{A_2}^{\lambda}(X) \supseteq \overline{A_1 \bigcup A_2}^{\lambda}(X)$. □

Proposition 2. Let (U, AT, V, f) be an IvIS, $\lambda \in [0.5, 1]$, $A_1, A_2 \subseteq AT$ and $A_2 \subseteq A_1$. $\forall X \subseteq U$, we have:

(1) $\underline{A_1 - A_2}^{\lambda}(X) \subseteq \underline{A_1}^{\lambda}(X)$;

(2) $\overline{A_1 - A_2}^{\lambda}(X) \supseteq \overline{A_1}^{\lambda}(X)$.

Proof. (1)$\forall x \in \underline{A_1 - A_2}^{\lambda}(X) \Rightarrow S_{A_1 - A_2}^{\lambda}(x) \subseteq X$. According to Lemma 1, we know that $S_{A_1}^{\lambda}(x) \subseteq S_{A_1 - A_2}^{\lambda}(x)$, so we have $S_{A_1}^{\lambda}(x) \subseteq X \Rightarrow x \in \underline{A_1}^{\lambda}(X)$. Finally, we have $\underline{A_1 - A_2}^{\lambda}(X) \subseteq \underline{A_1}^{\lambda}(X)$.

(2)$\forall x \in \overline{A_1}^{\lambda}(X) \Rightarrow S_{A_1}^{\lambda}(x) \cap X \neq \emptyset$, similar to (1), we have $S_{A_1 - A_2}^{\lambda}(x) \cap X \neq \emptyset \Rightarrow x \in \overline{A_1 - A_2}^{\lambda}(X)$. Finally, we have $\overline{A_1 - A_2}^{\lambda}(X) \supseteq \overline{A_1}^{\lambda}(X)$. □

3.1 Adding a New Attribute Set

Considering the addition of an attribute set A^+ in the IvIS, the following incremental mechanisms are proposed for updating rough approximations.

Proposition 3. Let (U, AT, V, f) be the IvIS. $\forall X \subseteq U$, the lower approximations of X with the addition of A^+ can be updated as follows:

$\underline{AT \cup A^+}^{\lambda}(X) = \underline{AT}^{\lambda}(X) \cup \underline{A^+}^{\lambda}(X) \cup Y_1$, where $Y_1 = \{x \in (X - \underline{AT}^{\lambda}(X)) \cap (X - \underline{A^+}^{\lambda}(X)) | S_{AT \cup A^+}^{\lambda}(x) \subseteq X\}$.

Proof. According to Proposition 1 and the definition of rough lower approximation, we have $\underline{AT}^{\lambda}(X) \cup \underline{A^+}^{\lambda}(X) \subseteq \underline{AT \cup A^+}^{\lambda}(X) \subseteq X$. Hence, we obtain that $\underline{AT \cup A^+}^{\lambda}(X) = \underline{AT}^{\lambda}(X) \cup \underline{A^+}^{\lambda}(X) \cup Y_1$, where $Y_1 = \{x \in (X - \underline{AT}^{\lambda}(X)) \cap (X - \underline{A^+}^{\lambda}(X)) | S_{AT \cup A^+}^{\lambda}(x) \subseteq X\}$. □

Proposition 4. *Let (U, AT, V, f) be the IvIS. $\forall X \subseteq U$, the upper approximations of X with the addition of A^+ can be updated as follows: $\overline{AT \cup A^+}^{\lambda}(X) = \overline{AT}^{\lambda}(X) \cap \overline{A^+}^{\lambda}(X) \setminus Z_1$, where $Z_1 = \{x \in (\overline{AT}^{\lambda}(X) \cap \overline{A^+}^{\lambda}(X) - X) | S^{\lambda}_{AT \cup A+}(x) \cap X = \emptyset\}$.*

Proof. According to Proposition 1 and the definition of rough upper approximation, we have $X \subseteq \overline{AT \cup A^+}^{\lambda}(X) \subseteq \overline{AT}^{\lambda}(X) \cap \overline{A^+}^{\lambda}(X)$. Therefore, we obtain that $\overline{AT \cup A^+}^{\lambda}(X) = \overline{AT + A^+}^{\lambda}(X) \setminus Z_1$, where $Z_1 = \{x \in (\overline{AT}^{\lambda}(X) \cap \overline{A^+}^{\lambda}(X) - X) | S^{\lambda}_{AT \cup A+}(x) \cap X = \emptyset\}$.   □

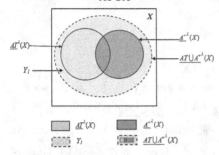

(a) The mechanism of incremental updating lower approximation

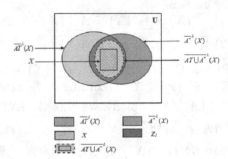

(b) The mechanism of incremental updating upper approximation

Fig. 1. The mechanisms of incremental updating approximations when adding an attribute set

Figure 1(a) and (b) present the incremental mechanisms for updating lower and upper approximations when adding an attribute set A^+, respectively. The dotted circles in Fig. 1 denotes the updated lower and upper approximations. From Fig. 1(a), it's easy to find that $(\underline{A^+}^{\lambda}(X) \setminus (\underline{A^+}^{\lambda}(X) \cap \underline{AT}^{\lambda}(X))) \cup Y_1$ is the incremental part of the updated lower approximation with respect to the original lower approximation $\underline{AT}^{\lambda}(X)$. From Fig. 1(b), we know that Z_1 is the incremental part of the update upper approximation.

3.2   Deleting an Attribute Set

Considering the deletion of an attribute set A^- from the IvIS, the following propositions are proposed for updating rough approximations, where AT' is defined by the updated attribute, i.e., $AT' = AT - A^-$.

Proposition 5. *Let (U, AT, V, f) be the IvIS. $\forall X \subseteq U$, the lower approximations of X with the deletion of A^- can be updated as follows: $\underline{AT'}^{\lambda}(X) = \underline{AT}^{\lambda}(X) - Y_2$, where $Y_2 = \{x \in \underline{AT}^{\lambda}(X) | S^{\lambda}_{AT'}(x) \nsubseteq X\}$.*

Proof. According to Proposition 2 and the definition of rough lower approximation, we have $\underline{AT'}^{\lambda}(X) \subseteq \underline{AT}^{\lambda}(X)$. Then we can obtain $\underline{AT'}^{\lambda}(X) = \underline{AT}^{\lambda}(X) - Y_2$, where $Y_2 = \{x \in \underline{AT}^{\lambda}(X) | S^{\lambda}_{AT'}(x) \nsubseteq X\}$.   □

Proposition 6. *Let* (U, AT, V, f) *be the IvIS.* $\forall X \subseteq U$, *the upper approxima-tions of* X *with the deletion of* A^- *can be updated as follows:*
$$\overline{AT'}^{\lambda}(X) = \overline{AT}^{\lambda}(X) \cup Z_2, \text{ where } Z_2 = \{x \in (U - \overline{AT}^{\lambda}(X)) | S^{\lambda}_{AT'}(x) \cap X \neq \emptyset\}.$$

Proof. According to Proposition 2 and the definition of rough upper approxima-tion, we have that $\overline{AT}^{\lambda}(X) \subseteq \overline{AT'}^{\lambda}(X)$. So we can get $\overline{AT'}^{\lambda}(X) = \overline{AT}^{\lambda}(X) \cup Z_2$, where $Z_2 = \{x \in (U - \overline{AT}^{\lambda}(X)) | S^{\lambda}_{AT'}(x) \cap X \neq \emptyset\}$.   □

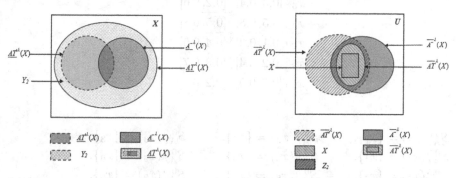

(a) The mechanism of incremental up-dating lower approximation

(b) The mechanism of incremental up-dating upper approximation

Fig. 2. The mechanisms of incremental updating approximations when deleting an attribute set

Figure 2(a) and (b) present the incremental mechanisms for updating lower and upper approximations when deleting an attribute set A^-, respectively. Simi-lar to Fig. 1, the updated lower and upper approximations are characterized with dotted circles. In the figures, it's clear that Y_2 and Z_2 are the incremental parts of lower and upper approximations, respectively, with respect to the original rough approximations $\underline{AT}^{\lambda}(X)$ and $\overline{AT}^{\lambda}(X)$ when an attribute set is deleted from IvIS. Here the two approximations $\underline{A}^{-\lambda}(X)$ and $\overline{A}^{-\lambda}(X)$ are shown for explanation, and they are not used to update rough approximations.

4   A Comparative Illustration

In this section, a comparative illustration is used to validate the feasibility of the proposed incremental methods.

Given the IvIS (U, AT, V, f) as shown in Table 1, we show the two aspects of changing attribute sets in IvIS as follows:
(1) The attribute set $A^+ = \{a_6, a_7\}$ shown in Table 2 is added into Table 1;
(2) The attribute set $A^- = \{a_2, a_3\}$ is removed from Table 1.

In the following, we suppose that $X = \{x_1, x_4, x_6\}$ similar to Example 3. From Table 1, we have

Table 2. The attribute set $A^+ = \{a_6, a_7\}$ is inserted into Table 1.

	a_6	a_7
x_1	$[0.1, 0.4]$	$[0.3, 0.5]$
x_2	$[0.2, 0.5]$	$[0.4, 0.7]$
x_3	$[0.4, 0.7]$	$[0.1, 0.5]$
x_4	$[0.6, 0.7]$	$[0.1, 0.4]$
x_5	$[0.3, 0.4]$	$[0.2, 0.6]$
x_6	$[0.5, 0.8]$	$[0.5, 0.8]$
x_7	$[0.1, 0.3]$	$[0.4, 0.5]$
x_8	$[0.2, 0.4]$	$[0.3, 0.7]$
x_9	$[0.7, 0.9]$	$[0.2, 0.5]$

$$S_{AT}^{0.5}(x_1) = \{x_1\}, \quad S_{AT}^{0.5}(x_2) = \{x_2\}, \quad S_{AT}^{0.5}(x_3) = \{x_3\},$$
$$S_{AT}^{0.5}(x_4) = \{x_4\}, \quad S_{AT}^{0.5}(x_5) = \{x_5\}, \quad S_{AT}^{0.5}(x_6) = \{x_6\},$$
$$S_{AT}^{0.5}(x_7) = \{x_7\}, \quad S_{AT}^{0.5}(x_8) = \{x_8\}, \quad S_{AT}^{0.5}(x_9) = \{x_9\}.$$

Then according to Definition 4, we have $\underline{AT}^{0.5}(X) = \{x_1, x_4, x_6\}$ and $\overline{AT}^{0.5}(X) = \{x_1, x_4, x_6\}$.

(1) When the attribute $A^+ = \{a_6, a_7\}$ is inserted into Table 1, we compute the lower approximation $\underline{AT \cup A^+}^{0.5}(X)$ and the upper approximation $\overline{AT \cup A^+}^{0.5}(X)$ by the non-incremental methods in the following:

First we obtain the 0.5-similarity classes about $AT \cup A^+$:

$$S_{AT \cup A^+}^{0.5}(x_1) = \{x_1\}, \quad S_{AT \cup A^+}^{0.5}(x_2) = \{x_2\}, \quad S_{AT \cup A^+}^{0.5}(x_3) = \{x_3\},$$
$$S_{AT \cup A^+}^{0.5}(x_4) = \{x_4\}, \quad S_{AT \cup A^+}^{0.5}(x_5) = \{x_5\}, \quad S_{AT \cup A^+}^{0.5}(x_6) = \{x_6\},$$
$$S_{AT \cup A^+}^{0.5}(x_7) = \{x_7\}, \quad S_{AT \cup A^+}^{0.5}(x_8) = \{x_8\}, \quad S_{AT \cup A^+}^{0.5}(x_9) = \{x_9\}.$$

Then according to Definition 4, we have $\underline{AT \cup A^+}^{0.5}(X) = \{x_1, x_4, x_6\}$ and $\overline{AT \cup A^+}^{0.5}(X) = \{x_1, x_4, x_6\}$.

We update the lower and upper approximations by Propositions 3 and 4 as follows:

From Table 2, we have

$$S_{A^+}^{0.5}(x_1) = \{x_1, x_7, x_8\}, \quad S_{A^+}^{0.5}(x_2) = \{x_2, x_8\}, \quad S_{A^+}^{0.5}(x_3) = \{x_3\},$$
$$S_{A^+}^{0.5}(x_4) = \{x_4\}, \quad S_{A^+}^{0.5}(x_5) = \{x_5, x_8\}, \quad S_{A^+}^{0.5}(x_6) = \{x_6\},$$
$$S_{A^+}^{0.5}(x_7) = \{x_1, x_7\}, \quad S_{A^+}^{0.5}(x_8) = \{x_1, x_2, x_5, x_8\}, \quad S_{A^+}^{0.5}(x_9) = \{x_9\}.$$

So we have $\underline{A^+}^{0.5}(X) = \{x_4, x_6\}$ and $\overline{A^+}^{0.5}(X) = \{x_1, x_4, x_6, x_7, x_8\}$. Then according to Propositions 3 and 4, we have $Y_1 = \emptyset$ and $Z_1 = \emptyset$, respectively. Thus $\underline{AT \cup A^+}^{0.5}(X) = \{x_1, x_4, x_6\}$ and $\overline{AT \cup A^+}^{0.5}(X) = \{x_1, x_4, x_6\}$, which are the same with the results of non-incremental methods.

(2) When the attribute $A^- = \{a_3, a_4\}$ is deleted from Table 1, we calculate the updated lower and upper approximations by Propositions 5 and 6, respectively as follows:

First, we know $AT' = A = \{a_1, a_2, a_5\}$ and the results of $S_{AT'}^{0.5}(x_i)$ for $i = 1, 2, \ldots, 9$ from Example 2.

Then according to Propositions 5 and 6, if $X = \{x_1, x_4, x_6\}$, we have $Y_2 = \{x_4\}$ and $Z_2 = \{x_2\}$, respectively.

Therefore, $\underline{AT'}^{0.5}(X) = \{x_1, x_6\}$ and $\overline{AT'}^{0.5}(X) = \{x_1, x_2, x_4, x_6\}$, which are the same with the results of Example 3, *i.e.* the non-incremental results.

5   Conclusion

IvIS is usually used to describe uncertain problems in the real-life applications. And the attribute set in an IvIS always changes over time. In this paper, by the similarity-based rough set theory in IvIS, we proposed the incremental methods for updating rough approximations when adding and removing attributes, respectively. A comparative example was presented to illustrate the effectiveness of the proposed methods. Our future work will focus on the development of the corresponding incremental algorithms and the experimental analysis of the proposed incremental methods in compare with the non-incremental methods.

Acknowledgements. This work is supported by the National Science Foundation of China (No. 61175047), NSAF (No. U1230117), the Young Software Innovation Foundation of Sichuan Province, China (No. 2014-046) and the Beijing Key Laboratory of Traffic Data Analysis and Mining (BKLTDAM2014001).

References

1. Inbarani, H.H., Azar, A.T., Jothi, G.: Supervised hybrid feature selection based on pso and rough sets for medical diagnosis. Comput. Methods Program. Biomed. **113**(1), 175–185 (2014)
2. Yao, J., Herbert, J.P.: Financial time-series analysis with rough sets. Appl. Soft Comput. **9**(3), 1000–1007 (2009)
3. Chen, Y., Cheng, C.: A soft-computing based rough sets classifier for classifying ipo returns in the financial markets. Appl. Soft Comput. **12**(1), 462–475 (2012)
4. Tseng, T.B., Jothishankar, M.C., Wu, T.T.: Quality control problem in printed circuit board manufacturingan extended rough set theory approach. J. Manufact. Syst. **23**(1), 56–72 (2004)
5. Wang, F.H.: On acquiring classification knowledge from noisy data based on rough set. Expert Syst. Appl. **29**(1), 49–64 (2005)
6. Bai, H.X., Ge, Y., Wang, J.F., Li, D.Y., Liao, Y.L., Zheng, X.Y.: A method for extracting rules from spatial data based on rough fuzzy sets. Knowl.-Based Syst. **57**, 28–40 (2014)
7. Pawlak, Z.: Information systems theoretical foundation. Inf. Syst. **3**(6), 205–218 (1981)

8. Chen, Z.C., Qin, K.Y.: Attribute reduction of interval-valued information system based on variable precision tolerance relation. Comput. Sci. **36**(3), 163–166 (2009)
9. Leung, Y., Fischer, M.M., Wu, W.Z., Mi, J.S.: A rough set approach for the discovery of classification rules in interval-valued information systems. Int. J. Approx. Reasoning **47**(2), 233–246 (2008)
10. Qian, Y.H., Liang, J.Y., Dang, C.Y.: Interval ordered information systems. Comput. Math. Appl. **56**(8), 1994–2009 (2008)
11. Miao, D.Q., Zhang, N., Yue, X.D.: Knowledge reduction in interval-valued information systems. In: Proceedings of the 8th IEEE International Conference on Congitive Informatics, pp. 320–327 (2009)
12. Yang, X.B., Yu, D.J., Yang, J.Y., Wei, L.H.: Dominance-based rough set approach to incomplete interval-valued information system. Data Knowl. Eng. **68**(11), 1331–1347 (2009)
13. Dai, J.H., Wang, W.T., Mi, J.S.: Uncertainty measurement for interval-valued information systems. Inf. Sci. **251**, 63–78 (2013)
14. Dai, J.H., Wang, W.T., Xu, Q., Tian, H.W.: Uncertainty measurement for interval-valued decision systems based on extended conditional entropy. Knowl.-Based Syst. **27**, 443–450 (2012)
15. Zhang, H.Y., Leung, Y., Zhou, L.: Variable-precision-dominance-based rough set approach to interval-valued information systems. Inf. Sci. **244**, 75–91 (2013)
16. Du, W.S., Hu, B.Q.: Approximate distribution reducts in inconsistent interval-valued ordered decision tables. Inf. Sci. **271**, 93–114 (2014)
17. Li, T.R., Ruan, D., Geert, W., Song, J., Xu, Y.: A rough Sets based characteristic relation approach for dynamic attribute generalization in data mining. Knowl.-Based Syst. **20**(5), 485–494 (2007)
18. Luo, C., Li, T.R., Chen, H.M., Liu, D.: Incremental approaches for updating approximations in set-valued ordered information systems. Knowl.-Based Syst. **50**, 218–233 (2013)
19. Luo, C., Li, T.R., Chen, H.M.: Dynamic maintenance of approximations in set-valued ordered decision systems under the attribute generalization. Inf. Sci. **257**, 210–228 (2014)
20. Luo, C., Li, T.R., Chen, H.M., Lu, L.X.: Fast algorithms for computing rough approximations in set-valued decision systems while updating criteria values. Inf. Sci. **299**, 221–242 (2015)
21. Li, S.Y., Li, T.R., Liu, D.: Incremental updating approximations in dominance-based rough sets approach under the variation of the attribute set. Knowl.-Based Syst. **40**, 17–26 (2013)
22. Zeng, A.P., Li, T.R., Liu, D., Zhang, J.B., Chen, H.M.: A fuzzy rough set approach for incremental feature selection on hybrid information systems. Fuzzy Sets Syst. **258**, 39–60 (2015)
23. Chen, H.M., Li, T.R., Luo, C., et al.: A Decision-theoretic Rough Set Approach for Dynamic Data Mining. IEEE Trans. Fuzzy Syst. (2015). doi:10.1109/TFUZZ.2014.2387877

Three-Way Decision

Methods and Practices of Three-Way Decisions for Complex Problem Solving

Hong Yu[1][(✉)], Guoyin Wang[1], Baoqing Hu[2], Xiuyi Jia[3], Huaxiong Li[4],
Tianrui Li[5], Decui Liang[6], Jiye Liang[7], Baoxiang Liu[8], Dun Liu[5],
Jianmin Ma[9], Duoqian Miao[10], Fan Min[11], Jianjun Qi[12], Lin Shang[4],
Jiucheng Xu[13], Hailong Yang[14], Jingtao Yao[15], Yiyu Yao[15],
Hongying Zhang[16], Yanping Zhang[17], and Yanhui Zhu[18]

[1] Chongqing University of Posts and Telecommunications, Chongqing 400065, China
yuhong@cqupt.edu.cn
[2] Wuhan University, Wuhan 430072, China
[3] Nanjing University of Science and Technology, Nanjing 210094, China
[4] Nanjing University, Nanjing 210093, China
[5] Southwest Jiaotong University, Chengdu 610031, China
[6] University of Electronic Science and Technology of China, Chengdu 610054, China
[7] Taiyuan Normal University, Taiyuan 030006, China
[8] North China University of Science and Technology, Tangshan 063009, China
[9] Chang'an University, Xi'an 710064, China
[10] Tongji University, Shanghai 201804, China
[11] Southwest Petroleum University, Chengdu 610500, China
[12] Xidian University, Xi'an 710071, China
[13] Henan Normal University, Xinxiang 453007, China
[14] Shaanxi Normal University, Xi'an 710119, China
[15] University of Regina, Regina, SK S4S 0A2, Canada
[16] Xi'an Jiaotong University, Xi'an 710049, China
[17] Anhui University, Hefei 230601, China
[18] Hunan University of Technology, Zhuzhou 412007, China

Abstract. A theory of three-way decisions is formulated based on the notions of three regions and associated actions for processing the three regions. Three-way decisions play a key role in everyday decision-making and have been widely used in many fields and disciplines. A group of Chinese researchers further investigated the theory of three-way decision and applied it in different domains. Their research results are highlighted in an edited Chinese book entitled "Three-way Decisions: Methods and Practices for Complex Problem Solving." Based on the contributed chapters of the edited book, this paper introduces and reviews most recent studies on three-way decisions.

1 Introduction

The essential ideas of three-way decisions are commonly used in everyday life and are widely applied in many fields and disciplines, including medical decision-making, social judgement theory, hypothesis testing in statistics, management

© Springer International Publishing Switzerland 2015
D. Ciucci et al. (Eds.): RSKT 2015, LNAI 9436, pp. 255–265, 2015.
DOI: 10.1007/978-3-319-25754-9_23

sciences, classifications, machine learning and data mining, and peer review processes. By extracting the common elements of decision-making methods in these fields, Yao [37] introduced the notion of three-way decisions, consisting of the positive, boundary and negative rules, as a new paradigm of decision-making. A key feature of three-way decisions is to take different actions and to make different decisions for the three regions. Three-way decision approaches have been applied to areas such as decision support systems, email spam filtering, clustering analysis and so on [6,17,18,41].

The main idea of three-way decisions is to divide a universe into three disjoint regions and to process the different regions by using different strategy. By using notations and terminology of rough set theory [26,27,38], we give a brief description of three-way decisions as follows [41].

Definition 1. *Suppose U is a finite nonempty set of objects or decision alternatives and C is a finite set of conditions. Each condition in C may be a criterion, an objective, or a constraint. The problem of three-way decisions is to divide, based on the set of conditions C, U into three pair-wise disjoint regions by a mapping f:*

$$f : U \longrightarrow \{\text{POS}, \text{BND}, \text{NEG}\}. \tag{1}$$

The three regions are called the positive, boundary, and negative regions, respectively.

By definition, the three regions POS, BND and NEG are subsets of U with $U = \text{POS} \cup \text{BND} \cup \text{NEG}$, $\text{POS} \cap \text{BND} = \emptyset$, $\text{BND} \cap \text{NEG} = \emptyset$, and $\text{NEG} \cap \text{POS} = \emptyset$. Because one or two of the three regions may be the empty set $\emptyset$, $\{\text{POS}, \text{BND}, \text{NEG}\}$ might not be a partition of U. In order to facilitate the discussion, we usually call $\{\text{POS}, \text{BND}, \text{NEG}\}$ a tripartition of U by slightly loosening the definition of a partition. Corresponding to these three regions, the complements are as follows: $\text{POS}^c = \text{BND} \cup \text{NEG}$, $\text{BND}^c = \text{POS} \cup \text{NEG}$, and $\text{NEG}^c = \text{POS} \cup \text{BND}$.

Depending on the construction and interpretation of the mapping f, there are qualitative three-way decisions and quantitative three-way decisions. In qualitative three-way decision models, the universe is divided into three regions based on a function f that is of a qualitative nature. Quantitative three-way decision models are induced by that is of a quantitative nature. An evaluation-based three-way decision model uses an evaluation function that measures the desirability of objects with reference the set of criteria.

It should be pointed out that we can have a more general description of three-way decisions by using more generic labels and names. For example, in an evaluation-based model of three-way decisions [37], we can use a pair of thresholds to divide a universe into three regions. If we arrange objects in an increasing order with lower values at left, then we can conveniently label the three regions as the left, middle, or right regions, respectively, or simply L, M, and R regions [39]. In a similar way, strategies for processing three regions can be described in more generic terms [40].

Originally, the concept of three-way decisions was proposed and used to interpret probabilistic rough set three regions. Further studies show that a theory of three-way decision can be developed by moving beyond rough set theory. In fact, many recent studies went far beyond rough sets. An edited Chinese book entitled "Three-way Decisions: Methods and Practices for Complex Problem Solving" provides a snapshot of such research efforts from a group of Chinese researchers. In order to go further insights into three-way decisions and promote further research, this paper gives a brief of results on three-way decisions reported in the edited book.

2   Studies on the Basic Issues of Three-Way Decisions

Cost-Sensitive Models Based on DTRS. Three-way decisions originate from the studies on the decision-theoretic rough set (DTRS) model. It presents a semantics explanation on how to decide a concept into positive, negative and boundary regions based on the minimization of the decision cost, rather than decision error. Therefore, three-way decisions can be viewed as a cost-sensitive decision. Li et al. incorporated the three-way decisions into cost-sensitive learning and proposed a three-region cost-sensitive classification [14]. The key content of three-region cost-sensitive classification is to classify an instance into boundary region when precise classification (decide positive instance and negative instance decisions) cannot be immediately decided [15]. It is evident that the boundary decision may achieve lower cost/risk than positive and negative decisions do, if available information for immediate decision is insufficient, which is consistent with human decision process [14,16].

Cost-Sensitive Sequential Three-Way Decisions. In real-world applications, the available information is always insufficient, or it may associate with extra costs to get available information, which leads to frequent boundary decision. However, if the available information continuously increases, the previous boundary decisions may be converted to positive or negative decisions, which forms a sequential decision process [35,36]. Li et al. proposed a cost-sensitive sequential three-way decision strategy [12,13], which simulates the human decisions on such dynamic sequential decision: from rough granule to precise granule process. The applications of cost-sensitive sequential three-way decisions in face recognition were investigated in [57], which indicated that cost sequential three-way decisions achieve lower misrecognition costs than traditional two-way decisions in the dynamic sequential decision process of face recognition.

Compared to two-way decisions approaches, three-way decisions approaches introduces deferment decision through a pair of thresholds (α, β). Therefore, for three-way decision models, a great challenge is acquirement of a set of pairs of thresholds (α, β).

Determination of the Thresholds Through Optimization. Shang and Jia [7,8] studied this problem from an optimization viewpoint. An optimization problem is constructed to minimize the decision cost. Through solving the optimization problem, the thresholds and corresponding cost functions for making three-way decisions can be learned from given data without any preliminary knowledge. An adaptive algorithm with computational complexity $O(n^2)$ and other evolutionary algorithms were also proposed. They also studied the problem of how to obtain a two-way decisions result through a three-way decisions procedure. A two-phase classification approach is proposed by importing ensemble learning method [19].

Determination of the Thresholds Through Gini Coefficients. Zhang and Yao [42] investigated the relationship between changes in rough set regions and their impacts on the Gini coefficients of decision regions. Effective decision regions can be obtained by satisfying objective functions of Gini coefficients of decision regions. Three different objective functions are discussed in the book. The example shows that effective decision regions can be obtained by tuning Gini coefficients of decision regions to satisfy a certain objective function. It is suggested that with the new approach more applicable decision regions and decision rules may be obtained.

Determination the Thresholds by a Constructive Covering Algorithm. Zhang and Zou et al. [44] proposed a cost-sensitive three-way decisions model based on constructive covering algorithm (CCA). Zhang and Xing et al. [43] introduced CCA to three-way decisions procedure and proposed a new three-way decisions model based on CCA. According to the samples, POS, NEG and BND are gotten automatically. The new model does not need any given parameters to form the regions. This model has three advantages: (1) it is easier to process multi-categories classification; (2) it can process discrete type data and continuous type data directly; (3) the most important one is that it provides a new method to form three regions automatically for three-way decisions.

Three-Way Decision Spaces. Hu [5] established three-way decision space based on the proposed decision measurement, decision condition and evaluation function. That is, we use fuzzy lattice (complete distributive lattice with an inverse order and involutive operator) as a decision measurement tool, decision condition is unified by a mapping from universe to fuzzy lattice and evaluation function is unified through three axioms, i.e. (E1) Minimum element axiom, (E2) Monotonicity axiom and (E3) Complement axiom. Thus, three-way decisions based on fuzzy sets, interval-valued fuzzy sets, random sets and rough sets are the special examples of three-way decision spaces. At the same time, multi-granulation three-way decisions space and its corresponding multi-granulation three-way decisions are also established. We also introduce novel dynamic two-way decisions and dynamic three-way decisions based on three-way decisions spaces and three-way decisions based on bi-evaluation functions.

Three-Way Attribute Reduction. Miao and his group constructed three-way weighted entropies from the concept level to classification level, and further explore three-way attribute reduction by a novel approach of Bayesian inference [45]; established region-based hierarchical attribute reduction [46] and reduction target structure-based hierarchical attribute reduction [48] for the two-category decision theoretic rough set model; and built an expanded double-quantitative model by logically integrating probabilities and grades and the relevant double-quantitative reduction by hierarchically preserving specific regions [47].

3   Three-Way Decisions with Rough Sets

Incremental Approaches with DTRS for Incomplete Data. Considering that incomplete data with missing values are very common in many data-intensive applications. Luo et al. [20] proposed an incremental approach for updating probabilistic rough approximations, i.e., positive, boundary and negative regions, with the variation of objects in an incomplete information system. Four different maintenance strategies of the three-way decision rules with an incremental object based on the rough approximations were implemented, respectively [21].

Dynamic Three-Way Decisions with DTRS. Considering the addition of new objects in an information system, some new attributes may appear simultaneously. Chen et al. [1] investigated the dynamic DTRS approach for updating rough approximations with respect to the variation of objects and attributes simultaneously. On consideration of the semantic explanations of three-way decisions, Liu et al. [22] proposed a "four-level" approach and integrated the existing probabilistic rough set models to a generalized research framework. In general, three-way decisions build a bridge to connect the rough sets and decision theory, and promote the development of the both two research fields. Liu et al. [23] considered the dynamic change of loss functions in DTRS with the time, and further proposed the dynamic three-way decision model. Liu et al. [24] investigated multiple-category classification problems with three-way decisions, and further proposed a dynamic two-stage method to choose the best candidate classification.

4   Three-Way Decisions with Other Theories

Three-Way Decisions with Dempster-Shafer Theory. Faced with a multi-criteria group decision making with more attributes and a great deal of alternatives, the traditional decision making methods usually have the intensive computations or too many parameters to determine. To overcome these limitations, Wang et al. [29] proposed a Dempster-Shafer theory based intelligent three-way group sorting method. The method constructs the decision evidences from computing the fuzzy memberships of an alternative belonging to the decision classes and aggregates these evidences by using the famous Dempster combination approach. The proposed method has two merits: only one parameter to determine and the decision evidences obtained from decision makers' assessments.

Three-Way Decisions with Fuzzy Sets. Zadeh had introduced the concept of fuzzy sets to address the uncertainty and fuzziness due to ambiguity and incomplete information in real world. Zhang and her team mainly study uncertainty measures, ranking and their applications of fuzzy sets and its generalized model, including interval-valued fuzzy sets and hesitant fuzzy sets [50,51]. A new axiomatic definition of entropy of interval-valued fuzzy sets was proposed and its relation with similarity measure was discussed in [50]. A hybrid monotonic inclusion measure was proposed in [51] to quantitative ranking of any two fuzzy sets. A general model of ranking of any two interval sets and the uncertainty and ambiguity measures were discussed in detail [49] and their applications to three-way decisions was presented. Yang et al. [30] had devoted to construct the basic frame of fuzzy three-way decisions by extending three-way decisions to fuzzy case based on fuzzy set theory. In fuzzy three-way decisions, one universe is divided into three fuzzy regions satisfying some conditions rather than three pair-wise disjoint crisp regions. Some research results about fuzzy three-way decisions have been contained in the published book [41]. They introduced notions of several kinds of evaluation functions based fuzzy three-way decisions by using fuzzy set theory, and illustrated the relationships among these notions. In future research, they will focus on how to apply fuzzy three-way decisions to real-world applications by using fuzzy mathematic theory and methods. Besides, three-way approximations of intuitionistic fuzzy sets is also this group's research issue.

Three-Way Decisions with Formal Concept Analysis. Qi et al. [28] proposed the three-way concept analysis based on combining three-way decisions [37] and formal concept analysis [2]. In the framework of formal concept analysis, given a formal context (U, V, R), a formal concept (X, A) means that X contains just those objects sharing all the attributes in A and the attributes in A are precisely those common to all the objects in X. This expresses the semantics of "*jointly possessed*" between an object subset and an attribute subset in a formal context. But such concept cannot express the semantics of "*jointly not possessed*" which also exists in a formal context. The extension (or/and intension) of a three-way concept is equipped with two parts: positive and negative ones, where the positive part is used to express "*jointly possessed*" and the negative one is used to express "*jointly not possessed*". For example, a three-way concept $(X, (A, B))$ implies that all the objects in X jointly possessed every attribute in A and jointly not possessed every attribute in B.

5   Applications of Three-Way Decisions

Clustering Analysis. Yu and her group studied overlapping clustering [31], determining the number of clusters [32], incremental clustering [34] and so on, based on the three-way decision theory. In their work, a cluster is represented by an interval set instead of a single set. They use three regions to represent a cluster. Objects in the POS region belong to the cluster definitely, objects in the NEG region do not belong to the cluster definitely, and objects in BND region are fringe elements of the cluster. The advantages of the representation are that it is not only show which objects just belong to this cluster but also show which objects might belong to the cluster intuitively. Through the further work on the BND region [33], we can know the degree of an object influences on the form of the cluster intuitively, which is very helpful in some practice of applications such as community evaluations.

Frequent Itemsets Mining. Min and his group applied three-way decisions to the incremental mining of frequent itemsets [25,52]. In these works, as new data added, the algorithm only need to check and update itemsets in the boundary region. All possible itemsets are divided into three regions, namely the positive, the boundary and the negative region. Itemsets in the positive region are already frequent. Itemsets in the boundary region are infrequent, however may be frequent after data increment in the near future. Itemsets in the negative region will not be frequent even after data increment. Therefore to keep the frequent itemsets up-to-date, one only need to check those in the boundary region, and the runtime is saved.

Text Classification. Shang and Jia combined the three-way decisions solution with text sentiment analysis to improve the performance of sentiment classification [53], and they also applied it to filter spam email to obtain a lower misclassification rate and a less misclassification cost [9]. Miao and his group established an instance-centric hierarchical classification framework based on decision-theoretic rough set model [10]. Furthermore, Zhu et al. [55] presented a two-stage three-way decision classifiers based on integration of three-way decision and traditional machine learning algorithms. Zhang et al. [56] applied three-way decisions to sentiment classification with sentiment uncertainty to deal with the problems of context-dependent sentiment classification and topic-dependent sentiment classification.

Image Processing, Video Analysis and Others. Shang and Jia have combined three-way decisions solution with fuzzy sets to propose a new fuzzy rough set model and applied it in image segmentation [3,4]; Miao and his group have proposed a novel algorithm for image segmentation with noise in the framework of decision-theoretic rough set model [11]. Shang and Jia have adopted three-way decisions to analyze video behaviors and detect video anomaly behaviors [54].

6 Conclusions

The notion of three-way decisions was introduced for meeting the needs to properly explain three regions of probabilistic rough sets. A theory of three-way decisions moves far beyond this original goal. We begin to see a more general theory that embraces ideas from many fields and disciplines. An edited Chinese book entitled "Three-way Decisions: Methods and Practices for Complex Problem Solving" marks a cornerstone in the development of three-way decisions. It provides a snapshot of most recent research on three-way decisions. This paper is a summary of the main results reported in the book. A reader is encourage to read the book for more detailed descriptions.

For studies on basic issues of three-way decisions, we review cost-sensitive models based on decision-theoretic rough sets, cost-sensitive sequential three-way decisions, approaches to computing the thresholds α and β, three-way decision spaces, and algorithms of three-way attribute reductions. In the context of rough set theory, we discuss multigranulation decision-theoretic rough set models, incremental approaches to rough set approximations, and dynamic three-way decisions. With respect to other theories, we examine three-way decisions with Dempster-Shafer theory of evidential reasoning, fuzzy sets and formal concept analysis. Finally, we touch upon applications of three-way decisions in clustering analysis, frequent itemsets mining, text classification, sentiment analysis, image processing and so on.

By introducing a sample of most recent studies on three-way decisions, we want to demonstrate the value and power, as well as the great potentials, of three-way decisions. Considering the significance and generality of three-way decisions, continued research efforts are needed. We welcome you to join us and to work on an emerging and exciting theory of three-way decisions. For additional information, please consult the homepage of three-way decisions at: http://www2.cs.uregina.ca/~twd/.

Acknowledgements. The authors would like to thank the organizers of IJCRS 2015 for inviting and encouraging them to present their results.

References

1. Chen, H.M., Li, T.R., Luo, C., Horng, S.J., Wang, G.Y.: A decision-theoretic rough set approach for dynamic data mining. IEEE Trans. Fuzzy Syst. (2015). doi:10.1109/TFUZZ.2014.2387877

2. Ganter, B., Wille, R.: Formal Concept Analysis. Mathematical Foundations. Springer, Berlin, Heidelberg (1999)

3. Guo, M., Shang, L.: Selecting the appropriate fuzzy membership functions based on user-demand in fuzzy decision-theoretic rough set model. In: Proceedings of FUZZ-IEEE, pp. 1–8 (2013)

4. Guo, M., Shang, L.: Color image segmentation based on decision-theoretic rough set model and fuzzy C-means algorithm. In: Proceedings of FUZZ-IEEE, pp. 229–236 (2014)

5. Hu, B.Q.: Three-way decisions space and three-way decisions. Inf. Sci. **281**, 21–52 (2014)

6. Jia, X.Y., Shang, L., Zhou, X.Z., Liang, J.Y., Miao, D.Q., Wang, G.Y., Li, T.R., Zhang, Y.P.: Theory of Three-Way Decisions and Application. Nanjing University Press, Nanjing, China (2012). (In Chinese)

7. Jia, X.Y., Li, W.W., Shang, L., Chen, J.J.: An optimization viewpoint of decision-theoretic rough set model. In: Yao, J.T., Ramanna, S., Wang, G.Y., Suraj, Z. (eds.) RSKT2011. LNCS, vol. 6954, pp. 457–465. Springer, Heidelberg (2011)

8. Jia, X.Y., Tang, Z.M., Liao, W.H., Shang, L.: On an optimization representation of decision-theoretic rough set model. Int. J. Approximate Reasoning **55**, 156–166 (2014)

9. Jia, X., Shang, L.: Three-way decisions versus two-way decisions on filtering spam email. In: Peters, J.F., Skowron, A., Li, T., Yang, Y., Yao, J.T., Nguyen, H.S. (eds.) Transactions on Rough Sets XVIII. LNCS, vol. 8449, pp. 69–91. Springer, Heidelberg (2014)
10. Li, W., Miao, D.Q., Wang, W.L.: Hierarchical rough decision theoretic framework for text classification. In: 9th IEEE International Conference on Cognitive Informatics, pp. 484–489. IEEE Press, New York (2010)
11. Li, F., Miao, D.Q., Liu, C.H.: An image segmentation algorithm based on decision-theoretic rough set model. CAAI Trans. Intell. Syst. **9**(2), 143–147 (2014)
12. Li, H.X., Zhou, X.Z., Huang, B.: Cost-sensitive sequential three-way decisions. In: Liu, D., Li, T.R., Miao, D.Q., Wang, G.Y., Liang, J.Y. (eds.) Three-Way Decisions Granul. Comput., pp. 42–59. Science Press, Beijing (2013). (in chinese)
13. Li, H., Zhou, X., Huang, B., Liu, D.: Cost-sensitive three-way decision: a sequential strategy. In: Lingras, P., Wolski, M., Cornelis, C., Mitra, S., Wasilewski, P. (eds.) RSKT 2013. LNCS, vol. 8171, pp. 325–337. Springer, Heidelberg (2013)
14. Li, H., Zhou, X., Zhao, J., Huang, B.: Cost-sensitive classification based on decision-theoretic rough set model. In: Li, T., Nguyen, H.S., Wang, G., Grzymala-Busse, J., Janicki, R., Hassanien, A.E., Yu, H. (eds.) RSKT 2012. LNCS, vol. 7414, pp. 379–388. Springer, Heidelberg (2012)
15. Li, H.X., Zhou, X.Z., Zhao, J.B.: Cost-sensitive classification based on three-way decision-theoretic rough sets. Theory of Three-way Decisions and Application, pp. 34–45. Nanjing University Press (2012) (in Chinese)
16. Li, H.X., Zhou, X.Z.: Risk decision making based on decision-theoretic rough set: a three-way view decision model. Int. J. Comput. Intell. Syst. **4**(1), 1–11 (2011)
17. Liu, D., Li, T.R., Miao, D.Q., Wang, G.Y., Liang, J.Y.: Three-Way Decisions and Granular Computing. Science Press, Beijing, China (2013). (In Chinese)
18. Li, H.X., Zhou, X.Z., Li, T.R., Wang, G.Y., Miao, D.Q., Yao, Y.Y.: Decision-Theoretic Rough Set Theory and Recent Progress. Science Press, Beijing, China (2011). (In Chinese)
19. Li, W., Huang, Z., Jia, X.: Two-phase classification based on three-way decisions. In: Lingras, P., Wolski, M., Cornelis, C., Mitra, S., Wasilewski, P. (eds.) RSKT 2013. LNCS, vol. 8171, pp. 338–345. Springer, Heidelberg (2013)
20. Luo, C., Li, T., Chen, H.: Dynamic maintenance of three-way decision rules. In: Miao, D., Pedrycz, W., Slezak, D., Peters, G., Hu, Q., Wang, R. (eds.) RSKT 2014. LNCS, vol. 8818, pp. 801–811. Springer, Heidelberg (2014)
21. Luo, C., Li, T.: Incremental three-way decisions with incomplete information. In: Cornelis, C., Kryszkiewicz, M., Ślęzak, D., Ruiz, E.M., Bello, R., Shang, L. (eds.) RSCTC 2014. LNCS, vol. 8536, pp. 128–135. Springer, Heidelberg (2014)
22. Liu, D., Li, T.R., Ruan, D.: Probabilistic model criteria with decision-theoretic rough sets. Inf. Sci. **181**, 3709–3722 (2011)
23. Liu, D., Li, T., Liang, D.: Three-way decisions in dynamic decision-theoretic rough sets. In: Lingras, P., Wolski, M., Cornelis, C., Mitra, S., Wasilewski, P. (eds.) RSKT 2013. LNCS, vol. 8171, pp. 291–301. Springer, Heidelberg (2013)
24. Liu, D., Li, T.R., Li, H.X.: A multiple-category classification approach with decision-theoretic rough sets. Fundamenta Informaticae **115**, 173–188 (2012)
25. Li, Y., Zhang, Z.H., Chen, W.B., Min, F.: Tdup: an approach to incremental mining of frequent itemsets with three-way-decision pattern updating. Int. J. Mach. Learn. Cybern. 1–13 (2015)
26. Pawlak, Z.: Rough sets. Int. J. Comput. Inf. Sci. **11**, 341–356 (1982)
27. Pawlak, Z.: Rough Sets. Theoretical Aspects of Reasoning about Data. Kluwer Academic Publishers, Dordrecht (1991)

28. Qi, J., Wei, L., Yao, Y.: Three-way formal concept analysis. In: Miao, D., Pedrycz, W., Slezak, D., Peters, G., Hu, Q., Wang, R. (eds.) RSKT 2014. LNCS, vol. 8818, pp. 732–741. Springer, Heidelberg (2014)

29. Wang, B., Liang, J.: A novel intelligent multi-attribute three-way group sorting method based on dempster-shafer theory. In: Miao, D., Pedrycz, W., Slezak, D., Peters, G., Hu, Q., Wang, R. (eds.) RSKT 2014. LNCS, vol. 8818, pp. 789–800. Springer, Heidelberg (2014)

30. Yang, H.L., Li, S.G., Wang, S.Y., Wang, J.: Bipolar fuzzy rough set model on two different universes and its application. Knowl. Based Syst. **35**, 94–101 (2012)

31. Yu, H., Wang, Y.: Three-way decisions method for overlapping clustering. In: Yao, J.T., Yang, Y., Słowiński, R., Greco, S., Li, H., Mitra, S., Polkowski, L. (eds.) RSCTC 2012. LNCS, vol. 7413, pp. 277–286. Springer, Heidelberg (2012)

32. Yu, H., Liu, Z.G., Wang, G.Y.: An automatic method to determine the number of clusters using decision-theoretic rough set. Int. J. Approximate Reasoning **55**(1), 101–115 (2014)

33. Yu, H., Jiao, P., Wang, G.Y., Yao, Y.Y.: Categorizing overlapping regions in clustering analysis using three-way decisions. In: Proceedings of IEEE/WIC/ACM International Joint Conferences on Web Intelligence (WI) and Intelligent Agent Technologies (IAT), vol. 2, pp. 350–357 (2014)

34. Yu, H., Zhang, C., Hu, F.: An incremental clustering approach based on three-way decisions. In: Cornelis, C., Kryszkiewicz, M., Ślęzak, D., Ruiz, E.M., Bello, R., Shang, L. (eds.) RSCTC 2014. LNCS, vol. 8536, pp. 152–159. Springer, Heidelberg (2014)

35. Yao, Y.: Granular computing and sequential three-way decisions. In: Lingras, P., Wolski, M., Cornelis, C., Mitra, S., Wasilewski, P. (eds.) RSKT 2013. LNCS, vol. 8171, pp. 16–27. Springer, Heidelberg (2013)

36. Yao, Y.Y., Deng, X.F.: Sequential three-way decisions with probabilistic rough sets. In: Proceedings of IEEE ICCI*CC 2011, pp. 120–125. IEEE (2011)

37. Yao, Y.: An outline of a theory of three-way decisions. In: Yao, J.T., Yang, Y., Słowiński, R., Greco, S., Li, H., Mitra, S., Polkowski, L. (eds.) RSCTC 2012. LNCS, vol. 7413, pp. 1–17. Springer, Heidelberg (2012)

38. Yao, Y.Y.: The two sides of the theory of rough sets. Knowl. Based Syst. **80**, 67–77 (2015)

39. Yao, Y.Y., Yu, H.: An introduction to three-way decisions. In: Yu, H., Wang, G.Y., Li, T.R., Liang, J.Y., Miao, D.Q., Yao, Y.Y. (eds.) Three-Way Decisions: Methods and Practices for Complex Problem Solving. Science Press, Beijing (2015). (In Chinese)

40. Yao, Y.Y.: Rough sets and three-way decisions. In: Ciucci, D., Wang, G.Y., Mitra, S., Wu, W.Z. (eds.) RSKT 2015. LNCS (LNA), vol. 9436, pp. 62–73. Springer, Heidelberg (2015)

41. Yu, H., Wang, G.Y., Li, T.R., Liang, J.Y., Miao, D.Q., Yao, Y.Y.: Three-Way Decisions: Methods and Practices for Complex Problem Solving. Science Press, Beijing, China (2015). (In Chinese)

42. Zhang, Y., Yao, J.T.: Determining three-way decision regions with gini coefficients. In: Cornelis, C., Kryszkiewicz, M., Ślęzak, D., Ruiz, E.M., Bello, R., Shang, L. (eds.) RSCTC 2014. LNCS, vol. 8536, pp. 160–171. Springer, Heidelberg (2014)

43. Zhang, Y., Xing, H., Zou, H., Zhao, S., Wang, X.: A three-way decisions model based on constructive covering algorithm. In: Lingras, P., Wolski, M., Cornelis, C., Mitra, S., Wasilewski, P. (eds.) RSKT 2013. LNCS, vol. 8171, pp. 346–353. Springer, Heidelberg (2013)

44. Zhang, Y., Zou, H., Chen, X., Wang, X., Tang, X., Zhao, S.: Cost-sensitive three-way decisions model based on CCA. In: Cornelis, C., Kryszkiewicz, M., Ślęzak, D., Ruiz, E.M., Bello, R., Shang, L. (eds.) RSCTC 2014. LNCS, vol. 8536, pp. 172–180. Springer, Heidelberg (2014)

45. Zhang, X., Miao, D.: Three-way weighted entropies and three-way attribute reduction. In: Miao, D., Pedrycz, W., Slezak, D., Peters, G., Hu, Q., Wang, R. (eds.) RSKT 2014. LNCS, vol. 8818, pp. 707–719. Springer, Heidelberg (2014)

46. Zhang, X.Y., Miao, D.Q.: Region-based quantitative and hierarchical attribute reduction in the two-category decision theoretic rough set model. Knowl. Based Syst. **71**, 146–161 (2014)

47. Zhang, X.Y., Miao, D.Q.: An expanded double-quantitative model regarding probabilities and grades and its hierarchical double-quantitative attribute reduction. Inf. Sci. **299**, 312–336 (2015)

48. Zhang, X.Y., Miao, D.Q.: Reduction target structure-based hierarchical attribute reduction for two-category decision-theoretic rough sets. Inf. Sci. **277**, 755–776 (2014)

49. Zhang, H.Y., Yang, S.Y.: Ranking Interval Sets Based on Inclusion Measures and Applications to Three-Way Decisions, Manuscript

50. Zhang, H.Y., Zhang, W.X.: Entropy of interval-valued fuzzy sets based on distance and its relationship with similarity measure. Knowl. Based Syst. **22**, 449–454 (2009)

51. Zhang, H.Y., Zhang, W.X.: Hybrid monotonic fuzzy inclusion measure and its use in measuring similarity and distance between fuzzy sets. Fuzzy Sets Syst. **51**, 56–70 (2009)

52. Zhang, Z.H., Li, Y., Chen, W.B., Min, F.: A three-way decision approach to incremental frequent itemsets mining. J. Inf. Comput. Sci. **11**, 3399–3410 (2014)

53. Zhou, Z., Zhao, W., Shang, L.: Sentiment analysis with automatically constructed lexicon and three-way decision. In: Miao, D., Pedrycz, W., Slezak, D., Peters, G., Hu, Q., Wang, R. (eds.) RSKT 2014. LNCS, vol. 8818, pp. 777–788. Springer, Heidelberg (2014)

54. Zhao, L., Shang, L., Gao, Y., Yang, Y., Jia, X.Y.: Video behavior analysis using topic models and rough sets applications. IEEE Comput. Intell. Mag. **8**(1), 56–67 (2013)

55. Zhu, Y., Tian, H., Ma, J., Liu, J., Liang, T.: An integrated method for micro-blog subjective sentence identification based on three-way decisions and naive bayes. In: Miao, D., Pedrycz, W., Slezak, D., Peters, G., Hu, Q., Wang, R. (eds.) RSKT 2014. LNCS, vol. 8818, pp. 844–855. Springer, Heidelberg (2014)

56. Zhang, Z., Wang, R.: Applying three-way decisions to sentiment classification with sentiment uncertainty. In: Miao, D., Pedrycz, W., Slezak, D., Peters, G., Hu, Q., Wang, R. (eds.) RSKT 2014. LNCS, vol. 8818, pp. 720–731. Springer, Heidelberg (2014)

57. Zhang, L.B., Li, H.X., Zhou, X.Z., Huang, B., Shang, L.: Cost-sensitive sequential three-way decision for face recognition. In: Miao, D., Pedrycz, W., Ślęzak, D., Peters, G., Hu, Q., Wang, R. (eds.) RSKT2014. LNCS, vol. 8818, pp. 375–383. Springer, Switzerland (2014)

A Multi-view Decision Model Based on CCA

Jie Chen[1,2,3], Shu Zhao[1,2,3]([✉]), and Yanping Zhang[1,2,3]

[1] Key Laboratory of Intelligent Computing and Signal Processing of Ministry
of Education, Hefei 230601, Anhui, People's Republic of China
[2] Center of Information Support and Assurance Technology, Anhui University,
Hefei 230601, Anhui, People's Republic of China
[3] School of Computer Science and Technology, Anhui University,
Hefei 230601, Anhui, People's Republic of China
zhaoshuzs2002@hotmail.com
http://ailab.ahu.edu.cn

Abstract. Three-way decision theory divides all samples into three
regions: positive region, negative region and boundary region. A lack
of detailed information may make a definite decision impossible for samples in boundary region. These samples may be further handled by using
new information. In this paper, we propose a method Multi-View Decision Model based on constructive three-way decision theory. Multi-view
Decision Model mines the global information of all samples for decision.
All samples firstly are decided by $MinCA$, which builds the min covers for each class. Then samples in boundary region are classified using
Multi-view information. Experiments have shown that in most cases,
Multi-View Decision Model is beneficial for reducing boundary region
and promoting classification precision.

Keywords: Boundary region · Multi-view information · Three-way
decision theory · $MinCA$

1 Introduction

In conventional two-way decision model, there are only two optional choices for
a decision: positive decision or negative decision regardless of lack of information or not. Thus, it may result in wrong decisions when the information is not
enough. To address this issue, Yao proposed Three-way decision model [1–4],
which extends two-way decision theory by incorporating an additional choice:
boundary decision. Three-way decision theory presents a sample as positive, negative and boundary region. Many researchers have done further research on it.
Yao and Zhao pointed out some characteristic which needs to stay the same
in attribute reduction of $DTRS$ and proposed the attribute reduction method;
Yao, Liu and Li et al. researched on the three-way decision semantic in $DTRS$
and proposed the three-way decision rough set model [2]. Ping Li et al. adopted
the idea of tri-training algorithm [5] and put forward a tri-training algorithm
based on three-way decisions to reduce the boundary region [6]. Yao proposed

© Springer International Publishing Switzerland 2015
D. Ciucci et al. (Eds.): RSKT 2015, LNAI 9436, pp. 266–274, 2015.
DOI: 10.1007/978-3-319-25754-9_24

sequential three-way decisions to make a definite decision of acceptance or rejection for some uncertain samples [7]. In recent years, the three-way decision was widely used in the real life, such as spam filtering [8,9], text classification [10], medical diagnosis [11], management theory [12], social judgment theory [13,14], paper review [15,16], risk preferences of decision-making [17,18], oil exploration decision [19,20], text classification [10], automatic clustering [21] and etc.

The main superiority of three-way decision compared to two-way decision is the utility of the boundary decision. In three-way decision theory, the boundary decision is regarded as a feasible choice of decision when the available information for decision is too limited to make a proper decision, which is similar to the human decision strategy in the practical decision problems. In this case, how to reduce the boundary region is a new problem [6].

Samples in boundary region with a non-commitment decision may be further investigated by using new mined information. For a practical decision problem, we may find diverse characteristics between the types of decisions. Different attitudes towards a decision can be seen among separate groups of people. Some people always take optimistic decision, while other people may take pessimistic decision or indifferent decision. As we can see, people will take different type decision according to their personal character. When considering the difference among different people, it is necessary to develop a flexible decision model in which diverse type of decision is embodied.

To solve this problem, we propose a method to mine more accurate information for decision. Before, multi-granular mining for boundary region in three-way decision theory had been put forward using extra information of boundary region. Now, we propose a new method, Multi-View Decision Model based on constructive three-way decision theory. Based on $MinCA$, we find detailed information about all samples and define multi-view attribute subsets. Then samples are classified according to multi-view information.

The paper is organized as follows: in Sect. 2, we introduce the related works. In Sect. 3, we introduce Multi-View Decision Model Based on constructive three-way decision theory in detail. In Sect. 4, we analyze the experimental results. We draw our conclusion in Sect. 5.

2   Related Work

2.1   An Overview of Three-Way Decisions Model Based on CCA

In three-way decision model, decision actions are given by $A = \{a_P, a_N, a_B\}$, representing POS, NEG and BND decisions, called the positive, negative, and boundary region respectively. The positive region POS consists of those objects that we accept as satisfying the conditions and the negative region NEG consists of those objects that we reject as satisfying the conditions. BND decisions are the third choice for decision which means we need to collect more information for further precise decision.

Three-Way Decisions Model Based on Constructive Covering Algorithm (CCA) was proposed by Zhang and Xing [22] which doesn't need threshold

α and β. Constructive Covering Algorithm (CCA) is a constructive supervised learning algorithm that maps all samples in the data set to an n-dimensional sphere S^n. The sphere neighborhoods are utilized to divide the samples. The CCA can construct neural networks (NNs) based on the samples' own characteristics.

Definition 1. *Constructive Covering Algorithm (CCA)*. *Given a training samples set* $X = \{x_1, x_2, \cdots, x_n\}$, $(i = 1, 2, \cdots, n)$, *which is the set in* n-*dimensional Euclidean space.* $A_i = (A_i^1, A_i^2, \cdots, A_i^m)$ *is* m-*dimensional characteristic attribute of the* i-*th sample. We assume* $C^j = \bigcup C_i^j$, $i = \{1, 2, \cdots\}$. C^j *represents all covers of the* j-*th category samples. We can define the distance between sample* i *and the farthest similar point as* $d_1(i)$, *the distance between sample* i *and the nearest other as* $d_2(i)$.

$$d_1(i) = \max_{k \in I(j)} \{d(x^i, x_k) | d(x^i, x_k) < d_2(i)\} \tag{1}$$

$$d_2(i) = \min_{k \notin I(j)} \{d(x^i, x_k)\} \tag{2}$$

$$d(i) = (d_1(i) + d_2(i))/2 \tag{3}$$

Then, C_i^j is the ith cover of class j which is constructed by x^i and $d(i)$. The center of C_i^j is x^i, the radius is $d(i)$.

After this, we can obtain a set of sphere neighborhoods (covers) such that $C(j) = \bigcup C_i^j$ covers every input $\forall x^l \in p(j)$ and does not cover any input $x^l \notin p(t)$. Therefore, $C(t) = \bigcup C(j), j = 1, \cdots, t$, which is the output of CCA.

The CCA transforms the design of the NNs to calculate the set of covers. This algorithm constructs NNs based on the samples' own characteristics that are suitable for large sets of data, and avoids the selection of the structure of NNs and local minimum point. The training speed is rapid.

For two categories y_1 and y_2, we define POS of C^{y_1}, namely, $POS(C^{y_1})$ by the difference of unions $\bigcup C_i^{y_1} - \bigcup C_j^{y_2}$, $NEG(C^{y_1})$ by $\bigcup C_j^{y_2} - \bigcup C_i^{y_1}$ and $BND(C^{y_1})$ by the rest, where $i = \{1, 2, \cdots\}$, $j = \{1, 2, \cdots\}$. That is to say, $POS(C^{y_1})$ is equal to $NEG(C^{y_2})$; $POS(C^{y_2})$ is equal to $NEG(C^{y_1})$; $BND(C^{y_1})$ is equal to $BND(C^{y_2})$. For various categories y_l, $l = 1, 2, \cdots$, we also define $POS(C^{y_l})$ by the difference of unions $\bigcup C_i^{y_l} - \bigcup C_j^{y_m}$, $NEG(C^{y_l})$ by $\bigcup C_j^{y_m} - \bigcup C_i^{y_l}$, $m \neq l$, and $BND(C^{y_l})$ by the rest.

2.2   Multi-granular Mining for Boundary Region

In CCA model, we use a single view to represent a sample. In real-word decision making, we may consider multi-view that eventually leads to two-way decisions. At each view, more information is acquired. So, we had presented a new way to formulate three-way decisions through the notion of multi-view of granularity, namely multi-granular three-way decision algorithm($MGTD$). The whole attribute set of samples is divided into different subsets and each subset is a view of granularity.

Definition 2. *The Pair of Heterogeneous Points(PHP). We assume that there are sample set* $X = \{x_1, x_2, \cdots, x_n\}$, n *is the number of samples, and attribute set* $A = \{A_1, A_2, \cdots, A_m\}$, m *is the number of attributes. For* $\forall x_i \in BND(X), \exists x_j \in BND(X),$ *satisfy:*

$$PHP(i,j) = \{(x_i, x_j)|MIN\{d(x_i, x_j)\}\} \tag{4}$$

where $MIN\{d(x_i, x_j)\}$ *means* $\forall k, x_k \in BND(X)$, *and* $y(x_i) \neq y(x_k)$, $d(x_i, x_j) \leq d(x_i, x_k)$. $y(x_i)$ *is the class of sample* x_i, $d(x_i, x_j)$ *denotes the Euclidean distance between* x_i *and* x_j. *So samples* x_i *and* x_j *are named as the pair of heterogeneous points.*

According to the definition of PHP, we can obtain the information of PHP in $BND(X)$. Then, we would deal with $x_i \in BND(X)$ through the notion of multi-view of granularity. In $MGTD$, the number of positive cover and negative cover is changed with thresholds k and t, which are rate that release samples from positive regions POS or negative regions NEG into boundary region $BND(X)$. The values of thresholds k and t are according to cost-sensitive classification strategy.

3 Multi-view Decision Model Based on Constructive Three-Way Decision Theory

In this section, we propose a Multi-View Decision Model based on constructive three-way decision theory and show its advantages over other three-way decision models.

In $MGTD$, thresholds k and t are also a great challenge for three-way decision which are choose based on experience of experts. Moreover, this algorithm mines information from samples in boundary region. It is local information about samples in boundary region, not all samples. In this paper, author proposes a multi-view decision model($MVDM$) based on $MinCA$ to deal with boundary region.

Definition 3. *MinCA. Given a training samples set* $X = \{x_1, x_2, \cdots, x_n\}$, $(i = 1, 2, \cdots, n)$, *which is the set in n-dimensional Euclidean space.* $A_i = (A_i^1, A_i^2, \cdots, A_i^m)$ *is m-dimensional characteristic attribute of the ith sample. We assume* $C^j = \bigcup C_i^j$, $i = \{1, 2, \cdots\}$. C^j *represents all Min covers of the jth category samples.*

$$d(i) = d_1(i) = \max_{k \in I(j)} \{d(x^i, x_k)|d(x^i, x_k) < d_2(i)\} \tag{5}$$

The center of Min cover C_i^j *is* x^i, *the radius is* $d(i)$.

$MinCA$ regards the max distance between the center and the similar points as the radius where the boundary does not have any dissimilar points [22].

The difference between $MinCA$ and CCA is shown in Fig. 1. The radius of CCA is $d(i)$, $d(i) = (d_1(i) + d_2(i))/2$, but the radius of $MinCA$ is $d_1(i)$.

In $MVDM$, we firstly divide samples into three regions:POS, NEG and BND based on $MinCA$, which is obtain most precise covers. Then multi-view information is obtained from all samples. The samples in BND will be decided again according to multi-view information of whole sample set.

The detail of $MVDM$ is presented as below.

Algorithm 1. Multi-View Decision Model based on constructive three-way decision theory.

Input: A set of objects $X = \{x_1, x_2, \cdots, x_n\}$.
   A set of attributes $A = \{A_1, A_2, \cdots, A_m\}$.
   Test set $Y = \{y_1^1, y_2^1, \cdots, y_1^2, y_2^2, \cdots\}$, where y_i^j means $y(y_i) = j$.
Output: Two regions POS and NEG.
   Initialize $C^{MV} = \emptyset$.
Step 1:
Train sample set X with attribute set A based on $MinCA$, and generate Min cover set $C^{min} = \{c_1^{min}, c_2^{min}, \cdots, c_{p^{min}}^{min}\}$, where p^{min} is the number of Min covers.
   $POS = \{x_i | y(x_i) = 1, x_i$ can be covered by $C^{min}\}$
   $NEG = \{x_i | y(x_i) = 2, x_i$ can be covered by $C^{min}\}$
   $BND = \{x_i | x_i$ can not be covered by $C^{min}\}$
Step 2:
   for all samples X and attribute set A **do**
      Define $PHPs$.
      Compute $bias_k = \sum\limits_{each PHP} |f_1(A_k) - f_2(A_k)|$.
      Sort $bias_k$, and select the top θ, named as new attribute set A_2.
      $BND \rightarrow X, A_2 \rightarrow A$.
      Obtain new cover set $C\prime$.
      $C^{MV} = C^{MV} \cup C\prime = \{c_1^{MV}, c_2^{MV}, \cdots, c_{p^{MV}}^{MV}\}$.
   until PHP is not found.
Step 3:
      $POS\prime = \{x_i | y(x_i) = 1, x_i$ can be covered by $C^{MV}, x_i \in BND\}$
      $NEG\prime = \{x_i | y(x_i) = 2, x_i$ can be covered by $C^{MV}, x_i \in BND\}$
      $POS = POS \bigcup POS\prime, NEG = NEG \bigcup NEG\prime$
   return POS and NEG.

In this algorithm, no threshold is need to become POS, NEG and BND regions. The value of θ is not sensitive because attribute subsets are selected over and over until all samples are covered. $f(A_k)$ is the value of attribute A_k, $k = 1, 2, \cdots, m$. So in experiments, we usually assume θ equals to two-thirds of all attribute number. All samples are firstly divided into POS, NEG and BND. Then samples in BND are secondly decided to POS or NEG based on global detail information from $PHPs$. Finally, BND is none.

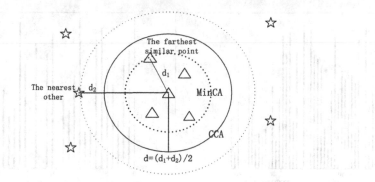

Fig. 1. The difference between $MinCA$ and CCA

4   Experiments

4.1   Experimental Settings

The experiments were performed on two data sets from *UCI Machine Learning Repository* [23]. Table 1 shows the details of the data sets.

Table 1. Data sets from UCI

Data	Number of data	Attributes	Classes
spambase	4601	58	2
chess	3196	36	2

We conduct four groups of comparative experiments to evaluate the performance of our algorithm. Those comparative algorithms are three-way decisions model based on decision-theoretic rough set ($DTRS$) [1], Cost-sensitive three-way decisions model based on $CCA(CCTDM)$ [24], and robustness three-way decisions model based on $CCA(RTDM)$ [25]. 10-fold cross-validation is used for all datasets in our experiments.

4.2   Experimental Results

The experimental results are shown in Figs. 2 and 3. $MVDM$ mines multi-view information of all samples to deal uncertain samples. Figure 2 shows the importance of each attribute in all subsets. The x-axis represents the serial number of the attribute. The y-axis represents the sequence number of each attribute, with larger numbers corresponding to more important attributes. From Fig. 2, we can see that some attributes are not important, such as attribute subset $\{4, 22, 38, 39, 43, \cdots, 58\}$ of *spambase* data set and attribute subset $\{14, 17, 19, 25, 27, 28, 29, 30, 33, \cdots, 37\}$ of *chess* data set. But attribute subset

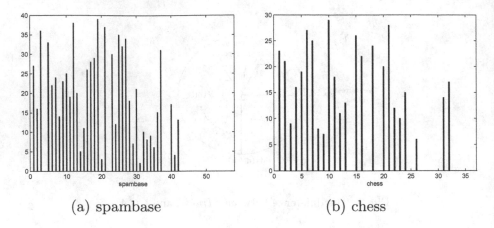

(a) spambase   (b) chess

Fig. 2. The importance order of each attribute in attribute subsets

$\{19, 12, 21, 3, 25, \cdots\}$ of *spambase* data set and attribute set $\{10, 21, 6, 15, 7, \cdots\}$ of *chess* data set are important to decision which are embodied in multi views. Hence, the multi-views of granularity in $MVDM$ reflect the importance of each attribute for decision-making.

Further, to evaluate the performance of our algorithm, we contrast $MVDM$ with $DTRS$, $CCTDM$ and $RTDM$. The experimental results are shown in Fig. 3. The x-axis is the group number of experiments. The y-axis is the whole classification rate of each three-way decision algorithm. It is equal to the number of correctly classified samples divided by the total number of samples. Experiments select eleven group thresholds for other algorithms, so the results of other algorithms is a curve and the results of $MVDM$ is a straight line.

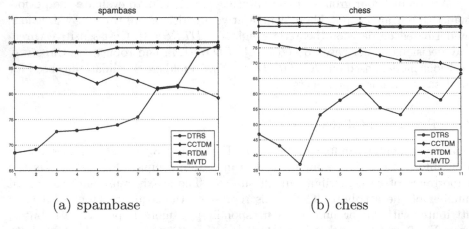

(a) spambase   (b) chess

Fig. 3. The whole classification rate of four three-way decision models on *spambase* and *chess*

In Fig. 3, $MVDM$ is superior to other three-way decision classifiers in most cases. To sum up, we can obtain:

- $MVDM$ is parameterless.
- $MVDM$ mines the global information of all samples to deal uncertain samples.

5   Conclusion

In this paper, we proposed a method $MDVM$ to reduce boundary region in three-way decision. We utilized $MinCA$ to obtain most precise covers and mine multi-view of all samples to deal uncertain samples. We conducted several groups of comparative experiments among $MVDM$, $DTRS$, $CCTDM$ and $RTDM$. The experimental results on *spambase* and *chess* showed that compared with other three-way decision classifiers, the classification precision of $MVDM$ is higher and boundary region are none.

Acknowledgments. This work is supported by the National Natural Science Foundation of China (No. 61175046, No. 61402006), supported by Provincial Natural Science Research Program of Higher Education Institutions of Anhui Province (No. KJ2013A016), and supported by Open Funding Project of Co-Innovation Center for Information Supply & Assurance Technology of Anhui University (No. ADXXBZ201410).

References

1. Yao, Y.Y.: The superiority of three-way decisions in probabilistic rough set models. Inf. Sci. **181**, 1080–1096 (2011)
2. Yao, Y.: Three-way decision: an interpretation of rules in rough set theory. In: Wen, P., Li, Y., Polkowski, L., Yao, Y., Tsumoto, S., Wang, G. (eds.) RSKT 2009. LNCS, vol. 5589, pp. 642–649. Springer, Heidelberg (2009)
3. Yao, Y.Y.: Three-way decisions with probabilistic rough sets. Inf. Sci. **180**, 341–353 (2010)
4. Yao, Y.Y.: Two semantic issues in a probabilistic rough set model. Fundamenta Informaticae **108**, 249–265 (2011)
5. Zhou, Z.H., Li, M.: Tri-training: exploiting unlabeled data using three classifiers. Knowl. Data Eng. **17**, 1529–1541 (2005)
6. Li, P., Shang, L., Li, H.: A method to reduce boundary regions in three-way decision theory. In: Miao, D., Pedrycz, W., Slezak, D., Peters, G., Hu, Q., Wang, R. (eds.) RSKT 2014. LNCS, vol. 8818, pp. 834–843. Springer, Heidelberg (2014)
7. Yao, Y.: Granular computing and sequential three-way decisions. In: Lingras, P., Wolski, M., Cornelis, C., Mitra, S., Wasilewski, P. (eds.) RSKT 2013. LNCS, vol. 8171, pp. 16–27. Springer, Heidelberg (2013)
8. Jia, X., Zheng, K., Li, W., Liu, T., Shang, L.: Three-way decisions solution to filter spam email: an empirical study. In: Yao, J.T., Yang, Y., Słowiński, R., Greco, S., Li, H., Mitra, S., Polkowski, L. (eds.) RSCTC 2012. LNCS, vol. 7413, pp. 287–296. Springer, Heidelberg (2012)

9. Zhou, B., Yao, Y.Y., Yao, J.T.: Cost-sensitive three-way email spam filtering. J. Intell. Inf. Syst. **42**, 1–27 (2014)
10. Li, W., Miao, D.Q., Wang, W.L., Zhang, N.: Hierarchical rough decision theoretic framework for text classification. In: Sun, F., Wang, Y., Lu, J., Zhang, B., Kinsner, W., Zadeh, L.A. (eds.) ICCI 2010, pp. 484–489. IEEE Press, Los Alamitos (2010)
11. Herbert, J.P., Yao, J.T.: Game-theoretic risk analysis in decision-theoretic rough sets. In: Wang, G., Li, T., Grzymala-Busse, J.W., Miao, D., Skowron, A., Yao, Y. (eds.) RSKT 2008. LNCS (LNAI), vol. 5009, pp. 132–139. Springer, Heidelberg (2008)
12. Liu, D., Li, T., Liang, D.: Three-way decisions in dynamic decision-theoretic rough sets. In: Lingras, P., Wolski, M., Cornelis, C., Mitra, S., Wasilewski, P. (eds.) RSKT 2013. LNCS, vol. 8171, pp. 291–301. Springer, Heidelberg (2013)
13. Li, H., Zhou, X., Zhao, J., Huang, B.: Cost-sensitive classification based on decision-theoretic rough set model. In: Li, T., Nguyen, H.S., Wang, G., Grzymala-Busse, J., Janicki, R., Hassanien, A.E., Yu, H. (eds.) RSKT 2012. LNCS, vol. 7414, pp. 379–388. Springer, Heidelberg (2012)
14. Jia, X.Y., Shang, L., Chen, J.J.: Attribute reduction based on minimum decision cost. J. Front. Comput. Sci. Technol. **5**, 155–160 (2011). (in Chinese)
15. Li, H., Zhou, X., Zhao, J., Liu, D.: Attribute reduction in decision-theoretic rough set model: a further investigation. In: Yao, J.T., Ramanna, S., Wang, G., Suraj, Z. (eds.) RSKT 2011. LNCS, vol. 6954, pp. 466–475. Springer, Heidelberg (2011)
16. Liu, D., Li, H.X., Zhou, X.Z.: Two decades research on decision-theoretic rough sets. In: Sun, F., Wang, Y., Lu, J., Zhang, B., Kinsner, W., Zadeh, L.A. (eds.) ICCI 2010, pp. 968–973. IEEE Press, Los Alamitos (2010)
17. Li, H.X., Zhou, X.Z.: Risk decision making based on decision-theoretic rough set: a three-way view decision model. Int. J. Comput. Intell. Syst. **4**, 1–11 (2011)
18. Zhou, X., Li, H.: A multi-view decision model based on decision-theoretic rough set. In: Wen, P., Li, Y., Polkowski, L., Yao, Y., Tsumoto, S., Wang, G. (eds.) RSKT 2009. LNCS, vol. 5589, pp. 650–657. Springer, Heidelberg (2009)
19. Yu, H., Chu, S., Yang, D.: Autonomous knowledge-oriented clustering using decision-theoretic rough set theory. Fundamenta Informaticae **115**, 141–156 (2012)
20. Yao, J.T., Herbert, J.P.: Web-based support systems with rough set analysis. In: Kryszkiewicz, M., Peters, J.F., Rybiński, H., Skowron, A. (eds.) RSEISP 2007. LNCS (LNAI), vol. 4585, pp. 360–370. Springer, Heidelberg (2007)
21. Yu, H., Liu, Z., Wang, G.: Automatically determining the number of clusters using decision-theoretic rough set. In: Yao, J.T., Ramanna, S., Wang, G., Suraj, Z. (eds.) RSKT 2011. LNCS, vol. 6954, pp. 504–513. Springer, Heidelberg (2011)
22. Zhang, Y., Xing, H., Zou, H., Zhao, S., Wang, X.: A three-way decisions model based on constructive covering algorithm. In: Lingras, P., Wolski, M., Cornelis, C., Mitra, S., Wasilewski, P. (eds.) RSKT 2013. LNCS, vol. 8171, pp. 346–353. Springer, Heidelberg (2013)
23. UCI Machine Learning Repository. http://archive.ics.uci.edu/ml/
24. Zhang, Y.P., Zou, H.J., Zhao, S.: Cost-sensitive three-way decisions model based on CCA. J. Nanjing Univ. **2**, 447–452 (2015)
25. Zhang, Y.P., Zou, H., Chen, X., Wang, X.Y., Tang, X.Q., Zhao, S.: Cost-sensitive three-way decisions model based on CCA. In: Miao, D.Q., et al. (eds.) RSKT 2014. LNCS(LNAI), vol. 8818, pp. 172–180. Springer, Heidelberg (2014)

A Teacher-Cost-Sensitive Decision-Theoretic Rough Set Model

Yu-Wan He, Heng-Ru Zhang, and Fan Min[✉]

School of Computer Science, Southwest Petroleum University,
Chengdu 610500, China
minfanphd@163.com

Abstract. Existing DTRS models use two states (success, fail) and three actions (accept, defer, reject) to describe the decision procedure. However, deferment provides a compromise instead of a solution. In this paper, we replace this action with *consult* which stands for consulting a teacher for correct classification. Naturally, the new action involves the cost of teacher, which is already known in many applications. Through computing the thresholds α and β with the misclassification cost and the teacher cost from the decision system, the positive, the boundary, and the negative regions are obtained. They correspond to positive rules for acceptance, boundary rules for consulting a teacher, and negative rules for rejection, respectively. We compare the new model with the Pawlak model and previous DTRS model through an example. This study indicates a new research direction of DTRS.

Keywords: Decision-theoretic rough sets · Delay cost · Misclassification cost · Teacher cost · Three-way decision

1 Introduction

Rough set theory was proposed by Pawlak [1] in the 1980s and has become a focus of data mining and knowledge discovery. Yao et al. [2] introduced Bayesian decision procedure to form Decision-Theoretic Rough Set (DTRS). Three-way decision originates from the researches on DTRS model and it has received much attention in decision-making and artificial intelligence communities. Zhang [3] built a three-way recommender system based on the misclassification cost and teacher cost. Yu [4] provided a three-way decision strategy for overlapping clustering based on the decision-theoretic rough set model. Its superiority compared with Pawlak ternary classification and probabilistic binary classification models were demonstrated in the paper [5].

In general, loss function analysis is an important theoretic topic on three-way decision. Zhao [6] proposed an optimal cost-sensitive granularization based on rough sets for variable costs. Li [7] investigated the relationship between risk preferences and loss functions. Liu [8] presented a Decision-theoretic rough sets with probabilistic distribution. Zhang [9] proposed a constructive covering-based

© Springer International Publishing Switzerland 2015
D. Ciucci et al. (Eds.): RSKT 2015, LNAI 9436, pp. 275–285, 2015.
DOI: 10.1007/978-3-319-25754-9_25

method to determine the thresholds. Jia [10] proposed a simulated annealing algorithm for learning thresholds.

However, in these extension DTRS models, a common content is to make deferment decision when precise decisions (positive and negative decisions) cannot be immediately decided. In many real-world scenarios, this defer decision is not reasonable. For example, considering a medical diagnose system, it is hard for a doctor to make an immediate decision due to knowledge limitation, but the patient is in emergency and hunger for treatment, what should the doctor do? Actually, he/she may consult a senior expert for solution at the cost of certain price.

This paper replaces action of deferment with consulting which stands for asking a teacher for correct classification without delay. The new action induces a new type of cast called teacher cost, which is more generalized and flexible than delay cost. With the change of loss function, the teacher-cost-sensitive DTRS model is formed. Then we focus on discussing the three-way decision rules by using probabilistic information based on this extension model. We provide a detail analysis via an example by comparison of three models at both micro level and macro level. Our results enhance the understanding of three-way decision and broaden the applications of DTRS models.

2   Three Types of Costs

Quality inspection is a very important step when a product is manufactured. It experiences a set of tests and is classified into qualified and unqualified. However, sometimes it is hard to classify a product by limited test results due to insufficient information. Decision maker may take different attitude and action towards the same case:

1. Make a random decision.
2. Wait and see.
3. Consult a teacher for correct solution.

Three issues arise in this scenario: how much is the misclassification? how much is the delay? how much is consulting a teacher? They are called misclassification cost, delay cost, teacher cost respectively. A hypothetical cost matrix is given by Table 1.

2.1   Misclassification Cost

The cost of classification errors [11] is the penalty of deciding that an object belongs to class J when its real class is K. Suppose there are two complementary states for a product: W_Q is a confirmed qualified state and W_U is a confirmed unqualified state, thus there are two actions regarding the two regions of the decision: a_P is to judge a product qualified, a_N is to judge a product unqualified. In many problems, the costs of all errors are not equal. Here, mistaking an unqualified product for a qualified one means penalty and has a cost of 17, while

Table 1. Three types of costs

(a) misclassification costs

	a_P (Qualified)	a_N (Unqualified)
W_Q (qualified)	0	15
W_U (unqualified)	17	0

(b) delay costs

	a_P (Qualified)	a_N (Unqualified)	a_B (Wait and Further test)
W_Q (qualified)	0	15	9
W_U (unqualified)	17	0	2

(c) teacher costs

	a_P (Qualified)	a_N (Unqualified)	a_T (Consult a teacher)
W_Q (qualified)	0	15	4
W_U (unqualified)	17	0	4

mistaking a qualified product for an unqualified one means loss on production and has a cost of 15. Note that costs may have any units. In this paper, costs represent dollars expended.

2.2 Delay Cost

Delay classification means to need further-investigation so it will result cost [11]. If there is not yet sufficient information to allow a decision, one would make a non-committed decision a_B, which means that a product needs additional tests for further observation. The loss of delay decision cost 9 and 2 for qualified product and unqualified one respectively. Here the costs for different category products are not assumed to be equal.

2.3 Teacher Cost

The teacher cost [11] is introduced which is derived from actively selecting cases for the teacher. A rational decision maker would calculate the expected cost of classifying the case by himself versus the cost of consulting a teacher in order to get the correct classification directly. Here we suppose that the cost of consulting a teacher for all products is equal.

3 Teacher-Cost-Sensitive DTRS Model

In DTRS model three types of decision rules are derived from the three regions. A positive rule makes a decision of acceptance, a negative rule makes a decision of rejection, and a boundary rule makes a decision of abstaining or delay. However, the delay or deferment is not a solution but a compromise. In this section, a new interpretation of rules in teacher-cost-sensitive DTRS model is introduced. One can make a whole another three-way decision: accept, consult and reject.

The three regions enable us to derive three types of decision rules, namely, positive rules for acceptance, boundary rules for consulting a teacher, and negative rules for rejection. Suppose the set of actions is given by $A = \{a_P, a_T, a_N\}$, where a_P denotes the action deciding $x \in POS(X)$, a_N means the action deciding $x \in NEG(X)$, a_T represents the action to consult a teacher by deciding $x \in BND(X)$.

The loss functions regarding the risk or cost of actions in different states are presented in Table 2. In the cost matrix, λ_{PP}, λ_{NP} denote the losses incurred for taking actions of a_P, a_N when an object belongs to X. Similarly, λ_{PN}, λ_{NN} denote the losses incurred for taking the same actions when the object belongs to $\overline{X}$. If it is not certain to decide an object belongs to X or $\overline{X}$, λ_{TT} denotes the loss incurred by consulting a teacher.

Table 2. Loss functions with a_P, a_T, a_N

	a_P	a_N	a_T
X	λ_{PP}	λ_{NP}	λ_{TT}
$\overline{X}$	λ_{PN}	λ_{NN}	λ_{TT}

Normally the cost for a right decision is less than that of boundary decision (consulting a teacher), and less than that of wrong decision, thus we have $\lambda_{PP} \leq \lambda_{TT} < \lambda_{NP}$ and $\lambda_{NN} \leq \lambda_{TT} < \lambda_{PN}$, the three parameters α, β and γ can be calculated as:

$$
\begin{aligned}
\alpha &= \frac{\lambda_{PN} - \lambda_{TT}}{\lambda_{PN} - \lambda_{PP}}, \\
\beta &= \frac{\lambda_{TT} - \lambda_{NN}}{(\lambda_{NP} - \lambda_{NN})}, \\
\gamma &= \frac{\lambda_{PN} - \lambda_{NN}}{(\lambda_{PN} - \lambda_{NN}) + (\lambda_{NP} - \lambda_{PP})}.
\end{aligned}
\tag{1}
$$

When $(\lambda_{PN} - \lambda_{TT})(\lambda_{NP} - \lambda_{TT}) < (\lambda_{TT} - \lambda_{PP})(\lambda_{TT} - \lambda_{NN})$, we have $0 \leq \beta < \gamma < \alpha \leq 1$. In this case, after tie-breaking, we obtained the following simplified rules and induce following decision rules:

(P) If $P(X|[x]) \geq \alpha$, decide $[x] \subseteq POS_\pi(X)$, make a acceptance decision.

(N)If $P(X|[x]) \leq \beta$, decide $[x] \subseteq NEG_\pi(X)$, make a rejection decision.

(B)If $\beta < P(X|[x]) < \alpha$, decide $[x] \subseteq BND_\pi(X)$, make a consulting decision.

The threshold γ is no longer needed.

As shown in Fig. 1, the unit interval $[0, 1]$ of probability values are divided into three regions, namely, the acceptance region $[\alpha, 1]$, the rejection region $[0, \beta]$, and the consulting region $[\beta, \alpha]$. Consider an acceptance decision characterized by the condition $P(X[x]) \geq \alpha$. The quantity $1 - \alpha$ presents the tolerance of

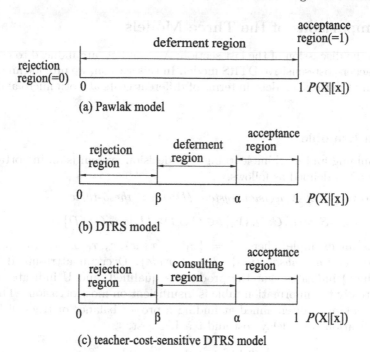

Fig. 1. Three classification model

the maximum error rate for incorrect acceptance. Similarly, β presents the tolerance of maximum error rate for incorrect rejection. One makes an acceptance or a rejection decision if the error rate is within acceptable levels of tolerance; Otherwise, one consults a teacher for correct solution.

In addition, the semantics differences of the three types of rules can be easily explained by their associated different costs:

$$
\begin{aligned}
positive\ rule &: conf * \lambda_{PP} + (1 - conf) * \lambda_{PN}, \\
boundary\ rule &: \lambda_{TT}, \\
negative\ rule &: conf * \lambda_{NP} + (1 - conf) * \lambda_{NN}.
\end{aligned}
\tag{2}
$$

where $conf = P(X|[x])$ for rule $Des([x]) \rightarrow \Lambda Des[X], \Lambda \in \{P, B, N\}$. In the special case where we assume zero cost for a correct classification, namely, $\lambda_{PP} = \lambda_{NN} = 0$. Costs associated with rules can be simplified to:

$$
\begin{aligned}
positive\ rule &: (1 - conf) * \lambda_{PN}, \\
boundary\ rule &: \lambda_{TT}, \\
negative\ rule &: conf * \lambda_{NP}.
\end{aligned}
\tag{3}
$$

They are much easier to understand in tradeoff between misclassification cost and teacher cost.

4   Comparisons of the Three Models

Through the discussion of the last section, λ_{TT} and a_T are induced to construct a new teacher-cost-sensitive DTRS model. In this section, we provide an example to compare the three models, in terms of different costs at both micro and macro levels [5].

4.1   An Example

In data mining and machine learning, the decision system is an important concept and often defined as follows.

Definition 1. *[12] A decision system (DS) S is the 5-tuple:*

$$S = (U, C, D, \{V_a | a \in C \cup D\}, \{I_a | a \in C \cup D\})\}. \tag{4}$$

Table 3 is an example where $U = \{x_1, x_2, x_3, x_4, x_5, x_6, x_7, x_8, x_9, x_{10}\}$, $C = \{Clay, Glaze, Waterabsorption\}$, $d = \{grade\}$. Decision attribute d denotes result: the Q indicates that the product is qualified, and U indicates unqualified. Actually the information table is insufficient on most occasions. Therefore, decision making is often aimed at finding a proper balance or tradeoff between misclassification cost, delay cost and teacher cost.

Table 3. A decision table

Chinaware	Clay	Glaze	Water absorption	Grade
x_1	high	high	good	Q
x_2	normal	high	bad	U
x_3	high	normal	good	Q
x_4	normal	high	bad	U
x_5	high	normal	good	U
x_6	high	high	good	Q
x_7	normal	high	bad	Q
x_8	normal	high	bad	U
x_9	high	high	good	Q
x_{10}	high	normal	good	Q

In Table 3, the set of condition attributes C determine an equivalence relation $\mathcal{R}$ [7]:

$$\mathcal{R} = \{(x, y) | I_a(x) = I_a(y), \forall a \in C\}$$

where $I_a(x)$ is the value of x on attribute a. A partition will be constructed with regard to equivalence relation $\mathcal{R}$:

$$U/\mathcal{R} = U/C = \{\{x_1, x_6, x_9\}, \{x_2, x_4, x_7, x_8\}, \{x_3, x_5, x_{10}\}$$
$$= \{[x_1]_{\mathcal{R}}, [x_2]_{\mathcal{R}}, [x_3]_{\mathcal{R}}\}$$

where $[x_i]_{\mathcal{R}}$ $(i = 1, 2, 3)$ is the equivalence class of x_i:

$$[x_i]_{\mathcal{R}} = \{y \in U | (x_i, y) \in \mathcal{R}\}$$

Similarly, the decision set $d = \{grade\}$ constructs a partition as: $U/d = \{\{x_1, x_3, x_6, x_7, x_9, x_{10}\}, \{x_2, x_4, x_5, x_8\}\} = \{X, \overline{X}\}$,

where $X = \{x_1, x_3, x_6, x_9\}$ represents the state Q and $\overline{X} = \{x_2, x_4, x_5, x_8\}$ represents the sate U.

Given description x, let $P(X|[x]_R)$ be the probability of an object in state Q given that the object is described by x.

In the following discussions, these conditional probabilities $P(X|[x]_R))$ are calculated by the following formula from the given information Table 3.

$$
\begin{aligned}
P(X|[x]) &= \frac{P(X) \cdot P([x]|X)}{P([x])} \\
&= \frac{P(X) \cdot P([x]|X)}{P(X) \cdot P([x]|X) + P(\overline{X}) \cdot P([x]|\overline{X})}
\end{aligned}
\tag{5}
$$

For each equivalence class in U/R, the conditional probability is calculated as follows:

$$
\begin{aligned}
P(X|[x_1]_{\mathcal{R}}) &= \frac{|[x_1]_{\mathcal{R}} \cap X|}{|[x_1]_{\mathcal{R}}|} \\
&= \frac{|\{x_1, x_6, x_9\} \cap \{x_1, x_3, x_6, x_7, x_9, x_{10}\}|}{|\{x_1, x_6, x_9\}|} \\
&= 1,
\end{aligned}
\tag{6}
$$

$$
\begin{aligned}
P(X|[x_2]_{\mathcal{R}}) &= \frac{|[x_2]_{\mathcal{R}} \cap X|}{|[x_2]_{\mathcal{R}}|} \\
&= \frac{|\{x_2, x_4, x_7, x_8\} \cap \{x_1, x_3, x_6, x_7, x_9, x_{10}\}|}{|\{x_2, x_4, x_7, x_8\}|} \\
&= 0.25,
\end{aligned}
\tag{7}
$$

$$
\begin{aligned}
P(X|[x_3]_{\mathcal{R}}) &= \frac{|[x_3]_{\mathcal{R}} \cap X|}{|[x_3]_{\mathcal{R}}|} \\
&= \frac{|\{x_3, x_5, x_{10}\} \cap \{x_1, x_3, x_6, x_7, x_9, x_{10}\}|}{|\{x_3, x_5, x_{10}\}|} \\
&= 0.67.
\end{aligned}
\tag{8}
$$

where $|\cdot|$ denotes the cardinality of a set. Then the decision table can be transformed to Table 4. Here, let X denote a subset of 10 objects. Giving the relationships between X and equivalence classes, where an integer in each cell of the second and third columns is the number of objects in the corresponding set. The number of cell of the last columns is a conditional probability, which has been obtained in Eqs. (6), (7) and (8).

The conditional probability $P(X|[x_i]_{\mathcal{R}})$ describes the probability of getting qualified under the description $[x_i]_{\mathcal{R}}$. The higher the conditional probability is, the more possible a produce is gotten qualified. Then the thresholds (α, β) are calculated.

$$\alpha = 0.63, \beta = 0.25, \gamma = 0.53.$$

According to these thresholds, for the Pawlak model and the DTRS model, X can be approximated, respectively as follows:

$$POS_{(1,0)}(X) = [x_1]_{\mathcal{R}},$$
$$BND_{(1,0)}(X) = [x_2]_{\mathcal{R}} \cup [x_3]_{\mathcal{R}},$$
$$NEG_{(1,0)}(X) = \emptyset.$$

$$POS_{(0.63,0.25)}(X) = [x_1]_{\mathcal{R}} \cup [x_3]_{\mathcal{R}},$$
$$BND_{(0.63,0.25)}(X) = \emptyset,$$
$$NEG_{(0.63,0.25)}(X) = [x_2]_{\mathcal{R}}.$$

In the same way, inserting the different cost values into Eq. (1), we obtain the following thresholds:

$$\alpha = 0.76, \beta = 0.27, \gamma = 0.53.$$

The X can be approximated as follows:

$$POS_{(0.76,0.27)}(X) = [x_1]_{\mathcal{R}},$$
$$BND_{(0.76,0.27)}(X) = [x_3]_{\mathcal{R}},$$
$$NEG_{(0.76,0.27)}(X) = [x_2]_{\mathcal{R}}.$$

4.2   Micro-Level Analysis

A micro-level analysis concerns decisions made for a particular equivalence class [13]. For the equivalence class $[x_i]_{\mathcal{R}}$, the Pawlak model, DTRS model and teacher-cost-sensitive DTRS model, one can make decisions by putting $[x_i]_{\mathcal{R}}$ into different region at different associated cost, which is denoted by $cost^p$, $cost^d$, $cost^t$ respectively (Table 4).

In Pawlak model the associated cost is calculated as follows:

$$cost^p_{[x_1]_{\mathcal{R}}} = \lambda_{PP} p(x|[x_1]) + \lambda_{PN} p(\overline{x}|[x_1])) = 0,$$
$$cost^p_{[x_2]_{\mathcal{R}}} = \lambda_{BP} p(x|[x_1]) + \lambda_{BN} p(\overline{x}|[x_1])) = 9*0.25 + 2*0.75 = 3.75,$$
$$cost^p_{[x_3]_{\mathcal{R}}} = \lambda_{BP} p(x|[x_1]) + \lambda_{BN} p(\overline{x}|[x_1])) = 9*0.67 + 2*0.33 = 6.69.$$

Table 4. A transformed decision table

	$X \cap Q$	$X \cap U$	$P(X|[x_i]_R)$	$P(\overline{X}|[x_i]_R)$
$[x_1]_{\mathcal{R}}$	3	0	1	0
$[x_2]_{\mathcal{R}}$	1	3	0.25	0.75
$[x_3]_{\mathcal{R}}$	2	1	0.67	0.33

In DTRS model the associated cost is calculated as follows:

$$cost^d_{[x_1]_\mathcal{R}} = \lambda_{PP} p(x|[x_1]) + \lambda_{PN} p(\overline{x}|[x_1])) = 0,$$

$$cost^d_{[x_2]_\mathcal{R}} = \lambda_{NP} p(x|[x_2]) + \lambda_{NN} p(\overline{x}|[x_2])) = 9*0.25 + 2*0.75 = 3.75,$$

$$cost^d_{[x_3]_\mathcal{R}} = \lambda_{PP} p(x|[x_3]) + \lambda_{PN} p(\overline{x}|[x_3])) = 17*0.33 = 5.61.$$

In teacher-cost-sensitive DTRS model the associated cost is given by:

$$cost^t_{[x_1]_\mathcal{R}} = \lambda_{PP} p(x|[x_1]) + \lambda_{PN} p(\overline{x}|[x_1])) = 0,$$

$$cost^t_{[x_2]_\mathcal{R}} = \lambda_{NP} p(x|[x_2]) + \lambda_{NN} p(\overline{x}|[x_2])) = 15*0.25 = 3.75,$$

$$cost^t_{[x_3]_\mathcal{R}} = \lambda_{TT}(p(x|[x_3]) + p(\overline{x}|[x_3]) = \lambda_{TT} = 4.$$

Table 5 summarizes the decisions and associated cost for all equivalence classes:

Table 5. Summarized decisions and associated cost table

	Pawlak model	DTRS model	teacher-cost-sensitive DTRS model
$[x_1]_\mathcal{R}$	acceptance: 0.00	acceptance: 0.00	acceptance: 0.00
$[x_2]_\mathcal{R}$	deferment: 3.75	rejection: 3.75	rejection: 3.75
$[x_3]_\mathcal{R}$	deferment: 6.69	acceptance: 5.61	consulting a teacher: 4

Thus, it can be seen that the teacher-cost-sensitive DTRS model is always the same as the lower cost of the Pawlak model and DTRS model under a certain condition (Table 5).

4.3 Macro-Level Analysis

At the macro level, one considers the collective performance of the set of rules induced by all equivalence classes [13]. Based on [5], the results of Pawlak classification are summarized by Table 6.

Finally, according to Table 6, we have

$$Cost^p = \frac{1}{10}\left[\lambda_{PP} n^P_{PP} + (\lambda_{BP} n^P_{BP} + \lambda_{BN} n^P_{BN}) + \lambda_{NN} n^P_{NN})\right]$$

$$= \frac{1}{10}(9*3 + 2*4) = 0.35,$$

$$Cost^d = \frac{1}{10}\left[(\lambda_{PP} n^d_{PP} + \lambda_{PN} n^d_{PN}) + (\lambda_{BP} n^d_{BP} + \lambda_{BN} n^d_{BN}) + (\lambda_{NP} n^d_{NP} + \lambda_{NN} n^d_{NN})\right]$$

$$= \frac{1}{10}(17*1 + 15*1) = 0.32,$$

$$Cost^t = \frac{1}{10}\left[(\lambda_{PP} n^t_{PP} + \lambda_{PN} n^t_{PN}) + \lambda_{TT}(n^d_{TP} + n^d_{TN}) + (\lambda_{NP} n^d_{NP} + \lambda_{NN} n^d_{NN})\right]$$

$$= \frac{1}{10}(0*17 + 4*3 + 15*1) = 0.27.$$

Again, under a certain condition, the total cost of teacher-cost-sensitive model is lower than both the cost of Pawlak model and the cost of the DTRS model (Table 6).

Table 6. The result of three different rough set models

(a) Pawlak model

	X(P):postive	$\overline{X}(P)$: negative	Total
a_P : accept	n_{PP}^p : 3	n_{PN}^p : 0	3
a_B : defer	n_{BP}^p : 3	n_{BN}^p : 4	7
a_N : reject	n_{NP}^p : 0	n_{NN}^p : 0	0
Total	6	4	10

(b) DTRS model

	X(P):postive	$\overline{X}(P)$: negative	Total
a_P : accept	n_{PP}^d : 5	n_{PN}^d : 1	6
a_B : defer	n_{BP}^d : 0	n_{BN}^d : 0	0
a_N : reject	n_{NP}^d : 1	n_{NN}^d : 3	4
Total	6	4	10

(c) teacher-cost-sensitive DTRS modle

	X(P):postive	$\overline{X}(P)$: negative	Total
a_P : accept	n_{PP}^d : 3	n_{PN}^d : 0	3
a_B : consult	n_{BP}^d : 2	n_{BN}^d : 1	3
a_N : reject	n_{NP}^d : 1	n_{NN}^d : 3	4
Total	6	4	10

5   Conclusions

By considering teacher cost in real decision procedures, the deferment decision in three-way decision can be replaced by the action of consulting a teacher. In the teacher-cost-sensitive DTRS model, we derive positive, negative, and boundary rules from corresponding regions. After comparison with the Pawlak model and classical DTRS model, the results denote that under a certain condition, the teacher-cost-sensitive model can get better classification at lower total cost than other two model. The teacher-cost-based analysis of three-way decision brings the rough set theory closer to real world applications. There is still an important problem needed for further investigation. In this new model, the boundary decision presents consulting a teacher for correct classification. However, with the available information increasing or obtaining an answer, the previous boundary decisions may be converted to positive or negative decisions. How to describe such dynamic sequential decision process is meaningful for future work.

Acknowledgements. This work is in part supported by National Science Foundation of China under Grant No. 61379089.

References

1. Pawlak, Z.: Rough sets. Int. J. Comput. Inf. Sci. **11**, 341–356 (1982)
2. Yao, Y.Y., Wong, S.: A decision theoretic framework for approximating concepts. Int. J. Man-Mach. Stud. **37**, 793–809 (1992)
3. Zhang, H.R., Min, F.: Three-way recommender systems based on random forests. Knowledge-Based Systems (2015). doi:10.1016/j.knosys.2015.06.019
4. Yu, H., Wang, Y., Jiao, P.: A three-way decisions approach to density-based overlapping clustering. In: Peters, J.F., Skowron, A., Li, T., Yang, Y., Yao, J.T., Nguyen, H.S. (eds.) Transactions on Rough Sets XVIII. LNCS, vol. 8449, pp. 92–109. Springer, Heidelberg (2014)
5. Yao, Y.: The superiority of three-way decisions in probabilistic rough set models. Inf. Sci. **181**, 1080–1096 (2011)
6. Zhao, H., Zhu, W.: Optimal cost-sensitive granularization based on rough sets for variable costs. Knowl.-Based Syst. **65**, 72–82 (2014)
7. Li, H.X., Zhou, X.Z.: Risk decision making based on decision-theoretic rough set: a three-way view decision model. Int. J. Comput. Intel. Syst. **4**(1), 1–11 (2011)
8. Liu, D., Li, T., Liang, D.: Decision-theoretic rough sets with probabilistic distribution. In: Li, T., Nguyen, H.S., Wang, G., Grzymala-Busse, J., Janicki, R., Hassanien, A.E., Yu, H. (eds.) RSKT 2012. LNCS, vol. 7414, pp. 389–398. Springer, Heidelberg (2012)
9. Zhang, Y., Xing, H., Zou, H., Zhao, S., Wang, X.: A three-way decisions model based on constructive covering algorithm. In: Lingras, P., Wolski, M., Cornelis, C., Mitra, S., Wasilewski, P. (eds.) RSKT 2013. LNCS, vol. 8171, pp. 346–353. Springer, Heidelberg (2013)
10. Yi, J.X., Lin, S.: A simulated annealing algorithm for learning thresholds in three-way decision-theoretic rough set model. J. Chin. Comput. Syst. **11**, 2603–2606 (2013). (in chinese)
11. Turney, P.D.: Types of cost in inductive concept learning. In: Proceedings of the Workshop on Cost-Sensitive Learning at the 17th ICML, pp. 1–7 (2000)
12. Min, F., He, H.P., Qian, Y.H., Zhu, W.: Test-cost-sensitive attribute reduction. Inf. Sci. **181**, 4928–4942 (2011)
13. Yao, Y.Y., Zhou, B.: Micro and macro evaluation of classification rules. In: 7th IEEE International Conference on Cognitive Informatics, ICCI 2008, pp. 441–448 (2008)

A Three-Way Decision Making Approach to Malware Analysis

Mohammad Nauman[1], Nouman Azam[1]([✉]), and JingTao Yao[2]

[1] National University of Computer and Emerging Sciences, Peshawar, Pakistan
{mohammad.nauman,nouman.azam}@nu.edu.pk
[2] Department of Computer Science, University of Regina, Regina S4S 0A2, Canada
jtyao@cs.uregina.ca

Abstract. Malware analysis techniques generally classify software behaviors as malicious (i.e., harmful) or benign (i.e., not harmful). Due to ambiguous nature of application behavior, there are cases where it may not be possible to confidently reach two-way conclusions. This may result in higher classification errors which in turn affect users trust on malware analysis outcomes. In this paper, we investigate a three-way decision making approach based on probabilistic rough set models, such as, information-theoretic rough sets and game-theoretic rough sets, for malware analysis. The essential idea is to add a third option of deferment or delaying a decision whenever the available information is not sufficient to reach certain conclusions. We demonstrate the applicability of the proposed approach with an example from system call sequences of a vulnerable Linux application.

1 Introduction

Recent years have seen an increasing rate of adoption in digital means of storing and sharing data. Despite many advantages and benefits, this also leads to several challenges and issues [3]. One such issue is the protection and security of sensitive data stored on digital devices [14]. Malware analysis techniques aim to develop and provide solutions for detecting illegal accesses to system resources, such as, digital data [5].

Several malware analysis techniques are in use today with reasonable classification performance [8]. The performance of these techniques are dependent on the ability to classify an application behavior as malicious (i.e., having harmful effects) or benign (i.e., having no harmful effects) [12]. There are situations where the application behavior is not convincing enough for deciding it as malicious or benign. Such hard to detect cases lead to errors having severe consequences [2]. For instance, consider a scenario where the malware analysis is accurate most of the time but occasionally may mistakenly raise a red flag on a benign application, warning the user of some malicious activity. Repeated cases of such misclassification may lead to a lack of trust on the analysis engine on part of the end user [2]. This may result in situations, where the users may ignore even correct results of malware analysis thus putting the integrity of user's system and

© Springer International Publishing Switzerland 2015
D. Ciucci et al. (Eds.): RSKT 2015, LNAI 9436, pp. 286–298, 2015.
DOI: 10.1007/978-3-319-25754-9_26

data at high risk. To address this issue, we propose a three-way decision making approach to malware analysis using probabilistic rough sets.

Rough set theory is a mathematical tool to analyze data characterized by uncertain and incomplete information [13]. It can be used to generate three regions while approximating a set (representing a certain concept). These three regions are called positive, negative and boundary regions which are sometimes interpreted and referred to as three-way decisions (or classification) in the form of acceptance, rejection and deferment, respectively [19]. We consider three-way decisions based on a generalized rough set model called probabilistic rough set model [18]. The three regions and the implied three-way decisions in the probabilistic rough set model are defined and controlled by a pair of thresholds [18]. Based on a specific interpretation of these thresholds, there are different forms of probabilistic rough set models. We consider two such models, i.e., information-theoretic rough sets (ITRS) [4] and game-theoretic rough sets (GTRS) [9] to examine the application of probabilistic rough sets in the context of malware analysis.

In this article, we extend a well known existing approach to malware analysis called sequence time-delay embedding by incorporating probabilistic rough sets based three-way decisions [7]. The target architecture of the malware analysis system is also outlined. The essential change in the time-delay embedding approach is the ability to generate a third possible result in the form of deferment. This enables the systems based on our approach to avoid hard to classify cases whenever the information is not sufficient to reach accurate decisions. A demonstrative example based on a vulnerable Linux applications is provided for generating three-way decisions using probabilistic rough set models, i.e., ITRS and GTRS. The results advocate for the use of three-way decisions in classifying applications behavior.

2   Malware Analysis

Malware-analysis deals with the detection of malicious operations or behavior by a software running on a system [5]. There are two generally used approaches for deciding upon the maliciousness of a software, i.e., signature-based approaches and heuristics-based approaches [5]. Signature-based approaches match the hash of an executable to that of known malware. In case of a positive hit, it is marked as being dangerous. This approach is, however, not feasible in the face of frequently changing executables and appearance of newer malware samples. Heuristics-based approaches are comparatively more dynamic in the sense that during the runtime, the behavior of an application is measured instead of its static executable [5].

A well-known heuristics-based approach to behavior measurement called sequence time-delay embedding was presented in [7]. This approach is based on the behavior of an application in terms of the system calls it generates. The rationale behind this approach is that any sensitive operation performed by an application has to be mediated by the operating system. Since applications only

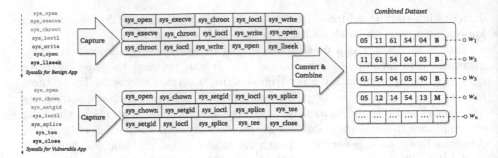

Fig. 1. System call based behaviors

communicate with the operating system through system calls [15], their behavior can be specified in terms of the system calls they generate. We further elaborate the details of this approach as it is being adopted in this study.

The sequence time-delay embedding approach is described visually in Fig. 1. The target application generates different system calls during its lifetime. Any modification or vulnerability exploit in the application will lead to a different sequence of system calls at the point of the vulnerability regardless of how it was exploited. It is possible to separately intercept and record these sequences through operating system constructs such as the Linux Security Modules [16] or the **strace** tool. These sequences are represented using windows of a fixed size, represented as n, which is set to $n = 5$ in the figure. The first 5 calls are captured in the first window. When the sixth call is executed, calls 2 to 5 are shifted left leaving a blank slot at the end of the second window where the sixth call is recorded. This leads to the notion of sliding windows that capture all n-sized unique sequences of system calls made by the application.

Once the sliding windows for malicious and benign applications are captured, they are associated with their respective classes and the resulting dataset is used as input to a classification model. Several models have been proposed in the past for this purpose. Generally speaking, all of these approaches are based on two-way classification decisions. This means that we are forced to identify an application as malicious or benign. However, it may not be always possible to reach such crisp conclusions due to unclear nature of application behavior.

In particular, there are many behaviors which are common in malicious and benign applications, i.e., a malicious application may behave sometimes like a benign application. The restriction of always identifying the application behavior as malicious or benign leads to many errors which significantly affect the user's trust on the system [17]. In some cases, this results in a negative feedback cycle where the user distrusts even accurate decisions. We propose a three-way decision making approach where hard to decide cases are deferred in the hope that useful information may evolve in future, which will make the decision making more evident and accurate.

3   Three-Way Decisions with Probabilistic Rough Sets

Three-way decisions are common in daily life and are being employed in different fields [11]. We focus on three-way decisions using probabilistic rough sets [18].

Probabilistic rough sets are an extension of Pawlak rough sets and are characterized by two main ingredients, i.e., conditional probability between a set and a concept and a pair of thresholds. The probabilistic rough sets make use of these two ingredients to define the three rough set regions as follows [18].

$$\text{POS}_{(\alpha,\beta)}(C) = \{x \in U | P(C|[x]) \geq \alpha\}, \tag{1}$$
$$\text{NEG}_{(\alpha,\beta)}(C) = \{x \in U | P(C|[x]) \leq \beta\}, \tag{2}$$
$$\text{BND}_{(\alpha,\beta)}(C) = \{x \in U | \beta < P(C|[x]) < \alpha\}, \tag{3}$$

where U is the universe of objects and $P(C|[x])$ is the evaluation of an object x to be in C, provided that $x \in [x]$. Moreover, $\text{POS}_{(\alpha,\beta)}(C), \text{NEG}_{(\alpha,\beta)}(C)$ and $\text{BND}_{(\alpha,\beta)}(C)$ are referred to as positive, negative and boundary regions, respectively. From decision making perspective, these three regions are frequently referred to as regions of acceptance, rejection and deferment decisions [18]. The determination of (α, β), which controls the three regions, is a fundamental challenge in the application of probabilistic rough sets [20]. We focus on two approaches in this regard, i.e., information-theoretic rough sets and game-theoretic rough sets. Each of these approaches provide a mechanism for determining thresholds which are used to obtain three-way decisions based on the three regions (defined in Eqs. (1)–(3)).

3.1   Game-Theoretic Rough Sets

Game-theoretic rough sets (GTRS) determine the thresholds by considering a tradeoff between multiple criteria [9]. A game is formulated for this purpose. A typical game may be defined as a tuple $\{P, S, u\}$ [10], where:

- P is a finite set of n players,
- $S = S_1 \times ... \times S_n$, where S_i is a finite strategies set for player i. Each $s = (s_1, s_2, ..., s_n) \in S$ is called a strategy profile where each player i plays strategy s_i, and
- $u = (u_1, ..., u_n)$ where $u_i : S \longmapsto \Re$ is a real-valued utility or payoff function for player i.

Nash game solution (also called Nash equilibrium) is commonly used in GTRS to determine possible game solutions. In order to define Nash solution, consider $s_{-i} = (s_1, s_2, ..., s_{i-1}, s_{i+1}, ..., s_n)$ as a strategy profile without player i strategy which means that $(s_1, s_2, ..., s_n)$ may be written as (s_i, s_{-i}). The strategy profile $(s_1, s_2, ..., s_n)$ is defined as a Nash equilibrium, if for every players i, s_i is the best response to s_{-i}. This is expressed mathematically as [10],

$$\forall i, \forall s_i' \in S_i, \quad u_i(s_i, s_{-i}) \geq u_i(s_i', s_{-i}), \quad \text{where } (s_i' \neq s_i) \tag{4}$$

Table 1. Payoff table for a two-player GTRS based game

		c_2		
		s_1	s_2	$\cdots$
c_1	s_1	$u_{c_1}(s_1, s_1),\ u_{c_2}(s_1, s_1)$	$u_{c_1}(s_1, s_2), u_{c_2}(s_1, s_2)$	$\cdots$
	s_2	$u_{c_1}(s_2, s_1),\ u_{c_2}(s_2, s_1)$	$u_{c_1}(s_2, s_2), u_{c_2}(s_2, s_2)$	$\cdots$
	$\cdots$	$\cdots$	$\cdots$	$\cdots$

Equation (4) intuitively suggests that Nash equilibrium constitutes a strategy profile in which no player would want to change his strategy, provided he has the knowledge of other players strategies.

GTRS formulates a game based on the above game description. The players are selected for highlighting different criteria or aspects of rough set based decision making, such as, accuracy or applicability of decision rules [1]. Suitable measures are used to evaluate these criteria. The strategies are considered as either direct or indirect modifications in the thresholds. Each criterion is affected in a different way based on different strategies within the game. The aim of the game is to find an acceptable solution based on the considered criteria which will be used to determine effective thresholds.

Table 1 represents a two-player game in GTRS. The players in this game are denoted as criteria c_1 and c_2, respectively. Each table cell corresponds to a particular strategy profile. For instance, the top right table cell corresponds to a strategy profile (s_1, s_1). Each table cell also corresponds to a certain threshold pair which is calculated based on the respective strategies. The values in the table cell contains a pair of utility functions which are calculated based on the respective strategy profile. The game solution, such as, the Nash equilibrium, is utilized to determine a possible game solution in terms of a strategy profile. The corresponding thresholds to the game solutions are the determined thresholds which are then used in the probabilistic rough set model to obtain the three regions and the implied three-way decisions.

3.2 Information-Theoretic Rough Sets

Information-theoretic rough sets determine the thresholds by minimizing the overall uncertainty due to probabilistic rough set regions. Specifically, it measures the uncertainty in the partition based on a concept C, i.e., $\pi_C = \{C, C^c\}$ with respect to the three probabilistic rough set regions, which are given as, [4]

$$
\begin{aligned}
H(\pi_C | \text{POS}_{(\alpha,\beta)}(C)) = {}& -P(C|\text{POS}_{(\alpha,\beta)}(C)) \log P(C|\text{POS}_{(\alpha,\beta)}(C)) \\
& -P(C^c|\text{POS}_{(\alpha,\beta)}(C)) \log P(C^c|\text{POS}_{(\alpha,\beta)}(C)), \quad (5)
\end{aligned}
$$

$$
\begin{aligned}
H(\pi_C | \text{NEG}_{(\alpha,\beta)}(C)) = {}& -P(C|\text{NEG}_{(\alpha,\beta)}(C)) \log P(C|\text{NEG}_{(\alpha,\beta)}(C)) \\
& -P(C^c|\text{NEG}_{(\alpha,\beta)}(C)) \log P(C^c|\text{NEG}_{(\alpha,\beta)}(C)), \quad (6)
\end{aligned}
$$

$$
\begin{aligned}
H(\pi_C | \text{BND}_{(\alpha,\beta)}(C)) = {}& -P(C|\text{BND}_{(\alpha,\beta)}(C)) \log P(C|\text{BND}_{(\alpha,\beta)}(C)) \\
& -P(C^c|\text{BND}_{(\alpha,\beta)}(C)) \log P(C^c|\text{BND}_{(\alpha,\beta)}(C)). \quad (7)
\end{aligned}
$$

The conditional probabilities in these equations, say, $P(C|\text{POS}_{(\alpha,\beta)}(C))$ is computed as, $P(C|\text{POS}_{(\alpha,\beta)}(C)) = \frac{|C \cap \text{POS}_{(\alpha,\beta)}(C)|}{|\text{POS}_{(\alpha,\beta)}(C)|}$. Conditional probabilities for other regions are similarly obtained.

The overall uncertainty is computed as an average uncertainty of the regions [4], i.e.,

$$
\begin{aligned}
H(\pi_C|\pi_{(\alpha,\beta)}) &= P(\text{POS}_{(\alpha,\beta)}(C))H(\pi_C|\text{POS}_{(\alpha,\beta)}(C)) \\
&\quad + P(\text{NEG}_{(\alpha,\beta)}(C))H(\pi_C|\text{NEG}_{(\alpha,\beta)}(C)) \\
&\quad + P(\text{BND}_{(\alpha,\beta)}(C))H(\pi_C|\text{BND}_{(\alpha,\beta)}(C)).
\end{aligned} \tag{8}
$$

Equation (8) was reformulated in a more readable form in [1]. Considering $\Delta_P(\alpha,\beta)$, $\Delta_N(\alpha,\beta)$ and $\Delta_B(\alpha,\beta)$ as the uncertainties of the positive, negative and boundary regions, respectively, i.e.,

$$\Delta_P(\alpha,\beta) = P(\text{POS}_{(\alpha,\beta)}(C))H(\pi_C|\text{POS}_{(\alpha,\beta)}(C)), \tag{9}$$

$$\Delta_N(\alpha,\beta) = P(\text{NEG}_{(\alpha,\beta)}(C))H(\pi_C|\text{NEG}_{(\alpha,\beta)}(C)), \tag{10}$$

$$\Delta_B(\alpha,\beta) = P(\text{BND}_{(\alpha,\beta)}(C))H(\pi_C|\text{BND}_{(\alpha,\beta)}(C)). \tag{11}$$

Using Eqs. (9)–(11), Eq. (8) is reexpressed as,

$$\Delta(\alpha,\beta) = \Delta_P(\alpha,\beta) + \Delta_N(\alpha,\beta) + \Delta_B(\alpha,\beta), \tag{12}$$

which represents the overall uncertainty with respect to a particular (α,β) threshold pair. The problem of finding optimum thresholds may be realized as the selection of (α,β) thresholds that minimizes Eq. (12).

4   A Three-Way Approach for Malware Analysis

In this section, we describe how three-way decision making can be used to augment malware analysis techniques. For the sake of demonstration, we extend the basic system call based behavior analysis technique presented in Sect. 2. It should however be noted that the approach is extensible to other analysis techniques.

In order to incorporate three-way decisions, we consider an extension of the system call mediation modules. For our proof-of-concept, we have chosen the Linux kernel due to its large user base and open source nature. We present the changes visually through Fig. 2. In Linux, whenever a sensitive system call is generated, it goes through several checks. The first one is the Discretionary Access Control (DAC) specified by the mode bits associated with specific resources. If this check is successful, control is passed to Linux Security Modules (LSM). There might be more than one LSMs in a single system. The Linux system also supports module stacking. Stacking means that each module will be called one after the other and if one of the modules denies the execution of the system call, the final result will be denied. How module stacking will be utilized is addressed later in the same section.

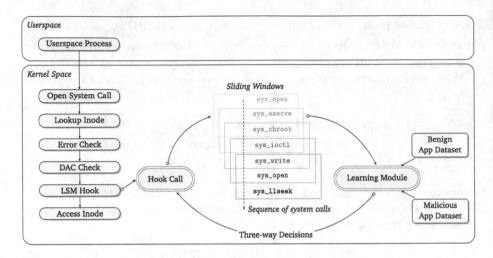

Fig. 2. Target architecture for Three-way decisions in malware analysis

A new LSM called System Call Behavior Monitor (SCBM) is created for implementing sliding windows. The SCBM captures each system call and based on the system call sequence, generates the sliding windows. In Fig. 2, the *hook call* component captures these calls and computes the sliding windows. The size of the sliding windows is configurable through the `menuconfig` utility during kernel compilation which remains fixed at runtime. The module keeps track of unique system calls on a per-application basis. (These windows can be seen in the `scbm` directory in the `securityfs` virtual filesystem at runtime). During each system call, the *learning module* within the SCBM will generate a three-way decision in the form of grant, deny or deferment. This decision is reached by utilizing previously collected data for the application identified as malicious and benign.

The three decisions are interpreted as follows. In case of a grant decision, access to the system call is permitted and the calling application proceeds normally. In case of a deny decision, a permission is denied and the result is sent to the application where it can be handled. The advantage of three-way decision is the addition of third kind of decision called deferment. For such decisions, we envision a modification to the kernel security module stacking to support true stack-based calling convention. In this approach, LSMs would be arranged in an ordered set instead of a bag arrangement currently in place. Each module would be called only if the preceding module resulted in a deferment. This would allow system administrators to have more efficient behavior measurement techniques where the most obvious malwares would be quickly recognized by the most efficient malware analysis technique. If the basic module fails to classify an application's behavior, the system would fall through to the next module which would be able to apply deeper analysis techniques. This modification to module stacking semantics forms part of our future work.

5   A Demonstrative Example

For our demonstrative example, we use the system call sequences available in the UNM dataset for the `xlock` application (available at http://www.cs.unm.edu/~immsec/data/xlock.html) [6]. The traces corresponding to malicious behavior are from the system calls of `xlock` exploited with the `CA-97.13.xlock` vulnerability. They are also retrieved from the UNM dataset. The system call sequences are converted to corresponding sliding windows using a custom bash script. Each of these windows are represented as rows in Table 2 indicated by $w_1, w_2, ..., w_{32}$. The sequence of system calls in each window are represented by $s_1, s_2, ..., s_5$, i.e., the columns of the Table 2. The values of the columns are the system call numbers as defined in the operating system. The final column of Table 2 indicate the identified behavior as malicious or benign represented as M and B, respectively.

Table 2 is essentially an information table since the set of objects (perceived as windows) are described by a set of attributes (considered as system calls). The concept of interest in this case is to determine the malicious behavior, i.e., Behaviour = M. We are unable to exactly specify this concept based on the equivalence classes $X_1, ..., X_9$. For instance, we are unable to tell whether or not X_3 belongs to this concept since 3 behaviors in X_3 are identified as malicious and the remaining 1 is identified as benign. We approximate this concept in the probabilistic rough set model. For this purpose, we calculated the conditional probability between an equivalence class X_i and a concept C, i.e., $P(C|X_i)$. Since the concept of interest in this case is Behaviour = M, the conditional probability is given by,

Table 2. System call sequences and their corresponding identification

Window	s_1	s_2	s_3	s_4	s_5	Behavior	Window	s_1	s_2	s_3	s_4	s_5	Behavior
w_1	106	106	106	106	106	M	w_2	125	5	5	3	90	B
w_3	125	5	5	3	90	M	w_4	106	5	90	6	5	B
w_5	106	106	106	106	106	M	w_6	125	5	5	3	90	M
w_7	3	90	90	90	6	B	w_8	3	90	90	90	6	M
w_9	106	5	90	6	5	M	w_{10}	125	5	5	3	90	M
w_{11}	125	5	5	3	90	M	w_{12}	5	108	3	19	6	B
w_{13}	108	3	19	6	33	B	w_{14}	108	3	19	6	33	M
w_{15}	3	90	90	90	6	M	w_{16}	3	90	90	90	6	M
w_{17}	106	5	90	6	5	M	w_{18}	3	6	5	108	3	B
w_{19}	5	108	3	19	6	B	w_{20}	5	108	3	19	6	M
w_{21}	125	5	5	3	90	B	w_{22}	45	45	5	108	45	B
w_{23}	45	45	5	108	45	M	w_{24}	3	6	5	108	3	B
w_{25}	3	6	5	108	3	B	w_{26}	3	6	5	108	3	M
w_{27}	125	5	5	3	90	B	w_{28}	6	5	108	3	19	B
w_{29}	3	6	5	108	3	M	w_{30}	45	45	5	108	45	B
w_{31}	5	108	3	19	6	B	w_{32}	6	5	108	3	19	B

Table 3. Equivalence classes based on Table 2

$X_1 = \{w_1, w_5\}$	$X_2 = \{w_7, w_8, w_{15}, w_{16}\}$
$X_3 = \{w_4, w_9, w_{17}\}$	$X_4 = \{w_2, w_3, w_6, w_{10}, w_{11}, w_{21}, w_{27}\}$
$X_5 = \{w_{13}, w_{14}\}$	$X_6 = \{w_{18}, w_{24}, w_{25}, w_{26}, w_{29}\}$
$X_7 = \{w_{22}, w_{23}, w_{30}\}$	$X_8 = \{w_{12}, w_{19}, w_{20}, w_{31}\}$
$X_9 = \{w_{28}, w_{32}\}$	

$$P(C|X_i) = P(\text{Behaviour} = \text{M}|X_i) = \frac{|\text{Behaviour} = \text{M} \cap X_i|}{|X_i|}. \tag{13}$$

The conditional probabilities of equivalence classes $X_1, ..., X_9$ based on Eq. (13) are calculated as 1.0, 0.75, 0.67, 0.57, 0.5, 0.4, 0.33 0.25 and 0.0, respectively. These conditional probabilities represent the level of agreement between similar patterns to be representative of malicious behavior. The probability of an equivalence class X_i is determined as $P(X_i) = |X_i|/|U|$ which means that the probability of X_1 is $|X_1|/|U| = 2/32 = 0.0625$. The probabilities of other equivalence classes $X_2, ..., X_9$ are similarly calculated as 0.125, 0.09375, 0.21875, 0.0625, 0.15625, 0.09375, 0.125 and 0.0625, respectively. In order to demonstrate the application of probabilistic rough set model for malware analysis, we consider two probabilistic rough set models, namely, ITRS and GTRS (explained in Sect. 3) for classifying application behaviours (Table 2).

5.1 Three-Way Decisions Based on Game-Theoretic Rough Sets

Game-theoretic rough sets provide three-way decisions by employing a game mechanism as outlined in Sect. 3.1. We consider a game in GTRS discussed in [1]. The players in the game are the measures of accuracy and generality. Each player can choose from three possible strategies, i.e., $s_1 = \alpha{\downarrow} = 25\,\%$ decrease in α, $s_2 = \beta{\uparrow} = 20\,\%$ increase in β and $s_3 = \alpha{\downarrow}\beta{\uparrow} = 20\,\%$ decrease in α and $20\,\%$ increase in β. We omit the details of the game for the sake of briefness. Interested reader is referred to reference [1].

$$Accuracy(\alpha, \beta) = \frac{|(\text{POS}_{(\alpha,\beta)}(C) \cap C) \bigcup (\text{NEG}_{(\alpha,\beta)}(C) \cap C^c)|}{|\text{POS}_{(\alpha,\beta)}(C) \bigcup \text{NEG}_{(\alpha,\beta)}(C)|}, \tag{14}$$

$$Generality(\alpha, \beta) = \frac{|\text{POS}_{(\alpha,\beta)}(C) \bigcup \text{NEG}_{(\alpha,\beta)}(C)|}{|U|}, \tag{15}$$

The accuracy measures the relative number of correct classification decisions for objects compared to the total classification decisions. The generality measures the number of objects for whom classification decisions can be made compared to all objects. In order to calculate these measures based on a certain thresholds, for instance, $(\alpha, \beta) = (0.8, 0.4)$, we first need to determine the three regions. The three regions in this case are given by $POS_{(0.8,0.4)}(C) = \bigcup\{X_1\}$, $BND_{(0.8,0.4)}(C) = \bigcup\{X_2, X_3, X_4, X_5\}$, and $NEG_{(0.8,0.4)}(C) = \{X_6, X_7, X_8, X_9\}$. The properties of accuracy and generality are calculated as,

$$Accuracy(\alpha, \beta) = \frac{|(X_1 \cap C) \bigcup ((X_6 \bigcup X_7 \bigcup X_8 \bigcup X_9) \cap C^c)|}{|X_1 \bigcup X_6 \bigcup X_7 \bigcup X_8 \bigcup X_9|}$$

$$= \frac{|\{w_1, w_5, w_{18}, w_{24}, w_{25}, w_{22}, w_{30}, w_{12}, w_{19}, w_{31}, w_{28}, w_{32}\}|}{|\{w_1, w_5, w_{18}, w_{24}, w_{25}, w_{26}, w_{29}, w_{22}, w_{23}, w_{30}, w_{12}, w_{19}, w_{20}, w_{31}, w_{28}, w_{32}\}|} = 0.75 \quad (16)$$

$$Generality(\alpha, \beta) = \frac{|(X_1 \bigcup X_6 \bigcup X_7 \bigcup X_8 \bigcup X_9)|}{|U|}$$

$$= \frac{|\{w_1, w_5, w_{18}, w_{24}, w_{25}, w_{26}, w_{29}, w_{22}, w_{23}, w_{30}, w_{12}, w_{19}, w_{20}, w_{31}, w_{28}, w_{32}\}|}{|\{w_1, w_2, ..., w_{32}\}|} = 0.5 \quad (17)$$

Table 4 represents the payoff table corresponding to the game which is determined based on the data in Table 2. The cell with bold values, i.e., **(0.8191,0.3438)**, represents the game solution based on Eq. (4). The strategies corresponding to this solution are (s_1, s_3) for the players which leads to thresholds $(\alpha, \beta) = (0.6, 0.2)$. This means that by setting the level of acceptance and rejection at 0.6 and 0.2, respectively, we are able to make 81.91 % correct decisions in 34.38 % of the cases.

The above example demonstrates the use of probabilistic rough sets based three-way decisions for determining malicious behaviour using two models, i.e., ITRS and GTRS. The essential difference between these two models may be explained based on the type and nature of the solution. The ITRS provides an optimization based solution in determining the thresholds and the GTRS provides a tradeoff based solution to obtain the thresholds. The other essential difference is that the ITRS is based on a single criterion, i.e., the uncertainty while the GTRS is based on multiple criteria. The question of which one to choose may depend on the intended requirements and usage of the application.

Table 4. Payoff table

		Generality		
		$s_1 = \alpha_\downarrow$	$s_2 = \beta_\uparrow$	$s_3 = \alpha_\downarrow\beta_\uparrow$
		= 20% dec. α	= 20% inc. β	= 20% (dec. α & inc. β)
Accuracy	$s_1 = \alpha_\downarrow$	(0.8191,0.3438)	(1.0,0.1250)	**(0.8191,0.3438)**
	= 20% dec. α			
	$s_2 = \beta_\uparrow$	(1.0000,0.1250)	(0.7500,0.5000)	(0.7500,0.5000)
	= 20% inc. β			
	$s_3 = \alpha_\downarrow\beta_\uparrow =$	(0.8191,0.3438)	(0.7500,0.5000)	(0.7400,0.7188)
	20% (dec. α & inc. β)			

5.2 Three-Way Decisions Based on Information-Theoretic Rough Sets

Information-theoretic rough sets determine thresholds for inducing three-way decisions based on minimization of the overall uncertainty. Considering the conditions $0 \leq \beta < 0.5 \leq \alpha \leq 1.0$ of majority oriented model, we have the domain of α as $D_\alpha = \{1.0, 0.7, 0.6, 0.5\}$, and the domain of β $D_\beta = \{0.0, 0.3, 0.4\}$. To evaluate a certain threshold pair, say $(\alpha, \beta) = (1.0, 0.0)$, we need to determine the positive, negative and boundary regions given in Eqs. (1)–(3). The $\text{POS}_{(1.0, 0.0)}(C) = \{X_1\}$, $\text{NEG}_{(1.0, 0.0)}(C) = \{X_9\}$ and $\text{BND}_{(1.0, 0.0)}(C) = \bigcup\{X_2, X_3, ..., X_8\}$. The probability of the positive, negative and boundary regions are $P(\text{POS}_{(\alpha, \beta)}(C)) = P(X_1) = 0.0625$, $P(\text{NEG}_{(\alpha, \beta)}(C)) = P(X_9) = 0.0625$ and $P(\text{BND}_{(\alpha, \beta)}(C)) = P(X_2) + P(X_3) + ... + P(X_8) = 0.875$, respectively. The conditional probability $P(C|\text{POS}_{(\alpha, \beta)}(C))$ is,

$$P(C|\text{POS}_{(1,0)}(C)) = \frac{\sum\limits_{i=1}^{1} P(C|X_i) * P(X_i)}{\sum\limits_{i=1}^{1} P(X_i)} = \frac{1 * 0.0625}{0.0625} = 1.0 \qquad (18)$$

The probability $P(C^c|\text{POS}_{(1,0)}(C))$ is determined as $1 - P(C|\text{POS}_{(1,0)}(C)) = 1 - 1 = 0$. The uncertainty due to the positive region based on Eq. (5) is calculated as,

$$H(\pi_C|\text{POS}_{(1,0)}(C)) = -1 * log1 - (0 * log0) = 0. \qquad (19)$$

According to Eq. (9), the average uncertainty of the positive region is $\Delta_P(1, 0) = P(\text{POS}_{(1,0)}(C))H(\pi_C|\text{POS}_{(1,0)}(C)) = 0$. In the same way, using Eqs. (10) and (11), the average uncertainties of the negative and boundary regions are determined as $\Delta_N(1, 0) = 0$ and $\Delta_B(1, 0) = 0.875$, respectively. The total uncertainty according to Eq. (12) is therefore $\Delta(1, 0) = 0.875$.

We can compute the overall uncertainties of all possible threshold pairs. The results are presented by the following matrix.

$$
\begin{array}{c}
\\
\beta = 0.0 \\
\beta = 0.3 \\
\beta = 0.4
\end{array}
\begin{array}{cccc}
\alpha = 1.0 & \alpha = 0.7 & \alpha = 0.6 & \alpha = 0.5 \\
\left(\begin{array}{cccc}
0.875 & 0.8680 & 0.8607 & 0.8606 \\
0.8682 & 0.8688 & 0.8661 & 0.8768 \\
\mathbf{0.8544} & 0.8665 & 0.8704 & 0.8937
\end{array} \right)
\end{array}
$$

We can search the minimum uncertainty in the table and the corresponding thresholds. In this example, the minimum value is **0.8544** corresponding to $(\alpha, \beta) = (1.0, 0.4)$.

6   Conclusion

We considered malware analysis techniques aiming at identifying potentially harmful softwares based on their behavior or interaction with the operating system. These techniques are typically based on some intelligent mechanism that consider two-way classification of software applications in the form of malicious or benign. It is argued that due to ambiguous nature of applications' behavior, it may not be possible to confidently reach such two-way conclusions. This seriously affects user trust and satisfaction on the malware analysis systems. We propose a three-way decision making approach for classifying application behaviors. The essential change is the addition of a deferment or delay decision. This is of particular interest when the available information is not convincing enough to reach certain conclusions. The proposed three-way approach for malware analysis is demonstrated by considering probabilistic rough set models. In particular, we examine two models, i.e., information-theoretic rough sets and game-theoretic rough sets on sequences of a vulnerable Linux application. A demonstrative example suggests that three-way approach to malware analysis can be useful alternative for classifying application behaviours.

Acknowledgment. This work was partially supported by a discovery grant from NSERC canada.

References

1. Azam, N., Yao, J.T.: Analyzing uncertainties of probabilistic rough set regions with game-theoretic rough sets. Int. J. Approx. Reasoning **55**(1), 142–155 (2014)
2. Biddle, R., van Oorschot, P.C., Patrick, A.S., Sobey, J., Whalen, T.: Browser interfaces and extended validation ssl certificates: an empirical study. In: Proceedings of the 2009 ACM Workshop on Cloud Computing Security, pp. 19–30 (2009)
3. Demchenko, Y., Ngo, C., de Laat, C., Membrey, P., Gordijenko, D.: Big security for big data: addressing security challenges for the big data infrastructure. In: Jonker, W., Petković, M. (eds.) SDM 2013. LNCS, vol. 8425, pp. 76–94. Springer, Heidelberg (2014)
4. Deng, X., Yao, Y.: An information-theoretic interpretation of thresholds in probabilistic rough sets. In: Li, T., Nguyen, H.S., Wang, G., Grzymala-Busse, J., Janicki, R., Hassanien, A.E., Yu, H. (eds.) RSKT 2012. LNCS, vol. 7414, pp. 369–378. Springer, Heidelberg (2012)
5. Egele, M., Scholte, T., Kirda, E., Kruegel, C.: A survey on automated dynamic malware-analysis techniques and tools. ACM Comput. Surv. **44**(2), 6 (2012)
6. Forrest, S., Hofmeyr, S.A., Somayaji, A., Longstaff, T.A.: UNM dataset. http://www.cs.unm.edu/immsec/data-sets.htm. Accessed 6 April 2015
7. Forrest, S., Hofmeyr, S.A., Somayaji, A., Longstaff, T.A.: A sense of self for unix processes. In: IEEE Symposium on Security and Privacy, pp. 120–128 (1996)
8. Gandotra, E., Bansal, D., Sofat, S.: Malware analysis and classification: a survey. J. Inf. Secur. **5**, 56–64 (2014)
9. Herbert, J.P., Yao, J.T.: Game-theoretic rough sets. Fundamenta Informaticae **108**(3–4), 267–286 (2011)
10. Leyton-Brown, K., Shoham, Y.: Essentials of Game Theory: A Concise Multidisciplinary Introduction. Morgan & Claypool Publishers, San Rafael (2008)
11. Liu, D., Yao, Y.Y., Li, T.R.: Three-way investment decisions with decision-theoretic rough sets. Int. J. Comput. Intel. Syst. **4**(1), 66–74 (2011)
12. Mehdi, B., Ahmed, F., Khayyam, S.A., Farooq, M.: Towards a theory of generalizing system call representation for in-execution malware detection. In: IEEE International Conference on Communications (ICC), pp. 1–5 (2010)
13. Pawlak, Z.: Rough sets. Int. J. Comput. Inf. Sci. **11**, 241–256 (1982)

14. Symantec: Internet security threat report, vol. 19. http://www.symantec.com/security_response/publications/threatreport.jsp. Accessed 12 April 2015
15. Tanenbaum, A.S., Woodhull, A.S.: Operating systems: design and implementation, vol. 2. Prentice-Hall, Englewood Cliffs (1987)
16. Wright, C., Cowan, C., Morris, J., Smalley, S., Kroah-Hartman, G.: Linux security module framework. In: Ottawa Linux Symposium, vol. 8032, pp. 6–16 (2002)
17. Yang, H., Li, T., Hu, X., Wang, F., Zou, Y.: A survey of artificial immune system based intrusion detection. The Scientific World Journal (2014)
18. Yao, Y.Y.: Probabilistic rough set approximations. Int. J. Approx. Reasoning **49**(2), 255–271 (2008)
19. Yao, Y.Y.: Three-way decisions with probabilistic rough sets. Inf. Sci. **180**(3), 341–353 (2010)
20. Yao, Y.Y.: Two semantic issues in a probabilistic rough set model. Fundamenta Informaticae **108**(3–4), 249–265 (2011)

Chinese Emotion Recognition Based on Three-Way Decisions

Lei Wang[1,2(✉)], Duoqian Miao[1,2], and Cairong Zhao[1,2]

[1] Department of Computer Science and Technology,
Tongji University, Shanghai, China
dragon_wlei@126.com
[2] The Key Laboratory of Embedded System and Service Computing,
Ministry of Education, Tongji University, Shanghai, China

Abstract. In recent years, affective computing has become a research hotspot in the area of natural language processing and Chinese emotion recognition is an important constituent. This paper proposes a method of Chinese emotion recognition based on three-way decisions. Given the emotion dictionary constructed firstly, the grammatical information of sentences, topic features of texts and three-way decisions are integrated and applied into Chinese emotion recognition, thus realizing the multi-label emotion recognition of sentences in Chinese texts. The results of experiments show that the method of Chinese emotion recognition, based on three-way decisions, has achieved excellent results in the emotion recognition of Chinese sentences.

Keywords: Three-way decisions · Probability topic · Emotion dictionary · Affective computing

1 Introduction

With the rapid development of the Internet, more and more common users enjoy expressing their own emotions on the Internet, making comments on product performance and discussing current affairs. It brings large quantities of online text information with subjective emotional such as personal blog, product comment, news comment and so on [1–3]. All these information reflect person's emotional tendency such as happiness, anger, sorrow and enjoy [4]. Through analyzing the emotion of the online information, we can understand the individual emotional state and the level of popular products which people likes or dislikes and get to understand the fondness degree of the users. However, all of this would not be achieved by solely relying on manual handling, which promotes the development of affective computing [5] and making it a research hotspot in the area of natural language processing.

The work is supported by the National Natural Science Foundation of China (No. 61273304), the Specialized Research Fund for the Doctoral Program of Higher Education of China (No. 2013007 213004) and partially supported by the National Science Foundation of China (No. 61203247).

© Springer International Publishing Switzerland 2015
D. Ciucci et al. (Eds.): RSKT 2015, LNAI 9436, pp. 299–308, 2015.
DOI: 10.1007/978-3-319-25754-9_27

Affective computing aims to promote the quality of the human-computer communication and obtain more valuable information by revealing the delicate human emotions [4]. After Picard put forward the concept of "Affective Computing", emotion intelligence analysis has drawn more attention from scientists and researchers in the fields of artificial intelligence and computational linguistics.

The text such as weblog, product reviews and news discussion plays a basic role among all the communication mediums in the Internet, so the analysis of text emotion is an important constituent of affective computing. According to the differences of the analysis particle, it can be subdivided into three layers: emotion analysis of words, emotion analysis of sentences and emotion analysis of texts. Emotion analysis of sentences plays the connecting role, which relies on the emotional analysis of words and in the meantime, it can achieve more abundant emotional elements, thus providing support for the emotion analysis of texts.

Emotion analysis of sentences mainly concentrates on the identification of various kinds of emotion object, the extraction of various kinds of emotion owners and judgment of the emotional polarity of sentences and so on. Considering the previous researches, there are two kinds of research: the approach based on emotion knowledge and the approach based on feature classification. The approach based on emotion knowledge [6, 7] mainly relies on the existing emotion dictionary, so it first identifies emotion words in the text, and then emotion polarity weight value is computed for recognizing emotion of the text. Generally, the emphasis of the approach is placed on the extraction and judgment of emotion words. Hu and Liu[1, 2] took advantage of the synonym and antonym relation of WordNet to identify the emotion of words, then recognized the emotion polarity of sentences according the weight of emotion words which have advantages in the sentence. The latter approach mainly adopts machine learning and chooses large numbers of features to achieve the emotion polarity classification of texts by utilizing corpus. Dave etc. [8] used the machine learning approach and studied the emotion analysis of sentences. They collected over 1000 pieces of comments which have been marked emotion polarity and counted the appearance frequency of n-gram combination in the documents; then, according to the proportion of appearance frequency, they graded the emotion polarity of these features and judged the emotion polarity according to these features and the grades.

This paper takes the emotion recognition of sentences in Chinese texts as the research topic, integrates the grammatical information of sentences, the topics of texts and Three-way decisions, and then brings them into the emotion recognition of sentences. A Chinese emotion recognition method based on Three-way decisions is proposed to recognize the emotion of sentences in texts and the experiment result shows that the approach has achieved good results.

The remainder of the paper is as follows: Sect. 2 briefly introduces relative studies including LDA Model, Three-way decisions theory. Section 3 explains in detail the Chinese emotion recognition approach based on Three-way decisions. In Sect. 4, we describe the Chinese emotion corpus (Ren_CECps), the experiment process and the analysis of results. Finally, we summarize the whole paper and some concluding remarks are presented in Sect. 5.

2 Related Work

2.1 Latent Dirichlet Model

LDA model [9] was proposed by Blei in 2003, and it is a three-layer Bayes generation model. The advantage of LDA is that the scale of parameter space is not related to corpus and is suitable to process large-scale corpus.

In LDA model, each document in the corpus can be expressed as a probability distribution composed by some topics, while each topic is a probability distribution composed by several words. As to each document in the corpus, the generative process of the documents in the LDA model is as follows:

1. To each text, choose a topic from topics distribution;
2. Choose a word from the corresponding word distribution of the chosen topic;
3. Repeat the above procedures until every word of the document has been chosen.

The graph model of the generative process is shown as Fig. 1:

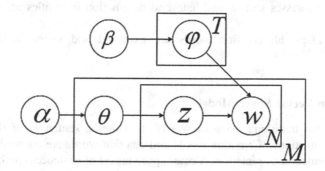

Fig. 1. Graph model representation of LDA

2.2 Three-Way Decisions Theory

Three-way decisions rough set model [10] is adopted to identify the multi-label emotion of sentences in Chinese texts. Compared with other various kinds of existing models, Three-way decisions rough set model takes a pair of threshold value on the basis of classic rough set theory and divide one set into three areas which do not intersect, thus bringing the third choice to delay decision-making so as to avoid the risks brought by direct decisions.

In Three-way decisions rough set, we divide the objects into corresponding positive area, negative area and boundaries by comparing the value of probability and the threshold and these areas correspond to the acceptance decision, rejection decision and delay decision respectively. The delay decision objects are in the boundaries and we need to collect further information to analyze and make corresponding decisions, which is acceptance or rejection. Three-way decisions theory can analyze data effectively and reduce wrong decisions so as to improve the accuracy of decisions.

3 Chinese Emotion Recognition

3.1 Emotion Dictionary

At present, some emotion dictionaries are developed for affective computing such as WordNet-Affect, SentiWordNet and the VSA Chinese-English Emotion Analysis Vocabulary. In this paper, emotion recognition research is completely based on Chinese emotion corpus (Ren_CECps) [11, 12]. In order to better achieve emotion recognition research, we extract all emotion words with emotion polarity and emotion intensity from Ren_CECps and conduct a new Chinese emotion dictionary to apply it into the experiments.

In Ren_CECps, each emotion word in the sentence is labeled for 8 fundamental emotion categories such as surprise, sorrow, love, joy, hate, expect, anxiety and anger. It has been labeled corresponding emotion intensity and expressed as an emotion vector $\vec{e}$. For example, emotion word "无私" ('selfless') is labeled as (0.0,0.0,0.7,0.0,0.0, 0.5,0.0,0.0), so it expresses love and expect and its emotion intensities are 0.7 and 0.5 respectively. The emotion word "战争"('war') is labeled as (0.0,0.5,0.0,0.0,0.5,0.0, 0.0,0.0), so it expresses sorrow and hate and its emotion intensities are 0.5 and 0.5 respectively.

In Ren_CECps, all emotion words are extracted and stored in the emotion dictionary.

3.2 Emotion Vector Space Model

According to the traditional "bag of words" hypothesis, sentences of the texts are treated as combinations of emotion words and emotion words are counted for helping recognize the emotion of sentences. Vector space model of sentences can be expressed as $\vec{S} = \{w_1, w_2, \cdots, w_n\}$, here n is the number of emotion words, w_i is the ith emotion word. For w_i, it can be expressed as an emotion vector $\vec{e_i} = (e_i^1, e_i^2, e_i^3, e_i^4, e_i^5, e_i^6, e_i^7, e_i^8)$. The emotion vector space model of the sentence can be further expressed as the following form:

$$\vec{S} = \{w_1, w_2, \ldots, w_n, \vec{e_1}, \vec{e_2}, \ldots, \vec{e_n}\} \qquad (1)$$

3.3 Emotion Vector Space Model Based on Topics

For all sentences of a text, it will separate the correlation of sentences and neglect the dependency of contexts only by using the emotion vector space model in Formula (1) to recognize the emotion of sentences. Suppose the sentence of a text describes the same topic information with the text, so the emotion of the sentence and the text also agree with and the emotion of sentences should be determined by emotion word which can better reflect the topic feature of sentence and the text.

Latent topic features are integrated into emotion vector space model of sentences. For document D, it applies LDA model and obtains T latent topic $\overrightarrow{T} = \{t_1, t_2, \ldots, t_T\}$ and the probability distribution $\vec{\phi}$ of topic-word, and then utilizes the probability distribution of "text-topic-word" to solve the dependency of the context. The author proposes an emotion vector space model based on topic, finds out the t_m with largest probability weight from T latent topic and further extends the emotion vector space model of sentences. The formula is as follows:

$$\overrightarrow{S} = \{w_1, w_2, \ldots, w_n, \overrightarrow{e_1}, \overrightarrow{e_2}, \ldots, \overrightarrow{e_n}, \varphi_{m1}, \varphi_{m2}, \ldots, \varphi_{mn}\} \tag{2}$$

w_i refers to the ith emotion word, $\vec{e}_i$ refers to the emotion vector of the ith emotion word, φ_{mi} means the probability distribution of the topic-word of the ith emotion word based on topic t_m.

In order to describe the emotion features of sentences more clearly, transform the formula further and we get emotion vector space model based on topics and the form is as follows:

$$\overrightarrow{S} = (\frac{1}{n}\sum_{i=1}^{n}(1+\varphi_{mi})e_i^1, \frac{1}{n}\sum_{i=1}^{n}(1+\varphi_{mi})e_i^2, \ldots, \frac{1}{n}\sum_{i=1}^{n}(1+\varphi_{mi})e_i^7, \frac{1}{n}\sum_{i=1}^{n}(1+\varphi_{mi})e_i^8)$$
$$\tag{3}$$

$\frac{1}{n}\sum_{i=1}^{n}(1+\varphi_{mi})e_i^j$ refers to the emotion intensity average of the jth emotion of all emotion words of the sentence. The emotion and emotion intensity of the sentence can be got by Formula (3).

There are parts of emotion words that could be defined by negative adverbs, it could weaken or change the emotion of the sentence. In the paper, the negative adverb just before or after the emotion words is considered and the emotion of the emotion words modified by negative adverbs is defined as 0. For example,

"一天十多小时的复习也需要健健康康的身体来支撑，否则万一晕倒在自习室里可就不妙了" (焦虑: 0.5).

"To review for over 10 h a day needs a healthy body. Otherwise, it would be no good if you faint in the self-study room" (anxiety: 0.5)

The emotion word "miao" (good) has the meaning of "love", but there is the negative adverb "no" before the emotion word which negates the emotion information. Therefore, the sentence does not have the "love" emotion and recognizes the sentence has the "anxiety" emotion according to other emotion words.

3.4 Chinese Emotion Recognition Based on Three-Way Decisions

After the initial multi-label recognition of the emotion of all sentences, each sentence in the text is labeled with certain emotions and intensity, and the emotions of each sentence is represented as an 8-dimension emotion vector which is treated as the input data of three-way decisions classifier.

According to the rules of three-way decisions, the threshold values α_t and β_t are set in advance to recognize whether the give testing sentence x has emotion k. The decision process is as follows:

(1) If $P(k|[x]) \geq \alpha_t$, then the sentence x has emotion k.
(2) If $P(k|[x]) \leq \beta_t$, then the sentence x does not have emotion k.
(3) If $\beta_t < P(k|[x]) < \alpha_t$, it means the sentence x may have emotion k, or may not have emotion k, and it needs further processing;

If sentences cannot be determined whether it has emotion k, a threshold value θ_t is set and the process is followed:

(1) If the number of the kth emotion words in the sentence x is equal or greater than θ_t, then sentence x is judged to have emotion k.
(2) If the number of the kth emotion words in the sentence x is smaller than θ_t, then sentence x is judged not to have emotion k.

3.5   Multi-label Emotion Recognition Framework

In the paper, the multi-label emotion recognition framework of sentence is shown as Fig. 2. The left is the training process, and the right is the test process, and we also apply emotion dictionary, LDA model and three-way decisions into it. The multi-label emotion recognition process of sentences is divided into 6 steps, and the specific description is as follows:

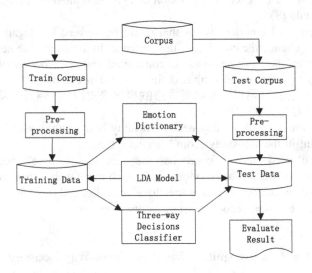

Fig. 2. Multi-label emotion recognition framework

Step 1: Draw the experimental data randomly from Ren_CECps corpus for training corpus and testing corpus;

Step 2: Pre-processing respectively the training corpus and the testing corpus. Remove small quantities of sentences without emotion polarity and the non-emotion words in sentences; only retain the emotion words in sentences.

Step 3: Utilize experiment corpus, make up emotion dictionary;

Step 4: Apply emotion dictionary and LDA model respectively on training data and testing data, make up the emotion vector space model based on topics features;

Step 5: Take advantage of training data to achieve all the threshold values needed by Three-way decisions classifiers;

Step 6: Apply Three-way Decision Classifier to the testing data and evaluate experiment result;

4 Experiment and Analysis

4.1 Experiment Data

The Ren_CECps Chinese emotion corpus is adopted for the experiments. The corpus contains 1487 Chinese blog which have altogether 11255 paragraphs, 35096 sentences and 878164 words, selected from Chinese websites as the initial text corpus.

In Ren_CECps Chinese emotion corpus, all the language information related to emotion expression is labeled manually and all emotions are divided into 8 basic emotion categories which are: surprise, sorrow, love, joy, hate, expect, anxiety and anger. The emotion category and intensity of texts, paragraphs and sentences are labeled as an 8-dimensional emotion vector, and is represented as follows:

$$\vec{e} = (e^1, e^2, e^3, e^4, e^5, e^6, e^7, e^8) \tag{4}$$

In the formula, e^i is labeled as the emotion intensity of a basic emotion category of 8 emotion category and the scope of the intensity is from 0.1 to 1.0.

In the experiment, 1000 blogs are selected randomly from Ren_CECps Chinese emotion corpus, which are altogether 21225 sentences. 10-fold cross-validation is performed on the whole data for each experiment.

4.2 Experiment Setting

The goal of the experiment is to recognize the multi-label emotion of sentences. BR (Binary Relevance) is a classic representative of multi-label classification algorithm and suitable to situations with small label quantity q. While there are only 8 types of emotion in Ren_CECps corpus, so BR algorithm is adopted as multi-label classification algorithm and Naïve Bayes is adopted as the basic classifier.

The paper adopts evaluation approach based on labels [13, 14] and the formula for the macro average accuracy of multi-label classification is as follows:

$$Macro - accuracy = \frac{1}{|K|} \sum_{k=1}^{|K|} \frac{tp_k + tn_k}{tp_k + tn_k + fp_k + fn_k} \tag{5}$$

tp_k denotes the correct positives label number, tn_k denotes the correct negatives label number, fp_k denotes the wrong label number, and fn_k denotes the wrong negatives label number.

In order to evaluate the accuracy degree of Chinese emotion recognition, the author also proposes three evaluation standards, which are single type of emotion identification, two types of emotion identification and all types of emotion identification, and the formulas are as follows:

$$One - match - accuracy = \frac{the\ number\ of\ sentences\ at\ least\ one\ emotion\ matched}{the\ total\ number\ of\ sentences}$$

(6)

$$Two - match - accuracy = \frac{the\ number\ of\ sentences\ at\ least\ two\ emotions\ matched}{the\ total\ number\ of\ sentences}$$

(7)

$$All - match - accuracy = \frac{the\ number\ of\ sentences\ all\ emotions\ matched}{the\ total\ number\ of\ sentences}$$

(8)

4.3   Analysis of Experiments Result

The emotion categories of this experiment are 8 types (surprise, sorrow, love, joy, hate, expect, anxiety and anger) and it does not consider sentences with neutral emotion. In the experiments, the parameter setting is as follows: $\alpha = 0.5$, $\beta = 0.1$, the type of emotions $L = 8$, the number of topics $T = 3$, $\alpha_t = 0.6$, $\beta_t = 0.3$, $\theta_t = 0.6$. All of the parameters are all empirical value and can be got from training set.

The NB algorithm is adopted as the benchmark and Table 1 gives the comparison of the result of the two methods.

Table 1. Comparison of the experiment

	NB approach	Approach based on three-way decisions
One-match-accuracy	0.040	0.601
Two-match-accuracy	0.029	0.306
All-match-accuracy	0.022	0.113
Macro-accuracy	0.807	0.752

Table 1 indicates that the approach based on three-way decisions is better than NB and can identify the emotions of the sentences accurately in single emotion identification, two emotion identification and all emotion identification. However, in macro accuracy, NB is a little better than the approach based on three-way decisions, which is because each sentence in the corpus only has several emotions and the value of tn_k is large, so the macro-accuracy of NB is higher, but it also indicates that the approach based on three-way decisions is better than NB approach.

Compare NB, the accuracy of 8 types of basic emotion is shown as Fig. 3. In Fig. 3, the method based on three-way decisions is better than NB for recognizing the emotion of joy, love, expect, while it is lower for other 5 type. Through studying the Ren_CECps corpus, we find that lots of sentences have positive emotion in the corpus such as joy, love and expect and the number of sentences containing negative emotion such as hate, sorrow, anger and surprise is few. So the value of m_k in formula 5 is great when recognizing the negative emotions and the result also indicates that the method based on three-way decisions has more advantages for recognizing the positive emotion of the sentences. It is the future work that how to further complete and enrich the Ren_CECps Chinese emotion corpus.

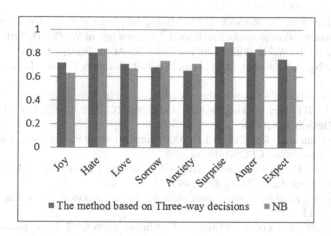

Fig. 3. Accuracy of 8 types of basic emotion recognition

5 Conclusion

The paper focuses on the multi-label emotion recognition of sentences and proposes the Chinese emotion recognition approach based on three-way decisions. The Ren_CECps Chinese emotion corpus is taken as the data of experiments, the context dependency between sentences is solved by using topic features and the emotion recognition of negative sentences is addressed by taking advantage of the negative adverbs. Lastly three-way decisions classifier is applied to recognize multi-label emotion of sentences, such as surprise, sorrow, love, joy, hate, expect, anxiety and anger.

The paper utilizes the negative adverbs, topic features and three-way decisions theory to study the multi-label emotion recognition of Chinese sentences and still has a lot to improve. It is a hot topic in the future how to improve the multi-label emotion recognition accuracy of sentences and how to recognize the emotion of texts by utilizing the emotion information of words and sentences.

References

1. Hu, M., Liu, B: Mining and summarizing customer reviews. In: Proceedings of the 10th International Conference on Knowledge Discovery and Data Mining, pp. 168–177 (2004)
2. Liu, B., Zhang, L.: A Survey on Opinion Mining and Sentiment Analysis. Mining Text Data. Springer, New York (2012)
3. Jo, Y., Oh, A.H.: Aspect and sentiment unification model for online review analysis. In: Proceedings of the 4th ACM International Conference on Web Search and Data Mining, pp. 815–824 (2011)
4. Ren, F.: Affective information processing and recognizing human emotion. J. Electron. Notes Theor. Comput. Sci. **225**, 39–50 (2009)
5. Picard, R.W.: Affective Computing. MIT Press, Cambridge (1997)
6. Taboada, M., Brooke, J., Tofiloski, M.: Lexicon-based methods for sentiment analysis. J. Comput. Linguist. **37**(2), 267–307 (2011)
7. Rao, D., Ravichandran, D.: Semi-supervised polity lexicon induction. In: Proceedings of the EACL, pp. 675–682 (2009)
8. Dave, K., Lawrence, S., Pennock, D.M.: Mining the peanut gallery: opinion extraction and semantic classification of product reviews. In: Proceedings of WWW-03, 12th International Conference on the World Wide Web, pp. 519–528. ACM, Budapest (2003)
9. Blei, D.M., Ng, A.Y., Jordan, M.I.: Latent Dirichlet allocation. J. Mach. Learn. Res. **3**, 993–1022 (2003)
10. Yao, Y.: An outline of a theory of three-way decisions. In: Yao, J., Yang, Y., Słowiński, R., Greco, S., Li, H., Mitra, S., Polkowski, L. (eds.) RSCTC 2012. LNCS, vol. 7413, pp. 1–17. Springer, Heidelberg (2012)
11. Ren, F.: Document for Ren-CECps 1.0 (2009). http://a1-www.is.tokushima-u.ac.jp/member/ren/Ren-CECps1.0/Ren-CECps1.0.html
12. Quan, C., Ren, F.: A blog emotion corpus for emotional expression analysis in Chinese. J. Comput. Speech Lang. **24**(4), 726–749 (2010)
13. Tsoumakas, G., Katakis, I.: Multi-label classification: an overview. Int. J. Data Warehouse. Min. **3**(3), 1–13 (2007)
14. Tsoumakas, G., Katakis, I., Vlahavas, I.: Mining multi-label Data, Data Mining and Knowledge Discovery Handbook, 2nd edn. Springer, Heidelberg (2010). Part 6

Statistical Interpretations of Three-Way Decisions

Yiyu Yao and Cong Gao[✉]

Department of Computer Science, University of Regina,
Regina, SK S4S 0A2, Canada
{yyao,gao266}@cs.uregina.ca

Abstract. In an evaluation based model of three-way decisions, one constructs three regions, namely, the left, middle, and right regions based on an evaluation function and a pair of thresholds. This paper examines statistical interpretations for the construction of three regions. Such interpretations rely on an understanding that the middle region consists of normal or typical instances in a population, while two side regions consist of, abnormal or untypical instances. By using statistical information such as median, mean, percentile, and standard deviation, two interpretations are discussed. One is based on non-numeric values and the other is based on numeric values. For non-numeric values, median and percentile are used to construct three pair-wise disjoint regions. For numeric values, mean and standard deviation are used. The interpretations provide a solid statistical basis of three-way decisions for applications.

Keywords: Statistical interpretations · Three-way decisions

1 Introduction

The basic ideas of three-way decisions may be explained in terms of the division and processing of a universe of objects by using three regions [41]. A two step trisecting-and-acting framework has been proposed [39]. In the first step, one divides the universe into three pair-wise disjoint regions, that is, a tripartition of the universe. In the second step, one designs most effective strategies to process three regions. Such ideas of dividing and processing the universe with three regions have been widely used in many fields, such as medicine [10,21,29], business [2], engineering and science [7,50]. Examples of three-way decision models include rough sets [22], interval sets [37], shadowed sets [23], three-way approximation of fuzzy sets [6], three-way classification [50], three-way clustering [43] and many more [4,13,18,20,24,27,45–47]. By integrating the results from these fields and exemplar models, a theory of three-way decisions has been proposed to study domain independent ways for fast and effective decision making and information processing [35]. There is growing interest in three-way decisions [1,8,9,11–17,19,31,33,39,42,44,49].

Typically, a division of a universe is based on an evaluation function and a pair of thresholds. The evaluation function assigns each object in the universe an

© Springer International Publishing Switzerland 2015
D. Ciucci et al. (Eds.): RSKT 2015, LNAI 9436, pp. 309–320, 2015.
DOI: 10.1007/978-3-319-25754-9_28

evaluation status value (ESV). The three regions are constructed by collecting, respectively, the set of objects whose ESV are equal or greater than one threshold, the set of objects whose ESV are equal or less than another threshold, and the set of objects whose ESV are between the two thresholds. Therefore, interpretation and determination of an evaluation function and a pair of thresholds is a fundamental issue of three-way decisions.

By generalizing Pawlak rough sets [22], Yao et al. [40] proposed a decision-theoretic rough set (DTRS) model by using probability as an evaluation function and a pair of thresholds on probability to derive three regions known as the probabilistic positive, negative and boundary regions. A pair of thresholds is determined and interpreted within Bayesian decision theory. JT Yao and Herbert [34] proposed a game-theoretic rough set (GTRS) model. A pair of thresholds is determined by designing a game. Deng and Yao [5] and Zhang [48] proposed and investigated information-theoretic rough set (ITRS) models. A pair of thresholds is determined by minimizing the information uncertainty of the three regions. These probabilistic models of rough sets can be considered as specific examples of three-way decisions. Their ideas of exploring statistical information may be applied to study three-way decision in general.

From a statistical point of view, the sequential hypothesis testing framework introduced by Wald [32] also influenced three-way decisions. If a test strongly supports a hypothesis, one accepts the hypothesis; if the test is strongly against the hypothesis, one rejects the hypothesis; otherwise, one performs further tests. Based on further tests, hypothesis originally cannot be accepted and rejected will be accepted or rejected. Due to a sequential feature of Wald's framework, one can either accept or reject some hypothesizes without the need for further testing. This sequential processing has benefits of efficiency and effectiveness. Yao [36] adopted such ideas to sequential three-way decisions and granular computing [23,38].

These studies on combining statistics and three-way decisions provide a promising research direction. The main objective of this paper is to further examine statistical interpretations of three-way decisions. In particular, we use statistical notions, such as mean, median, standard deviation, central tendency and dispersion, for constructing and interpreting an evaluation function and a pair of thresholds required by three-way decisions. Objects around the mean value form one region, and two tails form the other two regions. In other words, we search for statistical interpretations of three-way decisions that enable us to examine structures of data and to make inference about data.

The paper is organized as follows. Section 2 briefly reviews an evaluation based three-way decisions model. Section 3 presents the statistical interpretations of three-way decisions. Section 4 provides some concluding remarks.

2   Overview of Evaluation Based Three-Way Decisions

An essential task of three-way decisions is to divide a universal set of objects into three pair-wise disjoint regions, as illustrated in Fig. 1(a). Following Yao and Yu [41], we label the three regions as the L, M and R regions, respectively.

They are interpreted as the left, middle and right regions, as indicated by their relative positions in Fig. 1 (a). It is required that the three subsets L, M and R of U satisfy the following conditions:

$$(i) \quad L \cup M \cup R = U,$$
$$(ii) \quad L \cap M = \emptyset, L \cap R = \emptyset, M \cap R = \emptyset.$$

In general, some of the three regions may be empty. The family $\{L, M, R\}$ may not be a partition of U. In this paper, we call $\{L, M, R\}$ a tripartition of U, with an understanding that some regions may be empty.

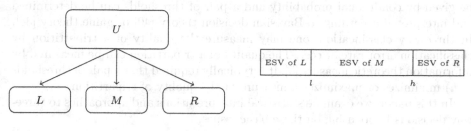

(a) Division of a set of objects U.   (b) Division of $(\mathbb{V}, \preceq)$.

Fig. 1. Illustration of three-way decisions.

To construct the three regions, we need to introduce the notion of an evaluation function. Let $v : U \to \mathbb{V}$ denote an evaluation function from U to a totally ordered set $(\mathbb{V}, \preceq)$ under a total order $\preceq$. For an object $x \in U$, $v(x) \in \mathbb{V}$ is the evaluation status value (ESV) of x. Given a pair of thresholds (l, h) with $l \prec h$, we divide $\mathbb{V}$ into three regions as shown in Fig. 1(b). The division of $\mathbb{V}$ immediately leads to a tripartition of U. If $v(x)$ is equal or less than the threshold l, then x is assigned to region L; if $v(x)$ is equal or greater than the threshold h, then x is assigned to region R; otherwise, x is assigned to region M.

We give a formal definition of a model of evaluation based three-way decisions [35].

Definition 1 (Evaluation based three-way decisions). *Suppose $(\mathbb{V}, \preceq)$ is a totally ordered set, where $\preceq$ is a total order. Suppose l and h are two elements in $\mathbb{V}$ with $l \prec h$, i.e., $l \preceq h \wedge \neg(h \preceq l)$. Given an evaluation function $v : U \to \mathbb{V}$, we can divide U into three regions as follows:*

$$L_{(l,h)}(v) = \{x \in U \mid v(x) \preceq l\},$$
$$M_{(l,h)}(v) = \{x \in U \mid l \prec v(x) \prec h\},$$
$$R_{(l,h)}(v) = \{x \in U \mid v(x) \succeq h\}, \tag{1}$$

where the subscripts (l, h) indicate that three regions are constructed based on a pair of thresholds (l, h).

In the rest of this paper, we only consider the task of dividing U into three regions, as formulated by Definition 1. For such a task, we must consider some fundamental issues, such as,

(1) construction and interpretation of a totally ordered set $(\mathbb{V}, \preceq)$,
(2) construction and interpretation of an evaluation function $v : U \to \mathbb{V}$,
(3) determination and interpretation of a pair of thresholds (l, h), and
(4) measurement of quality of a tripartition $\{L, M, R\}$.

These issues have been investigated in specific models of three-way decisions. For example, in decision theoretic rough set models, an evaluation function can be given by conditional probability and a pair of thresholds can be determined and interpreted according to Bayesian decision theory [40] or game theory [34]. In three-way classification, one may measure the quality of a tripartition by classification error rate or cost. The quality of a tripartition can be measured by information theoretic measure [5]. It is typically required that a pair of thresholds (l, h) minimizes or maximizes a measure of the quality of tripartition.

In this paper, we examine statistical interpretations and approaches to three-way decisions by focusing on these basic issues.

3   A Statistical Framework for Interpreting Three-Way Decisions

Depending on the structures of the set $\mathbb{V}$, we examine two specific statistical interpretations of three-way decisions.

3.1   General Considerations

In many applications, we typically have statistical information about objects in U. For example, we may have frequencies of measurement values with respect to a particular feature of objects. Such information may be used to construct both an evaluation function and a pair of thresholds. In the cases when an evaluation function is given, we may use a distribution of the evaluation status values (ESVs) to find a pair of thresholds.

In formulating three-way decisions, we use a totally ordered set $(\mathbb{V}, \preceq)$ as the set of evaluation status values (ESVs). According to Equation (1), we divide U into three regions. Region L consists of objects with low ESVs, region M with medium ESVs, and region R with high ESVs. Such a division in terms of low, medium and high values has been well used in many practical applications. For example, in blood pressure examination [21], the middle region means a normal blood pressure area, the left region is a hypotension area that is below the normal blood pressure, and the right region is hypertension area that is above the normal blood pressure. This example suggests one possible interpretation of three regions. The middle region M consists of normal or typical instances from a population, while regions L and R consist of, respectively, abnormal or untypical instances. In other words, blood pressure of a healthy person is expected to be fallen within a certain region, e.g., between 90 and 130 in systolic blood pressure [21]. An interesting question is how to interpret the intuitive notions of low, medium, and high values used in three-way decisions based on concepts from statistics.

In statistics, the concepts of median, mean, percentile, and standard deviation are used to describe distributional characteristics of a population. To establish a connection to three-way decisions, we may collect objects with ESVs around the median or mean value to form region M. The percentile or standard deviation may be used to calculate the positions of objects with ESVs away from the median or mean, which in turn determine a pair of thresholds. Two special cases of $\mathbb{V}$ may be considered. One is a set of non-numeric values and the other is a set of numeric values.

When $\mathbb{V}$ is a set of non-numeric values, we can perform comparisons based on the total order $\preceq$ and we may not carry out arithmetic operations such as addition and multiplication. In other words, we can only consider the ranking of values in $\mathbb{V}$ and the distribution of ESVs. The ordering enables us to locate median, that is, an object in the middle point of a ranked list. In addition, we can also use the frequency information to compute certain percentiles. Consequently, we use the median as the middle point of region M and use two percentiles to determine the size of region M. One percentile is used to calculate the left boundary of M, and the other percentile is used to calculate the right boundary of M. Region L is a set of objects with ESVs below the left boundary and region R is the set of objects with ESVs above the right boundary.

When $\mathbb{V}$ is a set of numeric ESVs, we can apply the median based interpretation by considering the ordering of objects induced by $\preceq$. Moreover, since we can perform arithmetic operations on $\mathbb{V}$, we can have a new interpretation by computing the mean and standard deviation. That is, we can use mean to set up the middle position, and use standard deviation to calculate the positions of two thresholds around the mean in terms of the distance away from the mean. Based on standard deviations, region M is the set of objects with ESVs around the mean, region L is the set of objects with ESVs much less than the mean, and region R is the set of objects with ESVs much greater than the mean.

The two interpretations make use of different types of statistical information. We discuss their detailed formulations in the rest of this section.

3.2 Interpretations Through Median and Percentile

When $\mathbb{V}$ is a set of non-numeric values, the ordering $\preceq$ only allows us to arrange objects in U into a ranked list according to their ESVs, as shown in Fig. 2. The median is the value at the middle position of this list and the positions of a pair of thresholds around the median are determined by two percentiles. A user can determine the three regions L, M, and R by the pair of percentiles. We use a simple example to demonstrate the main ideas.

Example 1. Suppose $U = \{e_1, e_2, e_3, e_4, e_5\}$ is a set of five eggs, and $\mathbb{V}$ is a set of words describing the size of eggs: $\{smallest, smaller, small, medium, large, larger, largest\}$ with the ordering $smallest \preceq smaller \preceq small \preceq medium \preceq large \preceq larger \preceq largest$. We want to divide U into three subsets according to their sizes. Given an evaluation function, suppose objects in U have the following ESVs: $v(e_1) = small, v(e_2) = smaller, v(e_3) = largest, v(e_4) = large, v(e_5) = larger$. We can arrange all objects in U into a ranked list based on their ESVs, e_2, e_1, e_4, e_5, e_3, according to the ordering $\preceq$. The object at the middle position 3 of this list is e_4 and its value is $large$, i.e., the median is $large$. Suppose we want the region M includes 60 % objects and each of regions L and R consists of 20 % objects. We use 20 % for computing the position of the left threshold and 20 % for computing the position of the right threshold. The position of the left threshold is 1 with object e_2 and ESV $smaller$, and the position of the right threshold is 4 with object e_3 and ESV $largest$. That is, $l = smaller$ and $h = largest$. Therefore, the tripartition is given $L = \{e_2\}$, $M = \{e_1, e_4, e_5\}$ and $R = \{e_3\}$.

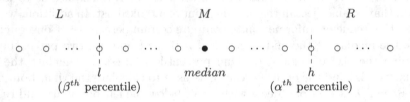

Fig. 2. Illustration of division on rank ordered list through median and percentile.

From the construction process of the example, we have an interpretation of three-way decisions through median and percentile, as depicted in Fig. 2. The ESV of the object denoted by the solid circle in the middle position is the median and (l, h) is a pair of thresholds based on the pair of percentiles.

The tripartition of three-way decisions can be constructed as follows. Suppose that the size of U is n. Step 1: arrange the set of objects into a ranked list according to their ESVs in ascending order, in which objects with the same ESV can be ranked in any order. In this way, we have a list of ESVs, $v_1, v_2, \cdots, v_n$, where v_1 is the smallest value and v_n the largest value. Step 2: we search for ESVs at β^{th} and α^{th} percentiles with $\beta < 50$ and $\alpha > 50$, we can calculate the pair of thresholds by:

$$l = v_{\lfloor \beta n/100 \rfloor},$$
$$h = v_{\lceil \alpha n/100 \rceil}, \tag{2}$$

where the floor function $\lfloor a \rfloor$ gives the largest integer that is not greater than a and the ceiling function $\lceil a \rceil$ gives the smallest integer that is not less than a. The floor and ceiling function used in l and h, respectively, are needed because $\beta n/100$ and $\alpha n/100$ may not be integers. As a result, three regions are constructed by:

$$L_{(\alpha,\beta)}(v) = \{x \in U \mid v(x) \preceq l\}$$
$$= \{x \in U \mid v(x) \preceq v_{\lfloor \beta n/100 \rfloor}\},$$
$$M_{(\alpha,\beta)}(v) = \{x \in U \mid l \prec v(x) \prec h\}$$
$$= \{x \in U \mid v_{\lfloor \beta n/100 \rfloor} \prec v(x) \prec v_{\lceil \alpha n/100 \rceil}\},$$
$$R_{(\alpha,\beta)}(v) = \{x \in U \mid v(x) \succeq h\}$$
$$= \{x \in U \mid v(x) \succeq v_{\lceil \alpha n/100 \rceil}\}. \tag{3}$$

In order to have pair-wise disjoint three regions, we require that $l \prec h$, i.e., l and h cannot be the same value in $\mathbb{V}$. This requires that the two percentiles must be chosen to satisfy the criterion.

Equation (3) provides an interpretation of three-way decisions through median and percentile. Such an interpretation has been widely used in many applications. For example, in boxplots [25], the values of l and h are obtained by first quartile and third quartile, and middle region M by interquartile range (IQR).

3.3   Interpretations Through Mean and Standard Deviation

When $\mathbb{V}$ consists of numeric values, statistical measures based on arithmetic operations such as mean and standard deviation can be applied. For simplicity, we assume that $\mathbb{V}$ is the set of real numbers. Suppose $v(x_1), v(x_2), \cdots, v(x_n)$ are the ESVs of objects in U, where n is the cardinality of U. The mean and standard deviation are calculated by:

$$\mu = \frac{1}{n} \sum_{i=1}^{n} v(x_i),$$

$$\sigma = \left(\frac{1}{n} \sum_{i=1}^{n} (v(x_i) - \mu)^2 \right)^{\frac{1}{2}}.$$

As shown by Fig. 3, we may interpret μ as the ESV for representing objects in M and σ as a unit to measure the positions of the two thresholds l and h. Suppose two non-negative numbers k_1 and k_2 represent the position of two thresholds away from the mean in terms of the number of times of the standard deviation. The pair of thresholds can be constructed as follows:

$$l = \mu - k_1 \sigma, \quad k_1 \geq 0,$$
$$h = \mu + k_2 \sigma, \quad k_2 \geq 0. \tag{4}$$

Generally, k_1 and k_2 do not need to be equal. According to l and h, three regions can be constructed by:

$$L_{(k_1,k_2)}(v) = \{x \in U \mid v(x) \leq l\}$$
$$= \{x \in U \mid v(x) \leq \mu - k_1 \sigma\},$$
$$M_{(k_1,k_2)}(v) = \{x \in U \mid l < v(x) < h\}$$

$$= \{x \in U \mid \mu - k_1\sigma < v(x) < \mu + k_2\sigma\},$$
$$R_{(k_1,k_2)}(v) = \{x \in U \mid v(x) \geq h\}$$
$$= \{x \in U \mid v(x) \geq \mu + k_2\sigma\}, \tag{5}$$

where $\leq$, $<$, and $\geq$ are standard relations on a set $\mathbb{V}$ of numeric values. It may be commented that k_1 and k_2 are related to z-score. Thus, we can interpret three-way decisions in terms of z-score.

Equation (5) makes no assumption of distribution of ESVs of objects. In many real applications, it is common that ESVs of objects in U satisfy a certain distribution. For example, Fig. 3 illustrates the tripartition based on two kinds of distributions. In Fig. 3(a), the normal distribution shows an unimodal and symmetric curve, in which the mean μ is obvious the normal or typical point of the distribution. While Fig. 3(b) shows a monotonic curve, the region around the mean represents the average area of the distribution, i.e., not too high and not too low.

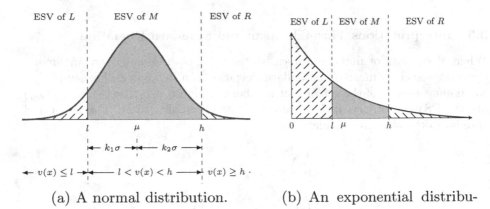

(a) A normal distribution.   (b) An exponential distribution.

Fig. 3. Illustration of three-way decisions on two kinds of distribution.

There are many applications based on this model. For example, Pater [21] suggested using $k_1 = k_2 = 2$ for blood pressure classifications. In other words, $M_{(2,2)}(v) = \{x \in U \mid \mu - 2\sigma < v(x) < \mu + 2\sigma\}$ is the region of normal blood pressure, while $L_{(2,2)}(v) = \{x \in U \mid v(x) \leq \mu - 2\sigma\}$ and $R_{(2,2)}(v) = \{x \in U \mid v(x) \geq \mu + 2\sigma\}$ regions are abnormal, that is, the hypotension and hypertension regions, respectively. The stanford-binet test [28] is the most often used approach to measure people's Intelligence Quotient (IQ), and $k_1 = k_2 = 2$ is usually used to classify people into three categories [28, 30]. People with IQ between $\mu - 2\sigma$ and $\mu + 2\sigma$ is considered as average, people with IQ greater than $\mu + 2\sigma$ is considered as above average, and people with IQ less than $\mu - 2\sigma$ is considered as below average.

The two statistical interpretations of three-way decisions are strongly related. Both people's IQ and blood pressure satisfy normal distribution [21, 28] and based on distributional properties, we know that the mean is equal to median and the middle region from $\mu - 2\sigma$ to $\mu + 2\sigma$ includes about 95 % ESVs, while both left and right regions include 2.5 % ESVs. This conversion from mean-deviation interpretation to median-percentile interpretation also can be done inversely. In real practices, the so-called criteria 68-95-99.7 rule [3] is widely used, which means the middle region is about 68 %, 95 %, and 99.7 % of all ESVs in a normal distribution when $k = 1$, $k = 2$, and $k = 3$, respectively. For non-symmetric distribution such as Fig. 3 (b), we also can use the mean and deviation to calculate three regions immediately, then the percentage of each region of all ESVs can be calculated by distribution parameters. Another way is that we can do standardization, the new standardized ESVs satisfy normal distribution with $\mu = 0$ and $\sigma^2 = 1$.

The interpretations through mean and standard deviation can also be described by z-score or standard score [26]. An object's z-score is calculated by $z = (x - \mu)/\sigma$, where x is object's ESV. The z-score integrates mean and deviation together and outcome only one value. This value is the ratio of the distance between object and mean and deviation. According to Equation (4), k_1 and k_2 are actually the two z-scores of the pair of thresholds. The z-score provides another measure to determine three regions.

4   Conclusion

Three-way decisions play an important role in many real world decision-making problems. Several interpretations of three-way decisions have been proposed. This paper examines statistical interpretations based on fundamental notions in statistics, including percentile, median, mean, and standard deviation.

Evaluation based three-way decisions use an evaluation function and a pair of thresholds to divide a universal set into three pair-wise disjoint regions. Statistical interpretations of three-way decisions construct three regions based on an understanding that the middle region M consists of normal or typical instances from a population, while regions L and R consist of, respectively, abnormal or untypical instances. Two special cases of statistical interpretation are proposed in this paper. One is a set of non-numeric values and the other is a set of numeric values. These interpretations are widely used in many applications, such as box-plots, IQ score classifications, and blood pressure classifications.

This paper considers only very simple ideas of using statistical information for three-way decisions. Such interpretations provide a way to determine the pair of required thresholds in three-way decisions. One may also explore other types of statistical information, such as mode, average absolute deviation, median absolute deviation, and z-score.

As future work, we will examine the process of constructing an evaluation function. We will also study three-way decisions to distinguish between normal and abnormal, and between typical and untypical instances by using specific classes of distributions. The criteria to determinate the pair of percentiles and the pair of thresholds for two interpretations will be studied based on concrete applications.

Acknowledgements. This work is partially supported by a Discovery Grant from NSERC, Canada and Sampson J. Goodfellow Scholarship.

References

1. Azam, N., Yao, J.T.: Analyzing uncertainties of probabilistic rough set regions with game-theoretic rough sets. Int. J. Approximate Reasoning **55**, 142–155 (2014)
2. Baram, Y.: Partial classification: the benefit of deferred decision. IEEE Trans. Pattern Anal. Mach. Intell. **20**, 769–776 (1998)
3. Czitrom, V., Spagon, P.D.: Statistical Case Studies for Industrial Process Improvement. SIAM, Philadelphia (1997)
4. Deng, X.F.: Three-Way Classification Models. Ph.D. Dissertation, Department of Computer Science, University of Regina (2015)
5. Deng, X.F., Yao, Y.Y.: A multifaceted analysis of probabilistic three-way decisions. Fundam. Informaticae **132**, 291–313 (2014)
6. Deng, X.F., Yao, Y.Y.: Decision-theoretic three-way approximations of fuzzy sets. Inf. Sci. **279**, 702–715 (2014)
7. Goudey, R.: Do statistical inferences allowing three alternative decision give better feedback for environmentally precautionary decision-making? J. Environ. Manage. **85**, 338–344 (2007)
8. Grzymala-Busse, J.W., Clarka, P.G., Kuehnhausena, M.: Generalized probabilistic approximations of incomplete data. Int. J. Approximate Reasoning **50**, 180–196 (2014)
9. Hu, B.Q.: Three-way decisions space and three-way decisions. Inf. Sci. **281**, 21–52 (2014)
10. Iserson, K.V., Moskop, J.C.: Triage in medicine, part I: concept, history, and types. Ann. Emerg. Med. **49**, 275–281 (2007)
11. Jia, X.Y., Shang, L., Zhou, X.Z., Liang, J.Y., Miao, D.Q., Wang, G.Y., Li, T.R., Zhang, Y.P. (eds.): Theory of Three-way Decisions and Applications (in Chinese). Nanjing University Press, Nanjing (2012)
12. Jia, X.Y., Tang, Z.M., Liao, W.H., Shang, L.: On an optimization representation of decision-theoretic rough set model. Int. J. Approximate Reasoning **55**, 156–166 (2014)
13. Li, H.X., Zhang, L.B., Huang, B., Zhou, X.Z.: Sequential three-way decision and granulation for cost-sensitive face recognition, Knowledge-Based Systems (2015). http://dx.doi.org/10.1016/j.knosys.2015.07.040
14. Li, H.X., Zhou, X.Z., Huang, B., Liu, D.: Cost-sensitive three-way decision: a sequential strategy. In: Lingras, P., Wolski, M., Cornelis, C., Mitra, S., Wasilewski, P. (eds.) RSKT 2013. LNCS, vol. 8171, pp. 325–337. Springer, Heidelberg (2013)
15. Liang, D.C., Liu, D.: Deriving three-way decisions from intuitionistic fuzzy decision-theoretic rough sets. Inf. Sci. **300**, 28–48 (2015)
16. Liang, D.C., Pedrycz, W., Liu, D., Hu, P.: Three-way decisions based on decision-theoretic rough sets under linguistic assessment with the aid of group decision making. Appl. Soft Comput. **29**, 256–269 (2015)
17. Liu, D., Li, T.R., Liang, D.C.: Incorporating logistic regression to decision-theoretic rough sets for classifications. Int. J. Approximate Reasoning **55**, 197–210 (2014)
18. Liu, D., Li, T.R., Liang, D.C.: Three-way government decision analysis with decision-theoretic rough sets. Int. J. Uncertainty Fuzziness Knowl.-Based Syst. **20**, 119–132 (2012)

19. Liu, D., Li, T.R., Miao, D.Q., Wang, G.Y., Liang, J.Y. (eds.): Three-way Decisions and Granular Computing (in Chinese). Science Press, Beijing (2013)
20. Liu, D., Liang, D.C., Wang, C.C.: A novel three-way decision model based on incomplete information system. Knowledge-Based Systems (2015). http://dx.doi.org/10.1016/j.knosys.2015.07.036
21. Pater, C.: The blood pressure "uncertainty range" - a pragmatic approach to overcome current diagnostic uncertainties (II). Curr. Controlled Trials Cardiovasc. Med. **6**, 5 (2005)
22. Pawlak, Z.: Rough sets. Int. J. Comput. Inf. Sci. **11**, 341–356 (1982)
23. Pedrycz, W., Skowron, A., Kreinovich, V.: Handbook of Granular Computing. Wiley, Chichester (2008)
24. Peters, J.F., Ramanna, S.: Proximal three-way decisions: theory and applications in social networks. Knowledge-Based Systems (2015). http://dx.doi.org/10.1016/j.knosys.2015.07.021
25. Rousseeuw, P.J., Ruts, I., Tukey, J.W.: The bagplot: a bivariate boxplot. Am. Stat. **53**, 382–387 (1999)
26. Sanders, D.H., Smidt, R.K., Adatia, A., Larson, G.A.: Statistics: A First Course. McGraw-Hill Ryerson, Toronto (2001)
27. Sang, Y.L., Liang, J.Y., Qian, Y.H.: Decision-theoretic rough sets under dynamic granulation. Knowledge-Based Systems (2015). http://dx.doi.org/10.1016/j.knosys.2015.08.001
28. Sattler, J.M.: Assessment of Children's Intelligence. W.B. Saunders Company, Philadelphia (1975)
29. Schechter, C.B.: Sequential analysis in a Bayesian model of diastolic blood pressure measurement. Med. Decis. Making **8**, 191–196 (1988)
30. Schofield, H.: Assess. Test. Introduction. Allen & Unwin, London (1972)
31. Shakiba, A., Hooshmandasl, M.R.: S-approximation spaces: a three-way decision approach. Fundam. Informaticae **39**, 307–328 (2015)
32. Wald, A.: Sequential Anal. Wiley, New York (1947)
33. Yao, J.T., Azam, N.: Web-based medical decision support systems for three-way medical decision making with game-theoretic rough sets. IEEE Trans. Fuzzy Syst. **23**, 3–15 (2014)
34. Yao, J.T., Herbert, J.P.: A game-theoretic perspective on rough set analysis. J. Chongqing Univ. Posts Telecommun. **20**, 291–298 (2008)
35. Yao, Y.Y.: An outline of a theory of three-way decisions. In: Yao, J.T., Yang, Y., Słowiński, R., Greco, S., Li, H., Mitra, S., Polkowski, L. (eds.) RSCTC 2012. LNCS, vol. 7413, pp. 1–17. Springer, Heidelberg (2012)
36. Yao, Y.Y.: Granular computing and sequential three-way decisions. In: Lingras, P., Wolski, M., Cornelis, C., Mitra, S., Wasilewski, P. (eds.) RSKT 2013. LNCS, vol. 8171, pp. 16–27. Springer, Heidelberg (2013)
37. Yao, Y.Y.: Interval-set algebra for qualitative knowledge representation. In: Proceedings of the 5th International Conference on Computing and Information (ICCI), pp. 370–374 (1993)
38. Yao, Y.Y.: Perspectives of granular computing. In: Proceedings of 2005 IEEE International Conference on Granular Computing, vol. 1, pp. 85–90 (2005)
39. Yao, Y.Y.: Rough sets and three-way decisions. In: Ciucci, D., Wang, G.Y., Mitra, S., Wu, W.Z. (eds.) RSKT 2015. LNCS (LNAI), vol. 9436, pp. 62–73. Springer International Publishing, Switzerland (2015)
40. Yao, Y.Y., Wong, S.K.M., Lingras, P.: A decision-theoretic rough set model. In: Proceedings of the 5th International Symposium on Methodologies for Intelligent Systems, pp. 17–25 (1990)

41. Yao, Y.Y., Yu, H.: An introduction of three-way decisions. In: Yu, H., Wang, G.Y., Li, T.R., Liang, J.Y., Miao, D.Q., Yao, Y.Y. (eds.) Three-Way Decisions: Methods and Practices for Complex Problem Solving, pp. 1–19. Science Press, Beijing (2015) (in Chinese)

42. Yu, H., Liu, Z.G., Wang, G.Y.: An automatic method to determine the number of clusters using decision-theoretic rough set. Int. J. Approximate Reasoning **55**, 101–115 (2014)

43. Yu, H., Su, T., Zeng, X.H.: A three-way decisions clustering algorithm for incomplete data. In: Miao, D., Pedrycz, W., Slezak, D., Peters, G., Hu, Q., Wang, R. (eds.) RSKT 2014. LNCS, vol. 8818, pp. 765–776. Springer, Heidelberg (2014)

44. Yu, H., Wang, G.Y., Li, T.R., Liang, J.Y., Miao, D.Q., Yao, Y.Y. (eds.): Three-Way Decisions: Methods and Practices for Complex Problem Solving. Science Press, Beijing (2015) (in Chinese)

45. Yu, H., Zhang, C., Wang, G.Y.: A tree-based incremental overlapping clustering method using the three-way decision theory. Knowledge-Based Systems (2015). http://dx.doi.org/10.1016/j.knosys.2015.05.028

46. Zhang, H.R., Min, F.: Three-way recommender systems based on random forests. Knowledge-Based Systems (2015). http://dx.doi.org/10.1016/j.knosys.2015.06.019

47. Zhang, H.Y., Yang, S.Y., Ma, J.M.: Ranking interval sets based on inclusion measures and applications to three-way decisions. Knowledge-Based Systems (2015). http://dx.doi.org/10.1016/j.knosys.2015.07.025

48. Zhang, Y.: Optimizing Gini coefficient of probabilistic rough set regions using game-theoretic rough sets. In: 26th Canadian Conference of Electrical And Computer Engineering (CCECE), pp. 1–4 (2013)

49. Zhou, B.: Multi-class decision-theoretic rough sets. Int. J. Approximate Reasoning **55**, 211–224 (2014)

50. Zhou, B., Yao, Y.Y., Luo, J.G.: Cost-sensitive three-way email spam filtering. J. Intell. Inf. Syst. **42**, 19–45 (2014)

Decision-Level Sensor-Fusion Based on DTRS

Bing Zhou[1][✉] and Yiyu Yao[2]

[1] Department of Computer Science, Sam Houston State University,
Huntsville, TX 77341, USA
zhou@shsu.edu
[2] Department of Computer Science, University of Regina,
Regina, SK S4S 0A2, Canada
yyao@cs.uregina.ca

Abstract. A decision-level sensor fusion based on decision-theoretic rough set (DTRS) model is proposed. Sensor fusion is the process of combining sensor readings from disparate resources such that the resulting information is more accurate and complete. Decision-level sensor fusion combines the detection results instead of raw data of different sensors, and it is most suitable when we have different types of sensors. Rough set theory offers a three-way decision approach to combine sensor results into three regions and reasoning under uncertain circumstances. Based on DTRS, we build a cost-sensitive sensor fusion model. A loss function is interpreted as the costs of making different classification decisions, the computation of required thresholds to define the three regions is based on the loss functions. Finally, an illustrative example demonstrates the framework's effectiveness and validity.

Keywords: Sensor fusion · Rough sets · Cost-sensitive · Uncertainty · Three-way

1 Introduction

Sensor fusion is the process of combining sensor readings from disparate resources such that the resulting information is more accurate and complete [4, 7]. Some of the common sensor types are Global Positioning System (GPS), infrared cameras, radars, and magnetic sensors. Sensor fusion is an example of information fusion. Fusion level defines the level at which the information of different sensors is combined. There are three different fusion levels: data-level fusion is the combination of raw sensor data; feature-level fusion is the combination of features extracted from different sensor data that represent object properties; decision-level fusion combines the detection results of different sensors. The choice of a suitable fusion level depends on the sensor types. For homogeneous sensors, one can use data-level fusion to take all unprocessed data into consideration. If the sensor types are very different, decision-level fusion is more suitable.

Rough set theory [11, 19] is a well-known mathematical tool for dealing with uncertain and imprecise information. It identifies decision rules and dependencies from data for decision making and classification. In sensor fusion problems,

© Springer International Publishing Switzerland 2015
D. Ciucci et al. (Eds.): RSKT 2015, LNAI 9436, pp. 321–332, 2015.
DOI: 10.1007/978-3-319-25754-9_29

since information is often incomplete, or is from unreliable resources, the sensor detection results based on these data may be uncertain. Rough set can be used to deal with this issue. In existing research of applying rough sets to sensor fusion, rough sets were mainly used for data preprocessing (e.g., constructing an attribute reduct) instead of decision making [9,13]. Pawlak et. al [12] proposed to use rough membership function to the fusion of homogeneous sensors. In their approach, different sensors are interpreted as attributes (columns) of an information table, sensor readings over a time interval are interpreted as rows, and the decision is the estimation of whether an agent tends to favor walking in a region (time interval). This approach serves as means of classifying sensors, that is, depending on the approximation of sensor values to a region defined by equivalence class, the rough integral computes the relevance of a sensor in a classification effort, and the identification of most relevant sensors provides a form of sensor fusion.

Fusion techniques can be seen as a classification problem with a discriminant function $f(a)$, where $f(a) \geq \gamma$ implies the positive result and $f(a) < \gamma$ implies negative result. However, it is often difficult to choose the best cut-off value (threshold) on the sensor output for decision. A higher threshold may be in favor of high precision but low recall, a lower threshold may produce the opposite results. In a three-way sensor fusion, a pair of thresholds is used to produce three decisions, the positive, the deferment, and the negative. By adding the deferment option, the misclassification errors from making the incorrect positive and negative decisions are reduced by introducing deferment errors. Thus, we can have a better precision and recall. In addition, Sensor fusion is a cost-sensitive task. For example, in landmine detection, misclassifying landmine to background is considered more costly than misclassifying background to landmine. Such characteristics have not been explicitly reflected and made clear in the existing sensor fusion methods. Cost-sensitive classification has received much attention in recent years. In the traditional classification task, minimizing error rate is used as the guideline of designing a classification system. The error rate assigns no loss to a correct decision, and assigns a unit loss to any error. Thus, all errors are equally costly. In real world applications, each error has different cost. Therefore, it is important to build a cost-sensitive classification system to incorporate different types of costs into the classification process.

In this paper, we introduce a cost-sensitive three-way decision approach to decision-level sensor fusion. More specifically, we adopt the systematic method from decision-theoretic rough set (DTRS) models to calculate a pair of thresholds based on well established Bayesian decision theory, with the aid of more practically operable notions such as cost, risk, benefit etc. [14]. A loss function is defined to state how costly each decision is, and a final decision is to choose the one for which the overall cost is minimum. Illustrative examples show that three-way classification has an overall lower coat compares to the traditional binary classification.

2   Sensor Fusion Techniques

In this section, the commonly used statistical fusion techniques are reviewed. This paper focuses on decision-level sensor fusion since all knowledge about the sensors can be applied separately. Each sensor can optimize the detection performance based on the information gathered.

The most common fusion technique is based on the well-known Bayesian decision theory, which is a fundamental statistical approach that makes decisions under uncertainty based on probabilities and costs associated with decisions [2]. For each sensor output a, a decision has to be made: a positive decision C or a negative decision C^c. Let $\lambda(C|C^c)$ denote the cost for making decision C when the true class is C^c. Let $Pr(C|a)$ be the posterior probability of a target being in class C given a. For a target with sensor output a, suppose decision C is made, the expected cost associated with making decision C is given by:

$$R(C|a) = \lambda(C|C^c)Pr(C^c|a) + \lambda(C|C)Pr(C|a). \tag{1}$$

Similarly, the expected cost associated with making decision C^c is given by:

$$R(C^c|a) = \lambda(C^c|C)Pr(C|a) + \lambda(C^c|C^c)Pr(C^c|a). \tag{2}$$

The posterior probability can be calculated using Bayes' theorem:

$$Pr(C|a) = \frac{Pr(C)Pr(a|C)}{Pr(a)}, \tag{3}$$

with $Pr(a|C)$ the likelihood of C given a, and $Pr(C)$ the a priori probability of class C.

For two class C and C^c, if the risk of making the decision that the class is C is lower than the risk of making the decision that the class is C^c given the description a, that is,

$$R(C|a) < R(C^c|a), \tag{4}$$

we make the decision that the class is C. This is equivalent to:

$$\frac{Pr(a|C)}{Pr(a|C^c)} \geq \frac{(\lambda(C|C^c) - \lambda(C^c|C^c))Pr(C^c)}{(\lambda(C^c|C) - \lambda(C|C))Pr(C)}. \tag{5}$$

The right-hand side of Eq. (5) is a constant value. The left-hand side of this equation is a ratio of two conditional probabilities and is called the likelihood ratio. This likelihood ratio is the optimal discriminant function for minimal expected cost. Estimating the conditional probabilities is the major challenge of classification.

Other common decision-level sensor fusion techniques include Naive Bayes' approach [2], Dempster-Shafer theory approach [1], fuzzy probability approach [3], and voting fusion approach [8]. These existing sensor fusion methods focus on working towards functions that give better estimations of sensor outputs. On the other hand, estimations of required threshold have not received much attention. There is a need for inferring these thresholds from a theoretical and practical basis.

3   Rough Set Theory

In Pawlak's rough set model [11], information about a finite set of objects are represented in an information table with a finite set of attributes. Formally, an information table can be expressed as:

$$S = (U, At, \{V_a \mid a \in At\}, \{I_a \mid a \in At\}),$$

where

U is a finite nonempty set of objects called the universe,

At is a finite nonempty set of attributes,

V_a is a nonempty set of values for $a \in At$,

$I_a : U \rightarrow V_a$ is an information function.

The information function I_a maps an object in U to a value of V_a for an attribute $a \in At$, that is, $I_a(x) \in V_a$. An equivalence relation can be defined with respect to a subset of attributes $A \subseteq At$, denoted as R_A, or simply R,

$$xRy \Longleftrightarrow \forall_{a \in A} I_a(x) = I_a(y)$$
$$\Longleftrightarrow I_A(x) = I_A(y).$$

The relation R is reflexive, symmetric and transitive. Two objects x and y in U are equivalent or indiscernible by the set of attributes A if and only if they have the same values on all attributes in A. The equivalence relation R induces a partition of U, denoted by U/R. The subsets in U/R are called equivalence classes, which are the building blocks to construct rough set approximations. The equivalence class containing x is defined as:

$$[x] = \{y \in U \mid xRy\}.$$

Consider an equivalence relation R on U. The equivalence classes induced by the partition U/R are the basic blocks to construct Pawlak's rough set approximations. For a subset $C \subseteq U$, the lower and upper approximations of C with respect to U/R are defined by:

$$\underline{apr}(C) = \{x \in U \mid [x] \subseteq C\}$$
$$= \bigcup \{[x] \in U/R \mid [x] \subseteq C\};$$
$$\overline{apr}(C) = \{x \in U \mid [x] \cap C \neq 0\}$$
$$= \bigcup \{[x] \in U/R \mid [x] \cap C \neq \emptyset\}. \qquad (6)$$

Based on the rough set approximations of C, one can divide the universe U into three pair-wise disjoint regions: the positive region POS(C) is the union of all the equivalence classes that is included in C; the negative region NEG(C) is the union of all equivalence classes that have an empty intersection with C;

and the boundary region $\text{BND}(C)$ is the difference between the upper and lower approximations:

$$\text{POS}(C) = \underline{apr}(C),$$
$$\text{NEG}(C) = U - \text{POS}(C) \cup \text{BND}(C),$$
$$\text{BND}(C) = \overline{apr}(C) - \underline{apr}(C). \tag{7}$$

It can be verified that $\text{NEG}(C) = \text{POS}(C^c)$, where C^c is the complement of C.

4 A Cost-Sensitive Three-Way Decision Approach

The definitions of positive, negative and boundary regions in rough sets lead to a three-way classification. In this section, we propose to use a specific model of rough set theory, called decision-theoretic rough set model for a cost-sensitive three-way decision-level sensor fusion.

4.1 The General Form

Sensor fusion can be considered as a classification problem with each sensor output as attributes and detection targets as objects. A discriminate function is used to differentiate positive and negative detection results on the basis of the outputs of different sensors. The output of each sensor is a measure of confidence in the presence of a positive result. The confidence level is a detection of a target given a certain sensor. The higher confidence level implies a higher probability of positive results. The discriminant function $f(a)$ is defined as:

$$f(a) \geq \gamma \longrightarrow \text{assign positive results,}$$
$$f(a) < \gamma \longrightarrow \text{assign negative results,}$$

where $a = \{a_1, a_2, ..., a_n\}$, $a_i \in [0, 1]$ is a sensor output vector with n the number of sensors and γ the threshold. $f(a) \geq \gamma$ indicates the positive detection result and $f(a) < \gamma$ indicates negative result.

Based on the sensor outputs, the discriminant function $f(a)$ can be interpreted differently for classification. For example, in naive Bayes classifier, $f(a)$ is the posterior probability or its monotonic transformations (e.g., the posterior odds). In SVM method, $f(a)$ is the distance between the given target and the decision hyperplane. In a recent paper of Yao and Deng [18], the discriminant function $f(a)$ is interpreted as a subsethood measure indicating the degree of inclusion between two sets. The statistical fusion technique based on Bayesian decision theory uses the likelihood ratio as the optimal discriminant function for minimal expected cost: $f(a) = \frac{Pr(a|C)}{Pr(a|C^c)}$.

In the classic decision-level sensor fusion, only one threshold $\gamma \in [0, 1]$ is used to compare with normalized $f(a)$ in order to determine whether or not the target is positive. Two classification regions, called the positive and negative regions, are produced for a given set of targets, that is,

$$POS_{(\gamma)}(C) = \{a \mid f(a) \geq \gamma\},$$
$$NEG_{(\gamma)}(C) = \{a \mid f(a) < \gamma\}. \tag{8}$$

The positive region $POS_{(\gamma)}(C)$ contains targets that are identified as positive, and the negative region $NEG_{(\gamma)}(C)$ contains targets that are identified as negative.

In three-way decision-level sensor fusion, although the true class is only binary (i.e., C or C^c), we make a three-way decision based on each sensor output. A pair of thresholds (α, β) with $0 \leq \beta < \alpha \leq 1$ is used to distinguish different value ranges of $f(a)$. The first threshold α determines the probability necessary for a re-examination, and the second threshold β determines the probability necessary for determining a negative results. The pair of thresholds produces three classification regions, called the positive, boundary, and negative regions as follows:

$$POS_{(\alpha,\beta)}(C) = \{a \mid f(a) \geq \alpha\},$$
$$BND_{(\alpha,\beta)}(C) = \{a \mid \beta < f(a) < \alpha\},$$
$$NEG_{(\alpha,\beta)}(C) = \{a \mid f(a) \leq \beta\}, \tag{9}$$

We determine a target as positive if $f(a)$ is greater than or equals to α. We determine a target as negative if $f(a)$ is less than or equals to β. We do not make an immediate decision if $f(a)$ is between α and β, instead, we make a decision of deferment.

4.2 Computing Thresholds Based on DTRS

In the context of rough set theory, the discriminate function of sensor fusion $f(a)$ can be interpreted as the conditional probability $Pr(C \mid [x])$:

$$f(a) = Pr(C \mid [x]), \tag{10}$$

in which the degrees of overlap between equivalence classes $[x]$ and a set C to be approximated are considered. In Eq. (7), an equivalence class is in the positive region if and only if it is fully contained in the set. This may be too restrictive to be practically useful in real applications. An attempt to use probabilistic information for approximations was suggested [5, 6, 10] to allow some tolerance of errors. A conditional probability is defined as:

$$Pr(C \mid [x]) = \frac{|C \cap [x]|}{|[x]|}, \tag{11}$$

where $|\cdot|$ denotes the cardinality of a set, and the conditional probability is written as $Pr(C \mid [x])$ representing the probability that an object belongs to C given that the object is described by $[x]$. The three regions in Eq. (7) can be equivalently defined by:

$$POS(C) = \{x \in U \mid Pr(C \mid [x]) = 1\},$$
$$BND(C) = \{x \in U \mid 0 < Pr(C \mid [x]) < 1\},$$
$$NEG(C) = \{x \in U \mid Pr(C \mid [x]) = 0\}. \tag{12}$$

They are defined by using the two extreme values, 0 and 1, of probabilities. They are of a qualitative nature; the magnitude of the value $Pr(C \mid [x])$ is not taken into account.

Yao et al. [14] introduced a more general probabilistic model, called decision-theoretic rough set (DTRS) model, in which a pair of thresholds α and β with $\alpha > \beta$ on the probability is used to define three probabilistic regions. The (α, β)-probabilistic positive, boundary and negative regions are defined by [14]:

$$\text{POS}_{(\alpha,\beta)}(C) = \{x \in U \mid Pr(C \mid [x]) \geq \alpha\},$$
$$\text{BND}_{(\alpha,\beta)}(C) = \{x \in U \mid \beta < Pr(C \mid [x]) < \alpha\},$$
$$\text{NEG}_{(\alpha,\beta)}(C) = \{x \in U \mid Pr(C \mid [x]) \leq \beta\}. \tag{13}$$

Probabilistic three regions may be interpreted in terms of costs of different types of classification decisions [15]. One obtains larger positive and negative regions by introducing classification errors in trade of a smaller boundary region so that the total classification cost is minimum [17]. Zhou and Yao proposed to estimate the conditional probability based on the naive probabilistic independence assumption which leads to analytical and computational implications [16].

The cost-sensitive nature of sensor fusion is reflected by working on the derivation of the required thresholds based on the systematic method provided in the DTRS models [14]. With respect to a set of detection targets, there are two classes C and C^c indicating that a target is in C (i.e., positive) or not in C (i.e., negative). To derive the three classification regions, the set of decisions is given by $\mathcal{A} = \{a_P, a_B, a_N\}$, where a_P, a_B, and a_N represent the three decisions in classifying a target x, namely, deciding $x \in \text{POS}(C)$, deciding $x \in \text{BND}(C)$, and deciding $x \in \text{NEG}(C)$, respectively. The loss function is given by a 3×2 matrix:

	C (P): positive	C^c (N): negative
a_P: positive	$\lambda_{PP} = \lambda(a_P|C)$	$\lambda_{PN} = \lambda(a_P|C^c)$
a_B: deferment	$\lambda_{BP} = \lambda(a_B|C)$	$\lambda_{BN} = \lambda(a_B|C^c)$
a_N: negative	$\lambda_{NP} = \lambda(a_N|C)$	$\lambda_{NN} = \lambda(a_N|C^c)$

In the matrix, λ_{PP}, λ_{BP} and λ_{NP} denote the losses incurred for making decisions a_P, a_B and a_N, respectively, when a target belongs to C, and λ_{PN}, λ_{BN} and λ_{NN} denote the losses incurred for making these decisions when the target does not belong to C. In particular, λ_{NP} is the loss incurred for mistakenly determining a positive target, and λ_{PN} is the loss incurred for mistakenly determining a negative target.

The expected losses associated with making different decisions for targets with description $[x]$ can be expressed as:

$$R(a_P|[x]) = \lambda_{PP} Pr(C|[x]) + \lambda_{PN} Pr(C^c|[x]),$$
$$R(a_B|[x]) = \lambda_{BP} Pr(C|[x]) + \lambda_{BN} Pr(C^c|[x]),$$
$$R(a_N|[x]) = \lambda_{NP} Pr(C|[x]) + \lambda_{NN} Pr(C^c|[x]). \tag{14}$$

The Bayesian decision procedure suggests the following minimum-risk decision rules:

(P) If $R(a_P|[x]) \leq R(a_B|[x])$ and $R(a_P|[x]) \leq R(a_N|[x])$, decide $x \in POS(C)$;

(B) If $R(a_B|[x]) \leq R(a_P|[x])$ and $R(a_B|[x]) \leq R(a_N|[x])$, decide $x \in BND(C)$;

(N) If $R(a_N|[x]) \leq R(a_P|[x])$ and $R(a_N|[x]) \leq R(a_B|[x])$, decide $x \in NEG(C)$.

Tie-breaking criteria should be added so that each target is put into only one region.

Based on the derivation results from DTRS [14], we can express three parameters using different loss functions:

$$\alpha = \frac{(\lambda_{PN} - \lambda_{BN})}{(\lambda_{PN} - \lambda_{BN}) + (\lambda_{BP} - \lambda_{PP})},$$

$$\beta = \frac{(\lambda_{BN} - \lambda_{NN})}{(\lambda_{BN} - \lambda_{NN}) + (\lambda_{NP} - \lambda_{BP})},$$

$$\gamma = \frac{(\lambda_{PN} - \lambda_{NN})}{(\lambda_{PN} - \lambda_{NN}) + (\lambda_{NP} - \lambda_{PP})}. \tag{15}$$

After tie-breaking, the following simplified rules are obtained:

(P1) If $Pr(C|[x]) \geq \alpha$, decide $x \in POS(C)$;

(B1) If $\beta < Pr(C|\mathbf{x}) < \alpha$, decide $x \in BND(C)$;

(N1) If $Pr(C|[x]) \leq \beta$, decide $x \in NEG(C)$.

The parameter γ is no longer needed. The (α, β)-probabilistic positive, negative and boundary regions are exactly same as in Eq. (9). The threshold α and β can be systematically calculated from loss functions based on Eq. (15).

5   An Example

Consider the use of multiple sensors for landmine detection as an example. Based on notations in Sect. 4.2, there are two true classes regarding a target: C denoting landmine and C^c denoting background. There are three actions: a_P for determining there is landmine, a_N for determining there is a background, and a_B for making a deferred decision.

A loss function is interpreted as the costs of making the corresponding decisions. Generally speaking, a higher cost occurs when misclassifying a landmine as background; it could result in life dangerous situation. On the other hand, misclassifying background to landmine brings unnecessary costs of defusing the landmine. Costs also occur when a deferment decision is made.

As shown in Table 1, there are four types of loss functions in the binary cost matrix. According to the minimum risk decision rules in Bayesian decision theory, we determine a landmine if the expected risk is smaller than determine a background, that is, $R(a_P|[x]) \leq R(a_N|[x])$. In the cost matrix for ternary

Table 1. Binary cost matrix

	C (P)(landmine)	C^c (N)(background)
a_P(positive)	$\lambda_{PP} = \lambda(a_P|C)$	$\lambda_{PN} = \lambda(a_P|C^c)$
a_N(negative)	$\lambda_{NP} = \lambda(a_N|C)$	$\lambda_{NN} = \lambda(a_N|C^c)$

Table 2. Ternary cost matrix

	C (P)(landmine)	C^c (N)(background)
a_P(positive)	$\lambda_{PP} = \lambda(a_P|C)$	$\lambda_{PN} = \lambda(a_P|C^c)$
a_B(deferment)	$\lambda_{BP} = \lambda(a_B|C)$	$\lambda_{BN} = \lambda(a_B|C^c)$
a_N (negative)	$\lambda_{NP} = \lambda(a_N|C)$	$\lambda_{NN} = \lambda(a_N|C^c)$

classification as shown in Table 2, there are six types of loss functions. In order to make a decision, we need to do a pairwise comparison of the three expected risks: $R(a_P|[x])$, $R(a_B|[x])$ and $R(a_N|[x])$. Each of these comparisons can be translated into the comparison between the a posteriori probability $Pr(C \mid [x])$ and a pair of thresholds α and β, where (α, β) can be calculated based on Eq. (15), and $Pr(C \mid [x])$ can be calculated based on the methods proposed in [16].

Tables 3 and 4 give two set of loss functions, respectively. It can be seen that mis-detecting a landmine considered more costly in Table 3 than in Table 4. The pair of thresholds α^1 and β^1 for Table 3 is calculated according to Eq. (15) as:

$$\alpha^1 = \frac{(\lambda^1_{PN} - \lambda^1_{BN})}{(\lambda^1_{PN} - \lambda^1_{BN}) + (\lambda^1_{BP} - \lambda^1_{PP})} = \frac{10 - 5}{(10 - 5) + (5 - 0)} = 0.50,$$

$$\beta^1 = \frac{(\lambda^1_{BN} - \lambda^1_{NN})}{(\lambda^1_{BN} - \lambda^1_{NN}) + (\lambda^1_{NP} - \lambda^1_{BP})} = \frac{5 - 0}{(5 - 0) + (90 - 5)} = 0.06.$$

The pair of thresholds α^2 and β^2 for Table 4 is calculated as:

$$\alpha^2 = \frac{(\lambda^2_{PN} - \lambda^2_{BN})}{(\lambda^2_{PN} - \lambda^2_{BN}) + (\lambda^2_{BP} - \lambda^2_{PP})} = \frac{8 - 5}{(8 - 5) + (5 - 0)} = 0.38,$$

$$\beta^2 = \frac{(\lambda^2_{BN} - \lambda^2_{NN})}{(\lambda^2_{BN} - \lambda^2_{NN}) + (\lambda^2_{NP} - \lambda^2_{BP})} = \frac{5 - 0}{(5 - 0) + (15 - 5)} = 0.33.$$

Table 3. Loss function set 1

	C (P)(landmine)	C^c (N)(background)
a_P(positive)	$\lambda^1_{PP} = 0$	$\lambda^1_{PN} = 10$
a_B(deferment)	$\lambda^1_{BP} = 5$	$\lambda^1_{BN} = 5$
a_N(negative)	$\lambda^1_{NP} = 90$	$\lambda^1_{NN} = 0$

Table 4. Loss function set 2

	$C\ (P)$(landmine)	$C^c\ (N)$(background)
a_P(positive)	$\lambda^2_{PP} = 0$	$\lambda^2_{PN} = 8$
a_B(deferment)	$\lambda^2_{BP} = 5$	$\lambda^2_{BN} = 5$
a_N(negative)	$\lambda^2_{NP} = 15$	$\lambda^2_{NN} = 0$

It follows that $\beta^1 < \beta^2 < \alpha^2 < \alpha^1$. As expected, the thresholds of Table 4 are within the thresholds of Table 3, which shows that the user of Table 3 is more critical than the user of Table 4 regarding both incorrect decisions. Consequently, Table 3 would have a larger group of deferred targets. In contract, Table 4 would have larger group of landmine and background targets but a smaller group of deferred targets.

Table 5. The sensor outputs for six targets. Here we consider three sensors s_1, s_2, s_3 and each column refers to the decision from a sensor

targets	s_1	s_2	s_3	class
t_1	0.4	0.3	0.7	background
t_2	0.8	0.65	0.9	landmine
t_3	0.1	0.3	0.45	background
t_4	0.75	0.5	0.6	landmine
t_5	0.95	0.8	0.7	landmine
t_6	0.2	0.35	0.4	background

For a new target t_7, suppose its probability of being landmine can be calculated from the training set as shown in Table 5, that is, $Pr(Landmine|[x]) = 0.3$. Base on loss functions in Table 3, t_7 will be classified into the deferment target group for further examination because $0.06 < Pr(Landmine|[x]) < 0.50$, but based on loss functions in Table 4, t_7 will be classified as background because $Pr(Landmine|[x]) \geq 0.38$. Different filtering options are tailored to meet individual requirements in terms of minimum overall cost based on our approach.

6   Conclusions and Future Work

Rough set theory is well-known for dealing with uncertainty and imprecise information. In this paper, we provide a three-way decision approach for decision-level sensor fusion based on DTRS. Compared to other commonly used sensor fusion techniques, a boundary region marked deferment is added to the classification results to convert potential misclassification errors into errors of deferment. The main advantage of our approach is that it allows the possibility of indecision, i.e., of refusing to make a decision. The undecided cases must be reexamined by collecting additional information. A pair of thresholds are used. The first threshold determines the point necessary for a re-examination, and the second threshold determines the point to reject. The required thresholds is systematically calculated based on DTRS. The cost associated with each decision is given by a loss function from the well-established Bayesian decision theory. The cost-sensitive characteristic is reflected by varying the values of loss functions.

References

1. Dempster, A.P.: Upper and lower probabilities induced by a multi-valued mapping. Ann. Math. Stat. **38**, 325–339 (1967)
2. Duda, R.O., Hart, P.E.: Pattern Classification and Scene Analysis. Wiley, New York (1973)
3. Freeling, A.N.S.: Possibility versus fuzzy probabilities - two alternatives. In: Zimmermann, H.J., Zadeh, L.A., Gaines, B.R. (eds.) Fuzzy Sets and Decision Analysis. Elsevier, Amsterdam (1984)
4. Gros, B., Bruschini, C.: Sensor technologies for the detection of antipersonnel mines: a survey of current research and system developments. In: Proceedings of the International Symposium on Measurement and Control in Robotics (ISMCR'96), Brussels, Belgium, pp. 509–518 (1996)
5. Grzymala-Busse, J.W.: Generalized probabilistic approximations. T. Rough Sets **16**, 1–16 (2013)
6. Grzymala-Busse, J.W., Clark, P.G., Kuehnhausen, M.: Generalized probabilistic approximations of incomplete data. Int. J. Approx. Reasoning **55**(1), 180–196 (2014)
7. Hall, D.L.: Mathematical Techniques in Multi-sensor Data Fusion. Artech House Inc, Norwood (1992)
8. Klein, L.A.: A Boolean algebra approach to multiple sensor voting fusion. IEEE Trans. Aerosp. Electron. Syst. **29**(2), 317–327 (1998)
9. Li, T., Fei, M.: Information fusion in wireless sensor network based on rough set. In: IEEE International Conference on Network Infrastructure and Digital Content, pp. 129–134 (2009)
10. Pawlak, Z., Wong, S.K.M., Ziarko, W.: Rough sets: probabilistic versus deterministic approach. Int. J. Man-Mach. Stud. **29**, 81–95 (1988)
11. Pawlak, Z.: Rough Sets, Theoretical Aspects of Reasoning about Data. Kluwer Academic Publishers, Dordrecht (1991)
12. Pawlak, Z., Peters, J.F., Skowron, A., Suraj, Z., Ramanna, S., Borkowski, M.: Rough measures, rough integrals and sensor fusion. Rough Set Theory Granular Comput. Stud. Fuzziness Soft Comput. **125**, 263–272 (2003)
13. Wang, H., Chen, Y.: Sensor data fusion using rough set for mobile robots system. In: Proceedings of the 2nd IEEE/ASME International Conference, pp. 1–5 (2006)
14. Yao, Y.Y., Wong, S.K.M., Lingras, P.: A decision-theoretic rough set model. In: Ras, Z.W., Zemankova, M., Emrich, M.L. (eds.) Methodologies for Intelligent Systems, New York, North-Holland, vol. 5, pp. 17–24 (1990)
15. Yao, Y.Y.: Three-way decisions with probabilistic rough sets. Inf. Sci. **180**, 341–353 (2010)

16. Yao, Y., Zhou, B.: Naive Bayesian rough sets. In: Yu, J., Greco, S., Lingras, P., Wang, G., Skowron, A. (eds.) RSKT 2010. LNCS, vol. 6401, pp. 719–726. Springer, Heidelberg (2010)
17. Yao, Y.Y.: The superiority of three-way decisions in probabilistic rough set models. Inf. Sci. **181**, 1080–1096 (2011)
18. Yao, Y.Y., Deng, X.F.: Quantitative rough sets based on subsethood measures. Inf. Sci. **267**, 306–322 (2014)
19. Yao, Y.Y.: The two sides of the theory of rough sets. Knowl.-Based Syst. **80**, 67–77 (2015)

Logic and Algebra

Antichain Based Semantics for Rough Sets

A. Mani[(✉)]

Department of Pure Mathematics, University of Calcutta,
9/1B, Jatin Bagchi Road, Kolkata 700029, India
a.mani.cms@gmail.com
http://www.logicamani.in

Abstract. The idea of using antichains of rough objects was suggested by the present author in her earlier papers. In this research basic aspects of such semantics are considered over general rough sets and general approximation spaces over quasi-equivalence relations. Most of the considerations are restricted to semantics associated with maximal antichains and their meaning. It is shown that even when the approximation operators are poorly behaved, some semantics with good structure and computational potential can be salvaged.

Keywords: Rough objects · Granular operator spaces · Maximal antichains · Antichains · Quasi equivalences · Axiomatic approach to granules · Granular rough semantics

1 Introduction

It is well known that sets of rough objects (in various senses) are quasi or partially orderable. In quasi or partially ordered sets, sets of mutually incomparable elements are called *antichains*. Some of the basic properties may be found in [1,2]. The possibility of using antichains of rough objects for a possible semantics was mentioned in [3,4] by the present author. In this research related semantics applicable for a large class of operator based rough sets including specific cases of RYS [5] and other general approaches like [6,7] are actually developed.

Let S be any set and l, u be lower and upper approximation operators on $\mathcal{S} \subseteq \wp(S)$ that satisfy monotonicity and $(\forall A \subseteq S) A \subseteq A^u$. An element $A \in \mathcal{S}$ will be said to be *lower definite* (resp. *upper definite*) if and only if $A^l = A$ (resp. $A^u = A$) and *definite*, when it is both lower and upper definite. The following are some of the the possible concepts of rough objects considered in the literature, (and all considerations will be restricted as indicated in the next definition):

- A non definite subset of S, that is A is a rough object if and only if $A^l \neq A^u$.
- Any pair of definite subsets of the form (A, B) satisfying $A \subseteq B$.
- Any pair of subsets of the form (A^l, A^u).
- Sets in an interval of the form (A^l, A^u).
- Sets in an interval of the form (A, B) satisfying $A \subseteq B$ and A, B being definite subsets.

D. Ciucci et al. (Eds.): RSKT 2015, LNAI 9436, pp. 335–346, 2015.
DOI: 10.1007/978-3-319-25754-9_30

- A non-definite element in a RYS, that is an x satisfying $\neg \mathbf{P}x^u x^l$
- An interval of the form, (A, B) satisfying $A \subseteq B$ and A, B being definite subsets.

Concepts of representation of objects necessarily relate to choice of semantic frameworks. In general, in most contexts, the order theoretic representations are of interest. In operator centric approaches, the problem is also about finding ideal representations. The central problem that is pursued in the present paper in relation to antichains of some types of rough objects is the following:

How good are representations of frameworks based on antichains of rough objects when the assumptions are minimal and sensible?

Set framework with operators will be used as all considerations will require quasi orders in an essential way. The evolution of the operators need not be induced by a cover or a relation (corresponding to cover or relation based systems respectively), but these would be special cases. The generalization to some rough Y-systems RYS (see [5] for definitions), will of course be possible as a result.

Definition 1. *A Granular Operator Space S will be a structure of the form $S = \langle \underline{S}, \mathcal{G}, l, u \rangle$ with $\underline{S}$ being a set, $\mathcal{G}$ an admissible granulation(defined below) over S and l, u being operators : $\wp(\underline{S}) \longmapsto \wp(\underline{S})$ satisfying the following:*

$$A^l \subseteq A \,\&\, A^{ll} = A^l \,\&\, A^u \subset A^{uu}$$

$$(A \subseteq B \longrightarrow A^l \subseteq B^l \,\&\, A^u \subseteq B^u)$$

$$\emptyset^l = \emptyset \,\&\, \emptyset^u = \emptyset \,\&\, S^l \subseteq S \,\&\, S^u \subseteq S.$$

Here, Admissible granulations are granulations $\mathcal{G}$ that satisfy the following three conditions (Relative RYS [5], $\mathbf{P} = \subseteq$, $\mathbb{P} = \subset$) and t is a term operation formed from set operations):

$$(\forall x \exists y_1, \ldots y_r \in \mathcal{G})\, t(y_1, y_2, \ldots y_r) = x^l$$

$$\text{and } (\forall x)\,(\exists y_1, \ldots y_r \in \mathcal{G})\, t(y_1, y_2, \ldots y_r) = x^u, \qquad \text{(Weak RA, WRA)}$$

$$(\forall y \in \mathcal{G})(\forall x \in \underline{S})\,(y \subseteq x \longrightarrow y \subseteq (x^l)), \qquad \text{(Lower Stability, LS)}$$

$$(\forall x, y \in \mathcal{G})(\exists z \in \underline{S})\, x \subset z,\, y \subset z \,\&\, z^l = z^u = z, \qquad \text{(Full Underlap, FU)}$$

On $\wp(\underline{S})$, the relation $\sqsubset$ is defined by

$$A \sqsubset B \text{ if and only if } A^l \subseteq B^l \,\&\, A^u \subseteq B^u.$$

The rough equality relation on $\wp(\underline{S})$ is defined via $A \approx B$ if and only if $A \sqsubset B \,\&\, B \sqsubset A$.

Regarding the quotient $\underline{S}| \approx$ as a subset of $\wp(\underline{S})$, the order $\Subset$ will be defined as per

$$\alpha \Subset \beta \text{ if and only if } \alpha^l \subseteq \beta^l \,\&\, \alpha^u \subseteq \beta^u.$$

Here α^l is being interpreted as the lower approximation of any of the elements of α and so on. $\Subset$ will be referred to as the *basic rough order*.

Definition 2. *By a roughly consistent object will be meant a set of subsets of* $\underline{S}$ *of the form* $H = \{A; (\forall B \in H)\, A^l = B^l, A^u = B^u\}$. *The set of all roughly consistent objects is partially ordered by the inclusion relation. Relative this maximal roughly consistent objects will be referred to as* rough objects. *By definite rough objects, will be meant rough objects of the form* H *that satisfy*

$$(\forall A \in H)\, A^{ll} = A^l \;\&\; A^{uu} = A^u.$$

Proposition 1. $\Subset$ *is a bounded partial order on* $\underline{S}|\approx$.

Proof. Reflexivity is obvious. If $\alpha \Subset \beta$ and $\beta \Subset \alpha$, then it follows that $\alpha^l = \beta^l$ and $\alpha^u = \beta^u$ and so antisymmetry holds.

If $\alpha \Subset \beta$, $\beta \Subset \gamma$, then the transitivity of set inclusion induces transitivity of $\Subset$. The poset is bounded by $0 = (\emptyset, \emptyset)$ and $1 = (S^l, S^u)$. Note that 1 need not coincide with (S, S).   □

Theorem 1. *Some known results relating to antichains and lattices are the following:*

1. *Let* X *be a partially ordered set with longest chains of length* r, *then* X *can be partitioned into* k *number of antichains implies* $r \leq k$.
2. *If* X *is a finite poset with* k *elements in its largest antichain, then a chain decomposition of* X *must contain at least* k *chains.*
3. *The poset* $AC_m(X)$ *of all maximum sized antichains of a poset* X *is a distributive lattice.*
4. *For every finite distributive lattice* L *and every chain decomposition* C *of* J_L *(the set of join irreducible elements of* L), *there is a poset* X_C *such that* $L \cong AC_m(X_C)$.

Proof. Proofs of the first three of the assertions can be found in [2,8] for example. Many proofs of results related to Dilworth's theorems are known in the literature and some discussion can be found in [8] (pages 126–135).

1. To prove the first, start from a chain decomposition and recursively extract the minimal elements from it to form r number of antichains.
2. This is proved by induction on the size of X across many possibilities.
3. See [2,8] for details.
4. In [9], the last connection between chain decompositions and representation by antichains reveals important gaps - there are other posets X that satisfy $L \cong AC_m(X)$. Further the restriction to posets is too strong and can be relaxed in many ways [10].   □

If R is a binary relation on a set $\underline{S}$, then the neighborhood generated by an $x \in \underline{S}$ will be

$$[x] = \{y \,:\, Ryx\}$$

A binary relation R on a set $\underline{S}$ is said to be a *Quasi-Equivalence* if and only if it satisfies:

$$(\forall x, y)\,([x] = [y] \leftrightarrow Rxy\,\&\,Ryx).$$

It is useful in algebras when it behaves as a good factor relation [11]. But the condition is of interest in rough sets by itself. *Note that* Rxy *is a compact form of* $(x, y) \in R$.

2   Anti Chains for Representation

It had been mentioned by the present author in [3], that anti-chains can be used for defining rough semantics in a meaningful way. But no related semantics were presented in the monograph. The simplest idea of dependence of rough objects corresponds to rough inclusion on the set of roughly equal objects.

Definition 3. $\mathbb{A}, \mathbb{B} \in \underline{S}| \approx$, *will be said to be* simply independent *(in symbols* $\Xi(\mathbb{A}, \mathbb{B}))$*if and only if*

$$\neg(\mathbb{A} \Subset \mathbb{B}) \ and \ \neg(\mathbb{B} \Subset \mathbb{A}).$$

A subset $\alpha \subseteq \underline{S}| \approx$ *will be said to be* simply independent *if and only if*

$$(\forall \mathbb{A}, \mathbb{B} \in \alpha) \, \Xi(\mathbb{A}, \mathbb{B}) \vee (\mathbb{A} = \mathbb{B}).$$

The set of all simply independent subsets shall be denoted by $\mathcal{SY}(S)$.

Definition 4. *A* maximal simply independent subset, *shall be a simply independent subset that is not properly contained in any other simply independent subset. The set of maximal simply independent subsets will be denoted by* $\mathcal{SY}_m(S)$.

On the set $\mathcal{SY}_m(S)$, $\ll$ *will be the relation defined by*

$$\alpha \ll \beta \ if \ and \ only \ if \ (\forall \mathbb{A} \in \alpha)(\exists \mathbb{B} \in \beta) \, \mathbb{A} \Subset \mathbb{B}.$$

Theorem 2. $\langle \mathcal{SY}_m(S), \ll \rangle$ *is a distributive lattice.*

Proof. This follows from the proof of order isomorphism between maximal order ideals and maximal antichains in general. Proofs of which can be found in [2] for example. □

Analogous to the above, it is possible to define essentially the same order on the set of maximal antichains of $\underline{S}| \approx$ denoted by $\mathfrak{S}$ with the $\Subset$ order. This order will be denoted by $\lessdot$ - this may also be seen to be induced by maximal ideals. But the more important issues are about definable operations on such independent sets. *Approximations of one system of independent sets by another system can be meaningful, but simple application of point-wise lower and upper approximation operators may end up violating closure in the first place.* The formal aspect is stated below.

Theorem 3. *If* $\alpha = \{\mathbb{A}_1, \mathbb{A}_2, \ldots, \mathbb{A}_n, \ldots\} \in \mathfrak{S}$, *and if* L *is defined by*

$$L(\alpha) = \{\mathbb{B}_1, \mathbb{B}_2, \ldots, \mathbb{B}_n, \ldots\}$$

with $X \in \mathbb{B}_i$ *if and only if* $X^l = \mathbb{A}_i^{ll} = \mathbb{B}_i^l$ *and* $X^u = \mathbb{A}_i^{lu} = \mathbb{B}_i^u$, *then* L *is a partial operation in general.*

Proof. The operation is partial because $L(\alpha)$ may not always be a maximal antichain. This can happen in general in which the properties $A^{ll} \subset A^l$ and/or $A^{ul} \subset A$ hold for some elements. The former possiblity is not possible by our assumptions, but the latter is scenario is permitted.

Specifically this can happen in bitten rough sets when the bitten upper approximation [12] operator is used in conjunction with the lower approximation. But many more examples are known in the literature (see [5]). □

This tells us that it is necessary to look elsewhere. If α, β are two sets of maximal antichains then they correspond to maximal sets of independent concepts in a sense. $\beta \cap \alpha$ represents the relative dependence or commonality of α relative β. Extensions of the antichain $\alpha \cap \beta$ to a maximal antichain ζ need not be unique in general. The strategy used here is to fix desired relation of the extension ζ to $\alpha \cap \beta$ and then define related operations of interest on α and β. Of the many possibilities, the most directly and non-trivially representable ones include those of extending $\alpha \cap \beta$ subject to cognitive dissonance of β (that is to require the commonality of the extension and β to be maximal) and that corresponding to the most radical extension. Formally,

Definition 5. *Let* $\chi(\alpha \cap \beta) = \{\xi; \ \xi$ *is a maximal antichain* $\& \alpha \cap \beta \subseteq \xi\}$ *be the set of all possible extensions of* $\alpha \cap \beta$. *The function* $\delta : \mathfrak{S}^2 \longmapsto \mathfrak{S}$ *corresponding to* extension under cognitive dissonance *will be defined as per* $\delta(\alpha, \beta) \in \chi(\alpha \cap \beta)$ *and (LST means* maximal subject to*)*

$$\delta(\alpha, \beta) = \begin{cases} \xi, & \textit{if } \xi \cap \beta \textit{ is a maximum subject to } \xi \neq \beta \textit{ and } \xi \textit{ is unique,} \\ \xi, & \textit{if } \xi \cap \beta \,\&\, \xi \cap \alpha \textit{ are LST } \xi \neq \beta, \alpha \textit{ and } \xi \textit{ is unique,} \\ \beta, & \textit{if } \xi \cap \beta \,\&\, \xi \cap \alpha \textit{ are LST } \&\, \xi \neq \beta, \alpha \textit{ but } \xi \textit{ is not unique,} \\ \beta, & \textit{if } \chi(\alpha \cap \beta) = \{\alpha, \beta\}. \end{cases}$$

Definition 6. *In the context of the above definition, the function* $\varrho : \mathfrak{S}^2 \longmapsto \mathfrak{S}$ *corresponding to* radical extension *will be defined as per* $\varrho(\alpha, \beta) \in \chi(\alpha \cap \beta)$ *and (MST means* minimal subject to*)*

$$\varrho(\alpha, \beta) = \begin{cases} \xi, & \textit{if } \xi \cap \beta \textit{ is a minimum under } \xi \neq \beta \textit{ and } \xi \textit{ is unique,} \\ \xi, & \textit{if } \xi \cap \beta \,\&\, \xi \cap \alpha \textit{ are MST } \xi \neq \beta, \alpha \textit{ and } \xi \textit{ is unique,} \\ \alpha, & \textit{if } (\exists \xi)\, \xi \cap \beta \,\&\, \xi \cap \alpha \textit{ are MST } \xi \neq \beta, \alpha \textit{ but } \xi \textit{ is not unique,} \\ \alpha, & \textit{if } \chi(\alpha \cap \beta) = \{\alpha, \beta\}. \end{cases}$$

Theorem 4. *The operations* ϱ, δ *satisfy all of the following:*

1. ϱ, δ *are groupoidal operations,*
2. $\varrho(\alpha, \alpha) = \alpha,$
3. $\delta(\alpha, \alpha) = \alpha,$
4. $\delta(\alpha, \beta) \cap \beta \subseteq \delta(\delta(\alpha, \beta), \beta) \cap \beta,$
5. $\varrho(\varrho(\alpha, \beta), \beta) \cap \beta \subseteq \varrho(\alpha, \beta) \cap \beta.$

Proof. 1. Obviously ϱ, δ are closed as the cases in their definition cover all possibilities. So they are groupoid operations. Associativity can be easily shown to fail through counterexamples.

2. Idempotence follows from definition.
3. Idempotence follows from definition.
4. By definition, $\alpha \cap \beta \subseteq \delta(\alpha, \beta)$ holds. The intersection with β of $\delta(\alpha, \beta)$ is a subset of $\delta(\delta(\alpha, \beta), \beta) \cap \beta$ by recursion.
5. Proof of this is similar to the above. □

In general, a number of possibilities (potential non-implications) like the following are satisfied by the algebra: $\alpha < \beta \,\&\, \alpha < \gamma \nrightarrow \alpha < \delta(\beta, \gamma)$. Given better properties of l and u, interesting operators can be induced on maximal antichains towards improving the properties of ϱ and δ.

Example 1 (Relation to Knowledge Interpretation).

In Pawlak's concept of knowledge in classical RST [13], if $\underline{S}$ is a set of attributes and P an indiscernibility relation on it, then sets of the form A^l and A^u represent clear and definite concepts. Extension of this to other kinds of RST have also been considered in [3,14,15] by the present author. In [15], the concept of knowledge advanced by her is that of union of pairwise independent granules (in set context corresponding to empty intersection) correspond to clear concepts. This granular condition is desirable in other situations too, but may not correspond to the approximations of interest. In real life, clear concepts whose parts may not have clear relation between themselves are too common. If all of the granules are not definite objects, then analogous concepts of knowledge may be graded or typed based on the properties satisfied by them [3]. Some examples of granular knowledge axioms are as follows:

1. Individual granules are atomic units of knowledge.
2. If collections of granules combine subject to a concept of mutual independence, then the result would be a concept of knowledge. The 'result' may be a single entity or a collection of granules depending on how one understands the concept of *fusion* in the underlying mereology. In set theoretic (ZF) setting the fusion operation reduces to set-theoretic union and so would result in a single entity.
3. Maximal collections of granules subject to a concept of mutual independence are admissible concepts of knowledge.
4. Parts common to subcollections of maximal collections are also knowledge.
5. All stable concepts of knowledge consistency should reduce to correspondences between granular components of knowledges. Two knowledges are *consistent* if and only if the granules generating the two have 'reasonable' correspondence.
6. Knowledge A is consistent with knowledge B if and only if the granules generating knowledge B are part of some of the granules generating A.

An antichain of rough objects is essentially a set of *some-sense mutually distinct rough concepts* relative that interpretation. Maximal antichains naturally correspond to represented rough knowledge that can be handled in a clear way in a context. The stress here should be on possible operations both within and over them. It is fairly clear that better the axioms satisfied by a concept of granular knowledge, better will be the nature of possible operations over sets of *some-sense mutually distinct rough concepts*. A clear picture of the connections will be part of future research.

From decision making perspectives, antichains of rough objects correspond to forming representative partitions of the whole and semantics relate to relation between different sets of representatives.

Example 2 (Micro-Fossils and Descriptively Remote Sets).

In the case study on numeric visual data including micro-fossils with the help of nearness and remoteness granules in [16], the difference between granules and approximations is very fluid as the precision level of the former can be varied.

The data set consists of values of probe functions that extract pixel data from images of micro-fossils trapped inside other media like amethyst crystals.

The idea of remoteness granules is relative a fixed set of nearness granules formed from upper approximations - so the approach is about reasoning with sets of objects which in turn arise from tolerance relations on a set. In [16], antichains of rough objects are not used, but the computations can be extended to form maximal antichains at different levels of precision towards working out the best antichains from the point of view of classification.

3   Quasi Equivalential Rough Set Theory

One of the most interesting type of granulation $\mathcal{G}$ in relational RST is one that satisfies

$$(\forall x, y)\, (\phi(x) = \phi(y) \leftrightarrow Rxy \,\&\, Ryx),$$

where $\phi(x)$ is the granule generated by $x \in \underline{S}$. This granular axiom says that if x is left-similar to y and y is left-similar to x, then the elements left similar to either of x and y must be the same. R is being read as *left-similarity* because it is directional and has effect similar to tolerances on neighborhood granules.

Reflexivity is not assumed as the present author's intention is to isolate the effect of the axiom alone.

For example, it is possible to find quasi equivalences that do not satisfy other properties from contexts relating to numeric measures. Let S be a set of variables such that Rxy if and only if $x \approx \kappa y \,\&\, y \approx \kappa' x \,\&\, \kappa, \kappa' \in (0.9, 1.1)$ for some interpretation of $\approx$.

Definition 7. *By a Quasi-Equivalential Approximation Space will be meant a pair of the form $S = \langle \underline{S}, R \rangle$ with R being a quasi equivalence. For an arbitrary subset $A \in \wp(S)$, the following can be defined:*

$$(\forall x \in \underline{S})\, [x] = \{y \,;\, y \in \underline{S} \,\&\, Rxx\}.$$

$$A^l = \bigcup \{[x] \,;\, [x] \subseteq A \,\&\, x \in \underline{S}\} \,\&\, A^u = \bigcup \{[x] \,;\, [x] \cap A \neq \emptyset \,\&\, x \in \underline{S}\}$$

$$A^{l_o} = \bigcup \{[x] \,;\, [x] \subseteq A \,\&\, x \in A\} \,\&\, A^{u_o} = \bigcup \{[x] \,;\, [x] \cap A \neq \emptyset \,\&\, x \in A\}$$

$$A^L = \{x \,;\, \emptyset \neq [x] \subseteq A \,\&\, x \in \underline{S}\} \,\&\, A^U = \{x \,;\, [x] \cap A \neq \emptyset \,\&\, x \in \underline{S}\}$$

$$A^{L_o} = \{x \,;\, [x] \subseteq A \,\&\, x \in A\} \,\&\, A^{U_o} = \{x \,;\, [x] \cap A \neq \emptyset \,\lor\, x \in A\}.$$

$$A^{L_1} = \{x \,;\, [x] \subseteq A \,\&\, x \in \underline{S}\} \,\&\, A^U = \{x \,;\, [x] \cap A \neq \emptyset \,\&\, x \in \underline{S}\}.$$

Note the requirement of non-emptiness of $[x]$ in the definition of A^L, but it is not necessary in that of A^{L_o}.

Theorem 5. *The following properties hold:*

1. *All of the approximations are distinct in general.*
2. *$(\forall A \in \wp(S))\, A^{L_o} \subseteq A^{l_o} \subseteq A^l \subseteq A$ and $A^{L_o} \subseteq A^L$.*
3. *$(\forall A \in \wp(S))\, A^{l_o l} = A^{l_o} \,\&\, A^{l l_o} \subseteq A^{l_o} \,\&\, A^{l_o l_o} \subseteq A^{l_o}$*

4. $(\forall A \in \wp(S))\, A^u = A^{ul} \subseteq A^{uu}$, but it is possible that $A \not\subseteq A^u$
5. It is possible that $A^L \not\subseteq A$ and $A \not\subseteq A^U$, but $(\forall A \in \wp(S))\, A^L \subseteq A^U$ holds. In general A^L would not be comparable with A^l and similarly for A^U and A^u.
6. $(\forall A \in \wp(S))\, A^{L \circ L \circ} \subseteq A^{L \circ} \subseteq A \subseteq A^{U \circ} \subseteq A^{U \circ U \circ}$. Further $A^U \subseteq A^{U \circ}$.

Proof. 1. The general example below can be evoked to prove the first assertion.
2. If $[x] \subseteq A^{l \circ}$, then it is not necessary that $[x] \subseteq A^{L \circ}$, but if $[x] \subseteq A^{L \circ}$, then $[x] \subseteq A^{l \circ}$. So $A^{L \circ} \subseteq A^{l \circ}$. For comparing A^l and $A^{l \circ}$, it suffices to look at the granules - it can happen that $[x] \subseteq A$, but $x \in \underline{S} \setminus A$. So $A^{l \circ} \subseteq A^l$, but the converse may fail.
3. All neighborhoods $[x] \subseteq A^{l \circ}$ need to be considered in the formation of the l-approximation of $A^{l \circ}$. So even if $x \in A \setminus A^{l \circ}$, $[x] \subseteq A^{l \circ}$ implies $[x] \subseteq A^{l \circ l}$. The converse is false in general.
　　 If $y \in A^{ll \circ}$, then $(\exists z \in A^l)\, y \in [z] \subseteq A^l \subseteq A$. But $z \in A^l$ implies $z \in A$. So $y \in A^{l \circ}$. If $a \in A^{l \circ}$, then $(\exists b \in A)\, a \in [b] \subseteq A$, but $[b] \subseteq A^l$ follows. So the conclusion $A^{ll \circ} = A^{l \circ}$ is justified.
4. Since R is not necessarily reflexive, it can happen that $x \notin [x]$ and $[x] \cap A \neq \emptyset$ and this specific x may not be related to any other y satisfying $[y] \cap A \neq \emptyset$. This situation will ensure $A \not\subseteq A^u$. This can happen even when R is a quasi-equivalence [17]. A^u is a union of neighborhood granules and so $A^{ul} = A^u$.
5. $A^L \not\subseteq A$ can happen precisely when a $x \in \underline{S} \setminus A$ satisfies $[x] \subseteq A$ and $x \notin [y]$ for all $[y] \cap A \neq \emptyset$. This ensures $x \notin [x]$. But if $y \in A^L$, then $y \in A^U$.
6. $A^{L \circ L \circ} \subset A^{L \circ}$ happens for example when $x \in A \setminus A^{L \circ}$ & $[x] \subset A^{L \circ}$ & $\neg(\exists y \in A)\, [y] = [x] \& y \neq x$. □

Clearly the operators l, u are granular approximations, but the latter is controversial as an upper approximation operator. The point-wise approximations L, U are more problematic - not only do they fail to satisfy representability in terms of neighborhood granules, but the lower approximation fails inclusion.

Example 3 (General)

$$\text{Let } \underline{S} = \{a, b, c, e, f, k, h, q\}$$

and let R be a binary relation on it defined via

$$
\begin{aligned}
R = \{&(a, a), (b, a), (c, a), (f, a),\\
&(k, k), (e, h), (f, c), (k, h)\\
(b, b),\, &(c, b), (f, b), (a, b), (c, e), (e, q)\}.
\end{aligned}
$$

The neighborhood granules $\mathcal{G}$ are then

$$
\begin{aligned}
&[a] = \{a, b, c, f\} = [b], [c] = \{f\}, [e] = \{c\},\\
&[k] = \{k\}, [h] = \{k, e\}, [f] = \emptyset \,\&\, [q] = \{e\}.
\end{aligned}
$$

So R is a quasi-equivalence relation.

If $A = \{a, k, q, f\}$, then

$$A^l = \{k, f\},\ A^u = \{a, b, c, f, k, e\},\ A^{uu} = \{a, b, c, f, k, e, h\}$$
$$A^{l_o} = \{k\},\ A^{u_o} = \{a, b, c, k, f\}.$$
$$A^L = \{k, f\},\ A^U = \{a, b, c, k, h, q\}.$$
$$A^{L_o} = \{q, k, f\},\ A^{U_o} = \{a, k, q, f, b, c, h\}.$$
$$A^{L_1} = \{k, c, f\},\ A^U = \{a, b, c, k, h, q\}.$$

Note that $A^{L_1} \not\subseteq A\ \&\ A^{L_1} \not\subseteq A^U\ \&\ A \not\subseteq A^U$.

4   Semantics of QE-Rough Sets

In this section a semantics of quasi-equivalential rough sets (QE-rough sets), using antichains generated from rough objects, is developed. Interestingly the properties of the approximation operators of QE-rough sets fall short of those of granular operator space. Denoting the set of maximal antichains of rough objects by $\mathfrak{S}$ and carrying over the operations $\ll, \varrho, \delta$, the following algebra can be defined.

Definition 8. *A maximal simply independent algebra Q of quasi equivalential rough sets shall be an algebra of the form*

$$Q = \langle \mathfrak{S}, \ll, \varrho, \delta \rangle$$

defined as in Sect. 2 with the approximation operators being l, u uniformly in all constructions and definitions.

Theorem 6. *Maximal simply independent algebras are well defined.*

Proof. None of the steps in the definition of the maximal antichains, or the operations ϱ or δ are problematic because of the properties of the operators l, u.   □

The above theorem suggests that it would be better to try and define more specific operations to improve the uniqueness aspect of the semantics or at least the properties of ϱ, δ. It is clearly easier to work with antichains as opposed to maximal antichains as more number of suitable operations are closed over the set of antichains as opposed to those over the set of maximal antichains.

Definition 9. *Let $\mathfrak{K}$ be the set of antichains of rough objects of S then the following operations $\mathfrak{L}, \mathfrak{U}$ and extensions of others can be defined:*

- *Let $\alpha = \{\mathbb{A}_1, \mathbb{A}_2, \dots, \mathbb{A}_n, \dots\} \in \mathfrak{K}$ with $\mathbb{A}_i$ being rough objects; the lower and upper approximation of any subset in $\mathbb{A}_i$ will be denoted by $\mathbb{A}_i^l$ and $\mathbb{A}_i^u$ respectively.*
- *Define $\mathfrak{L}(\alpha) = \{\mathbb{A}_1^l, \mathbb{A}_2^l, \dots, \mathbb{A}_r^l, \dots\}$ with duplicates being dropped*
- *Define $\mathfrak{U}(\alpha) = \{\mathbb{A}_1^u, \mathbb{A}_2^u, \dots, \mathbb{A}_r^u, \dots\}$ with duplicates being dropped*

- *Define*

$$\mu(\alpha) = \begin{cases} \alpha & \text{if } \alpha \in \mathfrak{S} \\ \text{undefined,} & \text{else.} \end{cases}$$

- *Partial operations* ϱ^*, δ^* *corresponding to* ϱ, δ *can also be defined as follows:* *Define*

$$\varrho^*(\alpha, \beta) = \begin{cases} \varrho(\alpha, \beta) & \text{if } \alpha, \beta \in \mathfrak{S} \\ \text{undefined,} & \text{else.} \end{cases}$$

$$\delta^*(\alpha, \beta) = \begin{cases} \delta(\alpha, \beta) & \text{if } \alpha, \beta \in \mathfrak{S} \\ \text{undefined,} & \text{else.} \end{cases}$$

The resulting partial algebra $\mathfrak{K} = \langle \mathfrak{K}, \mu, \vee, \wedge, \varrho^*, \delta^*, \mathfrak{L}, \mathfrak{U}, 0 \rangle$ *will be said to be a simply independent QE algebra*

Theorem 7. *Simply independent QE algebras are well defined and satisfy the following:*

- $\mathfrak{L}(\alpha) \vee \alpha = \alpha$.
- $\mathfrak{U}(\alpha) \vee \alpha = \mathfrak{U}(\alpha)$.

More aspects of this algebra will appear in a separate paper.

5 Aggregations of AC Semantics

Given a semantics in terms of antichains and the order $\ll$, then how is it possible to know that the semantics can be partitioned into multiple independent semantics?

This general class of problems is partially solved in this section from an essentially order theoretic perspective. Let $\mathcal{Y}(S)$ be the set of antichains of rough objects on the granular operator space S. A subset τ of antichains $\mathfrak{A}$ will be said to be *full* if and only if every interval of the form (α, β) in $\mathfrak{A}$ satisfies

$$(\alpha, \beta) \cap \tau \neq \emptyset \longrightarrow (\exists \gamma, \mu)\,(\alpha, \beta) \cap \tau = \{\gamma, \mu\}\,\&\,\gamma \neq \mu.$$

The condition is based in combinatorial aspects and is admittedly not intuitive.

Theorem 8. *Every full maximal antichain* α *can be partitioned into* $\{\alpha_1, \alpha_2\}$ *with* $\alpha_1 \cup \alpha_2 = \alpha$. *This in turn induces a partition* $\{\alpha_1 \uparrow, \alpha_2 \downarrow\}$ *of* $\mathcal{Y}(S)$ - $\alpha \uparrow$, $\alpha \downarrow$ *respectively being the order filter and order ideal generated by* α *respectively.*

Proof. Let $p : \alpha^2 \mapsto \alpha$ be a map such that

$$p(\alpha, \alpha) = \alpha,$$
$$p(\alpha, \beta) = \alpha \text{ or } \beta,$$
$$p(\alpha, \beta) = \alpha\,\&\,p(\beta, \gamma) = \beta \longrightarrow p(\alpha, \gamma) = \alpha,$$

$$\text{Let } \varphi(\alpha) = \beta \leftrightarrow (\forall \gamma)\,\alpha < \gamma\,\&\,\alpha < \beta\,\&\,p(\beta, \gamma) = \beta.$$

Defining $\alpha_1 = \{\varphi(x) : x \in \{z\,;\,(\exists a \in \alpha)\,z \ll a\}\}$ and by setting $\alpha_2 = \alpha \setminus \alpha_1$, it can be checked that $\{\alpha_1 \uparrow, \alpha_2 \downarrow\}$ is a partition of $\mathcal{Y}(S)$ through a contradiction argument. $\qquad \square$

Theorem 9. *The above theorem extends to maximal antichains directly.*

6 Further Directions and Remarks

In this research, general approaches to semantics of rough sets using antichains of rough objects and maximal antichains have been developed. The semantics is shown to be valid for a very large class of general rough set theories. This has been possible mainly because the objects of study have been taken to be antichains of rough objects as opposed to plain rough objects. From the meaning point of view, the objects studied are rough objects that appear to be pairwise independent. Both maximal antichains and antichains allow operators that are related to relative independence.

This research also motivates the following:

- Further study of specific cases.
- Research into connections with the rough membership function based semantics of [18] and extensions by the present author in a forthcoming paper. This is justified by advances in concepts of so-called cut-sets in antichains.
- Research into computational aspects as the theory is well developed for antichains.

Acknowledgement. The present author would like to thank the anonymous referees for useful remarks that led to improvement of the presentation of the paper.

References

1. Gratzer, G.: General Lattice Theory. Birkhauser, Berlin (1998)
2. Davey, B.A., Priestley, H.A.: Introduction to Lattices and Order, 2nd edn. Cambridge University Press, Cambridge (2002)
3. Mani, A.: Algebraic Semantics of Proto-Transitive Rough Sets. 1st edn., July 2014. arxiv:1410.0572
4. Mani, A.: Approximation dialectics of proto-transitive rough sets. In: Chakraborty, M.K., Skowron, A., Kar, S. (eds.) Facets of Uncertainties and Applications. Springer Proceedings in Math and Statistics, vol. 125, pp. 99–109. Springer, Heidelberg (2015)
5. Mani, A.: Dialectics of counting and the mathematics of vagueness. In: Peters, J.F., Skowron, A. (eds.) Transactions on Rough Sets XV. LNCS, vol. 7255, pp. 122–180. Springer, Heidelberg (2012)
6. Ciucci, D.: Approximation algebra and framework. Fundam. Informaticae **94**, 147–161 (2009)
7. Yao, Y.: Relational interpretation of neighbourhood operators and rough set approximation operators. Inf. Sci. **111**, 239–259 (1998)
8. Kung, J.P.S., Rota, G.C., Yan, C.H.: Combinatorics-The Rota Way. Cambridge University Press, Cambridge (2009)
9. Koh, K.: On the lattice of maximum-sized antichains of a finite poset. Algebra Univers. **17**, 73–86 (1983)

10. Siggers, M.: On the Representation of Finite Distributive Lattices. Mathematics **1**, 1–17 (2014). arXiv:1412.0011
11. Bosnjak, I., Madarasz, R.: On some classes of good quotient relations. Novisad J. Math. **32**(2), 131–140 (2002)
12. Mani, A.: Algebraic semantics of similarity-based bitten rough set theory. Fundam. Informaticae **97**(1–2), 177–197 (2009)
13. Pawlak, Z.: Rough Sets: Theoretical Aspects of Reasoning About Data. Kluwer Academic Publishers, Dodrecht (1991)
14. Mani, A.: Towards logics of some rough perspectives of knowledge. In: Suraj, Z., Skowron, A. (eds.) Rough Sets and Intelligent Systems - Professor Zdzisław Pawlak in Memoriam. Intelligent Systems Reference Library, vol. 43, pp. 342–367. Springer, Heidelberg (2013)
15. Mani, A.: Choice inclusive general rough semantics. Inf. Sci. **181**(6), 1097–1115 (2011)
16. Peters, J., Skowron, A., Stepaniuk, J.: Nearness of visual objects - application of rough sets in proximity spaces. Fundam. Informaticae **128**, 159–176 (2013)
17. Mani, A.: Esoteric rough set theory: algebraic semantics of a generalized VPRS and VPFRS. In: Peters, J.F., Skowron, A. (eds.) Transactions on Rough Sets VIII. LNCS, vol. 5084, pp. 175–223. Springer, Heidelberg (2008)
18. Chakraborty, M.K.: Membership function based rough set. Inf. Sci. **55**, 402–411 (2014)

Formalizing Lattice-Theoretical Aspects of Rough and Fuzzy Sets

Adam Grabowski[1]([✉]) and Takashi Mitsuishi[2]

[1] Institute of Informatics, University of Białystok,
Ciołkowskiego 1M, 15-245 Białystok, Poland
adam@math.uwb.edu.pl
[2] University of Marketing and Distribution Sciences,
Kobe, 3-1 Gakuen Nishimachi Nishi-ku, Kobe 655-2188, Japan
takashi_mitsuishi@red.umds.ac.jp

Abstract. Fuzzy sets and rough sets are well-known approaches to incomplete or imprecise data. In the paper we briefly report how these frameworks were successfully encoded with the help of one of the leading computer proof assistants in the world. Even though fuzzy sets are much closer to the set theory implemented within the Mizar library than rough sets, lattices as a basic viewpoint appeared a very feasible one. We focus on the lattice-theoretical aspects of rough and fuzzy sets to enable the application of external theorem provers like EQP or Prover9 as well as to translate them into TPTP format widely recognized in the world of automated proof search. The paper is illustrated with the examples taken just from one of the largest repositories of computer-checked mathematical knowledge – the Mizar Mathematical Library. Our formal development allows both for further generalizations, building on top of the existing knowledge, and even merging of these approaches.

1 Introduction

Through the years ordinary set theory appeared not to be feasible enough for modelling incomplete or imprecise information. Even if basically built on top of Zermelo-Fraenkel widely accepted by most mathematicians, fuzzy sets by Zadeh [27] proposed new view for membership functions, where degree of membership taken from the unit interval was considered rather than classical discrete bipolarity. Pawlak's alternative approach [21], although essentially of the same origin, was different – its probabilistic features were underlined. Also the focus was put rather on collective properties of clusters of objects than those of individuals as the latter can be hardly accessible.

Formalization is doing mathematics in a language formal enough to be understandable by computers [24] (of course, doing mathematics means also the act of proving theorems and correctness of definitions according to classical logic and Zermelo-Fraenkel set theory). This activity, obviously without the use of computers is dated back to Peano and Bourbaki as every mathematician uses more or less formal language [1]; but computer certification of mathematics can be

© Springer International Publishing Switzerland 2015
D. Ciucci et al. (Eds.): RSKT 2015, LNAI 9436, pp. 347–356, 2015.
DOI: 10.1007/978-3-319-25754-9_31

useful for many reasons – machines open new possibilities of information analysis and exchange, they can help to discover new proofs or to shed some light on approaches from various perspectives; with the help of such automated proof assistants one can observe deeper connections between various areas of mathematics. For example, lattice theory delivers interesting and powerful algebraic model – useful in quantum theory, logic, linear algebra, and topology, to list only most popular ones. Hence it is not very surprising that also rough [14] and fuzzy set theories [5] can be modelled in this way. Studying connections between theories can be also benefitting to lattice theory itself – see [4] for the use of upper and lower rough ideals (filters) in a lattice.

The structure of the paper is as follows: in the next section we briefly explain the issues concerned with the formalization of mathematics using computers (essentially concentrating on the topic of incomplete information about objects), then we introduce basic notions of lattice theory, also in mechanized setting. Sections 4 and 5 are devoted to concrete implementation of rough and fuzzy sets with Mizar system [20]. Section 6 summarizes our implementation and experiments with the Mizar Mathematical Library (MML in short), while in the last section we draw some concluding remarks and plans for future.

2 Lattice Theory – Formally and Informally

Lattices [13] are structures of the form

$$\langle L, \sqcup, \sqcap \rangle,$$

where L is a set (sometimes assumed to be non-empty), both binary operations $\sqcup$ and $\sqcap$ are commutative, associative, and satisfy the absorption laws. There is also an alternative definition of a lattice as $\langle L, \leq \rangle$, where $\leq$ is a partial ordering on L with the existence of suprema and infima for arbitrary pairs of elements of L. Essentially then, one can see a lattice as $\langle L, \sqcup, \sqcap, \leq \rangle$, where both parts are defined by one another.

It is worth noticing that lattices, especially those of them which have equational characterizations, can be automatically explored. Famous question on another axiomatization of Boolean algebras, known as the Robbins problem, was solved with EQP/OTTER system in 1996 after sixty years of unsuccessful human research. The first author provided also some proof developments in this problem [6], but with the use of another computer proof-assistant, namely the Mizar system [12]. It was created in the early seventies of the previous century in order to assist mathematicians in their work. Now the system consist of three main parts: the language in which all the mathematics can be expressed, close to the vernacular used by human mathematicians, which at the same time can be automatically verified, the software which verifies the correctness of formalized knowledge in the classical logical framework, and last but not least, the huge collection of certified mathematical knowledge – the MML.

In Mizar formalism [12], lattices are structures of the form

```
definition
  struct (/\-SemiLattRelStr, \/-SemiLattRelStr, LattStr)
  LattRelStr
            (# carrier -> set,
        L_join, L_meet -> (BinOp of the carrier),
            InternalRel -> Relation of the carrier #);
end;
```

introduced by the first author to benefit from using binary operations and orderings at the same time. However, defining a structure we give only information about the signature – arities of operations and their results; specific axioms are needed additionally. Even if all of them can be freely used in their predicative variant, definitional expansions proved their usefulness.

```
definition
  let L be non empty LattStr;
  attr L is meet-absorbing means                    :: LATTICES:def 8
    for a,b being Element of L holds (a "/\" b) "\/" b = b;
end;
```

The above is faithful translation of

$$\forall_{a,b\in L} (a \sqcap b) \sqcup b = b.$$

Continuing with all other axioms, we finally obtain lattices as corresponding structures with the collection of attributes under a common name Lattice-like.

```
definition
  mode Lattice is Lattice-like non empty LattStr;
end;
```

The alternative approach to lattices through the properties of binary relations is a little bit different, so the underlying structure is just the set with InternalRel, namely RelStr.

```
definition
  mode Poset is reflexive transitive antisymmetric RelStr;
end;
```

Boolean algebras, distributive lattices, and lattices with various operators of negation are useful both in logic and in mathematics as a whole, it is not very surprising that also rough set theory adopted some of the specific axiom sets – with Stone and Nelson algebras as most prominent examples.

The type LATTICE is, unlike the alternative approach where Lattice Mizar mode was taken into account, a poset with binary suprema and infima. Both approaches are in fact complementary, there is a formal correspondence between them shown, and even a common structure on which two of them are proved to be exactly the same. One can ask the question why to have both approaches available in the repository of computer verified mathematical knowledge available at the same time? The simplest reason is that the ways of their generalizations vary; in case of posets we could use relational structures based on the very general properties of binary relations (even not necessarily more general, but just different – including equivalence relations or tolerances); equationally defined lattices are good starting point to consider e.g. semilattices or lattices with various additional operators – here also equational provers can show their deductive power [17]. In this setting such important theorems as Stone's representation theorem for Boolean algebras was originally formulated.

The operations of supremum and infimum are "\/" and "/\", respectively – in both approaches. The natural ordering is <= in posets and [= in lattices equationally defined. Note that c= is set-theoretical inclusion. The viewpoint of posets was extensively studied in Mizar during the big formalization project – translating into Mizar the book *Compendium of Continuous Lattices* (CCL) by Gierz et al.

3   Incomplete Information and Set Theory

Recall that a fuzzy set A over a universe X is a set defined as

$$A = \{(x, \mu_A(x)) : x \in X\},$$

where $\mu_A(x) \in [0, 1]$ is membership degree of x in A. Because the notions in the MML make a natural hierarchy (as the base set theory is the Tarski-Grothendieck set theory, which is basically Zermelo-Fraenkel set theory with the Axiom of Choice where the axiom of infinity is replaced by Tarski's axiom of existence of arbitrarily large, strongly inaccessible cardinals) of the form: functions are relations, which are subsets of Cartesian product, and all these are just sets, so μ_A is a binary relation, i.e. subset of Cartesian product $X \times [0, 1]$.

Zadeh's approach assumes furthermore that μ_A is a function, extending a characteristic function χ_A. So, for arbitrary point x of the set A, the pair $(x, \mu_A(x))$ can be replaced just by the value of the membership function $\mu_A(x)$, which is in fact, formally speaking, the pair under consideration. Then all operations can be viewed as operations on functions, which appeared to be quite natural in the set-theoretical background taken in the MML as the base. All basic formalized definitions and theorems can be tracked under the address http:// mizar.org.

```
definition let C be non empty set;
  mode FuzzySet of C is Membership_Func of C;
end;
```

Of course, Membership_Func is not uniquely determined for C – the keyword mode denotes the shorthand for a type[1] in Mizar, that is, in fact C variable can be read from the corresponding function rather than vice versa.

---

[1] Most theorem provers are untyped, the Mizar language has types.

Table 1. Formalized notions and their formal translations (see also [8])

The notion	Formal counterpart
the membership function	Membership_Func of C
fuzzy set	FuzzySet of C
$\chi_A(x)$	chi(A,X).x
α-set	alpha-set C
supp C	support C
$F \cap G$	min (F,G)
$F \cup G$	max (F,G)
cF	1_minus F

We collected translations of selected formalized notions in Table 1. As we can read from this table, there are standard operations of fuzzy sets available, usually taken componentwise (note that F.x stands for the value of the function F on an argument x) [16]. The support of a membership function should not be defined because it was used already by the theory of formal power series.

Note that the Mizar repository extensively uses a difference between functions and partial functions (Function of X, Y and PartFunc of X, Y in Mizar formalism); because in case of partial functions only the inclusion of the domain in the set X is required, hence the earlier type expands to the latter automatically. Of course all such automation techniques are turned on after proving corresponding properties formally.

4   Rough Sets and Approximation Spaces

Main disadvantage behind formalization of rough set theory [21] was that we had to choose among two basic approaches to rough sets: either as classes of equivalence relations (with further generalizations into tolerances or even arbitrary binary relations) or as pairs consisting of the lower and the upper approximation (see [10] for detailed description of approximation operators in Mizar).

```
definition let X be Tolerance_Space, A be Subset of X;
  func RS A -> RoughSet of X equals         :: INTERVA1:def 14
    [LAp A, UAp A];
end;
```

The structure is defined as follows (_\/_ and _/_ are taken componentwise):

```
definition let X be Tolerance_Space;
  func RSLattice X -> strict LattStr means      :: INTERVA1:def 23
    the carrier of it = RoughSets X &
    for A, B being Element of RoughSets X,
       A1, B1 being RoughSet of X st A = A1 & B = B1 holds
```

```
        (the L_join of it).(A,B) = A1 _\/_ B1 &
        (the L_meet of it).(A,B) = A1 _/\_ B1;
end;
```

We have proved formally that these structures are Lattice-like, distributive, and complete (arbitrary suprema and infima exist, not only binary ones). The properties are expressed in the form allowing automatic treatment of such structures.

```
registration let X be Tolerance_Space;
  cluster RSLattice X -> bounded complete;
end;
```

Taking into account that [= is the ordering generated by the lattice operations, we can prove that it is just determined by the set-theoretical inclusion of underlying approximations.

```
theorem :: INTERVA1:71
  for X being Tolerance_Space, A, B being Element of RSLattice X,
      A1, B1 being RoughSet of X st A = A1 & B = B1 holds
    A [= B iff
      LAp A1 c= LAp B1 & UAp A1 c= UAp B1;
```

Detailed survey of the lattice-theoretical approach to rough sets is contained e.g. in Järvinen [14] paper, which enumerated some basic classes of lattices useful within rough set theory – e.g. Stone, de Morgan, Boolean lattices, and distributive lattices, to mention just the more general ones. All listed structures are well represented in the MML.

Flexibility of the Mizar language allows for defining new operators on already existing lattices without the requirement of repetitions of old structures, e.g. Nelson algebras are based on earlier defined de Morgan lattices (which are also of the more general interest), also defining various negation operators is possible in this framework. Furthermore, topological content widely represented in the Mizar Mathematical Library helped us to have illustrative projections into other areas of mathematics [26]. Here good example was (automatically discovered by us) linking with Isomichi classification of subsets of a topological space into three classes of subsets [9].

5   From Lattices of Fuzzy Sets to L-Fuzzy Sets

Exactly just like in the case of rough sets (where we tried to stick to equivalence relations instead of the more general case), first formalization in Mizar had to be tuned to better reflect Zadeh's mathematical idea [27]. The basic object RealPoset is a poset (a set equipped with the partial ordering, i.e. reflexive, transitive, and antisymmetric relation) – unit interval with the natural ordering defined of real numbers. We use Heyting algebras as they are naturally connected with L-fuzzy sets. We can define the product of the copies of the (naturally ordered) unit intervals (by the way, this was defined in Mizar article YELLOW_1 by the first author during the formalization of CCL).

```
definition let A be non empty set;
  func FuzzyLattice A -> Heyting complete LATTICE equals :: LFUZZY_0:def 4
    (RealPoset [. 0,1 .]) |^ A;
end;
```

The above can be somewhat cryptic, however the next theorem explains how elements of the considered structure look like – these are just functions from A into the unit interval [19].

```
theorem :: LFUZZY_0:14
  for A being non empty set holds
    the carrier of FuzzyLattice A = Funcs (A, [. 0, 1 .]);
```

Although the lattice operations are not stated explicitly here (which could be a kind of advantage here), the ordering determines it uniquely via the natural ordering of real functions. The underlying structure is a complete Heyting algebra. Of course, the type of this Mizar functor is not only declarative – it had to be proved[2].

```
theorem :: LFUZZY_0:19
  for C being non empty set,
      s,t being Element of FuzzyLattice C holds
  s"\/"t = max(@s, @t);
```

The functor @s returns for an element of the lattice of fuzzy sets the corresponding membership function (i.e. a fuzzy set). The so-called type cast is needed to properly recognize which operation max should be used [2] – in our case it is the operation defined for arbitrary functions as noted in Table 1. Of course, in the above definition, A is an arbitrarily chosen non-empty set; it can be replaced by the carrier of a lattice, and the definition of FuzzyLattice A can be tuned accordingly. This can show a wide area of applications of lattice theory within fuzzy sets [15] – from lattices of fuzzy sets to fuzzy sets on lattices.

6   Comparison of Both Developments

Even at the first sight, folklore, two basic objects are especially correlated with our area of research: partitions (resp. coverings in more general approach [29]) and intervals [25]. Querying MML, we discovered that lattices of partitions were formalized already pretty well, but to our big surprise, this was not the case of intervals, so we had to make some preparatory work by ourselves.

Instead of using the ordered pair of approximations, we can claim that both coordinates are just arbitrary objects, so that we can do just a little bit *reverse mathematics*. It happens even in the heart of rough set theory – correlation of properties of an indiscernibility relation with those of approximation operators in style of [28] or underlying lattice properties [14] could lead us to develop

---

[2] All the proofs are partially translated in the journal *Formalized Mathematics* and independently can be browsed at the homepage of the project.

Table 2. List of our submissions to MML concerning described topics

MML Identifier	Content
INTERVA1	algebra of intervals (including lattice of rough sets)
NELSON_1	Nelson algebras
YELLOW_1	products of posets
ROBBINS2	correspondence between relational structures and lattices
FUZZY_1	basic notions of fuzzy set theory [18]
LFUZZY_0	lattices of fuzzy sets
LFUZZY_1	transitive closure of fuzzy relations
FUZNUM_1	focus on fuzzy numbers, fundamentals revisited [7]
ROUGHS_1	introduction to rough sets [11]
ROUGHS_2	relational characterization of rough sets
ROUGHS_3	Zhu's paper [28] formalized (to appear)
ROUGHS_4	topological aspects of RST

some general theory of mathematical objects. One of the significant matchings discovered automatically with the help of our work was that classification of rough sets basically coincides with that of objects described by Isomichi – first class objects are precisely crisp sets, second class – rough sets with non-empty boundary, so third class just vanishes in equivalence-based approximation space. Details of this correspondence can be found in [9].

During our project of making both formal approaches to incomplete information accessible in the Mizar Mathematical Library, an extensive framework was created, besides lattices of rough sets and fuzzy sets. Its summary is shown in Table 2.

Second and third group in Table 2 lists our MML articles about fuzzy and rough sets, respectively, while the first group contains some preliminary notions and properties needed for smooth further work. We listed only MML items of which we are authors – among nearly 1250 files written by over 250 people. Not all were enumerated – the complete list could contain nearly 20 files containing about 40 thousand lines of Mizar code (so it is about 2 % of the MML). Not all of them are tightly connected with the theory of rough and fuzzy sets – they expand the theory of intervals, topologies, relational structures, and lattices – to list more notable areas of mathematics.

Following Grätzer's classical textbook [13] in Mizar, practically all chapters from Järvinen work are covered, maybe except Sect. 8 on information systems (reducts are poorly covered in Mizar, but e.g. the Mizar article about Armstrong systems on ordered sets was coauthored by Armstrong himself). Also the final section on lattices of rough sets is not sufficiently covered in the MML, because not all combinations of properties of binary relations were considered. Our personal perspective was that [14] was even better written from the viewpoint of a developer of MML. The fact is that the theory is much simpler than in *Compendium of Continuous Lattices* by Gierz et al. (and it is more self-contained), although a little bit painful – at least from the formal point of view – category theory was used in a similar degree.

7   Conclusions and Future Work

Essentially, in order to widen the formal framework we can formulate the queries on MML, we should construct more algebraic models both for rough sets and fuzzy sets [3]. Among our aims, we can point out merging fuzzy with rough sets to benefit having both approaches at the same time. Although we underlined the possibility of using external provers, we already applied some automatic tools available in the Mizar system aiming at discovering alternative proofs [22] and unnecessary assumptions to generalize the approach (similarly to generalizing from equivalence into tolerance relations in case of rough sets).

It is hard to describe all the formalized work which was done during this project in such a short form; however we tried to outline the very basic constructions. Of course, complete development (together with all the proofs) can be browsed in the Mizar Mathematical Library freely available for anyone interested under MML identifiers, e.g. listed in Table 2. Our work was very much inspired by Järvinen [14] and Zhu [28], although regular formalization of this topic started back in 2000, much before both these papers. Our short term goal is to provide annotated version of [14] where all items (definitions, propositions, and illustrative examples) will be mapped with their formal counterpart written in Mizar.

Mizar code enables also the information exchange with other proof assistants through its XML intermediate format. Even if the Mizar code is relatively well readable, LaTeX version and HTML script with expandable and fully hyperlinked proofs are available. All Mizar files are also translated into TPTP [23] (first order logic form which can serve as a direct input for Thousands of Problems for Theorem Provers).

References

1. Bryniarski, E.: Formal conception of rough sets. Fundamenta Informaticae **27**(2/3), 109–136 (1996)
2. Dubois, D., Prade, H.: Operations on fuzzy numbers. Int. J. Syst. Sci. **9**(6), 613–626 (1978)
3. Dubois, D., Prade, H.: Rough fuzzy sets and fuzzy rough sets. Int. J. Gen. Syst. **17**(2–3), 191–209 (1990)
4. Estaji, A.A., Hooshmandasl, M.R., Davvaz, B.: Rough set theory applied to lattice theory. Inf. Sci. **200**, 108–122 (2012)
5. Goguen, J.A.: L-fuzzy sets. J. Math. Anal. Appl. **18**(1), 145–174 (1967)
6. Grabowski, A.: Mechanizing complemented lattices within Mizar type system. J. Autom. Reasoning **55**(3), 211–221 (2015). doi:10.1007/s10817-015-9333-5
7. Grabowski, A.: The formal construction of fuzzy numbers. Formalized Math. **22**(4), 321–327 (2014)

8. Grabowski, A.: On the computer certification of fuzzy numbers. In: Ganzha, M., Maciaszek, L., Paprzycki, M. (eds.) Proceedings of Federated Conference on Computer Science and Information Systems, FedCSIS 2013, pp. 51–54 (2013)

9. Grabowski, A.: Automated discovery of properties of rough sets. Fundamenta Informaticae 128(1–2), 65–79 (2013)

10. Grabowski, A.: On the computer-assisted reasoning about rough sets. In: Dunin-Kęplicz, B., Jankowski, A., Szczuka, M. (eds.) Monitoring, Security and Rescue Techniques in Multiagent Systems. Advances in Soft Computing, vol. 28, pp. 215–226. Springer, Heidelberg (2005)

11. Grabowski, A., Jastrzębska, M.: Rough set theory from a math-assistant perspective. In: Kryszkiewicz, M., Peters, J.F., Rybiński, H., Skowron, A. (eds.) RSEISP 2007. LNCS (LNAI), vol. 4585, pp. 152–161. Springer, Heidelberg (2007)

12. Grabowski, A., Korniłowicz, A., Naumowicz, A.: Mizar in a nutshell. J. Formalized Reasoning 3(2), 153–245 (2010)

13. Grätzer, G.: General Lattice Theory. Birkhäuser, Boston (1998)

14. Järvinen, J.: Lattice theory for rough sets. In: Peters, J.F., Skowron, A., Düntsch, I., Grzymała-Busse, J.W., Orłowska, E., Polkowski, L. (eds.) Transactions on Rough Sets VI. LNCS, vol. 4374, pp. 400–498. Springer, Heidelberg (2007)

15. Kawahara, Y., Furusawa, H.: An algebraic formalization of fuzzy relations. Fuzzy Sets Syst. 101, 125–135 (1999)

16. Klir, G.J.: Fuzzy arithmetic with requisite constraints. Fuzzy Sets Syst. 91, 165–175 (1997)

17. Korniłowicz, A.: On rewriting rules in Mizar. J. Autom. Reasoning 50(2), 203–210 (2013)

18. Mitsuishi, T., Endou, N., Shidama, Y.: The concept of fuzzy set and membership function and basic properties of fuzzy set operation. Formalized Math. 9(2), 351–356 (2001)

19. Moore, R., Lodwick, W.: Interval analysis and fuzzy set theory. Fuzzy Sets Syst. 135(1), 5–9 (2003)

20. Naumowicz, A., Korniłowicz, A.: A brief overview of MIZAR. In: Berghofer, S., Nipkow, T., Urban, C., Wenzel, M. (eds.) TPHOLs 2009. LNCS, vol. 5674, pp. 67–72. Springer, Heidelberg (2009)

21. Pawlak, Z.: Rough Sets: Theoretical Aspects of Reasoning about Data. Kluwer, Dordrecht (1991)

22. Pak, K.: Methods of lemma extraction in natural deduction proofs. J. Autom. Reasoning 50(2), 217–228 (2013)

23. Urban, J., Sutcliffe, G.: Automated reasoning and presentation support for formalizing mathematics in mizar. In: Autexier, S., Calmet, J., Delahaye, D., Ion, P.D.F., Rideau, L., Rioboo, R., Sexton, A.P. (eds.) AISC 2010. LNCS, vol. 6167, pp. 132–146. Springer, Heidelberg (2010)

24. Wiedijk, F.: Formal proof - getting started. Not. AMS 55(11), 1408–1414 (2008)

25. Yao, Y.Y.: Two views of the rough set theory in finite universes. Int. J. Approximate Reasoning 15(4), 291–317 (1996)

26. Yao, Y., Yao, B.: Covering based rough set approximations. Inf. Sci. 200, 91–107 (2012)

27. Zadeh, L.: Fuzzy sets. Inf. Control 8(3), 338–353 (1965)

28. Zhu, W.: Generalized rough sets based on relations. Inf. Sci. 177(22), 4997–5011 (2007)

29. Zhu, W.: Topological approaches to covering rough sets. Inf. Sci. 177(6), 1499–1508 (2007)

Generalized Fuzzy Regular Filters
on Residuated Lattices

Yi Liu[1](✉) and Lianming Mou[2]

[1] Data Recovery Key Lab of Sichuan Province, Neijiang Normal University,
Neijiang 641000, Sichuan, People's Republic of China
liuyiyl@163.com
[2] College of Mathematics and Information Sciences, Neijiang Normal University,
Neijiang 641000, Sichuan, People's Republic of China

Abstract. The aim of this paper is further to develop the fuzzy filter theory of general residuated lattices. The concepts of $(\in, \in \vee q_k)$-fuzzy positive implicative filter, $(\in, \in \vee q_k)$-fuzzy MV filter and $(\in, \in \vee q_k)$-fuzzy regular filter are introduced; Their properties are investigated, and some equivalent characterizations of these generalized fuzzy filters are also derived.

1 Introduction

Non-classical logic have become as a formal and useful tool for computer science to deal with uncertain information. Many-valued logic, a great extension and development of classical logic, has always been a crucial direction in non-classical logic. Various logical algebras have been proposed as the semantical systems of non-classical logic systems, such as residuated lattices, lattice implication algebras(MV-algebras), BL-algebras, MTL-algebras, etc. Among these logical algebras, residuated lattices are very basic and important algebraic structure because the other logical algebras are all particular cases of residuated lattices.

The filter theory of a logical algebra [4–9,11,12] is related to the completeness of the corresponding non-classical logic. From a logical point of view, all kinds of filters correspond to various set of provable formulas. Still now, the filter theory of BL-algebras and residuated lattices have been widely investigated, and some of their characterizations and relations were presented. In additional, some new types of filter of BL-algebra and lattice implication algebras are derived. The concept of fuzzy set was introduced by Zadeh [10]. Rosenfeld inspired the fuzzification of algebraic structure and introduced the notion of fuzzy subgroup (Rosenfeld, 1971). The idea of fuzzy point and 'belongingness' and 'quasi-coincidence' with a fuzzy set were given by Pu and Liu [1]. A new type of fuzzy subgroup (viz $(\in, \in \vee q)$-fuzzy subgroup) was introduced [8].

This work is supported by National Natural Science Foundation of P.R.China (Grant no. 61175055, 61305074); The Application Basic Research Plan Project of Sichuan Province (No.2015JY0120), The Scientific Research Project of Department of Education of Sichuan Province (14ZA0245, 15ZB0270).

D. Ciucci et al. (Eds.): RSKT 2015, LNAI 9436, pp. 357–368, 2015.
DOI: 10.1007/978-3-319-25754-9_32

In fact, $(\in, \in \vee q)$-fuzzy subgroup is an important and useful generalization of Rosenfeld's fuzzy subgroup. The idea of fuzzy point and 'belongingness' and 'quasi-coincidence' with a fuzzy set have been applied some important algebraic system.

Therefore, it is meaningful to establish the fuzzy filter theory of general residuated lattice for studying the common properties of the above-mentioned logical algebras. This paper, as a continuation of above work, we extend the concept of quasi-coincidence and further investigate the $(\in, \in \vee q)$-fuzzy (implicative, positive implicative) filters of general residuated lattice, introducing the concept of $(\in, \in \vee q_k)$-fuzzy (implicative, positive implicative, MV, regular) filters, and some equivalent results are obtained respectively. Specially, these results still hold in other logical algebras (such as lattice implication algebras(MV-algebras), BL-algebras, MTL-algebras, etc.).

2   Basic Results on Residuated Lattices

In this section, we will list some relative concepts and properties on residuated lattices, they are used in later sections of this paper.

Definition 1. *[3] A residuated lattice is an algebraic structure*

$$L = (L, \vee, \wedge, \otimes, \rightarrow, 0, 1)$$

of type (2,2,2,2,0,0) satisfying the following axioms:
  (C1) $(L, \vee, \wedge, 0, 1)$ is a bounded lattice.
  (C2) $(L, \otimes, 1)$ is a commutative semigroup (with the unit element 1).
  (C3) $(\otimes, \rightarrow)$ is an adjoint pair.

In what follows, let L denoted a residuated lattice unless otherwise specified.

Proposition 1. *[3] In each residuated lattice L, the following properties hold for all $x, y, z \in L$:*
  (P1) $(x \otimes y) \rightarrow z = x \rightarrow (y \rightarrow z)$.
  (P2) $z \leq x \rightarrow y \Leftrightarrow z \otimes x \leq y$.
  (P3) $x \leq y \Leftrightarrow z \otimes x \leq z \otimes y$.
  (P4) $x \rightarrow (y \rightarrow z) = y \rightarrow (x \rightarrow z)$.
  (P5) $x \leq y \Rightarrow z \rightarrow x \leq z \rightarrow y$.
  (P6) $x \leq y \Rightarrow y \rightarrow z \leq x \rightarrow z, y' \leq x'$.
  (P7) $y \rightarrow z \leq (x \rightarrow y) \rightarrow (x \rightarrow z)$.
  (P8) $y \rightarrow x \leq (x \rightarrow z) \rightarrow (y \rightarrow z)$.
  (P9) $1 \rightarrow x = x, x \rightarrow x = 1$.
  (P10) $x^m \leq x^n$, $m, n \in N$, $m \geq n$.
  (P11) $x \leq y \Leftrightarrow x \rightarrow y = 1$.
  (P12) $0' = 1, 1' = 0, x' = x''', x \leq x''$.

Let L be a residuated lattice, $F \subseteq L$, and $x, y, z \in L$. We list some conditions which will be used in the following study:

(F1) $x, y \in F \Rightarrow x \otimes y \in F$.
(F2) $x \in F, x \leq F \Rightarrow y \in F$.
(F3) $1 \in F$.
(F4) $x \in F, x \to y \in F \Rightarrow y \in F$.
(F5) $z, z \to ((x \to y) \to x) \in F \Rightarrow x \in F$.
(F6) $z \to (x \to y), z \to x \in F \Rightarrow z \to y \in F$.

Definition 2. *[3, 11] (1) A non-empty subset F of a residuated lattice is called a **filter** of L if it satisfies (F1) and (F2).*

*(2) A non-empty subset F of a residuated lattice L is called an **implicative filter** of L if it satisfies (F3) and (F5).*

*(3) A non-empty subset F of a residuated lattice L is called a **positive implicative filter** of L if it satisfies (F3) and (F6).*

Proposition 2. *[3, 11] A non-empty subset F of a residuated lattice is called a **filter** of L if it satisfies (F3) and (F4).*

Let L be a residuated lattice, μ a fuzzy set of L and $x, y \in L$. For convenience, we list some conditions which will be used in the following study:
  (FF1) $\mu(x \otimes y) \geq min\{\mu(x), \mu(y)\}$.
  (FF2) $x \leq y \Rightarrow \mu(x) \leq \mu(y)$.
  (FF3) $\mu(1) \geq \mu(x)$.
  (FF4) $\mu(y) \geq min\{\mu(x \to y), \mu(x)\}$.
  (FF5) $\mu(x \to z) \geq min\{\mu(x \to (z' \to y)), \mu(y \to z)\}$.
  (FF6) $\mu(x \to z) \geq min\{\mu(x \to (y \to z)), \mu(x \to y)\}$.

Definition 3. *[3, 11] (1) A fuzzy set μ of a residuated lattice L is called a **fuzzy filter**, if it satisfies (FF1) and (FF2).*

*(2)A fuzzy set μ of a residuated lattice L is called a **fuzzy implicative filter**, if it satisfies (FF3) and (FF5).*

*(3)A fuzzy set μ of a residuated lattice L is called a **fuzzy positive implicative filter**, if it satisfies (FF3) and (FF6).*

Proposition 3. *[3, 11] A fuzzy set μ of a residuated lattice L is a **fuzzy filter**, if and only if it satisfies (FF3) and (FF4).*

A fuzzy set μ of a residuated lattice L of the form: $\mu(y) = t \in (0, 1]$ if $y = x$; Otherwise $\mu(y) = 0$. This fuzzy set is said to be a **fuzzy point** with support x and value t and is denoted by x_t.

A fuzzy point x_t is said to belong to (resp. be quasi-coincident with) a fuzzy set μ, written as $x_t \in \mu$ (resp. $x_t q \mu$) if $\mu(x) \geq t$ (resp. $\mu(x) + t > 1$). If $x_t \in \mu$ or (resp. and) $x_t q \mu$, then we write $x_t \in \vee q \mu$ (resp. $x_t \in \wedge q \mu$). The symbol $\overline{\in \vee q}$ means $\in \vee q$ doesn't hold. Using the notion of 'belongingness ($\in$)' and 'quasi-coincidence (q)' of fuzzy point with fuzzy subsets, the concept of (α, β)-fuzzy sub semigroup, where α and β are any one of $\{\in, \in \vee q, \in \wedge q\}$ with $\alpha \neq \in \wedge q$, was introduced in [2]. It is worthy to note that the most viable generalization of Rosenfeld's fuzzy subgroup is the notion of $(\in, \in \vee q)$-fuzzy subgroup. The detailed research with $(\in, \in \vee q)$-fuzzy subgroup has been considered in [1].

3   Main Results

In this section, we first to extend the concept of quasi-coincidence. In what follows, we let $k \in [0,1)$ unless otherwise specified. For a fuzzy point x_r and a fuzzy subset μ on L, we say that

(1) $x_r q_k \mu$ if $\mu(x) + r + k > 1$.
(2) $x_r \in \vee q_k \mu$ if $x_r \in \mu$ or $x_r q_k \mu$.
(3) $x_r \overline{\alpha} \mu$ if $x_r \alpha \mu$ doesn't hold for $\alpha \in \{\in, q_k, \in \vee q_k\}$.

3.1 $(\in, \in \vee q_k)$-Fuzzy Positive Implicative $(G\text{-})$ Filter

In this subsection, we will apply the ideas of quasi-coincidence to fuzzy (positive, positive implicative) filters of residuated lattices. The concepts of $(\in, \in \vee q_k)$-fuzzy positive implicative $(G\text{-})$ filter will be introduced and their properties will be investigated.

Definition 4. *A fuzzy subset μ on a residuated lattice L is called an $(\in, \in \vee q_k)$-fuzzy positive implicative filter, if it satisfies, for any $x, y, z \in L$, $t, r \in (0,1]$:*
  (FF11) $x_t \in \mu$ imply $1_t \in \vee q_k \mu$,
  (FF12) if $(x \rightarrow (y \rightarrow z))_t \in \mu$ and $(x \rightarrow y)_r \in \mu$, then $(x \rightarrow z)_{\min\{t,r\}} \in \vee q_k \mu$.

Example 1. Let $L = \{0, a, b, c, d, 1\}$, the Hasse diagram of L be defined as Fig. 1 and its implication operator $\rightarrow$ be defined as Table 1 and operator $\otimes$ be defined as Table 2. Then $\mathcal{L} = (L, \vee, \wedge, \otimes, \rightarrow, 0, 1)$ is a residuated lattice. $\mathcal{L}$ is also a regular residuated lattice.

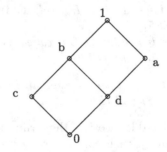

Fig. 1. Hasse diagram of L

We define a fuzzy subset B of $\mathcal{L}$

$$B(I) = 0.6, B(b) = B(c) = 0.7,$$

$$B(O) = B(d) = B(a) = 0.2.$$

It is routine to verify that B is an $(\in, \in \vee q_{0.2})$-fuzzy (implicative) filter. In fact, B is also an $(\in, \in \vee q_{0.2})$-fuzzy filter of $\mathcal{L}$.

Table 1. → of $\mathcal{L}$

→	0	a	b	c	d	1
0	1	1	1	1	1	1
a	c	1	b	c	b	1
b	d	a	1	b	a	1
c	a	a	1	1	a	1
d	b	1	1	b	1	1
1	0	a	b	c	d	1

Table 2. ⊗ of $\mathcal{L}$

⊗	0	a	b	c	d	1
0	0	0	0	0	0	0
a	0	a	d	0	d	a
b	0	d	c	c	0	b
c	0	0	c	c	0	c
d	0	d	0	0	0	d
1	0	a	b	c	d	1

Theorem 1. *Let μ be a fuzzy set of L. Then μ is an $(\in, \in \vee q_k)$-fuzzy positive implicative filter of L if and only if μ is an $(\in, \in \vee q_k)$-fuzzy filter and $\mu(x \to z) \geq min\{\mu(x \to (y \to z)), \mu(x \to y), \frac{1-k}{2}\}$, for any $x, y, z \in L$.*

Theorem 2. *Let μ be an $(\in, \in \vee q_k)$-fuzzy positive implicative filter of L if and only if $U(\mu; t)(\neq \emptyset)$ is a positive implicative filter of L for any $t \in (0, \frac{1-k}{2}]$.*

Theorem 3. *Let μ be fuzzy set of L and μ be an $(\in, \in \vee q_k)$-fuzzy filter of L. Then the following are equivalent:*

(1) μ is an $(\in, \in \vee q_k)$-fuzzy positive implicative filter;
(2) $\mu(x \to y) \geq min\{\mu(x \to (x \to y)), \frac{1-k}{2}\}$, for any $x, y \in L$;
(3) $\mu((x \to y) \to (x \to z)) \geq min\{\mu(x \to (y \to z)), \frac{1-k}{2}\}$, for any $x, y, z \in L$.

Proof. Assume that μ is an $(\in, \in \vee q_k)$-fuzzy positive implicative filter, we have $\mu(x \to y) \geq min\{\mu(x \to (x \to y)), \mu(x \to x), \frac{1-k}{2}\} = min\{\mu(x \to (x \to y)), \frac{1-k}{2}\}$. Thus (2) is valid.

Suppose that (2) holds. That is, assume that μ is an $(\in, \in \vee q_k)$-fuzzy filter of L and $\mu(x \to y) \geq min\{\mu(x \to (x \to y)), \frac{1-k}{2}\}$. Note that $x \to (y \to z) \geq x \to ((x \to y) \to (x \to z))$, it follows that

$$\mu((x \to y) \to (x \to z)) = \mu(x \to ((x \to y) \to z))$$
$$\geq min\{\mu(x \to (x \to ((x \to y) \to z))), \frac{1-k}{2}\}$$
$$= min\{\mu(x \to ((x \to y) \to (x \to z))), \frac{1-k}{2}\}$$

$$\geq min\{min\{\mu(x \to (y \to z))\frac{1-k}{2}\}, \frac{1-k}{2}\}$$

$$= min\{\mu(x \to (y \to z)), \frac{1-k}{2}\}.$$

And so $\mu(x \to z) \geq min\{\mu(x \to y), \mu((x \to y) \to (x \to z)), \frac{1-k}{2}\} \geq min\{\mu(x \to (y \to z)), \frac{1-k}{2}\} = min\{\mu(x \to y), \mu(x \to (y \to z)), \frac{1-k}{2}\}$. Therefore, μ is an $(\in, \in \vee q_k)$-fuzzy positive implicative filter of L.

$(1) \Leftrightarrow (3)$ is obvious from the above proof.

Definition 5. *Let μ be a fuzzy set of L. Then μ is called an $(\in, \in \vee q_k)$ fuzzy G filter of L if it is an $(\in, \in \vee q_k)$-fuzzy filter and $\mu(x \to y) \geq min\{\mu(x \to (x \to y)), \frac{1-k}{2}\}$, for any $x, y \in L$.*

Corollary 1. *Let μ be a fuzzy set of L. Then μ is an $(\in, \in \vee q_k)$-fuzzy positive implicative filter of L if and only if μ is an $(\in, \in \vee q_k)$-fuzzy G filter of L.*

Theorem 4. *Let μ be an $(\in, \in \vee q_k)$-fuzzy filter of L. Then μ is an $(\in, \in \vee q_k)$-fuzzy positive implicative filter if and only if $\mu(y \to x) \geq min\{\mu(z \to (y \to (y \to x))), \mu(z), \frac{1-k}{2}\}$, for any $x, y, z \in L$.*

Proof. Since μ is an $(\in, \in \vee q_k)$-fuzzy positive implicative filter, we have $\mu(y \to (y \to z)) \geq min\{\mu(z \to (y \to (y \to x))), \mu(z), \frac{1-k}{2}\}$. In fact, $y \to x = 1 \to (y \to x) = (y \to y) \to (y \to x)$, so $\mu(y \to x) = \mu((y \to y) \to (y \to x)) \geq min\{\mu(y \to (y \to x)), \frac{1-k}{2}\} \geq min\{\mu(z \to (y \to (y \to x))), \mu(z), \frac{1-k}{2}\}$.

Conversely, let μ be an $(\in, \in \vee q_k)$-fuzzy filter and satisfy the condition: $\mu(y \to x) \geq min\{\mu(z \to (y \to (y \to x))), \mu(z), \frac{1-k}{2}\}$, for any $x, y, z \in L$. And so

$$\mu(z \to x) \geq min\{\mu((z \to y) \to (z \to (z \to x))), \mu(z \to y), \frac{1-k}{2}\}. \quad (1)$$

Since $z \to (y \to x) = y \to (z \to x) \leq (z \to y) \to (z \to (z \to x))$, we have $\mu((z \to y) \to (z \to (z \to x))) \geq min\{\mu(z \to (y \to x)), \frac{1-k}{2}\}$. Together with (2), we have $\mu(z \to x) \geq min\{\mu((z \to y) \to (z \to (z \to x))), \mu(z \to y), \frac{1-k}{2}\} \geq min\{min\{\mu(z \to (y \to x)), \frac{1-k}{2}\}, \mu(z \to y), \frac{1-k}{2}\} = min\{\mu(z \to (y \to x)), \mu(z \to y), \frac{1-k}{2}\}$. Hence μ is an $(\in, \in \vee q_k)$-fuzzy positive implicative filter.

Theorem 5. *Let μ be an $(\in, \in \vee q_k)$-fuzzy positive implicative filter of L. Then μ is an $(\in, \in \vee q_k)$-fuzzy implicative filter of L if and only if $\mu((y \to x) \to x) \geq min\{\mu((x \to y) \to y), \frac{1-k}{2}\}$, for any $x, y \in L$.*

Proof. Assume that μ is an $(\in, \in \vee q_k)$-fuzzy implicative filter of L. We have, for any $x, y, z \in L$ $\mu((y \to x) \to x) \geq min\{\mu(z \to (((y \to x) \to x) \to y) \to ((y \to x) \to x)), \mu(z), \frac{1-k}{2}\}$. Taking $z = 1$ in (1), we have

$$\mu((y \to x) \to x) \geq min\{\mu((((y \to x) \to x) \to y) \to ((y \to x) \to x), \frac{1-k}{2}\}. \quad (2)$$

Since $(x \to y) \to y \leq (y \to x) \to ((x \to y) \to x) = (x \to y) \to ((y \to x) \to x) \leq (((y \to x) \to x) \to y) \to (y \to x) \to x$. Since μ is an $(\in, \in \vee q_k)$-fuzzy filter of L, we have $\mu((((y \to x) \to x) \to y) \to (y \to x) \to x) \geq min\{\mu((x \to y) \to y), \frac{1-k}{2}\}$. From (2), $\mu((y \to x) \to x) \geq min\{\mu((x \to y) \to y), \frac{1-k}{2}\}$, for any $x, y \in L$.

Conversely, since μ is an $(\in, \in \vee q_k)$-fuzzy filter of L, we have, for any $x, y, z \in L$, $\mu((x \to y) \to x) \geq min\{\mu(z \to ((x \to y) \to x)), \mu(z), \frac{1-k}{2}\}$. Since $(x \to y) \to x \leq (x \to y) \to ((x \to y) \to y)$, we have $\mu((x \to y) \to ((x \to y) \to y)) \leq min\{\mu((x \to y) \to x), \frac{1-k}{2}\}$. By hypotheses, $\mu((y \to x) \to x) \geq min\{\mu((x \to y) \to y), \frac{1-k}{2}\}$. Since μ is an $(\in, \in \vee q_k)$-fuzzy filter of L, we have $\mu((x \to y) \to y) \geq min\{\mu((x \to y) \to ((x \to y) \to y)), \frac{1-k}{2}\}$. It follows that $\mu((y \to x) \to x) \geq min\{\mu((x \to y) \to x), \frac{1-k}{2}\}$. Since $y \leq x \to y$ and $y \to x \leq z \to (y \to x)$, we get $(x \to y) \to x \leq y \to x \leq z \to (y \to x)$. It follows that $\mu(z \to (y \to x)) \geq min\{\mu((x \to y) \to x), \frac{1-k}{2}\}$. And $\mu(y \to x) \geq min\{\mu(z \to (y \to x)), \frac{1-k}{2}\}$. Hence $\mu(y \to x) \geq min\{\mu((x \to y) \to x), \mu(z), \frac{1-k}{2}\}$. And so

$$\mu(x) \geq min\{\mu((y \to x) \to x), \mu(y \to x), \frac{1-k}{2}\}$$

$$\geq min\{\mu((x \to y) \to x), \mu(y \to x), \frac{1-k}{2}\}$$

$$\geq min\{\mu((x \to y) \to x), \mu(z), \frac{1-k}{2}\}$$

$$\geq min\{min\{\mu(z), \mu(z \to ((x \to y) \to x)), \frac{1-k}{2}\}, \mu(z), \frac{1-k}{2}\}$$

$$= min\{\mu(z), \mu(z \to ((x \to y) \to x)), \frac{1-k}{2}\}.$$

Therefore μ is an $(\in, \in \vee q_k)$-fuzzy implicative filter of L.

3.2 $(\in, \in \vee q_k)$-Fuzzy MV (Fantastic) Filters

In this subsection, we will apply the ideas of quasi-coincidence to fuzzy MV-filters of residuated lattices. The concepts of $(\in, \in \vee q_k)$-fuzzy MV-filter will be introduced and their properties will be investigated.

Definition 6. *An $(\in, \in \vee q_k)$-fuzzy filter of L is called an $(\in, \in \vee q_k)$-fuzzy MV filter if it satisfies the condition:* $\mu(((x \to y) \to y) \to x) \geq min\{\mu(y \to x), \frac{1-k}{2}\}$.

Remark 1. In lattice implication algebras, BL-algebras, R_0-algebras, the MV filters are called fanstic filters.

Theorem 6. *A fuzzy set μ of L is an $(\in, \in \vee q_k)$-fuzzy MV filter if and only if it satisfies the following:*

(1) $\mu(1) \geq min\{\mu(x), \frac{1-k}{2}\}$, *for any $x \in L$;*
(2) $\mu(((x \to y) \to y) \to x) \geq min\{\mu(z), \mu(z \to (y \to x)), \frac{1-k}{2}\}$, *for any $x, y, z \in L$.*

Proof. Suppose that μ is an $(\in, \in \vee q_k)$-fuzzy MV filter, then (1) is trivial. By Definition 3.3, we have

$$\mu(((x \to y) \to y) \to x) \geq min\{\mu(y \to x), \frac{1-k}{2}\}$$

$$\geq min\{min\{\mu(z \to (y \to x)), \mu(z), \frac{1-k}{2}\}, \frac{1-k}{2}\}$$

$$= min\{\mu(z \to (y \to x)), \mu(z), \frac{1-k}{2}\}.$$

Conversely, suppose that μ satisfies the conditions (2). Taking $z = 1$ in (3), we have

$$\mu(((x \to y) \to y) \to x) \geq min\{\mu(y \to x), \frac{1-k}{2}\}. \tag{3}$$

Taking $y = 1$ in (2), $\mu(x) = \mu(((x \to 1) \to 1) \to x) \geq min\{\mu(z), \mu(z \to (1 \to x)), \frac{1-k}{2}\} = min\{\mu(z \to x), \mu(z), \frac{1-k}{2}\}$. Together with (1), we have μ is an $(\in, \in \vee q_k)$-fuzzy filter of L. Therefore, μ is an $(\in, \in \vee q_k)$-fuzzy MV filter.

Theorem 7. *Let μ be a fuzzy subset of L. Then μ is an $(\in, \in \vee q_k)$-fuzzy MV filter if and only if, for any $t \in (0, \frac{1-k}{2}]$, $U(\mu; t)(\neq \emptyset)$ is a MV filter of L.*

Theorem 8. *In a residuated lattice, every $(\in, \in \vee q_k)$-fuzzy implicative filter is an $(\in, \in \vee q_k)$-fuzzy MV filter.*

Proof. Assume that μ be an $(\in, \in \vee q_k)$-fuzzy implicative filter of L. Since $x \leq ((x \to y) \to y) \to x$, and so $(((x \to y) \to y) \to x) \to y \leq x \to y$. This implies that $((((x \to y) \to y) \to x) \to y) \to (((x \to y) \to y) \to x) \geq (x \to y) \to ((((x \to y) \to y) \to x)) = ((x \to y) \to y) \to ((x \to y) \to x) \geq y \to x$. It follows that $\mu(((((x \to y) \to y) \to x) \to y) \to (((x \to y) \to y) \to x)) \geq min\{\mu(y \to x), \frac{1-k}{2}\}$. Since μ be an $(\in, \in \vee q_k)$-fuzzy implicative filter of L, we have $\mu((((x \to y) \to y) \to x)) \geq min\{\mu(z \to (((((x \to y) \to y) \to x) \to y) \to (((x \to y) \to y) \to x))), \mu(z), \frac{1-k}{2}\}$. Taking $z = 1$, we get $\mu((((x \to y) \to y) \to x)) \geq min\{\mu((((((x \to y) \to y) \to x) \to y) \to (((x \to y) \to y) \to x)), \frac{1-k}{2}\} \geq min\{\mu(y \to x), \frac{1-k}{2}\}$. Hence, μ is an $(\in, \in \vee q_k)$-fuzzy MV filter.

Theorem 9. *A fuzzy set μ of L is an $(\in, \in \vee q_k)$-fuzzy implicative filter of L if and only if μ is both an $(\in, \in \vee q_k)$-fuzzy positive and an $(\in, \in \vee q_k)$-fuzzy MV filter of L.*

Proof. Assume that μ is an $(\in, \in \vee q_k)$-fuzzy implicative filter of L. We know μ is both an $(\in, \in \vee q_k)$-fuzzy positive and an $(\in, \in \vee q_k)$-fuzzy MV filter of L.

Conversely, suppose that μ is both an $(\in, \in \vee q_k)$-fuzzy positive filter and an $(\in, \in \vee q_k)$-fuzzy MV filter of L, we have $\mu((x \to y) \to y) \geq min\{\mu((x \to y) \to ((x \to y) \to y))\}$. Since $(x \to y) \to x \leq (x \to y) \to ((x \to y) \to y)$, it follows that $\mu((x \to y) \to ((x \to y) \to y)) \geq min\{\mu((x \to y) \to x), \frac{1-k}{2}\}$. And so $\mu((x \to y) \to y) \geq min\{\mu((x \to y) \to x), \frac{1-k}{2}\}$. On the other hand, since μ is an

$(\in, \in \vee q_k)$-fuzzy MV filter of L, we have $\mu(((x \to y) \to y) \to x) \geq min\{\mu(y \to x), \frac{1-k}{2}\}$. Since $(x \to y) \to x \leq y \to x$, $\mu(y \to x) \geq min\{\mu((x \to y) \to x), \frac{1-k}{2}\}$. Thus, $\mu(((x \to y) \to y) \to x) min\{\mu((x \to y) \to x), \frac{1-k}{2}\}$, and so $\mu(x) \geq min\{\mu(((x \to y) \to y) \to x), \mu((x \to y) \to y), \frac{1-k}{2}\} \geq min\{\mu((x \to y) \to x), \frac{1-k}{2}\}$, $\mu((x \to y) \to x) \geq min\{\mu(z \to ((x \to y) \to x)), \mu(z), \frac{1-k}{2}\}$. Hence $\mu(x) \geq min\{\mu(z \to ((x \to y) \to x)), \mu(z), \frac{1-k}{2}\}$, we have μ is an $(\in, \in \vee q_k)$-fuzzy implicative filter of L.

3.3 $(\in, \in \vee Q_k)$-Fuzzy Regular Filters

In this subsection, we will apply the ideas of quasi-coincidence to fuzzy regular filters of residuated lattices. The concepts of $(\in, \in \vee q_k)$-fuzzy regular filter will be introduced, their equivalent characterizations will also be investigated.

Definition 7. *An $(\in, \in \vee q_k)$-fuzzy filter μ of L is called an $(\in, \in \vee q_k)$-fuzzy regular filter if it satisfies the condition: $\mu(x'' \to x) \geq min\{\mu(1), \frac{1-k}{2}\}$ for any $x \in L$.*

Theorem 10. *Let μ be an $(\in, \in \vee q_k)$-fuzzy filter of L. Then following assertions are equivalent:*

(GFR1) μ is an $(\in, \in \vee q_k)$-fuzzy regular filter.
(GFR2) $\mu(y \to x) \geq min\{\mu(x' \to y'), \frac{1-k}{2}\}$ for any $x, y \in L$.
(GFR3) $\mu(y' \to x) \geq min\{\mu(x' \to y), \frac{1-k}{2}\}$ for any $x, y \in L$.

Proof. Suppose that μ is an $(\in, \in \vee q_k)$-fuzzy regular filter of L, satisfies the condition $\mu(x'' \to x) \geq min\{\mu(1), \frac{1-k}{2}\}$. For any $x \in L$, we have that $x' \to y' \leq y'' \to x'' \leq y \to x''$, and so $(x' \to y') \to (y \to x) \geq (y \to x'') \to (y \to x) \geq x'' \to x$. Since μ is an $(\in, \in \vee q_k)$-fuzzy filter, we have that $\mu(y \to x) \geq min\{\mu((x' \to y') \to (y \to x)), \mu(x' \to y'), \frac{1-k}{2}\} \geq min\{\mu(x'' \to x), \mu(x' \to y'), \frac{1-k}{2}\} \geq min\{min\{\mu(1), \frac{1-k}{2}\}, \mu(x' \to y'), \frac{1-k}{2}\} = min\{\mu(x' \to y'), \frac{1-k}{2}\}$. And so (GFR2) holds.

Assume that μ satisfies condition (GFR2) and let $x \in L$. Since $x''' = x'$, we have that $x' \to (x'')' = 1$, and so $\mu(x' \to x''') = \mu(1)$. By condition (GFR2), we have $\mu(x'' \to x) \geq min\{\mu(x' \to x'''), \frac{1-k}{2}\} = min\{\mu(1), \frac{1-k}{2}\}$. Therefore, (GFR1) is valid.

Suppose that μ is an $(\in, \in \vee q_k)$-fuzzy filter of L. Since $x' \to y \leq y' \to x''$, we have that $(x' \to y) \to (y' \to x) \geq (y' \to x'') \to (y' \to x) \geq x'' \to x$. Since μ is an $(\in, \in \vee q_k)$-fuzzy filter, we have that $\mu(y' \to x) \geq min\{\mu((x' \to y) \to (y' \to x)), \mu(x' \to y), \frac{1-k}{2}\} \geq min\{\mu(x'' \to x), \mu(x' \to y), \frac{1-k}{2}\} \geq min\{min\{\mu(1), \frac{1-k}{2}\}, \mu(x' \to y), \frac{1-k}{2}\} = min\{\mu(x' \to y), \frac{1-k}{2}\}$. Therefore, (GFR3) is valid.

Suppose that μ satisfies (GFR3). Since $x' \to x' = 1$ for any $x \in L$. It follows from (GFR3) that $\mu(x'' \to x) \geq min\{\mu(x' \to x'), \frac{1-k}{2}\} = min\{\mu(1), \frac{1-k}{2}\}$. Thus μ is an $(\in, \in \vee q_k)$-fuzzy regular filter of L.

Corollary 2. *If μ is an $(\in, \in \vee q_k)$-fuzzy regular filter of L, then $\mu(x'') \geq min\{\mu(x), \frac{1-k}{2}\}$ for any $x \in L$.*

Proof. For any $x \in L$, from (GFR3), we have $\mu(x^{''}) = \mu(x^{'} \to 0) \geq min\{\mu(0^{'} \to x), \frac{1-k}{2}\} = min\{\mu(1 \to x), \frac{1-k}{2}\} = min\{\mu(x), \frac{1-k}{2}\}$.

Theorem 11. *A fuzzy set μ of L is an $(\in, \in \vee q_k)$-fuzzy regular filter if and only if it satisfies*

(GF1) $\mu(1) \geq min\{\mu(x), \frac{1-k}{2}\}$ *for any $x \in L$;*

(GFR5) $\mu(y^{'} \to x) \geq min\{\mu(z \to (x^{'} \to y)), \mu(z), \frac{1-k}{2}\}$ *for any $x, y, z \in L$.*

Proof. Suppose that μ is an $(\in, \in \vee q_k)$-fuzzy regular filter of L. Then μ is an $(\in, \in \vee q_k)$-fuzzy filter of L, thus, (GF1) holds and $\mu(x^{'} \to y) \geq min\{\mu(z \to (x^{'} \to y)), \mu(z), \frac{1-k}{2}\}$. By (GFR3), we have $\mu(y^{'} \to x) \geq min\{\mu(x^{'} \to y), \frac{1-k}{2}\}$. And so, $\mu(y^{'} \to x) \geq min\{\mu(x^{'} \to y), \frac{1-k}{2}\} \geq min\{min\{\mu(z \to (x^{'} \to y)), \mu(z), \frac{1-k}{2}\}, \frac{1-k}{2}\} = min\{\mu(z \to (x^{'} \to y)), \mu(z), \frac{1-k}{2}\}$.

Conversely, assume that μ satisfies (GF1) and (GFR5). Taking $z = 1$ in (GFR5), we have

$$\mu(y^{'} \to x) \geq min\{\mu(x^{'} \to y), \frac{1-k}{2}\}. \tag{4}$$

Next, we show μ satisfies Theorem 3.2 (2). Let $x, y \in L$, since $x \to y = x \to (1 \to y) = x \to (0^{'} \to y)$, it follows from (GFR5) that $\mu(y^{''}) = \mu(y^{'} \to 0) \geq min\{\mu(x \to (0^{'} \to y)), \mu(x), \frac{1-k}{2}\} = min\{\mu(x \to y), \mu(x), \frac{1-k}{2}\}$. Therefore, μ is an $(\in, \in \vee q_k)$-fuzzy filter. From (4), we have μ is an $(\in, \in \vee q_k)$-fuzzy regular filter of L.

Theorem 12. *A fuzzy set μ of L is an $(\in, \in \vee q_k)$-fuzzy regular filter if and only if it satisfies*

(GF1) $\mu(1) \geq min\{\mu(x), \frac{1-k}{2}\}$ *for any $x \in L$;*

(GFR6) $\mu(y \to x) \geq min\{\mu(z \to (x^{'} \to y^{'})), \mu(z), \frac{1-k}{2}\}$ *for any $x, y, z \in L$.*

Denote V_{tk} by the set $\{x \in L | x_t \in \vee q_k \mu\}$. It is called an $\in \vee q_k$-level set of a fuzzy set μ.

Theorem 13. *For any fuzzy set μ in L, then μ is an $(\in, \in \vee q_k)$-fuzzy regular filter if and only if $V_{tk} (\neq \emptyset)$ is a regular filter of L for any $t \in (0, 1]$.*

Proof. Assume that μ is an $(\in, \in \vee q_k)$-fuzzy regular filter, then μ is an $(\in, \in \vee q_k)$-fuzzy filter. From Theorem 3.20 in [10], we have V_{tk} is a filter of L. Therefore, $1 \in V_{tk}$ i.e. $1_t \in \vee q_k \mu$, we have $\mu(1) \geq t$ or $\mu(1) + t + k > 1$. Since μ is an $(\in, \in \vee q_k)$-fuzzy regular filter of L, It follows from Definition 3.4 that $\mu(x^{''} \to x) min\{\mu(1), \frac{1-k}{2}\}$.

Case I: If $\mu(1) < \frac{1-k}{2}$, then $\mu(x^{''} \to x) \geq \mu(1)$.

(a) If $\mu(1) \geq t$, then $\mu(x^{''} \to x) \geq t$. That is, $(x^{''} \to x)_t \in \mu$, hence $x^{''} \to x \in V_{tk}$.

(b) If $\mu(1) + t + k > 1$, then $\mu(x^{''} \to x) + t + k > \mu(1) + t + k > 1$, i.e. $(x^{''} \to x)_t q_k \mu$, hence $x^{''} \to x \in V_{tk}$.

Case II: If $\mu(1) \geq \frac{1-k}{2}$, then $\mu(x^{''} \to x) \geq \frac{1-k}{2}$.

(a) If $t \in (0, \frac{1-k}{2}]$, then $\mu(x^{''} \rightarrow x) \geq \frac{1-k}{2} \geq t$, i.e., $x^{''} \rightarrow x \in V_{tk}$.

(b) If $t \in (\frac{1-k}{2}, 1]$, then $\mu(x^{''} \rightarrow x) + t + k \geq \frac{1-k}{2} + t + k \geq \frac{1-k}{2} + \frac{1-k}{2} + k = 1$.

Hence $(x^{''} \rightarrow x)_t q_k \mu$, i.e., $x^{''} \rightarrow x \in V_{tk}$.

Therefore $V_{tk}(\neq \emptyset)$ is a regular filter for any $t \in (0, 1]$.

Conversely, let μ be a fuzzy set in L such that $V_{tk}(\neq \emptyset)$ is a regular filter of L for any $t \in (0, 1]$. If there is $a \in L$ such that $\mu(1) < min\{\mu(a), \frac{1-k}{2}\}$, then $\mu(1) < r \leq min\{\mu(a), \frac{1-k}{2}\}$ for some $r \in (0, \frac{1-k}{2}]$. It follows that $1_r \notin \mu$. Also $\mu(1) + r + k < \frac{1-k}{2} + r + k \leq \frac{1-k}{2} + \frac{1-k}{2} + k = 1$, hence $1_t \overline{q_k} \mu$. Therefore $1_t \overline{\in \vee q_k} \mu$, i.e., $1 \notin V_{rk}$, which contradicts with V_{tk} is a regular filter of L. Therefore $\mu(1) \geq min\{\mu(1), \frac{1-k}{2}\}$ for any $x \in L$. Suppose that there exist $a, b, c \in L$ such that $\mu(b^{'} \rightarrow a) < min\{\mu(c \rightarrow (a^{'} \rightarrow b)), \mu(c), \frac{1-k}{2}\}$. Then $\mu(b^{'} \rightarrow a) < r \leq min\{\mu(c \rightarrow (a^{'} \rightarrow b)), \mu(c), \frac{1-k}{2}\}$ for some $t \in (0, \frac{1-k}{2}]$, it follows that $(c \rightarrow (a^{'} \rightarrow b))_r \in \mu$ and $c_r \in \mu$, of course, $(c \rightarrow (a^{'} \rightarrow b))_r \in \vee q_k \mu$ and $c_r \in \vee q_k \mu$. Since V_{tk} is a regular filter for any $t \in (0, 1]$, we have $b^{'} \rightarrow a \in V_{rt}$. But $(b^{'} \rightarrow a)_r \notin \mu$. And $\mu(b^{'} \rightarrow a) + r + k < \frac{1-k}{2} + r + k \geq \frac{1-k}{2} + \frac{1-k}{2} + k = 1$, hence $(b^{'} \rightarrow a)_r \overline{q_k} \mu$. Thus $(b^{'} \rightarrow a)_r \overline{\in \vee q_k} \mu$, that is $b^{'} \rightarrow a \notin V_{rk}$, this contradicts with V_{tk} is a regular filter of L. Therefore $\mu(y^{'} \rightarrow x) < min\{\mu(z \rightarrow (x^{'} \rightarrow y)), \mu(z), \frac{1-k}{2}\}$ for any $x, y, z \in L$. We have μ is an $(\in, \in \vee q_k)$-fuzzy regular filter of L.

Theorem 14. *For any fuzzy set μ in L, then μ is an $(\in, \in \vee q_k)$-fuzzy (implicative, positive implicative, MV) filter if and only if $V_{tk}(\neq \emptyset)$ is an (implicative, positive implicative, MV) filter of L for any $t \in (0, 1]$.*

4   Conclusions

In this paper, we develop the fuzzy filter theory of general residuated lattices. Mainly, we give some new characterizations of $(\in, \in \vee q_k)$-fuzzy positive implicative filters in general residuated lattices. We introduced the $(\in, \in \vee q_k)$-fuzzy congruence, study the relation between $(\in, \in \vee q_k)$-fuzzy filter, obtain a new residuated lattice, which induced by $(\in, \in \vee q_k)$-fuzzy filter. The theory can be used in lattice implication algebras(MV-algebras), BL-algebras, MTL-algebras, etc.

In our future work, we will continue investigating the relation between $(\overline{\in}, \overline{\in} \vee \overline{q_k})$-fuzzy filter and $(\overline{\in}, \overline{\in} \vee \overline{q_k})$-fuzzy congruence in general residuated lattices, at the same time, we shall establish generalized fuzzy filter theory and general fuzzy congruence theory based on interval valued fuzzy set in general residuated lattices. For more details, we shall give them out in the future paper. We hope that it will be of great use to provide theoretical foundation to design intelligent information processing systems.

References

1. Bhakat, S.K., Das, P.: $(\in, \in \vee q)$-fuzzy subgroups. Fuzzy Sets Syst. **80**, 359–368 (1996)
2. Bhakat, S.K.: $(\in \vee q)$-level subset. Fuzzy Sets Syst. **103**, 529–533 (1999)
3. Dilworth, R.P., Ward, M.: Residuated lattices. Trans. Am. Math. Soc. **45**, 335–354 (1939)
4. Jun, Y.B., Song, S.Z., Zhan, J.M.: Generalizations of of $(\in, \in \vee q)$-Fuzzy Filters in R_0 algebras. Int. J. Math. Math. Sci. **2010**, 1–19 (2010)
5. Kondo, M., Dudek, W.A.: Filter theory of BL-algebras. Soft. Comput. **12**, 419–423 (2008)
6. Liu, L.Z., Li, K.T.: Fuzzy implicative and Boolean filters of R_0-algebras. Inform. Sci. **171**, 61–71 (2005)
7. Ma, X.L., Zhan, J.M., Jun, Y.B.: On $(\in, \in \vee q)$-fuzzy filters of R_0-algebras. Math. Log. Quart. **55**, 452–467 (2009)
8. Liu, Y., Xu, Y.: Inter-valued (α, β)-fuzzy implication subalgebras. Comput. Sci. **38**(4), 263–266 (2011)
9. Liu, Y., Xu, Y., Qin, X.Y.: Interval-valued T-fuzzy filters and interval-valued T-fuzzy congruences on residuated lattices. J. Intell. Fuzzy Syst. **26**, 2021–2033 (2014)
10. Zadeh, L.A.: Fuzzy sets. Inf. Control **8**, 338–353 (1965)
11. Zhu, Y.Q., Xu, Y.: On filter theory of residuated lattices. Inf. Sci. **180**, 3614–3632 (2010)
12. Zhu, Y., Zhan, J., Jun, Y.B.: On $(\bar{\in}, \bar{\in} \vee \bar{q})$-fuzzy filters of residuated lattices. In: Cao, B., Wang, G., Chen, S., Guo, S. (eds.) Quantitative Logic and Soft Computing 2010. AISC, vol. 82, pp. 631–640. Springer, Heidelberg (2010)

Approximations on Normal Forms in Rough-Fuzzy Predicate Calculus

B.N.V. Satish[✉] and G. Ganesan

Department of Mathematics, Adikavi Nannaya University,
Rajahmundry, Andhra Pradesh, India
bnvsathish@gmail.com, prof.ganesan@yahoo.com

Abstract. Considering the importance of Pawlak's Rough Set Model in information systems, in 2005, G. Ganesan discussed the rough approximations on fuzzy attributes of the information systems. In 2008, G. Ganesan et al., have introduced Rough-Fuzzy connectives confining to the information system as a logical system with fuzzy membership values. In this paper, a two way approach on normal forms through the Rough-Fuzzy connectives using lower and upper literals is discussed.

Keywords: Rough sets · Rough-fuzzy connectives · Lower and upper literals · Normal forms

1 Introduction

The theory of rough sets [10], introduced by Z. Pawlak, finds significant applications in knowledge discovery in information systems. Considering the importance of this theory, the concepts of rough logic had been studied by several researchers [6,7].

Now, consider the following information system with the attributes P_1, $P_2, \ldots, P_m$ and the records $a_1, a_2, \ldots, a_n$.

	P_1	P_2	$\ldots P_i$	$\ldots P_m$
a_1	$P_1(a_1)$	$P_2(a_1)$	$\ldots P_i(a_1)$	$\ldots P_m(a_1)$
a_2	$P_1(a_2)$	$P_2(a_2)$	$\ldots P_i(a_2)$	$\ldots P_m(a_2)$
$\vdots$	$\vdots$	$\vdots$	$\vdots$ $\vdots$	$\vdots$ $\vdots$
a_j	$P_1(a_j)$	$P_2(a_j)$	$\ldots P_i(a_j)$	$\ldots P_m(a_j)$
$\vdots$	$\vdots$	$\vdots$	$\vdots$ $\vdots$	$\vdots$ $\vdots$
a_n	$P_1(a_n)$	$P_2(a_n)$	$\ldots P_i(a_n)$	$\ldots P_m(a_n)$

One may notice that each entity $P_i(a_j)$ represents the specific value or item of the record a_j under the attribute P_i. Considering the fuzzy membership values for each $P_i(a_j)$, in 2008, G. Ganesan et al., introduced R-F Predicate Calculus [2] by defining the rough connectives between the predicates. In 2014, G. Ganesan et al., have derived implication rules and equivalent forms in R-F Predicates. In this paper, we introduce various types of normal forms similar to [5].

© Springer International Publishing Switzerland 2015
D. Ciucci et al. (Eds.): RSKT 2015, LNAI 9436, pp. 369–380, 2015.
DOI: 10.1007/978-3-319-25754-9_33

The connectives AND, OR, IF-THEN, IF AND ONLY IF are commonly in use (along with Negation) in propositional logic. Since the said logic is closely associated with digital logic hardware designing, and the simple logic circuits are to be constructed only with the gates AND, OR, NEGATION, NAND, NOR (and XOR), the study of normal forms have been under great importance. In propositional logic, any Boolean function can be put into the canonical disjunctive normal form (CDNF) or minterm canonical form and its dual canonical conjunctive normal form (CCNF) or maxterm canonical form. These forms are useful for the simplification of Boolean functions, which is of great importance in the digital circuits to minimize the number of gates, to minimize the settling time, etc.

This paper is organized into four sections. In section two, we discuss fuzzy predicates and the rough connectives on them. Also, this section narrates all equivalence rules of R-F Predicate Calculus. In section three, we introduce the various kinds of normal forms using these equivalences.

2   Mathematical Aspects

In this section, we discuss the implications and equivalence rules in R-F Predicate Calculus. Fuzziness [9], Zadeh's invention plays an important role in logic. The predicates, which do not have precise logical value, are called fuzzy predicates.

For any two fuzzy predicates P and Q [with arguments a and b], the fuzzy conjunction ($\wedge$), fuzzy disjunction ($\vee$), fuzzy negation (neg), fuzzy implication ($\rightarrow$) and fuzzy bi implication ($\leftrightarrow$) are defined [4] as follows.

Fuzzy Conjunction ($\wedge$)	$:P(a) \wedge Q(b) = \min(P(a), Q(b))$
Fuzzy Disjunction ($\vee$)	$:P(a) \vee Q(b) = \max(P(a), Q(b))$
Fuzzy Negation (neg)	$:negP(a) = 1 - P(a)$
Fuzzy implication ($\rightarrow$)	$:P(a){\rightarrow}Q(b)=\max(1 - P(a), Q(b))$
Fuzzy bi-implication ($\leftrightarrow$)	$:P(a){\leftrightarrow}Q(b)=\min(P(a) \rightarrow Q(b), Q(b) \rightarrow P(a))$

2.1   Rough-Fuzzy Connectives

For a given threshold α chosen from the complement of a set of all critical points and complements [1], the lower and upper rough approximations of a given predicate P with argument a are defined by $P_\alpha(a) = \cup\{[x] : x \in X$ and $[x] \subseteq P(a)[\alpha]\}$ and $P^\alpha(a) = \cup\{[x] : x \in X$ and $[x] \cap P(a)[\alpha] \neq \emptyset\}$ respectively, where $[\alpha]$ represents strong α cut [1,4]. Let τ denote negation in usual predicate calculus.

For the given fuzzy predicates $P_i(x)$ and $P_j(y)$ are defined [2,3,8] as follows:

- **Rough conjunction** $(\wedge, \overset{\alpha}{\wedge}) \Rightarrow P_i(x) \underset{\alpha}{\wedge} P_j(y) = P_{i,\alpha}(x) \wedge P_{j,\alpha}(y)$ and $P_i(x) \overset{\alpha}{\wedge} P_j(y) = P_i^\alpha(x) \wedge P_j^\alpha(y)$ respectively.
- **Rough disjunction** $(\vee, \overset{\alpha}{\vee}) \Rightarrow P_i(x) \underset{\alpha}{\vee} P_j(y) = P_{i,\alpha}(x) \vee P_{j,\alpha}(y)$ and $P_i(x) \overset{\alpha}{\vee} P_j(y) = P_i^\alpha(x) \vee P_j^\alpha(y)$ respectively.

- **Rough implication** $(\xrightarrow{}, \xrightarrow{\alpha}) \Rightarrow P_i(x) \xrightarrow{\alpha} P_j(y) = P_i^{1-\alpha}(x) \longrightarrow P_{j,\alpha}(y)$
  and $P_i(x) \xrightarrow{\alpha} P_j(y) = P_{i,1-\alpha}(x) \longrightarrow P_j^{\alpha}(y)$ respectively.
- **Rough bi-implication** $(\xleftrightarrow{}, \xleftrightarrow{\alpha}) \Rightarrow P_i(x) \xleftrightarrow{\alpha} P_j(y) = [P_i(x) \xrightarrow{\alpha} P_j(y)] \wedge$
  $[P_j(y) \xrightarrow{\alpha} P_i(x)]$ and $P_i(x) \xleftrightarrow{\alpha} P_j(y) = [P_i(x) \xrightarrow{\alpha} P_j(y)] \wedge [P_j(y) \xrightarrow{\alpha} P_i(x)]$
  respectively.
- **Rough negation** $(\tau_\alpha, \tau^\alpha) \Rightarrow \tau_\alpha P_i(x) = (negP_i(x))_\alpha = \tau(P_i^{1-\alpha}(x))$ and
  $\tau^\alpha P_i(x) = (negP_i(x))^\alpha = \tau(P_{i,1-\alpha}(x))$ respectively.

Also, these connectives satisfy the properties
(i) $(negP_i(x))_\alpha = \tau(P_i^{1-\alpha}(x))$, (ii) $(negP_i(x))^\alpha = \tau(P_{i,1-\alpha}(x))$

The equivalence rules are shown in Tables 1 and 2.

3   Rough-Fuzzy Normal Forms

Normal forms are standard ways of constructing circuits from the tables defining Boolean functions. The role of disjunctive and conjunctive normal forms in classical logic and in boolean algebra of functions is well known. Both forms are used for standard representation of elements. In this section, a two way method of reducing a formula equivalent to a given formula called a normal form of fuzzy predicates similar to normal forms [5] in propositional calculus are introduced. The two way approximating on each kind of normal forms, which is the lower and upper approximations on connectives of fuzzy predicates are introduced in this section. Therefore, four kinds of normal forms *disjunctive normal form (DNF), conjunctive normal form (CNF), principal disjunctive normal form (PDNF) and principal conjunctive normal form (PCNF)* are discussed and some examples are solved.

In this section, the word *lower sum* is used instead of *lower disjunction* $(\underset{\alpha}{\vee})$, and the word *lower product* is used instead of *lower conjunction* $(\underset{\alpha}{\wedge})$ for convenience. Similarly, *upper sum, upper product* are used instead of *upper disjunction* $(\overset{\alpha}{\vee})$ and *upper conjunction* $(\overset{\alpha}{\wedge})$ respectively.

Definition 1. *For any fuzzy predicate $P(x)$ and α is a threshold,*

1. *The lower literal of $P_\alpha(x)$: $P_\alpha(x)$ or $\tau(P_\alpha(x))$ or $(negP(x))_\alpha$ or $\tau(negP(x))^{1-\alpha})$*
2. *The upper literal of $P^\alpha(x)$: $P^\alpha(x)$ or $\tau(P^\alpha(x))$ or $(negP(x))^\alpha$ or $\tau(negP(x))_{1-\alpha})$*

Definition 2. *The elementary sum of lower (upper) literals is called the lower (upper) elementary sum.*

Definition 3. *The elementary product of lower (upper) literals is called the lower (upper) elementary product.*

Table 1. Equivalence rules

Double negation laws			
LE_1	$\tau(\tau(P_\alpha(x))) = P_\alpha(x)$	UE_1	$\tau(\tau(P^\alpha(x))) = P^\alpha(x)$
Idempotent laws			
LE_2	$(P(x) \underset{\alpha}{\wedge} P(x)) = P_\alpha(x)$	UE_2	$(P(x) \overset{\alpha}{\wedge} P(x)) = P^\alpha(x)$
LE_3	$(P(x) \underset{\alpha}{\vee} P(x)) = P_\alpha(x)$	UE_3	$(P(x) \overset{\alpha}{\vee} P(x)) = P^\alpha(x)$
Commutative laws			
LE_4	$P(x) \underset{\alpha}{\wedge} Q(y) = Q(y) \underset{\alpha}{\wedge} P(x)$	UE_4	$P(x) \overset{\alpha}{\wedge} Q(y) = Q(y) \overset{\alpha}{\wedge} P(x)$
LE_5	$P(x) \underset{\alpha}{\vee} Q(y) = Q(y) \underset{\alpha}{\vee} P(x)$	UE_5	$P(x) \overset{\alpha}{\vee} Q(y) = Q(y) \overset{\alpha}{\vee} P(x)$

Some examples of lower (upper) elementary sum and lower (upper) elementary product are given below:

1. Lower elementary sum: $P_\alpha(x) \vee P_\alpha(x)$, $P_\alpha(x) \vee \tau(P_\alpha(x))$, $P_\alpha(x) \vee (negP(x))_\alpha$, $P_\alpha(x) \vee \tau((negP(x))^{1-\alpha})$, $(negP(x))_\alpha \vee (negP(x))_\alpha$

2. Upper elementary sum: $P^\alpha(x) \vee P^\alpha(x)$, $P^\alpha(x) \vee \tau(P^\alpha(x))$, $P^\alpha(x) \vee (negP(x))^\alpha$, $P^\alpha(x) \vee \tau((negP(x))_{1-\alpha})$, $(negP(x))^\alpha \vee (negP(x))^\alpha$

3. Lower elementary product: $P_\alpha(x) \wedge P_\alpha(x)$, $P_\alpha(x) \wedge \tau(P_\alpha(x))$, $P_\alpha(x) \wedge (negP(x))_\alpha$, $P_\alpha(x) \wedge \tau((negP(x))^{1-\alpha})$, $(negP(x))_\alpha \wedge (negP(x))_\alpha$

4. Upper elementary product: $P^\alpha(x) \wedge P^\alpha(x)$, $P^\alpha(x) \wedge \tau(P^\alpha(x))$, $P^\alpha(x) \wedge (negP(x))^\alpha$, $P^\alpha(x) \wedge \tau((negP(x))_{1-\alpha})$, $(negP(x))^\alpha \vee (negP(x))^\alpha$

In similar manner, lower (upper) elementary sum and lower (upper) elementary product for two or more fuzzy predicates are defined.

3.1   Disjunctive Normal Forms (DNF)

Disjunctive normal forms have studied in propositional logic so far. In this section, the disjunctive normal forms are introduced through two way approximating the connectives using a threshold α similar to the disjunctive normal forms [5] which are discussed in propositional calculus. Every propositional formula can be converted into an equivalent formula that is in DNF using both approximations. This transformation is based on rules about logical equivalences: the double negative law, De Morgan's laws, and the distributive law. Therefore, two kinds of disjunctive normal forms are used namely, *lower disjunctive normal forms* (L-DNF), *upper disjunctive normal forms* (U-DNF) of fuzzy predicates using a threshold α are defined in this section and also, some examples are solved by using these definitions.

Table 2. Equivalence rules

Associative laws			
LE_6	$(P(x) \underset{\alpha}{\wedge} Q(y)) \underset{\alpha}{\wedge} R(z) =$ $P(x) \underset{\alpha}{\wedge} (Q(y) \underset{\alpha}{\wedge} R(z))$	UE_6	$(P(x) \overset{\alpha}{\wedge} Q(y)) \overset{\alpha}{\wedge} R(z) =$ $P(x) \overset{\alpha}{\wedge} (Q(y) \overset{\alpha}{\wedge} R(z))$
LE_7	$(P(x) \underset{\alpha}{\vee} Q(y)) \underset{\alpha}{\vee} R(z) =$ $P(x) \underset{\alpha}{\vee} (Q(y) \underset{\alpha}{\vee} R(z))$	UE_7	$(P(x) \overset{\alpha}{\vee} Q(y)) \overset{\alpha}{\vee} R(z) =$ $P(x) \overset{\alpha}{\vee} (Q(y) \overset{\alpha}{\vee} R(z))$
Distributive laws			
LE_8	$P(x) \underset{\alpha}{\wedge} (Q(y) \underset{\alpha}{\vee} R(z))$ $= (P(x) \underset{\alpha}{\wedge} Q(y)) \underset{\alpha}{\vee} (P(x) \underset{\alpha}{\wedge} R(z))$	UE_8	$P(x) \overset{\alpha}{\wedge} (Q(y) \overset{\alpha}{\vee} R(z)) =$ $(P(x) \overset{\alpha}{\wedge} Q(y)) \overset{\alpha}{\vee} (P(x) \overset{\alpha}{\wedge} R(z))$
LE_9	$P(x) \underset{\alpha}{\vee} (Q(y) \underset{\alpha}{\wedge} R(z))$ $= (P(x) \underset{\alpha}{\vee} Q(y)) \underset{\alpha}{\wedge} (P(x) \underset{\alpha}{\vee} R(z))$	UE_9	$P(x) \overset{\alpha}{\vee} (Q(y) \overset{\alpha}{\wedge} R(z)) =$ $(P(x) \overset{\alpha}{\vee} Q(y)) \overset{\alpha}{\wedge} (P(x) \overset{\alpha}{\vee} R(z))$
Laws of contradiction			
LE_{10}	$P(x) \underset{\alpha}{\wedge} (negP(x)) = .f.$	UE_{10}	$P(x) \overset{\alpha}{\wedge} (negP(x)) = .f.$
Laws of excluded middle			
LE_{11}	$P(x) \underset{\alpha}{\vee} (negP(x)) = .t.$	UE_{11}	$P(x) \overset{\alpha}{\vee} (negP(x)) = .t.$
Domination laws			
LE_{12}	$P(x) \underset{\alpha}{\wedge} .f. = .f.$	UE_{12}	$P(x) \overset{\alpha}{\wedge} .f. = .f.$
LE_{13}	$P(x) \underset{\alpha}{\vee} .t. = .t.$	UE_{13}	$P(x) \overset{\alpha}{\vee} .t. = .t.$
Identity			
LE_{14}	$P(x) \underset{\alpha}{\wedge} .t. = P_\alpha(x)$	UE_{14}	$P(x) \overset{\alpha}{\wedge} .t. = P^\alpha(x)$
LE_{15}	$P(x) \underset{\alpha}{\vee} .f. = P_\alpha(x)$	UE_{15}	$P(x) \overset{\alpha}{\vee} .f. = P^\alpha(x)$
Contrapositive			
LE_{16}	$P(x) \underset{\alpha}{\longrightarrow} Q(y) =$ $(negQ(y)) \underset{\alpha}{\longrightarrow} (negP(x))$	UE_{16}	$P(x) \overset{\alpha}{\longrightarrow} Q(y) =$ $(negQ(y)) \overset{\alpha}{\longrightarrow} (negP(x))$
ME_1	$\tau(P(x) \underset{\alpha}{\wedge} Q(y)) = (negP(x) \overset{1-\alpha}{\vee} negQ(y))$	De Morgan's laws	
ME_2	$\tau(P(x) \overset{\alpha}{\wedge} Q(y)) = (negP(x) \underset{1-\alpha}{\vee} negQ(y))$		
ME_3	$\tau(P(x) \underset{\alpha}{\vee} Q(y)) = (negP(x) \overset{1-\alpha}{\wedge} negQ(y))$		
ME_4	$\tau(P(x) \overset{\alpha}{\vee} Q(y)) = (negP(x) \underset{1-\alpha}{\wedge} negQ(y))$		
ME_5	$\tau(P(x) \underset{\alpha}{\longrightarrow} Q(y)) = (P(x) \overset{1-\alpha}{\wedge} negQ(y))$		
ME_6	$\tau(P(x) \overset{\alpha}{\longrightarrow} Q(y)) = (P(x) \underset{1-\alpha}{\wedge} negQ(y))$		
ME_7	$P(x) \underset{\alpha}{\longrightarrow} (Q(y) \underset{\alpha}{\longrightarrow} R(z)) = (P(x) \overset{1-\alpha}{\wedge} Q(y)) \underset{\alpha}{\longrightarrow} R(z)$		
ME_8	$P(x) \overset{\alpha}{\longrightarrow} (Q(y) \overset{\alpha}{\longrightarrow} R(z)) = (P(x) \underset{1-\alpha}{\wedge} Q(y)) \overset{\alpha}{\longrightarrow} R(z)$		
ME_9	$P(x) \underset{\alpha}{\longleftrightarrow} negQ(y) = \tau(P(x) \overset{1-\alpha}{\longleftrightarrow} Q(y))$		
ME_{10}	$P(x) \overset{\alpha}{\longleftrightarrow} negQ(y) = \tau(P(x) \underset{1-\alpha}{\longleftrightarrow} Q(y))$		

Lower Disjunctive Normal Forms (L-DNF). An L-DNF is an elementary sum of lower elementary products.

Example 1. Find L-DNF of $(negP(x) \underset{\alpha}{\vee} Q(y))$.

Solution: Now

$$
\begin{aligned}
(negP(x) \underset{\alpha}{\vee} Q(y)) &= ((negP(x))_\alpha \vee Q_\alpha(y)) \\
&= ((negP(x))_\alpha \wedge .t.) \vee (Q_\alpha(y) \wedge .t.) \\
&= ((negP(x))_\alpha \wedge (Q(y) \underset{\alpha}{\vee} negQ(y))) \\
&\qquad \vee (Q_\alpha(y) \wedge (P(x) \underset{\alpha}{\vee} negP(x))) \\
&= ((negP(x))_\alpha \wedge Q_\alpha(y)) \vee ((negP(x))_\alpha \wedge (negQ(y))_\alpha) \\
&\qquad \vee (Q_\alpha(y) \wedge P_\alpha(x)) \vee (Q_\alpha(y) \wedge (negP(x))_\alpha) \\
&= ((negP(x))_\alpha \wedge Q_\alpha(y)) \vee ((negP(x))_\alpha \wedge (negQ(y))_\alpha) \\
&\qquad \vee (Q_\alpha(y) \wedge P_\alpha(x))
\end{aligned}
$$

Therefore the given form is in L-DNF.

Upper Disjunctive Normal Forms (U-DNF). A U-DNF is an elementary sum of upper elementary products.

Example 2. Find U-DNF of $((negP(x)) \overset{\alpha}{\vee} Q(y))$.

Solution: Now

$$
\begin{aligned}
((negP(x)) \overset{\alpha}{\vee} Q(y)) &= ((negP(x))^\alpha \vee Q^\alpha(y)) \\
&= ((negP(x))^\alpha \wedge .t.) \vee (Q^\alpha(y) \wedge .t.)) \\
&= ((negP(x))^\alpha \wedge (Q(y) \overset{\alpha}{\vee} negQ(y))) \\
&\qquad \vee (Q^\alpha(y) \vee (P(x) \overset{\alpha}{\vee} negP(x))) \\
&= ((negP(x))^\alpha \wedge Q^\alpha(y)) \vee ((negP(x))^\alpha \wedge (negQ(y))^\alpha) \\
&\qquad \vee (Q^\alpha(y) \wedge P^\alpha(x)) \vee (Q^\alpha(y) \wedge (negP(x))^\alpha) \\
&= ((negP(x))^\alpha \wedge Q^\alpha(y)) \vee ((negP(x))^\alpha \wedge (negQ(y))^\alpha) \\
&\qquad \vee (Q^\alpha(y) \wedge P^\alpha(x))
\end{aligned}
$$

Therefore the given form is in U-DNF.

Similarly we can prove that

1. Find L-DNF of $P_\alpha(x) \wedge (P(x) \xrightarrow{\alpha} Q(y))$
2. Find U-DNF of $P^\alpha(x) \wedge (P(x) \xrightarrow{\alpha} Q(y))$
3. Find L-DNF of $\tau(P(x) \underset{\alpha}{\vee} Q(y)) \Longleftrightarrow (P(x) \underset{\alpha}{\wedge} Q(y))$
4. Find U-DNF of $\tau(P(x) \overset{\alpha}{\vee} Q(y)) \Longleftrightarrow (P(x) \overset{\alpha}{\wedge} Q(y))$

3.2 Conjunctive Normal Forms (CNF)

Conjunctive normal forms have studied in propositional logic so far. As a normal form, it is useful in automated theorem proving. It is similar to the product of sums form used in circuit theory. In this section, the conjunctive normal forms are introduced through two way approximating the connectives using a threshold α similar to the conjunctive normal forms [5] which are discussed in propositional calculus. Every propositional formula can be converted into an equivalent formula that is in CNF using both approximations. This transformation is based on rules about logical equivalences: the double negative law, De Morgan's laws, and the distributive law. Therefore, two kinds of conjunctive normal forms are used namely, *lower conjunctive normal forms* (L-CNF), *upper conjunctive normal forms* (U-CNF) of fuzzy predicates using a threshold α are defined in this section and also, some examples are solved by using these definitions.

Proposition 1. *For every lower approximated formula, there is an equivalent formula in L-CNF (and also an equivalent formula in L-DNF).*

Proof. For any fuzzy predicates $P(x), Q(y), R(z)$ and α is a threshold, let us consider the case of L-CNF and propose a naive algorithm.
Apply the following rules as long as possible:
Step 1: Eliminate equivalences:
$$(P(x) \underset{\alpha}{\longleftrightarrow} Q(y)) \Longrightarrow (P(x) \underset{\alpha}{\longrightarrow} Q(y)) \wedge (Q(y) \underset{\alpha}{\longrightarrow} P(x))$$
Step 2: Eliminate implications:
$$(P(x) \underset{\alpha}{\longrightarrow} Q(y)) \Longrightarrow (negP(x) \underset{\alpha}{\vee} Q(y))$$
Step 3: Push negations downward:
$$\tau(P(x) \underset{\alpha}{\wedge} Q(y)) \Longrightarrow (negP(x) \overset{1-\alpha}{\vee} negQ(y))$$
Step 4: Eliminate multiple negations:
$$\tau(\tau(P_\alpha(x))) \Longrightarrow P_\alpha(x)$$
Step 5: Push disjunctions downward:
$$P(x) \underset{\alpha}{\vee} (Q(y) \wedge R(z)) \Longrightarrow (P(x) \underset{\alpha}{\vee} Q(y)) \wedge (P(x) \underset{\alpha}{\vee} R(z))$$
Step 6: Eliminate .t. and .f.:

(i) $P(x) \underset{\alpha}{\wedge} (negP(x)) \Longrightarrow .f.$ (ii) $P(x) \underset{\alpha}{\vee} (negP(x)) \Longrightarrow .t.$

(iii) $P(x) \underset{\alpha}{\wedge} .f. \Longrightarrow .f.$ (iv) $P(x) \underset{\alpha}{\vee} .t. \Longrightarrow .t.$

(v) $P(x) \underset{\alpha}{\wedge} .t. \Longrightarrow P_\alpha(x)$ (vi) $P(x) \underset{\alpha}{\vee} .f. \Longrightarrow P_\alpha(x)$

(vii) $\tau(.f.) \Longrightarrow .t.$ (viii) $\tau(.t.) \Longrightarrow .f.$

Therefore, the resulting formula is equivalent to the original one and in L-CNF.

Similarly, conversion of a formula to L-DNF works in the same way, except that conjunctions have to be pushed downward in step 5. Also, for every upper approximated formula, there is an equivalent formula in U-CNF and an equivalent in U-DNF.

Lower Conjunctive Normal Forms (L-CNF). An L-CNF is an elementary product of lower elementary sums.

Example 3. Find L-CNF of $Q_\alpha(y) \vee (P(x) \underset{\alpha}{\wedge} negQ(y)) \vee (negP(x) \underset{\alpha}{\wedge} negQ(y))$.

Solution: Now

$Q_\alpha(y) \vee (P(x)\underset{\alpha}{\wedge}negQ(y)) \vee (negP(x) \underset{\alpha}{\wedge} negQ(y))$

$$= Q_\alpha(y) \vee (P_\alpha(x) \wedge (negQ(y))_\alpha) \vee ((negP(x))_\alpha \wedge (negQ(y))_\alpha)$$
$$= Q_\alpha(y) \vee [(P_\alpha(x) \vee (negP(x))_\alpha) \wedge (negQ(y))_\alpha]$$
$$= [Q_\alpha(y) \vee (P_\alpha(x) \vee (negP(x))_\alpha)] \wedge [Q_\alpha(y) \vee (negQ(y))_\alpha]$$

Therefore the given form is in L-CNF.

Example 4. Find L-CNF of $\tau(P(x) \underset{\alpha}{\vee} Q(y)) \Longleftrightarrow (P(x) \underset{\alpha}{\wedge} Q(y))$.

Solution: Now

$$\tau(P(x)\underset{\alpha}{\vee}Q(y)) \Longleftrightarrow (P(x) \underset{\alpha}{\wedge} Q(y))$$
$$= [\tau(P(x) \underset{\alpha}{\vee} Q(y)) \longrightarrow (P(x) \underset{\alpha}{\wedge} Q(y))]$$
$$\wedge [(P(x) \underset{\alpha}{\wedge} Q(y)) \longrightarrow \tau(P(x) \underset{\alpha}{\vee} Q(y))]$$
$$= [(P(x) \underset{\alpha}{\vee} Q(y)) \vee (P(x) \underset{\alpha}{\wedge} Q(y))]$$
$$\wedge [\tau(P(x) \underset{\alpha}{\wedge} Q(y)) \vee ((negP(x))^{1-\alpha} \wedge (negQ(y))^{1-\alpha})]$$
$$= [(P(x) \underset{\alpha}{\vee} Q(y)) \vee P_\alpha(x)] \wedge [(P(x) \underset{\alpha}{\vee} Q(y)) \vee Q_\alpha(y)]$$
$$\wedge [((negP(x))^{1-\alpha} \vee (negQ(y))^{1-\alpha}) \vee (negP(x))^{1-\alpha}]$$
$$\wedge [((negP(x))^{1-\alpha} \vee (negQ(y))^{1-\alpha}) \vee (negQ(y))^{1-\alpha}]$$
$$= [(P(x) \underset{\alpha}{\vee} Q(y)) \vee P_\alpha(x)] \wedge [(P(x) \underset{\alpha}{\vee} Q(y)) \vee Q_\alpha(y)]$$
$$\wedge [(\tau(P_\alpha(x)) \vee \tau(Q_\alpha(y))) \vee \tau(P_\alpha(x))]$$
$$\wedge [(\tau(P_\alpha(x)) \vee \tau(Q_\alpha(y))) \vee \tau(Q_\alpha(y))]$$

Therefore the given form is in L-CNF.

Upper Conjunctive Normal Forms (U-CNF). A U-CNF is an elementary product of upper elementary sums.

Example 5. Find U-CNF of $Q^\alpha(y) \vee (P(x) \overset{\alpha}{\wedge} negQ(y)) \vee (negP(x) \overset{\alpha}{\wedge} negQ(y))$.

Solution: Now

$Q^\alpha(y) \vee (P(x)\overset{\alpha}{\wedge}negQ(y)) \vee (negP(x) \overset{\alpha}{\wedge} negQ(y))$

$$= Q^\alpha(y) \vee (P^\alpha(x) \wedge (negQ(y))^\alpha) \vee ((negP(x))^\alpha \wedge (negQ(y))^\alpha)$$
$$= Q^\alpha(y) \vee [(P^\alpha(x) \vee (negP(x))^\alpha) \wedge (negQ(y))^\alpha]$$
$$= [Q^\alpha(y) \vee (P^\alpha(x) \vee (negP(x))^\alpha)] \wedge [Q^\alpha(y) \vee (negQ(y))^\alpha]$$

Therefore the given form is in U-CNF.

Example 6. Find U-CNF of $\tau(P(x) \overset{\alpha}{\vee} Q(y)) \Longleftrightarrow (P(x) \overset{\alpha}{\wedge} Q(y))$.
Solution: Now

$$\tau(P(x)\overset{\alpha}{\vee}Q(y)) \Longleftrightarrow (P(x) \overset{\alpha}{\wedge} Q(y))$$

$$= [\tau(P(x) \overset{\alpha}{\vee} Q(y)) \longrightarrow (P(x) \overset{\alpha}{\wedge} Q(y))]$$
$$\wedge [(P(x) \overset{\alpha}{\wedge} Q(y)) \longrightarrow \tau(P(x) \overset{\alpha}{\vee} Q(y))]$$

$$= [(P(x) \overset{\alpha}{\vee} Q(y)) \vee (P(x) \overset{\alpha}{\wedge} Q(y))]$$
$$\wedge [\tau(P(x) \overset{\alpha}{\wedge} Q(y)) \vee ((negP(x))_{1-\alpha} \wedge (negQ(y))_{1-\alpha})]$$

$$= [(P(x) \overset{\alpha}{\vee} Q(y)) \vee P^{\alpha}(x)] \wedge [(P(x) \overset{\alpha}{\vee} Q(y)) \vee Q^{\alpha}(y)]$$
$$\wedge [((negP(x))_{1-\alpha} \vee (negQ(y))_{1-\alpha}) \vee (negP(x))_{1-\alpha}]$$
$$\wedge [((negP(x))_{1-\alpha} \vee (negQ(y))_{1-\alpha}) \vee (negQ(y))_{1-\alpha}]$$

$$= [(P(x) \overset{\alpha}{\vee} Q(y)) \vee P^{\alpha}(x)] \wedge [(P(x) \overset{\alpha}{\vee} Q(y)) \vee Q^{\alpha}(y)]$$
$$\wedge [(\tau(P^{\alpha}(x)) \vee \tau(Q^{\alpha}(y))) \vee \tau(P^{\alpha}(x))]$$
$$\wedge [(\tau(P^{\alpha}(x)) \vee \tau(Q^{\alpha}(y))) \vee \tau(Q^{\alpha}(y))]$$

Therefore the given form is in U-CNF.

3.3   Principal Disjunctive Normal Forms (PDNF)

In this section, two way approach of principal disjunctive normal forms on approximating the connectives using a threshold α are introduced similar to the principal disjunctive normal forms [5] which are discussed in propositional calculus. Therefore, two kinds of principal disjunctive normal forms are used namely, *lower principal disjunctive normal form* (L-PDNF), *upper principal disjunctive normal form* (U-PDNF) of fuzzy predicates using a threshold α are defined in this section and also, some examples are solved by using these definitions.

Definition 4. *For any two fuzzy predicates $P(x)$ and $Q(y)$,*

1. $P_{\alpha}(x) \wedge Q_{\alpha}(y), \tau(P_{\alpha}(x)) \wedge Q_{\alpha}(y), P_{\alpha}(x) \wedge \tau(Q_{\alpha}(y)), \tau(P_{\alpha}(x)) \wedge \tau(Q_{\alpha}(y)),$
   $(negP(x))_{\alpha} \wedge Q_{\alpha}(y), P_{\alpha}(x) \wedge (negQ(y))_{\alpha}, (negP(x))_{\alpha} \wedge (negQ(y))_{\alpha}$ *are called lower min terms.*
2. $P^{\alpha}(x) \wedge Q^{\alpha}(y), \tau(P^{\alpha}(x)) \wedge Q^{\alpha}(y), P^{\alpha}(x) \wedge \tau(Q^{\alpha}(y)), \tau(P^{\alpha}(x)) \wedge \tau(Q^{\alpha}(y)),$
   $(negP(x))^{\alpha} \wedge Q^{\alpha}(y), P^{\alpha}(x) \wedge (negQ(y))^{\alpha}, (negP(x))^{\alpha} \wedge (negQ(y))^{\alpha}$ *are called upper min terms.*

Lower Principal Disjunctive Normal Forms (L-PDNF). A formula is in L-PDNF of a given formula, if it is a sum of lower min terms.

Example 7. Obtain L-PDNF $(negP(x) \underset{\alpha}{\vee} Q(y))$.

Solution: Now

$$(negP(x) \underset{\alpha}{\vee} Q(y)) = ((negP(x))_\alpha \vee Q_\alpha(y))$$

$$= [(negP(x))_\alpha \wedge (Q(y) \underset{\alpha}{\vee} negQ(y))]$$

$$\vee [Q_\alpha(y) \wedge (P(x) \underset{\alpha}{\vee} negP(x))]$$

$$= [(negP(x))_\alpha \wedge Q_\alpha(y)] \vee [(negP(x))_\alpha \wedge (neg(Q(y))_\alpha]$$

$$\vee [Q_\alpha(y) \wedge P_\alpha(x)] \vee [Q_\alpha(y) \wedge (negP(x))_\alpha]$$

$$= [(negP(x))_\alpha \wedge (Q_\alpha(y))] \vee [(negP(x))_\alpha \wedge (negQ(y))_\alpha]$$

$$\vee [Q_\alpha(y) \wedge P_\alpha(x)]$$

Therefore the given form is in L-PDNF.

Upper Principal Disjunctive Normal Forms (U-PDNF).

A formula is in U-PDNF of a given formula, if it is a sum of upper min terms.

Example 8. Obtain U-PDNF $(negP(x) \overset{\alpha}{\vee} Q(y))$.

Solution: Now

$$(negP(x) \overset{\alpha}{\vee} Q(y)) = ((negP(x))^\alpha \vee Q^\alpha(y))$$

$$= [(negP(x))^\alpha \wedge (Q(y) \overset{\alpha}{\vee} negQ(y))]$$

$$\vee [Q^\alpha(y) \wedge (P(x) \overset{\alpha}{\vee} negP(x))]$$

$$= [(negP(x))^\alpha \wedge Q^\alpha(y)] \vee [(negP(x))^\alpha \wedge (neg(Q(y))^\alpha]$$

$$\vee [Q^\alpha(y) \wedge P^\alpha(x)] \vee [Q^\alpha(y) \wedge (negP(x))^\alpha]$$

$$= [(negP(x))^\alpha \wedge (Q^\alpha(y))] \vee [(negP(x))^\alpha \wedge (negQ(y))^\alpha]$$

$$\vee [Q^\alpha(y) \wedge P^\alpha(x)]$$

Therefore the given form is in U-PDNF.

3.4 Principal Conjunctive Normal Forms (PCNF)

In this section, two way approach of principal conjunctive normal forms on approximating the connectives using a threshold α are introduced similar to the principal conjunctive normal forms [5] which are discussed in propositional calculus. Therefore, two kinds of principal conjunctive normal forms are used namely, *lower principal conjunctive normal forms* (L-PCNF), *upper principal conjunctive normal forms* (U-PCNF) of fuzzy predicates using a threshold α are defined in this section and also, some examples are solved by using these definitions.

Definition 5. *For any two fuzzy predicates $P(x)$ and $Q(y)$,*

1. $P_\alpha(x) \vee Q_\alpha(y), \tau(P_\alpha(x)) \vee Q_\alpha(y), P_\alpha(x) \vee \tau(Q_\alpha(y)), \tau(P_\alpha(x)) \vee \tau(Q_\alpha(y)),$
$(negP(x))_\alpha \vee Q_\alpha(y), P_\alpha(x) \vee (negQ(y))_\alpha, (negP(x))_\alpha \vee (negQ(y))_\alpha$ *are called lower max terms.*

2. $P^\alpha(x) \vee Q^\alpha(y), \tau(P^\alpha(x)) \vee Q^\alpha(y), P^\alpha(x) \vee \tau(Q^\alpha(y)), \tau(P^\alpha(x)) \vee \tau(Q^\alpha(y)),$
$(negP(x))^\alpha \vee Q^\alpha(y), P^\alpha(x) \vee (negQ(y))^\alpha, (negP(x))^\alpha \vee (negQ(y))^\alpha$ *are called upper max terms.*

Lower Principal Conjunctive Normal Forms (L-PCNF). A formula is in L-PCNF of a given formula, if it is a product of lower max terms.

Example 9. Obtain L-PCNF $(negP(x) \xrightarrow{\alpha} R(z)) \wedge (Q(y) \xleftrightarrow{\alpha} P(x))$.

Solution: Now

$(negP(x) \xrightarrow{\alpha} R(z)) \wedge (Q(y) \xleftrightarrow{\alpha} P(x))$

$\qquad = (P(x) \underset{\alpha}{\vee} R(z)) \wedge [(Q(y) \xrightarrow{\alpha} P(x)) \wedge (P(x) \xrightarrow{\alpha} Q(y))]$

$\qquad = ((P(x) \underset{\alpha}{\vee} R(z)) \vee .f.) \wedge [(negQ(y) \underset{\alpha}{\vee} P(x)) \wedge (negP(x) \underset{\alpha}{\vee} Q(y))]$

$\qquad = [(P(x) \underset{\alpha}{\vee} R(z)) \vee (Q(y) \underset{\alpha}{\wedge} negQ(y))]$

$\qquad\qquad \wedge [(negQ(y) \underset{\alpha}{\vee} P(x)) \vee (R(z) \underset{\alpha}{\wedge} negR(z))]$

$\qquad\qquad \wedge [(negP(x) \underset{\alpha}{\vee} Q(y)) \vee (R(z) \underset{\alpha}{\wedge} negR(z))]$

$\qquad = [P(x) \underset{\alpha}{\vee} R(z)) \vee Q_\alpha(y)] \wedge [(P(x) \underset{\alpha}{\vee} R(z)) \vee (negQ(y))_\alpha]$

$\qquad\qquad \wedge [(negQ(y) \underset{\alpha}{\vee} P(x)) \vee R_\alpha(z)] \wedge [(negQ(y) \underset{\alpha}{\vee} P(x)) \vee (negR(z))_\alpha]$

$\qquad\qquad \wedge [(negP(x) \underset{\alpha}{\vee} Q(y)) \vee R_\alpha(z)] \wedge [(negP(x) \underset{\alpha}{\vee} Q(y)) \vee (negR(z))_\alpha]$

Therefore the given form is in L-PCNF.

Upper Principal Conjunctive Normal Forms (U-PCNF). A formula is in U-PCNF of a given formula, if it is a product of upper max terms.

Example 10. Obtain U-PCNF $\tau(P(x) \xrightarrow{\alpha} Q(y))$.

Solution: Now

$\tau(P(x) \xrightarrow{\alpha} Q(y)) = (P(x) \overset{1-\alpha}{\wedge} negQ(y))$

$\qquad = [(P^{1-\alpha}(x) \vee (Q(y) \overset{1-\alpha}{\wedge} negQ(y))]$

$\qquad\qquad \wedge [(negQ(y))^{1-\alpha} \vee (P(x) \overset{1-\alpha}{\wedge} negP(x))]$

$\qquad = [(P^{1-\alpha}(x) \vee Q^{1-\alpha}(y)] \wedge [P^{1-\alpha}(x) \vee (negQ(y))^{1-\alpha}]$

$\qquad\qquad \wedge [(negQ(y))^{1-\alpha} \vee P^{1-\alpha}(x)] \wedge [(negQ(y))^{1-\alpha} \vee negP(x))^{1-\alpha}]$

Therefore the given form is in U-PCNF.

Similarly, we can prove the following

1. Obtain U-PCNF $(negP(x) \xrightarrow{\alpha} R(z)) \wedge (Q(y) \xleftrightarrow{\alpha} P(x))$.
2. Obtain L-PCNF $\tau(P(x) \xrightarrow{\alpha} Q(y))$.

4 Conclusion

In this paper, we have derived normal forms of rough connectives of fuzzy predicates using Zadeh's min and max operators with illustrations. In future, it is planned to extend this work on generalized S and T norms and also it is aimed to implement these normal forms into designing (approximated) logical gates and to derive a dynamic key based encryption.

References

1. Ganesan, G., Raghavendra Rao, C.: Rough set: analysis of fuzzy sets using thresholds. Comput. Math, pp. 81–87. Narosa Publishers, New Delhi, India (2005)
2. Ganesan, G., Rao, C.R.: Rough connectives of fuzzy predicates. Int. J. Comput. Math. Sci. Appl. 1(2), 189–196 (2008)
3. Ganesan, G., Satish, B.N.V.: Naïve properties on rough connectives under fuzziness. Int. J. Recent Innov. Trends Comput. Commun. 2(2), 216–221 (2014)
4. George, J.K., Yuan, B.: Fuzzy Sets and Fuzzy Logic Theory and Applications. Prentice-Hall of India Pvt Ltd, Upper Saddle River (1997)
5. Tremblay, J.P., Manohar, R.: Discrete Mathematical Structures with Applications to Computer Science. McGraw-Hill International Edition, New York (1987)
6. Banerjee, M., Chakraborty, M.K.: Rough sets through algebraic logic. Fundamenta Informaticae 28, 211–221 (1996)
7. Liu, Q., Liu, L.: Rough logic and its reasoning. In: Gavrilova, M.L., Tan, C.J.K., Wang, Y., Yao, Y., Wang, G. (eds.) Transactions on Computational Science II. LNCS, vol. 5150, pp. 84–99. Springer, Heidelberg (2008)
8. Satish, B.N.V., Ganesan, G.: Approximations on intuitionistic fuzzy predicate calculus through rough computing. J. Intell. Fuzzy Syst. 27(4), 1873–1879 (2014)
9. Zadeh, L.: Fuzzy sets. J. Inf. Control 8, 338–353 (1965)
10. Pawlak, Z.: Rough Sets-Theoretical Aspects and Reasoning About Data. Kluwer Academic Publications, Dordrecht (1991)

Clustering

Combining Rough Clustering Schemes as a Rough Ensemble

Pawan Lingras$^{(\boxtimes)}$ and Farhana Haider

Mathematics and Computing Science, Saint Marys University, Halifax, Canada
pawan@cs.smu.ca, farhanahdr@gmail.com

Abstract. One of the challenges of big data is to combine results of data mining obtained from a distributed dataset. The objective is to minimize the amount data transfer with minimum information loss. A generic combination process will not necessarily provide an optimal ensemble of results. In this paper, we describe a rough clustering problem that leads to a natural ordering of clusters. These ordered rough clusterings are then combined while preserving the properties of rough clustering. A time series dataset of commodity prices is clustered using two different representations to demonstrate the ordered rough clustering process. The information from the ordering of clusters is shown to help us retain salient aspects of individual rough clustering schemes.

Keywords: Clustering · Ensemble · Rough sets · Granular computing · Financial time series · Volatility

1 Introduction

Since clustering is an unsupervised learning process, different algorithms may create different grouping of objects. To preserve such groupings Fred [3] proposed the concept of clustering ensemble. Further research by Strehl and Ghosh [19], and subsequently by Gionis et al. [6] helped to formalize the clustering ensemble process. Most of the early clustering ensemble techniques created conventional crisp clusters from clustering of the objects generated using different algorithms. Crisp clustering assigns an object to one and only one cluster. However, given the fact that ensemble clustering was devised to reconcile different cluster assignments of an object, it may make sense to allow for an object to belong to more than one cluster when there is a strong disagreement between the two clustering schemes.

It is also possible that an object is represented using different information granules depending on the context or point of view. If these two representations of the same group of objects are clustered, we will get two different clustering schemes. We demonstrate such a multi-granular view for a time series of commodity prices. A trader finds a daily price pattern interesting when it is volatile. The higher the fluctuations in prices, the more volatile the pattern. Black Scholes index is a popular measure to quantify volatility of a pattern. We can segment

© Springer International Publishing Switzerland 2015
D. Ciucci et al. (Eds.): RSKT 2015, LNAI 9436, pp. 383–394, 2015.
DOI: 10.1007/978-3-319-25754-9_34

daily patterns based on values of the Black Scholes index. This segmentation is essentially a clustering of one dimensional representation (Black Scholes index) of the daily pattern. Black Scholes index is a single concise index to identify volatility in a daily pattern.

However, a complete distribution of prices during the day can provide a more elaborate information on the volatility during the day. While a distribution consisting of frequency of different prices is not a concise description for a single day, it can be a very useful representation of daily patterns for clustering based on volatility. That means, we will have two different ways of grouping daily patterns. Instead of using one grouping or the other, clustering ensemble can be used to come up with a consensual grouping. Traditionally, clustering ensemble applied to crisp clustering schemes creates another crisp clustering scheme. In crisp clustering schemes, an object is assigned to one and only one cluster. There is no room for ambiguity in such a clustering. However, while combining results from two clustering schemes, there will be situations when the two clustering schemes do not agree with each other. When the two clustering schemes do not agree with each other on the assignment of an object, it should belong to two different clusters in the resulting clustering schemes. That means, the clusters will need to overlap.

Recently, Lingras and Haider [9] proposed creation of rough clustering ensembles by combining multiple crisp clustering schemes based on rough set theory that allows assignment of objects to multiple clusters. This paper extends the rough clustering ensemble technique proposed by Lingras and Haider for combining initial rough clustering schemes. Use of rough clustering instead of crisp, ensures that the objects that may belong to multiple clusters in any clustering scheme are not forced to a single cluster. Thus, the final clustering ensembles will preserve the ambivalent membership of these objects. The rough clustering schemes that are assembled in this paper are based on the same set of objects represented by different attributes. While this is a novel application of rough clustering ensemble for grouping daily trading patterns, application of rough set theory to financial time series data is not new. Yao and Herbert demonstrated a rough set data analysis model for decision rule discovery from financial time series data [21]. Jia and Han proposed a prediction system for stock price using support vector machine that uses rough set theory for pre-processing [7]. Nair et al. proposed a method that presents the design and performance evaluation of a hybrid rough set based decision tree for predicting the future trends in a stock exchange [12].

2   Rough Set and Clustering

We will first define mathematical notations that will be used throughout the paper. Let $X = \{x_1, ..., x_n\}$ be a finite set of objects. Assume that the objects are represented by m-dimensional vectors. A clustering scheme groups n objects into k clusters $C = \{c_1, ..., c_k\}$.

Rough sets were proposed using equivalence relations by Pawlak [13]. However, it is possible to define a pair of lower and upper bounds $(\underline{A}(Y), \overline{A}(Y))$

or a Rough Set for every set $Y \subseteq X$ as long as the properties specified by Pawlak [13,14] are satisfied [17,18,22]. Let us consider a clustering scheme C based on a hypothetical relation R

$$X/R = C = \{c_1, c_2, \ldots, c_k\} \tag{1}$$

that partitions the set X based on certain criteria. The actual values of c_i are not known. Let us assume that due to insufficient knowledge it is not possible to precisely describe the sets $c_i, 1 \leq i \leq k$, in the partition. However, it is possible to define each set $c_i \in X/R$ using its lower $\underline{A}(c_i)$ and upper $\overline{A}(c_i)$ bounds based on the available information.

We are considering the upper and lower bounds of only a few subsets of X. Therefore, it is not possible to verify all the properties of the Rough Sets [13,14]. However, the family of upper and lower bounds of $c_i \in X/R$ are required to follow some of the basic rough set properties such as:

(PR1) An object x can be part of at most one lower bound

(PR2) $\mathbf{x} \in \underline{A}(c_i) \Longrightarrow \mathbf{x} \in \overline{A}(c_i)$

(PR3) An object x is not part of any lower bound

$\Updownarrow$

x belongs to two or more upper bounds.

The rough K-means [10] and its various extensions [11,15] have been found to be effective in distance based clustering. A comparative study of crisp, rough and evolutionary clustering depicts how rough clustering outperforms crisp clustering [8]. Peters, et al. [16] provide a good comparison of rough clustering and other conventional clustering algorithms.

3   Ensemble Clustering

Though there are a number of clustering algorithms, no single clustering algorithm is capable of delivering sound solutions for all data sets. Combining clustering results or creating cluster ensembles is an alternate approach.

For a given set of objects, cluster ensemble uses two major steps; namely, generation and consensus. In the generation stage, population of diverse multiple clustering partitions are made by generative mechanisms using different feature subsets, clustering algorithms, parameter initialization, projection to subspaces or subsets of objects. In consensus stage, partitions are aggregated based on objects co-occurrence using relabeling and voting, co-association matrix, graph and hyper-graph etc. In addition, median partition using genetic algorithms, kernel methods etc. or probabilistic models can also be used for consensus selection [5]. The most expected properties of an ensemble clustering result are robustness, consistency, novelty and stability [4,19,20].

More formally, for our dataset X of n objects, let P be a set of n partitions of objects in X. Thus $P = \{P_1, P_2, \cdots, P_n\}$. Each partition in P is a set of disjoint and nonempty clusters $P_i = \{c_1^i, c_2^i, \cdots, c_{K(i)}^i\}$, $X = c_1^i \cup c_2^i \cup \cdots \cup c_{k(i)}^i$, and for any P_i, $k(i)$ is the number of clusters in the i-th clustering partition. The problem of clustering ensemble is to find a new partition $C = \{c_1, c_2, , c_k\}$ of data X, given the partitions in P, such that the final clustering solution is better than any individual clustering partition [4]. Fern and Lin [2] suggested a clustering scheme C maximizing summation of normalized mutual information (SNMI), that maximizes the information it shares with all the clusterings in the ensemble. Thus it can be considered to be the best one having general trend in the ensemble. Depending on the application, it may be necessary to define a different criteria or definition for determining a cluster ensemble. For example, in this paper we will be looking at clustering schemes, where clusters can be ordered based on a criteria such as volatility. The objective of the clustering ensemble would be to preserve different definitions of volatility obtained from different granular points of view.

4   Study Data and Initial Rough Clustering

4.1   Dataset and Knowledge Representation

Volatility of financial data series is an important indicator used by traders. The fluctuation in prices create trading opportunities. Volatility is a measure for variation of price of a financial instrument over time. Black Scholes index of volatility can be a good way to measure this. The equation of volatility index is an extension from the Black Scholes model which estimates the price of the option over time. This model is widely used by the options market participants. The key idea behind the model is to hedge the option by buying and selling the underlying asset in just the right way and, as a consequence, to eliminate risk. The instantaneous log returns of the stock price considered in this formula is an infinitesimal random walk with drift or more precisely is a geometric Brownian motion. The equation to estimate volatility using this model is:

$$Volatility = \sqrt{LogPriceRelativeVariance} \times (Observations - 1), \qquad (2)$$

where $LogPriceRelativeVariance = \sum (LogPriceRelative - Mean)^2$ [1]. In our data set, this single value represents volatility in prices of a financial instrument over a 6.5 hours of trading period divided into 39 ten-minute intervals. The trading patterns with similar daily volatilities are grouped together. For instance, the patterns with higher volatility index will form one cluster, if the difference between those volatilities are small enough to fit them in one cluster. Similarly, patterns with lower volatility index will form another cluster. The number of clusters is predetermined by analyzing distances between clusters and within cluster scatters.

While Black Scholes index is a concise measure, distribution of prices during the day can provide a more elaborate description of price fluctuations. We propose the use of five percentile values; 10 %, 25 %, 50 %, 75 % and 90 % to represent the price distribution. 10 % of the prices are below the 10th percentile value, 25 % of the prices are below the 25th percentile value and so on.

Our data set contains average prices at 10 min interval of 223 instruments transacted on 121 days comprising a total of 27,012 records. Each daily pattern has 39 intervals. This data set is used to create two representations of the daily patterns. The first one is a five dimensional pattern, which represents 10, 25, 50, 75 and 90 percentile values of the prices. The prices are normalized by the opening price so that a commodity selling for $100 has the same pattern as the one that is selling for $10. Afterwards, natural logarithm of the five percentiles are calculated. The second representation is one dimensional Black Scholes volatility for the day.

4.2 Individual Ordered Rough Clustering Using Two Knowledge Representations

Optimal number of clusters is an important measure in determining an appropriate clustering scheme. Lingras and Haider [9] used a two stage process that included first plotting scatter in clusters followed by the use of Davies-Bouldin (DB) index that minimizes the scatter within clusters and maximizes separation between clusters. Based on these two criteria, they chose five as a reasonable number of clusters. We will use the same number of clusters in our rough clustering.

In many cases, the groups generated by a clustering process have an implicit ordering. For example, clusters of customers in a retail store could be ordered based on their average spending and loyalty (their propensity to visit). Or products could be ordered based on their revenues, profits, and popularity (how many customers buy it). Similarly, the clusters of financial instruments (stocks) could be ordered based on their volatility. The volatility is an important indicator. A volatile daily pattern in a stock makes it more interesting to an aggressive trader and less interesting to a conservative trader.

Once we have obtained the rough clustering using two representations, we will study the patterns and number the clusters based on their increasing volatility. Let $cpr = \{cpr_1, cpr_2, cpr_3, cpr_4, cpr_5\}$ be the rough clustering scheme based on percentile values and $cdvr = \{cdvr_1, cdvr_2, cdvr_3, cdvr_4, cdvr_5\}$ be the rough clustering scheme based on the Black Scholes volatility. Each cluster cpr_i will be represented by a lower bound $\underline{PR}(cpr_i)$ and an upper bound $\overline{PR}(cpr_i)$, where PR is an approximation space for percentile rankings. Similarly, each cluster cvr_i will be represented by a lower bound $\underline{VR}(cdvr_i)$ and an upper bound $\overline{VR}(cdvr_i)$, where VR is an approximation space for Black Scholes volatility rankings. Since an object x can belong to one or more clusters, we cannot assign a single ranking for the object, but rather an interval of ranks. An object x can belong to multiple upper bounds $\overline{PR}(cpr_i)$ and its ranking will be represented as an interval $[\underline{cpr}(x), \overline{cpr}(x)]$, where $\underline{cpr}(x) = min(\{i \| x \in \overline{PR}(cpr_i)\})$ and $\overline{cpr}(x) = max(\{i \| x \in \overline{PR}(cpr_i)\})$. Similarly, the object x can also belong to multiple upper bounds $\overline{VR}(cdvr_i)$ and its ranking will be represented as an interval $[\underline{cdvr}(x), \overline{cdvr}(x)]$, where $\underline{cdvr}(x) = min(\{i \| x \in \overline{VR}(cdvr_i)\})$ and $\overline{cdvr}(x) = max(\{i \| x \in \overline{VR}(cdvr_i)\})$.

While there will be a general consensus on the membership of objects in both the clustering schemes, there will be some disagreements. We need to create a cluster ensemble that will preserve the ordering of clusters by accommodating both the volatility rankings. We propose the use of rough set theory for creating such an ensemble as described in the next section.

5   Ensemble of Two Rough Ordered Clustering Scheme

Let us recall the concept of lower and upper bounds of a cluster. If available information does not make it possible to represent an arbitrary subset $Y \subseteq X$, we can create lower $\underline{A}(Y)$ and upper approximations $\overline{A}(Y)$ of the set such that $\underline{A}(Y) \subseteq Y \subseteq \overline{A}(Y)$.

In our case, we have two different rough clustering schemes $cdvr$ and cpr, which use the approximation spaces VR and PR to represent the clusters as rough sets. We do not know the clustering scheme that will result from combining the two clustering schemes, since there are disagreements about the memberships of some of the objects in X. Therefore, we will build a rough clustering scheme with a combined approximation space $VR \otimes PR$ that satisfies the properties (PR1)-(PR3) discussed previously.

Since our clusters are ordered from $1, 2, \cdots, k$. we want to create clusters based on consensus in ordering. Let us use $cdvr$ and cpr as the two clustering schemes that will be combined to form a cluster ensemble. For an object $x \in X$, we have previously defined the ranks as $cdvr(x)$ and $cpr(x)$ from the two clustering schemes. Let $C = cdvr \otimes cpr = \{C_1, C_2, \cdots, C_k\}$ be the rough cluster ensemble, and rank of an object $x \in X$ will be defined as an interval:

$$cdvr \otimes cpr(x) = [min(\underline{cdvr}(x), \underline{cpr}(x)), max(\overline{cdvr}(x), \overline{cpr}(x))] \qquad (3)$$

We can then define each cluster $c_i \in C$ as rough sets with upper and lower bounds as follows:

$$\underline{VR \otimes PR}(c_i) = \underline{VR}(cdvr_i) \cap \underline{PR}(cpr_i) \qquad (4)$$

$$\overline{VR \otimes PR}(c_i) = \overline{VR}(cdvr_i) \cup \overline{PR}(cpr_i) \qquad (5)$$

We can easily verify that the lower and upper approximations given by Eqs. 4 and 5 satisfy the properties for rough clustering (PR1)-(PR3).

The following subsection describes the complete algorithm.

5.1   Algorithm: Rough Ensemble Clustering

Procedure: roughEnsembleClustering(rEnClusLo, rEnClusUp, clSet1, clSet2, noCls)

rEnClusLo ← Empty
rEnClusUp ← Empty

1. for cnt = 1 to noCls do step 2 to 3
2.   makeClusterLo(rEnClusLo[cnt], clSet1[cnt], clSet2[cnt])
3.   makeClusterUp(rEnClusUp[cnt], clSet1[cnt], clSet2[cnt])

Procedure: makeClusterLo(rEnClus, cls1, cls2)

1. for each i = 1 to size(cls1) do step 2
2.   for each j = 1 to size(cls2) do step 3
3.   if (clSet1[i] = clSet2[j]) do step 4
4.   rEnClus ← clSet1[i]

Procedure: makeClusterUp(rEnClus, cls1, cls2)

1. for each i = 1 to size(cls1) do step 2
2.   rEnClus ← clSet1[i]
3. for each j = 1 to size(cls2) do step 4 to 6
4.   for each k = 1 to size(rEnClus) do step 5
5.   if (rEnClus[k] = clSet2[j]) continue step 3
6.   rEnClus ← clSet2[j]

6   Discussion and Analysis of the Resulting Rough Cluster Ensemble

In this section, we will present the initial rough clustering of stocks based on percentile distribution and Black Scholes index followed by their ordering. We will then apply the proposed rough clustering ensemble to study the resulting combined clustering scheme. The section will conclude with a theoretical analysis of the robustness of the proposed approach followed by a comparison with the existing clustering ensemble techniques.

6.1   Initial Rough Clustering Using Percentile Values and Black Scholes Index

As mentioned before, using the scatter in clusters and Davies-Bouldin index, we determined that the appropriate number of clusters for both the clustering schemes was five.

We applied rough K-means algorithm to generate five rough clusters. Table 1 shows the number of patterns in each cluster for the two clustering schemes. We can see that the cardinalities of clusters are more or less similar with a small amount of disagreement. Figure 1 shows the plots of cluster means for the two clustering schemes. There is a clear ranking of the clusters in terms of the volatility. For example, the top graph of percentile values is the most volatile, while the one at the bottom is the least volatile. Similarly, the highest values of Black Scholes volatility indicate high volatility. We number the clusters based on the increasing volatility. Thus, $cpr = \{cpr_1, cpr_2, cpr_3, cpr_4, cpr_5\}$ is the clustering scheme based on percentile values and $cdvr = \{cdvr_1, cdvr_2, cdvr_3, cdvr_4, cdvr_5\}$

Table 1. Cluster cardinalities

Cluster number	1	2	3	4	5
cpr	15320	5833	1758	304	19
$\overline{cpr}$	17815	9385	3075	551	29
$\underline{cdvr}$	13972	7011	2384	557	80
$\overline{cdvr}$	15871	9757	3456	819	117

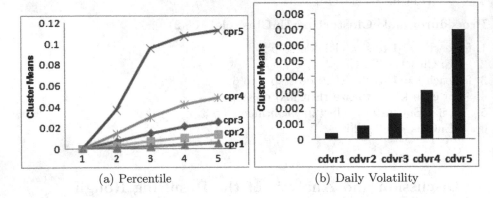

(a) Percentile   (b) Daily Volatility

Fig. 1. Means of 5 clusters after ranking

is the clustering scheme based on the Black Scholes volatility. Each cluster cpr_i will be represented by a lower bound $\underline{PR}(cpr_i)$ and an upper bound $\overline{PR}(cpr_i)$, where PR is an approximation space for percentile rankings. Similarly, each cluster $cdvr_i$ will be represented by a lower bound $\underline{VR}(cdvr_i)$ and an upper bound $\overline{VR}(cdvr_i)$, where VR is an approximation space for Black Scholes volatility rankings. Since an object x can belong to one or more clusters, we cannot assign a single ranking for the object, rather an interval of ranks can be used. An object x can belong to multiple upper bounds $\overline{PR}(cpr_i)$ and its ranking will be represented as an interval $[\underline{cpr}(x), \overline{cpr}(x)]$, where $\underline{cpr}(x) = min(\{i \| x \in \overline{PR}(cpr_i)\})$ and $\overline{cpr}(x) = max(\{i \| x \in \overline{PR}(cpr_i)\})$. Similarly, the object x can also belong to multiple upper bounds $\overline{VR}(cdvr_i)$ and its ranking will be represented as an interval $[\underline{cdvr}(x), \overline{cdvr}(x)]$, where $\underline{cdvr}(x) = min(\{i \| x \in \overline{VR}(cdvr_i)\})$ and $\overline{cdvr}(x) = max(\{i \| x \in \overline{VR}(cdvr_i)\})$.

The plots of rough cluster means visually confirmed that there is a reasonable matching for clusters. In order to understand the disagreement between the two clustering schemes, we plotted the patterns from the overlaps of clusters. Table 2 shows the intersections of clustering schemes $cdvr$ and cpr. There is reasonable agreement for the cluster 1. However the others clusters also agrees with respect to cluster memberships with some disagreements.

We will use the intersection tables shown in Table 2 to describe the lower and upper approximations for our example. Let us look at the tables as a pair

Table 2. Cluster intersections

(a) Lower Bound Regions	cdvr1	cdvr2	cdvr3	cdvr4	cdvr5
cpr1	**10972**	2675	441	57	3
cpr2	1895	2339	663	70	2
cpr3	56	702	589	123	2
cpr4	0	4	108	119	23
cpr5	0	0	0	1	17

(b)Upper Bound Regions	cdvr1	cdvr2	cdvr3	cdvr4	cdvr5
cpr1	**13162**	5033	1027	172	22
cpr2	3793	4967	1711	295	33
cpr3	294	1555	1283	394	45
cpr4	0	51	267	253	67
cpr5	0	0	0	6	25

of two dimensional matrix $\underline{table}$ and $\overline{table}$. Here, $\underline{table}[i][j]$ is the cell in row i and column j for the intersections of lower bounds and $\overline{table}[i][j]$ is the cell in row i and column j for the intersections of upper bounds. All the objects in the diagonal of the lower bound intersection correspond to the lower bounds, that is the cardinality of the lower approximations,

$$|\underline{VR \otimes PR}(c_i)| = \underline{table}[i][i]. \qquad (6)$$

That means, $\underline{VR \otimes PR}(c_1)$ has 10972 objects, $\underline{VR \otimes PR}(c_2)$ has 2339 objects, and so on. For the cardinality of upper approximation we will take all the distinct objects of the upper bound region of the cluster under consideration:

$$|\overline{VR \otimes PR}(c_i)| = \overline{table}[i][i] + \sum_{i \neq j} distinct(\overline{table}[i][j]) + \sum_{i \neq j} distinct(\overline{table}[j][i])$$

$$(7)$$

That means all the unique objects in row i and column j belong to $\overline{VR \otimes PR}(c_i)$. Taking unique objects ensure that we do not count the same object more than once. We use the condition $i \neq j$ in the sums because we do not want to count the diagonal twice.

In our table that means $\overline{VR \otimes PR}(c_1)$ has:

13162+distinct(5033+1027+172+22+3793+294+0+0)
= 13162+distinct(10341)=13162+7362 =20524 objects.

On the other end of the spectrum,$\overline{VR \otimes PR}(c_5)$ has:

25+distinct(0+0+0+6+22+33+45+67)
=25+distinct(173)=25+ 96=121 objects.

We can verify that the ranking in the rough cluster ensemble is consistent with the ranking from the initial rough clustering schemes.
For all $x \in \underline{VR \otimes PR}(c_i)$, $cdvr(x) = cpr(x) = i$.
While for all $x \in \overline{VR \otimes PR}(c_i)$, $min(cdvr(x), cpr(x)) \leq i \leq max(cdvr(x), cpr(x))$.
We have managed to preserve the original ordering of clusters in the cluster ensemble with the help of interval values.

6.2   Comparison with Conventional Clustering Ensemble Techniques

The relatively new concept of the clustering ensemble was first proposed a little more than a decade ago by Fred [3]. The early research focused on the crisp clustering ensemble, that ignores the noise or genuine outliers in real-world data. As for instance, in our dataset, if a pattern was shown as an outlier in one of the clustering schemes and part of a cluster in another clustering scheme, a conventional cluster ensemble will force the outlier in a cluster. On the other hand, the rough clustering ensemble can reconcile such a disagreement by putting the outlier in the boundary region. While our dataset does not have noticeable outliers, the most volatile fifth cluster $cdvr_5$ has a cardinality of 19 in the lower and 29 in the upper approximations. While corresponding cpr_5 has a cardinality of 80 in lower and 117 in upper approximations. These represent less than 1 % of the population. They are on the extreme end of the spectrum. The disagreement between the clustering scheme leads to only 17 in the intersections of the two lower approximations and 25 in the intersections of the upper approximations of the most volatile clusters. The rest of them are scattered among boundary regions of the clusters with lower volatility. The lower bound of the clusters in our rough ensemble will always form the core of the corresponding cluster like any of the crisp ensembles, since there is no disagreement between the individual clustering schemes. The crisp clustering ensemble forces the non-diagonal objects in Table 2 in one of the clusters. Our proposal will not disagree with any of the conflict resolution strategies proposed by different clustering ensemble methods, since our upper bounds of the clusters will include the corresponding common clusters. Therefore, our proposal is consistent with any of the conventional crisp clustering ensemble approaches. In addition, it will include the objects residing in the boundary region, which are common members as well as the outliers of upper approximation region in both the clustering schemes. One of the unique aspects of our proposal is an ability to preserve an implicit ranking between the clusters obtained by different clustering schemes. To the best of our knowledge, there is no other clustering ensemble proposal that addresses the issue of ordered clusters. The uniqueness of our approach is also a limitation, as this proposed technique is based on the combination of ordered clustering schemes. However, the experience gathered from this research will be useful in creating a more general rough clustering ensemble technique.

7   Conclusion

This paper describes the use of rough clustering to create ordered ranking of objects. In order to group daily price patterns based on volatility, we represent daily patterns using a single dimensional information granule using Black Scholes index. Another grouping is based on an information granule that consists of a more elaborate distribution of prices during the day. The rough clusters within these two groupings can be ordered based on their volatility. While these rough groupings tend to have a general consensus on volatility for most of the daily patterns, they disagree on a small number of patterns, even when the clusters are represented using rough sets. A closer inspection of the patterns suggests that both points of view have some merits. That means, we have a certain amount of order ambiguity in the resulting clustering ensemble. Lingras and

Haider [9] had proposed the use of rough sets to combine crisp clustering schemes. This paper proposes an extension of this previous proposal to combine ordered rough clustering schemes. The proposed ensemble method uses the preservation of cluster ordering as a guiding principle for creating the combined clustering.

References

1. Black, F., Scholes, M.: The pricing of options and corporate liabilities. J. Polit. Econ. **81**(3), 637–654 (1973)
2. Fern, X.Z., Lin, W.: Cluster ensemble selection. Stat. Anal. Data Mining **1**(3), 128–141 (2008)
3. Fred, A.: Finding consistent clusters in data partitions. In: Kittler, J., Roli, F. (eds.) MCS 2001. LNCS, vol. 2096, p. 309. Springer, Heidelberg (2001)
4. Gao, C., Pedrycz, W., Miao, D.: Rough subspace-based clustering ensemble for categorical data. Soft Comput. **5**(3), 1–16 (2013)
5. Ghosh, J., Acharya, A.: Cluster ensembles. Wiley Interdisc. Rev. Data Mining Knowl. Discov. **1**(4), 305–315 (2011)
6. Gionis, A., Mannila, H., Tsaparas, P.: Clustering aggregation. ACM Trans. Knowl. Discov. Data **1**(1), 4 (2007)
7. Jia, Z., Han, J.: An improved model of executive stock option based on rough set and support vector machines. In: 2008 Pacific-Asia Workshop on Computational Intelligence and Industrial Application, PACIIA 2008, **1**, pp. 256–261. IEEE (2008)
8. Joshi, M., Lingras, P.: Evolutionary and iterative crisp and rough clustering II: experiments. In: Chaudhury, S., Mitra, S., Murthy, C.A., Sastry, P.S., Pal, S.K. (eds.) PReMI 2009. LNCS, vol. 5909, pp. 621–627. Springer, Heidelberg (2009). http://dx.doi.org/10.1007/978-3-642-11164-8
9. Lingras, P., Haider, F.: Rough ensemble clustering. In: Intelligent Data Analysis, Special Issue on Business Analytics in Finance and Industry (2014)
10. Lingras, P., West, C.: Interval set clustering of web users with rough k-means. J. Intell. Inf. Sys. **23**(1), 5–16 (2004)
11. Mitra, S.: An evolutionary rough partitive clustering. Pattern Recogn. Lett. **25**(12), 1439–1449 (2004)
12. Nair, B.B., Mohandas, V., Sakthivel, N.: A decision treerough set hybrid system for stock market trend prediction. Int. J. Comput. Appl. **6**(9), 1–6 (2010)
13. Pawlak, Z.: Rough sets. Int. J. Comput. Inf. Sci. **11**(5), 341–356 (1982)
14. Pawlak, Z.: Fuzzy Logic for the Management of Uncertainty. Rough sets: a new approach to vagueness. Wiley, Newyork (1992)
15. Peters, G.: Some refinements of rough k-means clustering. Pattern Recogn. **39**(8), 1481–1491 (2006)
16. Peters, G., Crespo, F., Lingras, P., Weber, R.: Soft clustering-fuzzy and rough approaches and their extensions and derivatives. Int. J. Approximate Reasoning **54**(2), 307–322 (2013)
17. Polkowski, L., Skowron, A.: Rough mereology: a new paradigm for approximate reasoning. Int. J. Approximate Reasoning **15**(4), 333–365 (1996)
18. Skowron, A., Stepaniuk, J.: Information granules in distributed environment. In: Zhong, N., Skowron, A., Ohsuga, S. (eds.) RSFDGrC 1999. LNCS (LNAI), vol. 1711, pp. 357–366. Springer, Heidelberg (1999)
19. Strehl, A., Ghosh, J.: Cluster ensembles - a knowledge reuse framework for combining multiple partitions. J. Mach. Learn. Res. **3**, 583–617 (2003)

20. Vega-Pons, S., Ruiz-Shulcloper, J.: A survey of clustering ensemble algorithms. Int. J. Pattern Recogn. Artif. Intell. **25**(03), 337–372 (2011)
21. Yao, J., Herbert, J.P.: Financial time-series analysis with rough sets. Appl. Soft Comput. **9**(3), 1000–1007 (2009)
22. Yao, Y., Li, X., Lin, T., Liu, Q.: Representation and classification of rough set models. In: Proceeding of Third International Workshop on Rough Sets and Soft Computing. pp. 630–637 (1994)

Water Quality Prediction Based on a Novel Fuzzy Time Series Model and Automatic Clustering Techniques

Hui Meng, Guoyin Wang$^{(\boxtimes)}$, Xuerui Zhang, Weihui Deng, Huyong Yan, and Ruoran Jia

Big Data Mining and Applications Center, Chongqing Institute of Green and Intelligent Technology, Chinese Academy of Sciences, Chongqing 400714, China
wangguoyin@cigit.ac.cn

Abstract. In recent years, Fuzzy time series models have been widely used to handle forecasting problems, such as forecasting exchange rates, stock index and the university enrollments. In this paper, we present a novel method for fuzzy forecasting designed with the use of the two key techniques, namely automatic clustering and the probabilities of trends of fuzzy trend logical relationships. The proposed method mainly utilizes the automatic clustering algorithm to partition the universe of discourse into different lengths of intervals, and calculates the probabilities of trends of fuzzy trend logical relationships. Finally, it performs the forecasting based on the probabilities that were obtained in the previous stages. We apply the presented method for forecasting the water temperature and potential of hydrogen of the Shing Mun River, Hong Kong. The experimental results show that the proposed method outperforms Chen's and the conventional methods.

Keywords: Water quality prediction · Fuzzy sets · Fuzzy time series · Fuzzy logical relationship · Automatic clustering algorithm

1 Introduction

Water is the source of life, also an indispensable material resources for human survival and development. Deterioration of water quality has initiated serious management efforts in many countries [1]. In order to effectively control water pollution, promote the harmonious development between human beings and nature, information management of water environment has become especially important. Water quality prediction plays an important role in information management of water resources, also is the basic work of water resources development and utilization. It is usually used to infer the water quality change trends, supply the credible decision support for accidental water pollution. Accurate predictions of future phenomena are the lifeblood of optimizing water resources management in a watershed. In general, two kinds of approaches have been extensively employed to simulate and predict water quality. One is mechanism

© Springer International Publishing Switzerland 2015
D. Ciucci et al. (Eds.): RSKT 2015, LNAI 9436, pp. 395–407, 2015.
DOI: 10.1007/978-3-319-25754-9_35

water quality model, and the other is non-mechanism water quality model. The mechanism water quality model takes into account the factors that have impact on change of water quality, so the concept of model is clear. Unfortunately, the mechanism water quality model is very complex, and need a lot of data to be constructed. With the development of modern stochastic mathematics, fuzzy theory, statistical learning theory, machine leaning and artificial intelligence, the non-mechanism water quality models have been widely applied in water environment, such as, exponential smoothing [2], autoregressive integrated moving average model (ARIMA) [3], multivariable linear regression model, artificial neural networks (ANN) [4], support vector machine (SVM) [5], grey system theory [6], fuzzy time series model [7], etc.

Fuzzy time series (FTS) model has successfully been used to deal with various forecasting problems, such as to forecast the temperature, the exchange rate, and the stock index. Fuzzy time series was proposed by Song and Chissom based on the fuzzy set theory [8]. They developed two kinds of fuzzy time series to forecast the enrollments of the University of Alabama, i.e., the time invariant fuzzy time series model and the time variant fuzzy time series model [9,10]. Chen used an simple arithmetic operation to forecast the enrollments of the University of Alabama [11]. Chen and Hwang presented two factors time variant fuzzy time series to deal with forecasting problems [12]. Huarng pointed out that the effective length of the intervals in the universe of discourse can affect the forecasting accuracy rate [13]. In other words, the appropriate length of intervals can improve the forecasting results. In [14], Huarng presented a method using a heuristic function to forecast the enrollments of the University of Alabama and the Taiwan Futures Exchange. In recent years, some fuzzy forecasting methods based on fuzzy time series have been presented, such as [15–20]. Chen and Kao presented a method for forecasting based on fuzzy time series, particle swarm optimization techniques and support vector machines [21]. Chen et al. proposed an approach for forecasting the Taiwan Stock Exchange Capitalization Weighted Stock Index (TAIEX) based on two-factors second-order fuzzy-trend logical relationship groups and the probabilities of trends of fuzzy logical relationships [22].

In this paper, we present an improved fuzzy time series forecasting method for water quality prediction based on Chen et al.'s model since they provided a better forecasting framework and more accurate forecasting results [22]. The proposed method can effectively deal with the uncertainty and obtain the probabilities of trends of fuzzy-trend logical relationships. Furthermore, it can determine the appropriate interval length to increase the precision of forecast. In what follows, the mainly steps of this method will be introduced:

Firstly, we apply an automatic clustering algorithm to generate the clusters and then transform the clusters into different lengths of intervals. Then the proposed method fuzzifies the historical training data of the main factor and the secondary factor into fuzzy sets, respectively, to form two-factors second-order fuzzy logical relationship. Then, it groups the obtained two-factors second-order fuzzy logical relationships into two-factors second-order fuzzy-trend logical relationship groups. Then, it calculates the probability of the down-trend, the probability of the equal-trend, and the probability of the up-trend of the two-factors second-order fuzzy logical relationships in each two-factors second-order fuzzy-trend logical relationship group, respectively. Finally, it performs the forecasting based on the probabilities of the down-trend, the equal-trend, and the up-trend of the two-factors second-order fuzzy logical relationships in each two-factors second-order fuzzy-trend logical relationship group, respectively. We apply the proposed method to forecast the water temperature and potential of hydrogen of the Shing Mun River, Hong Kong. The experimental results show that the proposed method outperforms Chen's and the conventional methods.

The remaining content of this paper is organized as follows. In Sect. 2, we briefly review basic concepts of fuzzy sets, fuzzy time series etc. In Sect. 3, we review an automatic clustering algorithm for clustering historical data. In Sect. 4, we present a new method for forecasting the water temperature of the Shing Mun River, Hong Kong, based on two-factors second-order fuzzy-trend logical relationship groups, the probabilities of trends of fuzzy logical relationships and automatic clustering techniques. In Sect. 5, we make a comparison among the proposed method with Chen's [22] and the conventional methods according to the experimental results. The conclusions are discussed in Sect. 6.

2   Preliminaries

In this section we briefly review some basic concepts of fuzzy sets [8], fuzzy time series proposed by Song and Chissom [9,10], fuzzy logical relationships [11], and a two-factors high-order fuzzy time series forecasting model [16].

Let U be the universe of discourse, where $U = \{u_1, u_2, \ldots, u_n\}$. A fuzzy set A of the universe of discourse can be defined as follows:

$$A = f_A(u_1)/u_1 + f_A(u_2)/u_2 + \ldots + f_A(u_n)/u_n$$

where f_A is the membership function of the fuzzy set A, $f_A : U \to [0,1]$, $f_A(u_i)$ denotes the grade of membership of u_i in the fuzzy set A, and $1 \le i \le n$.

Let $Y_t(t = \cdots, 0, 1, 2, \cdots)$, a subset of R, be the universe of discourse in which fuzzy sets $f_A(t)(i = 1, 2, \cdots)$ are defined. Let F_t be a collection of $f_i(t)(i = 1, 2, \cdots)$. Then $F(t)$ is called a fuzzy time series on $Y(t)(t = \cdots, 0, 1, 2, \cdots)$.

When $F(t - 1)$ and $F(t)$ are fuzzy sets, assume that there exists a fuzzy logical relationship $R(t, t - 1)$ such that $F(t) = F(t - 1) \circ R(t, t - 1)$, where the symbol " $\circ$ " is max-min composition operator. Then, $F(t)$ is said to be caused by $F(t - 1)$, and the relationship between $F(t)$ and $F(t - 1)$ is denoted by a fuzzy logical relationship $F(t - 1) \to F(t)$, where $F(t)$ and $F(t - 1)$ refer to the current state and the next state of fuzzy time series, respectively.

Let $F_1(t), F_2(t)$ and $F(t)$ be fuzzy time series $(t = \cdots, 0, 1, 2, \cdots)$. If $F(t)$ is caused by $(F_1(t - 1), F_2(t - 1)), (F_1(t - 2), F_2(t - 2)), \cdots, (F_1(t - n), F_2(t - n))$. Hence, this fuzzy logical relationship can be represented by $((F_1(t - n), F_2(t - n)), \cdots, (F_1(t - 2), F_2(t - 2)), (F_1(t - 1), F_2(t - 1))) \to F(t)$, and it is called the two-factors nth order fuzzy time series forecasting model. If $F_1(t - 2) = A_i, F_2(t - 2) = B_j, F_1(t - 1) = A_k, F_2(t - 1) = B_l$ and $F(t) = A_m$, where

A_i, B_j, A_k, B_l, and A_m are fuzzy sets. Then we can represent the two-factors second-order fuzzy logical relationship as follows:

$$(A_i, B_j), (A_k, B_l) \rightarrow A_m$$

where $(A_i, B_j), (A_k, B_l)$, and A_m are called the current state and the next state of the fuzzy logical relationship, respectively [16].

3   An Automatic Clustering Algorithm

In this section, we present an improved automatic clustering algorithm (AC) for clustering numerical data [19]. The algorithm is now presented as the following steps:

Step 1: Sort the numerical data into an ascending order. Assume that ascending numerical data sequence is shown as follows:

$$d_1, d_2, \cdots, d_i, \cdots, d_n$$

where n denotes the total number of the datum, d_1 denotes the smallest datum, and d_n denotes the largest datum among the n numerical data, respectively. Then calculate the average difference $average\_diff$ and the standard deviation difference dev_diff between any two adjacent data, shown as follows:

$$average\_diff = \frac{\sum_{i=1}^{n-1}(d_{i+1} - d_i)}{n-1} \tag{1}$$

$$dev\_diff = \sqrt{\frac{\sum_{i=1}^{n-1}(d_{i+1} - d_i - average\_diff)^2}{n-1-1}} \tag{2}$$

Step 2: Based on the dev_diff, calculate the maximum distance max_data_dist, shown as follows:

$$max\_data\_dist = c * dev\_diff \tag{3}$$

where c denotes a positive constant, which affects the partition of the universe of discourse. Then we cluster for the first datum and let the cluster to be the current cluster. We determine whether the next datum can be put into the current cluster or create a new cluster for it. Assume that the current condition is shown as follows:

$$\cdots, \{\cdots, d_i\}, d_{i+1}, d_{i+2}, d_{i+3}, \cdots, d_n$$

if $d_{i+1} - d_i \leq max\_data\_dist$, then put d_{i+1} into the current cluster in which d_i belongs. Otherwise, create a new cluster for d_{i+1} and let the new cluster be the current cluster. Repeatedly, check the datum until all the data have been clustered. Assume that the clustering result is shown as follows:

$$\{d_{11}, \cdots, d_{1a}\}, \{d_{21}, \cdots, d_{2b}\}, \cdots, \{d_{i1}, \cdots, d_{in}\}, \{d_{j1}, \cdots, d_{jm}\}, \cdots, \{d_{p1}, \cdots, d_{pl}\}$$

Step 3: Let the number of intervals equal the number of clusters and transform all the clusters into the intervals according to the following rules:

Rule 1: If the interval is the first interval, then the lower bound of it can be calculated as follows:

$$interval\_lower_1 = d_1 - max\_data\_dist \qquad (4)$$

Rule 2: If the interval is the last interval, then the upper bound of it can be calculated as follows:

$$interval\_upper_p = d_n + max\_data\_dist \qquad (5)$$

Rule 3: The lower bound and the upper bound of the common intervals can be calculated as follows:

$$interval\_lower_i = interval\_upper_{i-1} \qquad (6)$$

$$interval\_upper_i = \frac{d_{in} + d_{j1}}{2} \qquad (7)$$

Based on the lower bound and the upper bound of the intervals, we can calculate the middle value for each interval and the average interval length. For example, the middle value of the ith interval and the average interval length can be calculated as follows:

$$mid\_value_i = \frac{interval\_lower_i + interval\_upper_i}{2} \qquad (8)$$

$$avg\_interval\_length = \frac{interval\_upper_p - interval\_lower_1}{p} \qquad (9)$$

where p, $interval\_upper_p$, and $interval\_lower_1$ denote the number of the intervals the last interval's upper bound, and the first interval's lower bound, respectively.

4   A Novel Method for Water Quality Prediction Based on Fuzzy Time Series Model and Automatic Clustering Techniques

In this section, we present a novel method for forecasting the water temperature and potential of hydrogen of the Shing Mun River, Hong Kong, based on two-factors second-order fuzzy-trend logical relationship groups, the probabilities of trends of fuzzy logical relationships and automatic clustering techniques. We totally divide the method into four stages. In what follows, all of the stages composing this model are going to be described in details.

Stage 1: Apply the automatic clustering algorithm introduced in Sect. 3 to generate different lengths of intervals from the training data for each factor. We hypothesize that the universe of discourse of the main factor is partitioned into p intervals, denoted as $u_1, u_2, \cdots, u_p$. Similarly, assume that the universe of discourse of the second factor is partitioned into m intervals, denoted as $v_1, v_2, \cdots, v_m$.

Stage 2: In this stage, we have to define the fuzzy sets and fuzzify the training data for each factor, establish two-factors second-order fuzzy logical relationship and group the fuzzy logical relationships into fuzzy-trend logical relationship groups. Therefore, this stage can be divided into four sub-steps, shown as follows:

Stage 2.1: Define the fuzzy sets $A_1, A_2, \cdots, A_p$ of the main factor based on the intervals $u_1, u_2, \cdots, u_p$ shown as follows:

$$A_1 = 1/u_1 + 0.5/u_2 + 0/u_3 + 0/u_4 + \cdots + 0/u_{p-1} + 0/u_p$$
$$A_2 = 0.5/u_1 + 1/u_2 + 0.5/u_3 + 0/u_4 + \cdots + 0/u_{p-1} + 0/u_p$$

$$\vdots$$

$$A_p = 0/u_1 + 0/u_2 + 0/u_3 + 0/u_4 + \cdots + 0.5/u_{p-1} + 1/u_p$$

Define the fuzzy sets $B_1, B_2, \cdots, B_m$ of the second factor based on the intervals $v_1, v_2, \cdots, v_m$, shown as follows:

$$B_1 = 1/v_1 + 0.5/v_2 + 0/v_3 + 0/v_4 + \cdots + 0/v_{p-1} + 0/v_m$$
$$B_2 = 0.5/v_1 + 1/v_2 + 0.5/v_3 + 0/v_4 + \cdots + 0/v_{p-1} + 0/v_m$$

$$\vdots$$

$$B_m = 0/v_1 + 0/v_2 + 0/v_3 + 0/v_4 + \cdots + 0.5/v_{p-1} + 1/v_m$$

Stage 2.2: Fuzzify each historical datum into a fuzzy set. If the datum of the main factor belongs to the interval $u_i, 1 \leq i \leq p$, then the datum is fuzzified into A_i. If the datum of the second factor belongs to the interval $v_k, 1 \leq k \leq m$, then the datum is fuzzified into B_k.

Stage 2.3: Based on the fuzzified historical training data of the main factor and the secondary factor obtained in Stage 2.2, respectively, construct two-factors second-order fuzzy logical relationships. If the fuzzified historical data of the main factor on time $t-2, t-1$ and t are A_{i2}, A_{i1} and A_k, respectively, where A_{i2}, A_{i1} and A_k are fuzzy sets. Likewise, if the fuzzified historical training data of the secondary factor on time $t-2, t-1$ are B_{j2} and B_{j1}, respectively, where B_{j2} and B_{j1} are fuzzy sets. Then we can construct the two-factors second-order fuzzy logical relationship, shown as follows:

$$(A_{i2}, B_{j2}), (A_{i1}, B_{j1}) \rightarrow A_k$$

Stage 2.4: Construct two-factors second-order fuzzy-trend logical relationship groups [22] by using the two-factors second-order fuzzy logical relationships obtained in Stage 2.3. Each two-factors second-order fuzzy-trend logical relationship group is represented by a "two-tuples", where the first element of the

Table 1. Nine groups of fuzzy-trend logical relationship for the two-factors second-order fuzzy logical relationship $(A_i, B_j), (A_k, B_l) \rightarrow A_m$ [23].

Group number	Fuzzy logical relationships	
	Between i and k	Between j and l
Group 1	Down	Down
Group 2	Down	Equal
Group 3	Down	Up
Group 4	Equal	Down
Group 5	Equal	Equal
Group 6	Equal	Up
Group 7	Up	Down
Group 8	Up	Equal
Group 9	Up	Up

two-tuples denotes the trend of adjacent fuzzy sets of the main factor and the second element of the two-tuples denotes the trend of adjacent fuzzy sets of the secondary factor. For example, the nine groups of fuzzy-trend logical relationships for the two-factors second-order fuzzy logical relationship $(A_i, B_j), (A_k, B_l) \rightarrow A_m$ are show in Table 1.

Stage 3: In this stage, we calculate the probability of the down-trend, the probability of the equal-trend, and the probability of the up-trend of the two-factors second-order fuzzy-trend logical relationships in each two-factors second-order fuzzy-trend logical relationship group, respectively.

Stage 3.1: For each two-factors second-order fuzzy-trend logical relationship group obtained in stage 2.4, we construct a table that containing the current status and the next status of each fuzzy logical relationship. For each two-factors second-order fuzzy logical relationship in the constructed two-factors second-order fuzzy-trend logical relationship groups, the fuzzified historical training data of the main factor on time $t-1$ and t are called the current status and the next status of the fuzzy logical relationship, respectively. For instance, we might as well consider the first two-factors second-order fuzzy logical relationship is shown in Table 2:

$$(A_{i2}, B_{j2}), (A_{i1}, B_{j1}) \rightarrow A_{k1}$$

we can see that the current status and the next status of the two-factors second-order fuzzy logical relationship are A_{i1} and A_{k1}, respectively.

Stage 3.2: Calculate the number $N_{D(i)}$ of the down-trend, the number $N_{E(i)}$ of the equal-trend, and the number $N_{U(i)}$ of the up-trend between the current status and the next status of each two-factors second-order fuzzy logical relationship in the ith two-factors second-order fuzzy logical relationship group *Group i*. For example, let's consider the following two-factors second-order fuzzy logical relationship in the ith two factors second-order fuzzy-trend logical relationship group *Group i*:

Table 2. Current status and the next status of each fuzzy logical relationship.

Fuzzy Logical Relationship	Current Status	Next Status
$(A_{i2}, B_{j2}), (A_{i1}, B_{j1}) \rightarrow A_{k1}$	A_{i1}	A_{k1}
$(A_{e2}, B_{f2}), (A_{e1}, B_{f1}) \rightarrow A_{k2}$	A_{e1}	A_{k2}
$(A_{p2}, B_{q2}), (A_{p1}, B_{q1}) \rightarrow A_{k3}$	A_{p1}	A_{k3}
$(A_{c2}, B_{d2}), (A_{c1}, B_{d1}) \rightarrow A_{k4}$	A_{c1}	A_{k4}
$\vdots$	$\vdots$	$\vdots$
$(A_{m2}, B_{n2}), (A_{m1}, B_{n1}) \rightarrow A_{kr}$	A_{m1}	A_{kr}

Table 3. The number $N_{D(i)}$ of the down-trend, the number $N_{E(i)}$ of, the equal-trend, and the number $N_{U(i)}$ of the up-trend in each fuzzy-trend logical relationship group.

Group Number	Relationships between the current status and the next status		
	Down-trend	Equal-trend	Up-trend
Group 1	$N_{D(1)}$	$N_{E(1)}$	$N_{U(1)}$
Group 2	$N_{D(2)}$	$N_{E(2)}$	$N_{U(2)}$
Group 3	$N_{D(3)}$	$N_{E(3)}$	$N_{U(3)}$
Group 4	$N_{D(4)}$	$N_{E(4)}$	$N_{U(4)}$
$\vdots$	$\vdots$	$\vdots$	$\vdots$
Group 8	$N_{D(8)}$	$N_{E(8)}$	$N_{U(8)}$
Group 9	$N_{D(9)}$	$N_{E(9)}$	$N_{U(9)}$

$$(A_i, B_j), (A_k, B_l) \rightarrow A_m$$

where A_k and A_m are called the current status and the next status of the two-factors second-order fuzzy logical relationship, respectively. If $k > m$, then increase the number $N_{D(i)}$ of the down-trend by one; If $k = m$, then increase the number $N_{E(i)}$ of the equal-trend by one; If $k < m$, then increase the number $N_{U(i)}$ of the up-trend by one. Table 3 shows the number $N_{U(i)}$ of the up-trend in each two-factors second-order fuzzy-trend logical relationship group *Group i*, where $1 \leq i \leq 9$.

Stage 3.3: According to Table 3, we can figure out the probability of the down-trend, the probability of the equal-trend and the probability of the up-trend for each two-factors second-order fuzzy-trend logical relationship group. For example, the probability of the down-trend of the two-factors second-order fuzzy logical relationship in the ith two-factors second-order fuzzy-trend logical relationship group *Group i* can be worked out by: $P_{D(i)} = N_{D(i)}/(N_{D(i)} + N_{E(i)} + N_{U(i)})$, where $(N_{D(i)} + N_{E(i)} + N_{U(i)}) \neq 0$. In the same way, we can get the probabilities of the equal-trend and up-trend for each fuzzy-trend logical relationship group, respectively.

Stage 4: Finally, it performs the forecasting based on the probabilities that obtained in the previous stage and the current actual value. Assume that the fuzzified historical testing data of the main factor on time $t-2$ and $t-1$ are A_i and A_k respectively. Suppose that the fuzzified historical testing data of the secondary factor on time $t-2$ and $t-1$ are B_j and B_l, respectively, where $i > k$ and $j > l$, and assume that we want to forecast the value of the main factor of time t, then from Table 1, we can see its corresponding two-factors second-order fuzzy-trend logical relationship group *Group 1* (i.e., the down-and-down group). When $(N_{D(i)} + N_{E(i)} + N_{U(i)}) \neq 0$, then $P_{D(i)}, P_{E(i)}$, and $P_{U(i)}$ denote the probabilities of the down-trend, the equal-trend and the up-trend of the two-factors second-order fuzzy logical relationship in the two-factors second-order fuzzy-trend logical relationship group *Group i*, respectively. Therefore, the forecasted value $F(t)$ of the main factor on time t is calculated as follows:

$$F(t) = [(R_{(t-1)} - C_1) \times P_{D(i)}] + (R_{(t-1)} \times P_{E(i)}) + [(R_{(t-1)} + C_2) \times P_{U(i)}] \quad (10)$$

$$C_1 = k_1 * avg\_interval\_length \ (k_1 \in R^+) \quad (11)$$

$$C_2 = k_2 * avg\_interval\_length \ (k_2 \in R^+) \quad (12)$$

where $R_{(t-1)}$ denotes the actual value of the main factor on time $t-1$. From (11), (12) we can conclude C_1 and C_2 denote k_1 and k_2 times of the average internal length in the universe of discourse of the main factor, respectively. The main idea of the (10) is that the forecasted value $F(t)$ of the main factor on time t belongs the interval $[R_{(t-1)} - C_1, R_{(t-1)} + C_2]$. Therefore, according to the (10), we can calculate the forecasting value of the main factor on time t. Otherwise, if $(N_{D(i)} + N_{E(i)} + N_{U(i)}) = 0$, then we let the forecasting value $F(t)$ of the main factor on time t be equal to the actual datum of the main factor on time $t-1$.

5   Experimental Results

In this section, we evaluate the effectiveness of the proposed forecasting model and conduct performance comparison with by conducting the subsequent experiments, including water temperature forecasting and potential of hydrogen forecasting of Shing Mun River, Hong Kong. All of the data sets which the paper used were published on the website of Environmental Protection Department of the Government of the Hong Kong Special Administrative Region of the People's Republic of China [24]. Furthermore, we use the Root Mean Squared Error (RMSE), the Mean Absolute Percentage Error (MAPE) and the Mean Absolute Error (MAE) to evaluate the forecasting performance of the proposed method, defined as follows:

$$RMSE = \sqrt{\frac{\sum_{i=1}^{n}(forecasted\_value_i - actual\_value_i)^2}{n}} \quad (13)$$

$$MAPE = \frac{1}{n}\sum_{i=1}^{n} \left| \frac{forecasted\_value_i - actual\_value_i}{actual\_value_i} \right| \quad (14)$$

$$MAE = \frac{1}{n}\sum_{i=1}^{n} |forecasted\_value_i - actual\_value_i| \quad (15)$$

Table 4. A comparison among the proposed method with Chen's and the conventional methods using MAE, RMSE and MAPE for forecasting water temperature.

Dates	Actual values	ARIMA	Chen's method	BPNN	SVM	The proposed method
2012/12/13	22	24.973	21.416	21.543	21.613	21.208
2013/01/18	19.2	19.653	22.018	22.349	22.236	21.811
2013/02/21	19	18.585	19.140	20.217	19.890	18.035
2013/03/15	22.6	19.564	19.016	20.185	19.789	22.574
2013/04/19	24.3	20.469	22.703	23.150	22.893	22.574
2013/05/30	29.4	24.251	24.308	24.392	24.746	21.429
2013/06/06	27.1	26.376	29.410	27.214	27.413	29.287
2013/07/18	26.6	27.429	27.108	26.009	26.868	26.878
2013/08/19	26.5	27.666	26.614	25.568	26.612	26.407
2013/09/18	26.9	27.229	26.585	25.485	26.554	25.343
2013/10/21	24.3	27.879	26.916	25.830	26.772	25.756
2013/11/14	21.5	25.831	24.307	24.392	24.746	23.076
2013/12/09	20.8	23.272	21.514	21.669	21.714	21.214
MAE		2.232	1.784	1.672	1.604	1.487
RMSE		2.771	2.330	2.122	2.139	1.942
MAPE(%)		9.347	7.470	7.130	6.911	6.162

where n denotes the number of dates needed to be forecasted, "$forecasted\_value_i$" denotes the forecasted value on time i, "$actual\_value_i$" denotes the actual value on time i, where $1 \leq i \leq n$.

5.1   Water Temperature Forecasting

In what follows, we use the proposed method to forecast the water temperature from December 13, 2012 to December 9, 2013, where we let the water temperature of the Shing Mun River, Hong Kong, as the main factor and let the water temperature of the Tai Wai Nullah, Hong Kong, as the second factor. The data from January 29, 1986 to November 29, 2012 are used as the training data, and the data from December 13, 2012 to December 9, 2013 are used as the testing data. It is worth mentioning that in this experiment we set the parameters $c = 1.5$ and $C_1 = C_2 = 12$, which were discussed in Sects. 3 and 4. In Table 3, we make a comparison between the RMSE, MAPE, and MAE obtained by the proposed model and the most representative forecasting models. From Table 4, it can be observed that MAE, MAPE, and RMSE of the proposed model is smaller than the conventional forecasting model (ARIMA model, SVM model, BPNN model), and the classical fuzzy time series model (Chen's model [22]).

Table 5. A comparison among the proposed method with Chen's and the conventional methods using MAE, RMSE and MAPE for forecasting the potential of hydrogen.

Dates	Actual values	ARIMA	BPNN	SVM	Chen's method	The proposed method
2013/01/18	8.2	7.861	7.904	7.877	8.062	8.137
2013/02/21	7.6	7.938	7.954	7.924	8.150	8.094
2013/03/15	7.7	8.097	7.695	7.746	7.568	7.492
2013/04/19	7.8	7.779	7.714	7.771	7.764	7.839
2013/05/30	7.8	7.870	7.760	7.804	7.773	7.696
2013/06/06	7.5	7.865	7.760	7.804	7.788	7.697
2013/07/18	7.5	7.932	7.689	7.730	7.454	7.418
2013/08/19	7.6	7.696	7.689	7.730	7.562	7.638
2013/09/18	7.8	7.781	7.695	7.746	7.611	7.541
2013/10/21	7.7	7.873	7.760	7.804	7.779	7.700
2013/11/14	7.5	7.754	7.714	7.771	7.650	7.598
2013/12/09	7.7	7.763	7.689	7.730	7.521	7.633
MAE		0.214	0.168	0.154	0.154	0.137
RMSE		0.261	0.201	0.196	0.209	0.190
MAPE(%)		2.790	2.170	2.011	2.010	1.790

5.2   The Potential of Hydrogen Forecasting

In this section, we also apply the presented method to predict the potential of hydrogen (ph) of the Shing Mun River, Hong Kong, where we let the potential of hydrogen of the Shing Mun River trunk stream as the main factor, and let the potential of hydrogen of the Shing Mun River tributary as the second factor. The data from January 29, 1986 to November 29, 2012 are used as the training data, and the data from January 18, 2013 to December 9, 2013 are used as the testing data. From Table 5, we can see that the proposed method get the smallest RMSE, MAPE and MAE than the conventional forecasting model(ARIMA model, SVM model, BPNN model), and the classical fuzzy time series model(Chen's model). Especially, in this experiment we set the parameters $c = 0.8$ and $C_1 = C_2 = 4$.

6   Conclusions

In this paper, we have successfully applied the proposed model to forecast the water temperature and potential of hydrogen of the Shing Mun River, Hong Kong based on two-factors second-order fuzzy trend logical relationship groups, the probabilities of trends of fuzzy-trend logical relationships and automatic clustering techniques. In particular, an improved automatic clustering algorithm has been used to partition the universe of discourse into different lengths of intervals according to the statistic distribution of historical data, rather than simply divide the universe of discourse into several static length of intervals [22]. The drawback of the static length of intervals is that the historical data are roughly put into the intervals, even if the variance of the historical data is not high [20]. The experiment results show that the proposed method outperforms Chen's and the traditional methods. However, we only use one secondary factor for fuzzy time series forecasting, and rely on empirical method to determine the parameters (e.g., k_1 and k_2). In the future, we will explore the impacts of the multiple secondary factors on fuzzy forecasting and utilize the suitable algorithms to find the optimal parameters like k_1 and k_2.

Acknowledgments. This work is supported by the National Science and Technology Major Project (2014ZX07104-006) and the Hundred Talents Program of CAS (NO. Y21Z110A10)

References

1. Faruk, D.Ö.: A hybrid neural network and ARIMA model for water quality time series prediction. Eng. Appl. Artif. Intell. **23**, 586–594 (2010)
2. Sbrana, G., Silvestrini, A.: Random switching exponential smoothing and inventory forecasting. Int. J. Prod. Econ. **156**, 283–294 (2014)
3. Shumway, R.H., Stoffer, D.S.: Time Series Analysis and its Applications: with R examples. Springer, Heidelberg (2010)
4. Han, H.G., Chen, Q.-L., Qiao, J.-F.: An efficient self-organizing RBF neural network for water quality prediction. Neural Netw. **24**, 717–725 (2011)
5. Vapnik, V.N.: The Nature of Statistical Learning Theory. Statistics for Engineering and Information Science. Springer-Verlag, New York (2000)
6. Julong, D.: The grey control system. J. Huazhong Univ. Sci. Technol. **3**, 18 (1982)
7. Song, Q., Chissom, B.S.: Fuzzy time series and its models. Fuzzy Sets Sys. **54**, 269–277 (1993)
8. Zadeh, L.A.: Fuzzy sets. Inf. Control **8**, 338–353 (1965)
9. Song, Q., Chissom, B.S.: Forecasting enrollments with fuzzy time seriespart I. Fuzzy Sets Sys. **54**, 1–9 (1993)
10. Song, Q., Chissom, B.S.: Forecasting enrollments with fuzzy time seriespart II. Fuzzy Sets Sys. **62**, 1–8 (1994)
11. Chen, S.M.: Forecasting enrollments based on fuzzy time series. Fuzzy Sets Sys. **81**, 311–319 (1996)
12. Chen, S.M., Hwang, J.R.: Temperature prediction using fuzzy time series. IEEE Trans. Sys. Man Cybern. Part B Cybern. **30**, 263–275 (2000)
13. Huarng, K.: Effective lengths of intervals to improve forecasting in fuzzy time series. Fuzzy Sets Sys. **123**, 387–394 (2001)
14. Huarng, K.: Heuristic models of fuzzy time series for forecasting. Fuzzy Sets Sys. **123**, 369–386 (2001)
15. Chen, S.M.: Forecasting enrollments based on high-order fuzzy time series. Cybern. Sys. **33**, 1–16 (2002)
16. Lee, L.W., Wang, L.H., Chen, S.M., Leu, Y.H.: Handling forecasting problems based on two-factors high-order fuzzy time series. IEEE Trans. Fuzzy Sys. **14**, 468–477 (2006)

17. Chen, S.M., Wang, N.Y., Pan, J.S.: Forecasting enrollments using automatic clustering techniques and fuzzy logical relationships. Expert Sys. Appl. **36**, 11070–11076 (2009)
18. Bang, Y.K., Lee, C.H.: Fuzzy time series prediction using hierarchical clustering algorithms. Expert Sys. Appl. **38**, 4312–4325 (2011)
19. Chen, S.M., Tanuwijaya, K.: Fuzzy forecasting based on high-order fuzzy logical relationships and automatic clustering techniques. Expert Sys. Appl. **38**, 15425–15437 (2011)
20. Chen, S.M., Tanuwijaya, K.: Multivariate fuzzy forecasting based on fuzzy time series and automatic clustering techniques. Expert Sys. Appl. **38**, 10594–10605 (2011)
21. Chen, S.M., Kao, P.Y.: TAIEX forecasting based on fuzzy time series, particle swarm optimization techniques and support vector machines. Inf. Sci. **247**, 62–71 (2013)
22. Chen, S.M., Chen, S.W.: Fuzzy forecasting based on two-factors second-order fuzzy-trend logical relationship groups and the probabilities of trends of fuzzy logical relationships. IEEE Trans. Cybern. **45**, 405–417 (2015)
23. Chen, S.M., Manalu, G.M.T., Pan, J.S., Liu, H.C.: Fuzzy forecasting based on two-factors second-order fuzzy-trend logical relationship groups and particle swarm optimization techniques. IEEE Trans. Cybern. **43**, 1102–1117 (2013)
24. Environmental Protection Department of the Government of the Hong Kong Special Administrative Region. http://epic.epd.gov.hk/EPICRIVER/river

Clustering Algorithm Based on Fruit Fly Optimization

Wenchao Xiao, Yan Yang$^{(\boxtimes)}$, Huanlai Xing, and Xiaolong Meng

School of Information Science and Technology, Southwest Jiaotong University,
Chengdu 610031, People's Republic of China
{xwc,xlmeng}@my.swjtu.edu.cn,
{yyang,hxx}@swjtu.edu.cn

Abstract. The swarm intelligence optimization algorithms have been widely applied in the fields of clustering analysis, such as ant colony algorithm, artificial immune algorithm and so on. Inspired by the idea of fruit fly optimization algorithms, this paper presents Fruit Fly Optimization Clustering Algorithm (FOCA) based on fruit fly optimization. The algorithm extends the space which fruit fly from two-dimension to three, in order to find the global optimum in each iteration. Besides, for the purpose of getting the optimize clusters centers, each fruit fly flies step by step, and every flight is a stochastic search in its own region. Compared with the other clustering algorithms of swarm intelligence, the proposed algorithm is simpler and with fewer parameters. The experimental results demonstrate that our algorithm outperforms some of state-of-the-art algorithms regarding to the accuracy and convergence time.

Keywords: Swarm intelligence · Clustering analysis · Fruit fly optimization · Convergence

1 Introduction

Clustering analysis is a class of discovery process which divides data into subsets. Each subset represents a cluster, with the intra-cluster similarity maximized and the inter-cluster similarity minimized [1]. It has been widely used in many application areas, such as business intelligence, image pattern recognition, web search, biology, security, and etc. In recent years, some scholars combine swarm intelligence optimization algorithms and clustering analysis, the fundamental strategy of which convert the clustering problem into an optimization problem; and carry out stochastic search by simulating the intelligent behavior of swarms to find the best objective function value of clustering division. Deneubourg et al. are the first researchers who applied the artificial ant colony to the clustering problem in 1991 [2]. Omran et al. proposed an image classification method using particle swarm optimization in 2002 [3]. Both algorithms have become well-known and been extensively used in the fields of clustering analysis. Recently, Tsang and Lau used multi-objective immune optimization evolutionary algorithm for clustering [4]. Amiri et al. have applied shuffled frog-leaping algorithm on clustering [5]. Zhang and Ning proposed an artificial bee colony approach for clustering [6]. Yazdani et al. presented a new algorithm based on an improved artificial fish swarm algorithm for data clustering [7].

© Springer International Publishing Switzerland 2015
D. Ciucci et al. (Eds.): RSKT 2015, LNAI 9436, pp. 408–419, 2015.
DOI: 10.1007/978-3-319-25754-9_36

However, all of the above clustering algorithms have deficiencies. For example, the ant colony clustering algorithm has too much parameters and its search quality is highly depend on the initial parameter selection [8]. Artificial immune clustering method is featured with grave complexity and poor generalization ability [9]. Clustering approach based on shuffled frog leaping algorithm is easily trapped into local optimum with a slow convergence [10]. Traditional bee colony clustering algorithm has disadvantages regarding to the random selection of initial swarm and search step [11]. Moreover, clustering problem using fish swarm approach has higher time complexity compared with the above methods [12].

This paper presents a novel clustering algorithm named Fruit Fly Optimization Clustering Algorithm (FOCA), based on the Fruit Fly Optimization Algorithm (FOA) [13], which is proposed by Taiwan scholar Pan in 2011. It is a swarm intelligence optimization algorithm based on the foraging behavior of the fruit fly. This algorithm has advantages such as simple computational process, less parameters, high precision, and ease of understanding [14]. The experimental results demonstrate that the algorithm outperforms some of state-of-the-art algorithms regarding to the accuracy and convergence time.

The remaining of this paper is organized as follows: Sect. 2 introduces the Fruit Fly Optimization Algorithm. Section 3 discusses the clustering analysis process using Fruit Fly Optimization Algorithm. Section 4 reports the experiments results. The final Section provides conclusions and future work.

2 Fruit Fly Optimization Algorithm

Fruit Fly Optimization Algorithm is a novel approach for global optimization based on the foraging behavior of drosophilas. The drosophilas has better sensing and perception than other species, in particular the sense of smell and sight. The olfactory organ of the drosophila can collect floating odors well in the air, and even can perceive the food source 40 km away. When a drosophila is close to the location of food source, it use its sensitive vision to find the food source and others drosophilas would fly towards the same direction [13].

The algorithm consists of several steps as below:

Step 1. Randomly initiate the location of fruit fly group, as shown in Fig. 1.

$$Init\ X\_axis$$
$$Init\ Y\_axis$$

Step 2. Give the random direction and distance to the search of food by osphresis of each fruit fly files.

$$X_i = X\_axis + Random\ Value$$
$$Y_i = Y\_axis + Random\ Value \tag{1}$$

Step 3. Since the location of food is unknown, the distance to the origin is estimated first (*Dist*). Then the smell concentration judgment value (*S*) is calculated, which is the reciprocal of the distance.

$$Dist(i) = \sqrt{X_i^2 + Y_i^2}$$
$$S(i) = 1/Dist(i)$$

(2)

Step 4. Substitute the smell concentration judgment value (*S*) into smell concentration judgment function (or called *Fitness Function*) so as to find the smell concentration (*Smell*) of the individual location of the fruit fly.

$$Smell(i) = Function(S(i))$$

(3)

Step 5. Find out the fruit fly with the maximal smell concentration (i.e. finding the optimal value) among the fruit fly swarm.

$$[BestSmell\ BestIndex] = max(Smell)$$

(4)

Step 6. Keep the best smell concentration value and *X_axis*, *Y_axis* coordinates. The fruit fly swarm group uses the above vision to fly towards that location.

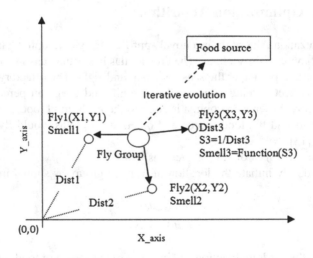

Fig. 1. Schematic diagram of the group iterative food searching of fruit fly

$$SmellBest = BestSmell$$
$$X\_axis = X(BestIndex) \qquad (5)$$
$$Y\_axis = Y(BestIndex)$$

Step 7. Repeat steps 2–5. If the smell concentration is superior to the previous iterative smell concentration, proceed step 6.

3 FOCA Approach

3.1 Motivation of FOCA

The traditional clustering algorithms based on swarm intelligence optimization algorithm suffer from a complex computational process and some drawbacks described above. Fruit Fly Optimization Algorithm is a novel swarm intelligence optimization algorithm proposed in 2011 [13], which has advantages such as simple computational process, few parameters, high precision, and ease of understanding. Fortunately, it is reasonable to tackle the clustering problem using Fruit Fly Optimization Algorithm.

3.2 Implementation of FOCA

Since the basic Fruit Fly Optimization Algorithm easily falls into premature convergence, Li et al. proposed a modified fly optimization algorithm for designing the self-turning proportional integral derivative controller [14]. The basic idea is to extend the two-dimensional coordinate system to three-dimensional X, Y and Z, and add a shock factor (δ) in the smell concentration to reduce the possibility of premature convergence. This paper is based on this improved version of fruit fly optimization algorithm.

As mentioned in the Sect. 2, the value of the best smell concentration may be found by iterative evolution according to the value of X, Y and Z. However, X, Y and Z just determine a single value (S). According to this feature, and considering that the cluster centers of an N-dimensional dataset are vectors consisting of N values of S ($center_j = \{S_i, i = 1, 2, 3, \ldots, N\}$), and each value of S is composed of $X = \{X_i, i = 1, 2, 3, \ldots, N\}$, $Y = \{Y_i, i = 1, 2, 3, \ldots, N\}$ and $Z = \{Z_i, i = 1, 2, 3, \ldots, N\}$. If the number of N-dimensional cluster centers is k ($k >= 2$), the one-dimensional matrix of X, Y and Z should be converted into two-dimensional matrix. Then the two-dimensional matrix of S ($S = \{S_{i,j}, i, j = 1, 2, 3, \ldots, N\}$) is obtained, where i represents the i-th cluster center and j represents an attribute value of the cluster center.

Since the attribute values of the cluster centers may be significantly different, the range of the initial value of X, Y and Z should be unified in Step 1, so that a better clustering result is obtained. To address this problem, firstly, the dataset is normalized. Then the range of the attribute value of cluster centers is between 0 and 1.

Data Normalized: Let A represent a numeric attribute consisting of N observations $v_1, v_2, \ldots, v_N$. The new attribute is calculated as:

$$v_i' = \frac{v_i - min(A)}{max(A) - min(A)} \tag{6}$$

where $min(A)$ and $max(A)$ represents the minimum value and the maximum value of Attribute A, respectively.

Some issues regarding the implementation of FOCA are discussed in the following subsections.

3.2.1   The Optimum Range Value of Initial Position

As we know, the attribute value of cluster centers is in the range [0, 1]. The final attribute value of cluster centers is determined by $1/\sqrt{X^2 + Y^2 + Z^2}$. Therefore, the range value of X, Y and Z should neither too big nor too small.

10 datasets (shown in Table 1) are given to test the best range value of initial position with five values are listed in Fig. 2 (each color represents a value). Each value denotes a range value of initial position. For example, if the value is 5, the range value of initial position is [0, 5]. The X axis represents the number of iterations. The Y axis is the value of Fitness Function. In general, the smaller the value of Fitness Function, the better the clustering performance.

It is observed from Fig. 2 that for the value 5 (range value of initial position) it is the best among the others, as well as approaches convergence before the iterations approach 50. Figure 3 gives the explanation in detail. Assume matrix S represent the entire solution space. When the range value of initial position is too small, the solution space is S1, where the best solution is excluded. Thus, none of the values are close to the best solution in the early stage of the evolution. On the contrary, if the range value of initial position is too large, the solution space is S3, where the best solution resides. Because the solution space is too large, the volatility of the solutions is very large in the early stage of the evolution. The solution space S2 is neither too large nor too small. Therefore, the solution is relatively stable and more close to the best solution in the early stage of the evolution.

Based on the above analysis, our algorithm chooses 5 to be the range value of initial position.

3.2.2   The Optimum Range Value of Step Size

In Step 2 of FOCA, the random direction and distance are given for the search of food using osphresis by each individual fruit fly, i.e., a random value (step size) is added to the X_axis, Y_axis and Z_axis. If the step size is too small, the ability of local search becomes stronger while the ability of global search becomes poor. If the step size is too big, the local search ability is weak while the global search ability turns to be significant. Hence, it is important to select an appropriate range value for the step size.

As mentioned in the previous subsection, 10 datasets (Table 1) are used to determine an appropriate range value of step size, with five values listed in Fig. 4. Each value represents a range value of step size. For example, if the value is 1, the range value of step size is [−1, 1].

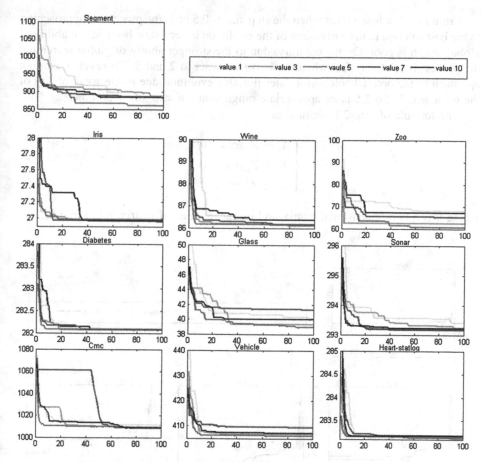

Fig. 2. Iterative evolution with different values of initial position

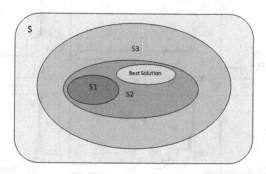

Fig. 3. Solution space

From Fig. 4, it is seen that when the step size is 0.5 or 1, the process of approaching to the best solution in the early stage of the evolution is very slow because the ability of global search is poor. On the contrary, due to the stronger ability of global searching, the process is relatively fast when the step size is set to 2 and 3. However, it can not approach to the optimal solution in later iterative evolution due to the weak ability of the local search. So 1.5 is an appropriate range value of step size than others.

The formula of step 2 is defined as

$$\begin{cases} X_i = X\_axis + \theta \\ Y_i = Y\_axis + \theta \\ Z_i = Z\_axis + \theta \end{cases} \tag{7}$$

where θ is a two-dimensional matrix and its range value of matrix is $[-1.5, 1.5]$.

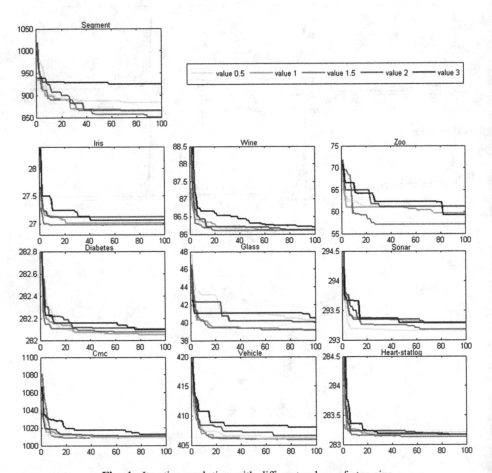

Fig. 4. Iterative evolution with different values of step size

3.2.3 Shock Factor

In [14], the shock factor (δ) is calculated as follows

$$\delta = max(Dist) * (0.5 - rand()) \tag{8}$$

where $rand()$ represent a random value [0, 1].

Since the range value of cluster centers is [0, 1], if the shock factor (δ) is calculated by formula (7), δ is then negative. In FOCA, the δ is computed as shown in Eq. (9).

$$\delta = \varepsilon * value * (0.5 - rand())$$
$$value = min(S) \tag{9}$$

where $value$ represents the minimum value of vector S, and ε represents the scale factor. If ε is too large, the change of smell concentration is too large, and ε will not contribute in the next generation. The formula of smell concentration $S(i)$ of our algorithm is defined as

$$S(i) = (1/Dist(i)) + \delta$$
$$Dist(i) = \sqrt{X_i^2 + Y_i^2 + Z_i^2} \tag{10}$$

3.2.4 Fitness Function

Let $R = \{R_i, i = 1, 2, \ldots, M\}$ represent a dataset, where M is the size of dataset R and R_i represents a sample of dataset R. In order to maximize the intra-cluster similarity and minimize the inter-cluster similarity, the fitness function is defined as

$$Fitness\ Funtion = \frac{InterCS}{1 + \alpha * IntraCS/InterCS} \tag{11}$$

where $InterCS$ and $IntraCS$ are defined as

$$InterCS = \sum_{j=1}^{k} \sum_{R_i \in \pi_j} d(R_i, \overline{R^{(\pi_j)}})$$
$$IntraCS = \sum_{i=1}^{k} \sum_{j=1}^{k} d(\overline{R^{(\pi_i)}}, \overline{R^{(\pi_j)}}) \tag{12}$$

where $InterCS$ and $IntraCS$ represent the sum of distance of inter-cluster and that of intra-cluster respectively. Symbol k is the number of cluster centers. $\overline{R^{(\pi_j)}}$ represents the j-th of cluster centers. $d(R_i, \overline{R^{(\pi_j)}})$ denotes the Euclidean distance of the sample corresponding to the cluster center. $d(\overline{R^{(\pi_i)}}, \overline{R^{(\pi_j)}})$ represents the Euclidean distance between cluster center i and cluster center j. α is a scale factor determining importance of J_2 on clustering results, where α is usually set to 1.

A smaller Fitness Function value results into better clustering results. The calculation process of Fitness Function is compared of several steps, as described below:

1. Use the value of Smell as the initial cluster centers. Then determine the partition of the dataset according to the nearest neighbor rule.
2. According to the division of the dataset, recalculate the cluster centers, then calculate the value of Fitness Function by formula (9) and formula (10).

```
Algorithm 1. FOCA
Input: {dataset, k(the number of clusters)}
Output: {C(clustering labels)}
Data normalization;
Initialize X_axis, Y_axis randomly;
Begin loop until the termination condition is sat-
isfied;
    Update the location of each fruit fly X_i,Y_i,Z_i
according o the formula (7);
    Calculate Dist(i), S(i) according to the for-
mula (10);
    Calculate Function value according to the for-
mula (11) and formula (12);
    If Function value is optimal;
        Save the best Function value and Cluster-
ing labels;
        Update the optimal location and fruit fly
swarm fly towards that location;
End Loop.
```

4   Experimental Study

In this section, the performance of FOCA is evaluated and also be compared with a number of existing state-of-the-art algorithms, including K-means [15], genetic algorithm (GA) [16], particle swarm optimization algorithm (PSO) [17], ant colony algorithm (ACO) [18], and artificial bee colony approach for clustering (ABC) [6]. 10 datasets are given to experiment, *F-measure* [19] (one of external evaluation) and *Dunn's index* [20] (one of relative evaluation) are used to evaluate the clustering results. Generally speaking, the larger the *F-measure* and *Dunn's index* are, the better the clustering performance is. All algorithms were run on Window OS computer with core I5, 2.5GHZ, and 8 GB ROM and Matlab2012 installed.

4.1   Datasets

10 datasets are used for this experiment, which are summarized in Table 1. All datasets are from UCI Data Repository, where Cmc and Heart represent Contraceptive-method-choice and Heart-statlog respectively.

Table 1. Related information of experimental datasets

Dataset	Instances	Features	Classes
Iris	150	4	3
Zoo	101	16	7
Glass	214	9	6
Sonar	208	60	2
Wine	178	14	3
Cmc	1473	9	3
Diabetes	768	8	2
Segment	2310	19	7
Vehicle	846	18	4
Heart	270	13	2

4.2 Results

In order to get more illustrative results, the datasets are normalized before the experiment. The number of population is set to 100. The experiment is conducted 20 times repeatedly with the same conditions. K-means is not compared with others in contrast to convergence time. The 6 clustering algorithms are compared in terms of the average value and the standard deviation. The average value reflects the clustering performance in general and the standard deviation reflects the degree of stability. Tables 2, 3 and 4 show all the results regarding *F-measure, Dunn's index* and convergence time respectively (the first number of column represents the average value and the last number of column is the standard deviation, respectively).

It is noted that the FOCA has 6 highest *F-measure* values on 10 datasets in Table 2 while 4 highest *Dunn's index* values in Table 3. Thus FOCA has a better performance in general among 6 algorithms in terms of the accuracy. Table 4 shows that FOCA has 7 lowest convergence time (seconds) in the 10 datasets. It demonstrates that the FOCA has a better convergence. It is also noted that the standard deviation of FOCA is smaller, which indicates the higher stability of FOCA.

Table 2. The average *F-measure* values of 6 algorithms on 10 datasets

K-means	ACO	GA	PSO	ABC	FOCA
0.8461±0.0826	0.8711±0.0371	0.8620±0.0109	0.8853±0.0001	0.8801±0.0061	**0.8905±0.0051**
0.7735±0.0747	0.8077±0.0286	0.7238±0.0811	0.7770±0.0414	0.7987±0.0655	**0.8461±0.0601**
0.4947±0.0461	0.4599±0.0355	0.4233±0.0314	0.4782±0.0176	0.4553±0.0154	**0.5033±0.0104**
0.5600±0.0066	0.5333±0.0079	0.5487±0.0011	0.5423±0.0000	0.5257±0.0021	0.5413±0.0045
0.9524±0.0031	0.9074±0.0321	0.8868±0.0122	0.9661±0.0000	0.8627±0.0219	**0.9683±0.0029**
0.4351±0.0125	0.4288±0.0224	0.3972±0.0070	0.4308±0.0105	0.4013±0.0175	**0.4384±0.0128**
0.6656±0.0000	0.6423±0.0053	0.6195±0.0089	0.6692±0.0010	**0.6721±0.0009**	0.6706±0.0011
0.6541±0.0372	0.5534±0.1258	0.5103±0.1035	0.6471±0.0432	0.6020±0.0701	**0.6968±0.0188**
0.4269±0.0052	0.4122±0.0044	0.3782±0.0127	**0.4369±0.0014**	0.4051±0.0071	0.4330±0.0051
0.7359±0.1027	0.7707±0.0215	0.7880±0.0056	**0.7895±0.0010**	0.7554±0.0102	0.7883±0.0019
1	0	0	2	1	6
2	0	6	0	2	0

Table 3. The average *Dunn's* values of 6 algorithms on 10 datasets

K-means	ACO	GA	PSO	ABC	FOCA
2.0019±0.4327	2.1354±0.0071	1.8558±0.1226	2.2072±0.0000	2.1170±0.0042	**2.2072±0.0000**
0.9144±0.0753	1.0778±0.0987	1.0331±0.1300	0.9675±0.0990	0.9977±0.0054	**1.1388±0.1575**
0.6765±0.1273	0.7114±0.0255	0.5211±0.0147	0.7091±0.1464	0.6588±0.1022	**0.8875±0.0378**
0.9097±0.0092	0.9305±0.0201	0.9101±0.0054	**0.9169±0.0033**	0.9003±0.0008	0.9165±0.0004
1.3574±0.0013	1.2886±0.0044	1.1407±0.0023	1.3356±0.0000	1.2071±0.0052	1.3450±0.0195
1.0340±0.0340	1.0489±0.0073	1.0644±0.0251	1.0509±0.0032	**1.1459±0.1548**	1.1275±0.1492
1.1536±0.0048	**1.2022±0.0064**	1.0031±0.0255	1.1191±0.0103	1.1885±0.0122	1.1160±0.0074
1.1757±0.1671	1.0887±0.0443	1.0016±0.1753	**1.2596±0.0221**	1.0007±0.0543	1.1991±0.1036
1.1363±0.0589	0.9794±0.0127	0.9100±0.0180	1.0885±0.0038	0.9878±0.0020	1.0807±0.0130
0.9902±0.0686	1.0101±0.0551	1.0045±0.0113	1.0396±0.0087	1.0001±0.0090	**1.0404±0.0022**
2	1	0	3	1	4
3	0	5	0	2	0

Table 4. The average time of convergence (seconds)

Dataset	ACO	GA	PSO	ABC	FOCA
iris	43.4210±5.0071	105.3132±18.0121	30.3367±3.6523	29.1301±0.5131	**16.3697±2.3008**
zoo	80.8250±15.5547	122.7850±8.8870	32.2312±5.8099	27.3350±3.7018	**23.4549±3.9152**
glass	93.1147±10.0205	301.9879±50.1107	55.0254±10.4810	66.6440±9.9987	**46.1081±6.3859**
sonar	75.4433±5.3343	158.3100±20.1704	32.7342±3.3548	**28.9871±2.5563**	32.0125±4.4453
wine	68.5734±12.5001	217.2511±35.2101	**21.7001±1.2936**	38.8525±3.3171	23.2001±3.3529
cmc	301.8447±18.2312	588.7714±55.5512	182.1221±15.7000	225.9943±22.2005	**141.1114±11.6677**
diabetes	211.3389±17.5690	452.9903±34.1245	89.2679±10.0928	87.5604±20.2000	**63.0125±15.6675**
segment	1187.4400±50.6971	1714.2239±197.5509	801.4751±70.2438	859.6541±97.3321	**655.4333±37.5395**
vehicle	288.7740±19.0007	500.0034±61.1177	153.3001±13.2687	201.2145±24.3343	**103.5567±15.8625**
Heart	56.0012±7.4411	89.4521±10.0244	19.5321±5.4962	**18.5573±4.0014**	25.7783±3.5018
Best	0	0	1	2	7
Worst	0	10	0	0	0

5 Conclusion

In his paper, a novel clustering algorithm called FOCA. Since the Fruit Fly Optimization Algorithm was proposed in 2011 [13], it's the first time to be used in clustering analysis. FOCA has advantages such as simple computational process, less parameters, high precision, and ease of understanding. The experimental results demonstrate that FOCA has advantages, such as high accuracy and short convergence time than other clustering algorithm based on swarm intelligence optimization algorithm.

In the future, we will extend this algorithm with semi-supervised model to improve its performance.

Acknowledgements. This work is supported by the National Science Foundation of China (Nos. 61170111, 61134002 and 61401374) and the Fundamental Research Funds for the Central Universities (No. 2682014RC23).

References

1. Han, J., Kamber, M.: Data Mining: Concepts and Techniques, 2nd edn. Morgan Kaufmann, Los Altos (2006)
2. Deneubourg, J.L., Goss,S., Franks, N., et al.: The dynamics of collective sorting: robot-like ant and ant-like robots. In: The First Conference on Simulation of Adaptive Behavior: From Animals to Animals, pp. 356–365 (1991)
3. Omran, M., Salman, A., Engelbrecht, A.P.: Image classification using particle swarm optimization. In: The 4th Asia-Pacific Conference on Simulated Evolution and Learning, pp. 370–374 (2002)
4. Tsang, W.W., Lau, H.Y.: Clustering-based multi-objective immune optimization evolutionary algorithm. In: Coello Coello, C.A., Greensmith, J., Krasnogor, N., Liò, P., Nicosia, G., Pavone, M. (eds.) ICARIS 2012. LNCS, vol. 7597, pp. 72–85. Springer, Heidelberg (2012)
5. Amiri, B., Fathian, M., Maroosi, A.: Application of shuffled frog-leaping algorithm on clustering. Adv. Manuf. Technol. **45**(2), 199–209 (2009)
6. Zhang, C.S., Ning, J.X.: An artificial bee colony approach for clustering. Expert Syst. Appl. **37**(7), 4761–4767 (2010)
7. Yazdani, D., Saman, B., Sepas, A., et al.: A new algorithm based on improved artificial fish swarm algorithm for data clustering. Artif. Intell. **13**(11), 170–192 (2013)
8. Yang, J.G., Zhuang, Y.B.: An improved ant colony optimization algorithm for solving a complex combinatorial optimization problem. Appl. Soft Comput. **10**(2), 653–660 (2010)
9. Li, Z.H., Zhang, Y.N., Tan, H.Z.: IA-AIS: an improved adaptive artificial immune system applied to complex optimization problem. Appl. Soft Comput. **11**(8), 4692–4700 (2011)
10. Wang, L., Fang, C.: An effective shuffled frog-leaping algorithm for multi-mode resource-constrained project scheduling problem. Inf. Sci. **181**(20), 4804–4822 (2011)
11. Akray, B., Karabog, D.: A modified artificial bee colony algorithm for real-parameter optimization. Inf. Sci. **192**(1), 120–142 (2012)
12. Tsai, H.C., Lin, Y.H.: Modification of the fish swarm algorithm with particle swarm optimization formulation and communication behavior. Appl. Soft Comput. **11**(8), 5367–5374 (2011)
13. Pan, W.C.: A new fruit fly optimization algorithm: taking the financial distress model as an example. Knowl. Based Syst. **26**, 69–74 (2012)
14. Li, C., Xu, S., Li, W., et al.: A novel modified fly optimization algorithm for designing the self-tuning proportional integral derivative controller. J. Convergence Inf. Technol. **16**(7), 69–77 (2012)
15. Xie, J.Y., Jiang, S., Xie, W.X., et al.: An efficient global k-means clustering algorithm. J. Comput. **6**(2), 271–279 (2011)
16. Mualik, U., Bandyopadhyay, S.: Genetic algorithm based clustering technique. Pattern Recogn. **33**(9), 1455–1465 (2000)
17. Omran, M., Salman, A., Engelbrecht, A.P.: Particle swarm optimization method for image clustering. Pattern Recog. Artif. Intell. **19**(3), 297–321 (2005)
18. Shelokar, P.S., Jayaraman, V.K., Kulkarni, B.D.: An ant colony approach for clustering. Anal. Chim. Acta **509**(2), 187–195 (2004)
19. Yang, Y., Kamel, M.S.: An aggregated clustering approach using multi-ant colonies algorithms. Pattern Recogn. **39**(7), 1278–1289 (2006)
20. Maulik, U., Bandyopadhyay, S.: Performance evaluation of some clustering algorithm and validity indices. Pattern Anal. Mach. Intell. **24**(12), 1650–1654 (2002)

Rough Sets and Graphs

Rough Set Theory Applied to Simple Undirected Graphs

Giampiero Chiaselotti[2], Davide Ciucci[1]([⊠]), Tommaso Gentile[2], and Federico Infusino[2]

[1] DISCo, University of Milano – Bicocca, Viale Sarca 336/14, 20126 Milano, Italy
ciucci@disco.unimib.it
[2] Department of Mathematics and Informatics, University of Calabria,
Via Pietro Bucci, Cubo 30B, 87036 Arcavacata di Rende, CS, Italy
giampiero.chiaselotti@unical.it, {gentile,f.infusino}@mat.unical.it

Abstract. The incidence matrix of a simple undirected graph is used as an information table. Then, rough set notions are applied to it: approximations, membership function, positive region and discernibility matrix. The particular cases of complete and bipartite graphs are analyzed. The symmetry induced in graphs by the indiscernibility relation is studied and a new concept of generalized discernibility matrix is introduced.

Keywords: Undirected graphs · Neighborhood · Discernibility matrix · Complete graphs · Bipartite graphs · Symmetry

1 Introduction

There are several possibilities to mix graph theory with rough set theory. First of all, we can prove the formal equivalence between a class of graphs and some concept related to rough sets [2,10], and then use the techniques of one theory to study the other. On the other hand, we can study rough sets using graphs [9] or vice versa study graphs using rough set techniques [1,4,6]. Here, we are following this last line. Our approach is very general and intuitive. Indeed, we consider the incidence matrix of a simple undirected graph as a Boolean Information Table and apply to it standard rough set techniques. We will see that the indiscernibility relation introduces a new kind of symmetry in graphs: roughly speaking, two vertices are similar if they behave in the same way with respect to all other vertices. In particular, the cases of complete and complete bipartite graphs are fully developed, by considering the corresponding of approximations, membership function, positive region and discernibility matrix on them. Some results on n-cycle and n-path graphs will also be given and a generalized notion of discernibility matrix introduced in order to characterize a new symmetry on graphs.

2 Basic Notions

The basic notions on rough set theory and graphs are given.

© Springer International Publishing Switzerland 2015
D. Ciucci et al. (Eds.): RSKT 2015, LNAI 9436, pp. 423–434, 2015.
DOI: 10.1007/978-3-319-25754-9_37

2.1   Rough Set Notation

A *partition* π on a finite set X is a collection of non-empty subsets $B_1, \ldots, B_M$ of X such that $B_i \cap B_j = \emptyset$ for all $i \neq j$ and such that $\bigcup_{i=1}^{M} B_i = X$. The subsets $B_1, \ldots, B_M$ are called *blocks* of π and we write $\pi := B_1 | \ldots | B_M$ to denote that π is a set partition having blocks $B_1, \ldots, B_M$. If $Y \subseteq B_i$, for some index i, we say that Y is a *sub-block* of π and we write $Y \preccurlyeq \pi$. If $x \in X$, we denote by $\pi(x)$ the (unique) block of π which contains the element x.

We assume the reader familiar with rough set theory, hence here we just fix the basic notations and refer to [7] for further details. An *information table* is denoted as $\mathcal{I} = \langle U, Att, Val, F \rangle$, where U are the *objects*, Att is the *attribute set*, Val is a non empty set of *values* and $F : U \times Att \to Val$ is the *information map*, mapping (object,attribute) pairs to values. If $Val = \{0, 1\}$ the table is said *Boolean*. The *indiscernibility* relation $\equiv_A$ is an equivalence relation between objects and depending on a set of attributes A: $u \equiv_A u'$ if $\forall a \in A$, $F(u, a) = F(u', a)$. Equivalence classes with respect to $\equiv_A$ are denoted as $[u]_A$ and the *indiscernibility partition* of $\mathcal{I}$ is the collection of all equivalence classes: $\pi_{\mathcal{I}}(A) := \{[u]_A : u \in U\}$. Given a set of objects Y and a set of attributes A, the lower and upper approximations of Y with respect to the knowledge given by A are:

$$\mathbf{l}_A(Y) := \{x \in U : [x]_A \subseteq Y\} = \bigcup\{C \in \pi_{\mathcal{I}}(A) : C \subseteq Y\} \tag{1}$$

$$\mathbf{u}_A(Y) := \{x \in U : [x]_A \cap Y \neq \emptyset\} = \bigcup\{C \in \pi_{\mathcal{I}}(A) : C \cap Y \neq \emptyset\}. \tag{2}$$

The subset Y is called A-exact if and only if $\mathbf{l}_A(Y) = \mathbf{u}_A(Y)$ and A-rough otherwise. Other notions that will be useful in the following are:

- The *rough membership function* $\mu_Y^A : U \to [0, 1]$: $\mu_Y^A(u) := \frac{|[u]_A \cap Y|}{|[u]_A|}$.
- If $B \subseteq Att$ is such that $\pi_{\mathcal{I}}(B) = \{Q_1, \ldots, Q_N\}$, the *positive region* is

$$POS_A(B) := \bigcup_{i=1}^{N} \mathbf{l}_A(Q_i) = \{u \in U : [u]_A \subseteq [u]_B\} \tag{3}$$

- The A-*degree dependency* of B is the number $\gamma_A(B) := \frac{|POS_A(B)|}{|U|}$

Let us notice that, since $Q_1, \ldots, Q_N$ are pairwise-disjoint, we have $|POS_A(B)| = \sum_{i=1}^{N} |\mathbf{l}_A(Q_i)|$.

2.2   Graphs

We denote by $G = (V(G), E(G))$ a finite simple (i.e., no loops and no multiple edges are allowed) undirected graph, with vertex set $V(G) = \{v_1, \ldots, v_n\}$ and edge set $E(G)$. If $v, v' \in V(G)$, we will write $v \sim v'$ if $\{v, v'\} \in E(G)$ and $v \nsim v'$ otherwise. We denote by $Adj(G)$ the adjacency matrix of G, defined as the $n \times n$ matrix (a_{ij}) such that $a_{ij} := 1$ if $v_i \sim v_j$ and $a_{ij} := 0$ otherwise. If $v \in V(G)$, we set

$$N_G(v) := \{w \in V(G) : \{v, w\} \in E(G)\} \tag{4}$$

$N_G(v)$ is usually called *neighborhood* of v in G. If $A \subseteq V(G)$ we set $N_G(A) := \bigcup_{v \in A} N_G(v)$.

Graphs of particular interest for our discussion will be complete and bipartite ones. If A and B are two vertex subsets of G we denote by $A \triangle B$ the symmetric difference between A and B in $V(G)$, that is $A \triangle B := (A \setminus B) \cup (B \setminus A)$.

Definition 2.1. *The* complete graph *on n vertices, denoted by K_n, is the graph with vertex set $\{v_1, \ldots, v_n\}$ and such that $\{v_i, v_j\}$ is an edge, for each pair of indexes $i \neq j$.*

Definition 2.2. *A graph $B = (V(B), E(B))$ is said* bipartite *if there exist two non-empty subsets B_1 and B_2 of $V(B)$ such that $B_1 \cap B_2 = \emptyset$, $B_1 \cup B_2 = V(B)$ and $E(B) \subseteq O(B) := \{\{x, y\} : x \in B_1, y \in B_2\}$. In this case the pair (B_1, B_2) is called a* bipartition *of B and we write $B = (B_1|B_2)$. It is said that $B = (B_1|B_2)$ is a* complete bipartite graph *if $E(B) = O(B)$. If $|B_1| = p$ and $|B_2| = q$ we denote by $K_{p,q}$ the complete bipartite graph having bipartition (B_1, B_2).*

Definition 2.3. *Let n be a positive integer. The n-cycle C_n is the graph having n vertices $v_1, \ldots, v_n$ and such that $E(C_n) = \{\{v_1, v_2\}, \{v_2, v_3\}, \ldots, \{v_{n-1}, v_n\}, \{v_n, v_1\}\}$. The n-path P_n is the graph having n vertices $v_1, \ldots, v_n$ and such that $E(P_n) = \{\{v_1, v_2\}, \{v_2, v_3\}, \ldots, \{v_{n-1}, v_n\}\}$.*

We will denote by I_n the $n \times n$ identity matrix and by J_n the $n \times n$ matrix having 1 in all its entries.

3   The Adjacency Matrix as an Information Table

We interpret now a simple undirected graph G as a Boolean information table denoted by $\mathcal{I}[G]$. Let $V(G) = V = \{v_1, \ldots, v_n\}$. We assume that the universe set and the attribute set of $\mathcal{I}[G]$ are both the vertex set V and we define the information map of $\mathcal{I}[G]$ as follows: $F(v_i, v_j) := 1$ if $v_i \sim v_j$ and $F(v_i, v_j) := 0$ if $v_i \nsim v_j$. If $A \subseteq V(G)$, we will write $\pi_G(A)$ instead of $\pi_{\mathcal{I}[G]}(A)$. The equivalence relation $\equiv_A$ is in relation with the notion of neighborhood as follows.

Theorem 3.1. *[4]. Let $A \subseteq V(G)$ and $v, v' \in V(G)$. The following conditions are equivalent:*

(i) $v \equiv_A v'$.
(ii) For all $z \in A$ it results that $v \sim z$ if and only if $v' \sim z$.
(iii) $N_G(v) \cap A = N_G(v') \cap A$.

The previous theorem provides us a precise geometric meaning for the indiscernibility relation $\equiv_A$: we can consider the equivalence relation $\equiv_A$ as a type of symmetry relation with respect to the vertex subset A. In fact, by part *(ii)* of Theorem 3.1 we can see that two vertices v and v' are A-indiscernible between them iff they have the same incidence relation with respect to all vertices $z \in A$. We say therefore that v and v' are A-*symmetric vertices* if $v \equiv_A v'$. Hence the A-indiscernibility relation in $\mathcal{I}[G]$ becomes an A-*symmetry relation* in $V(G)$ and the A-granule $[v]_A$ becomes the A-*symmetry block* of v.

Example 3.1. Consider the graph in the next picture:

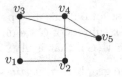

We have $N_G(v_1) = \{v_2, v_3\}$, $N_G(v_2) = \{v_1, v_4\}$, $N_G(v_3) = \{v_1, v_4, v_5\}$, $N_G(v_4) = \{v_2, v_3, v_5\}$ and $N_G(v_5) = \{v_3, v_4\}$. If we fix the vertex subset $A \doteq \{v_1, v_5\}$, we have $v_1 \equiv_A v_5$, so $\pi_G(A) = v_1 v_5 | v_2 | v_3 | v_4$.

As a direct consequence of Theorem 3.1 we can provide the general form for the indiscernibility partitions when $G = K_n$ and $G = K_{p,q}$.

Proposition 3.1. *[4] (i) Let $n \geq 1$ and let $A = \{v_{i_1}, \dots, v_{i_k}\}$ be a generic subset of $V(K_n) = \{v_1, \dots, v_n\}$. Then $\pi_{K_n}(A) = v_{i_1} | v_{i_2} | \dots | v_{i_k} | A^c$.*
(ii) Let p and q be two positive integers. Let $K_{p,q} = (B_1 | B_2)$. Then $\pi_{K_{p,q}}(A) = B_1 | B_2$ for each subset $A \subseteq V(K_{p,q})$ such that $A \neq \emptyset$.

In the next results we deal with the indiscernibility partitions of C_n and P_n.

Lemma 3.1. *Let $G = C_n$ or $G = P_n$ and let $V = V(G) = \{v_1, \dots, v_n\}$. Fix a subset $A \subseteq V$, $A = \{v_{i_1}, \dots, v_{i_k}\}$, and let $v_i, v_j \in V$, with $i < j$. Then $v_i \equiv_A v_j$ if and only if $N_G(v_i) \cap A = N_G(v_j) \cap A = \emptyset$ or $N_G(v_i) \cap A = N_G(v_j) \cap A = \{v_{i+1}\} = \{v_{j-1}\}$.*

Proof. At first, we observe that for each $i \in \{1, \dots, n\}$, $N_{C_n}(v_i) = \{v_{i-1}, v_{i+1}\}$, where the index sums are all taken $\mathrm{mod}(n)$, and

$$N_{P_n}(v_i) = \begin{cases} \{v_2\} & \text{if } i = 1 \\ \{v_{n-1}\} & \text{if } i = n \\ \{v_{i-1}, v_{i+1}\} & \text{otherwise.} \end{cases}$$

The proof follows easily by observing that, since $v_i \neq v_j$, then $|N_G(v_i) \cap N_G(v_j)| \leq 1$ and the equality holds if and only if $j = i + 2$. It follows that, if $N_G(v_i) \cap A = N_G(v_j) \cap A$, then $N_G(v_i) \cap A = (N_G(v_i) \cap N_G(v_j)) \cap A \subseteq N_G(v_i) \cap N_G(v_j)$. By Theorem 3.1, $v_i \equiv_A v_j$ if and only if $N_G(v_i) \cap A = N_G(v_j) \cap A$. Thus $|N_G(v_i) \cap A| = |N_G(v_j) \cap A| \leq 1$ and the equality holds if and only if $j = i + 2$ and $v_{i+1} = v_{j-1} \in A$. This proves the thesis. $\square$

We give now a complete description of the indiscernibility partition for both the two graphs C_n and P_n.

Proposition 3.2. *Let G and A as in Lemma 3.1. We set $B_G(A) := (N_G(A))^c$ and*

$$C_G(A) := \begin{cases} \{v_i \in A : v_{i-2} \notin A \wedge v_{i+2} \notin A\} & \text{if } G = C_n, \\ \{v_i \in A : 3 \leq i \leq n - 2 \wedge v_{i-2} \notin A \wedge v_{i+2} \notin A\} & \text{if } G = P_n. \end{cases}$$

Then, if $v_{s_1}, \ldots, v_{s_l}$ are the vertices in $V(G) \setminus [B_G(A) \cup N_G(C_G(A))]$ and $v_{j_1}, \ldots,$
v_{j_h} are the vertices in $C_G(A)$, we have

$$\pi_G(A) = B_G(A)|v_{j_1-1}v_{j_1+1}| \cdots |v_{j_h-1}v_{j_h+1}|v_{s_1}| \cdots |v_{s_l}|.$$

Proof. The proof follows directly by Lemma 3.1. In fact, let v_i, $v_j \in V(G)$, with
$i < j$ and $v_i \equiv_A v_j$. By the previous lemma, then either (a) $N_G(v_i) \cap A =$
$N_G(v_j) \cap A = \emptyset$ or (b) $N_G(v_i) \cap A = N_G(v_j) \cap A = v_{i+1} = v_{j-1}$. But (a) is
equivalent to say that v_i, $v_j \in B_A$, (b) that $\{v_i, v_j\} = N_G(v)$, for some $v \in C_A$.
The proposition is thus proved. □

Example 3.2. Let $n = 7$ and let $G = C_7$ or $G = P_7$ be, respectively, the 7-cycle
and the 7-path on the set $V = \{v_1, \ldots, v_7\}$. Let $A = \{v_3, v_4, v_7\}$. Then,

$$B_G(A) = \begin{cases} \{v_7\} & \text{if } G = C_7, \\ \{v_1, v_7\} & \text{if } G = P_7, \end{cases} \qquad C_G(A) = \begin{cases} \{v_3, v_4, v_7\} & \text{if } G = C_7, \\ \{v_3, v_4\} & \text{if } G = P_7, \end{cases}$$

and

$$V(G) \setminus [B_G(A) \cup N_G(C_G(A))] = \begin{cases} \emptyset & \text{if } G = C_7, \\ \{v_6\} & \text{if } G = P_7. \end{cases}$$

Thus,

$$\pi_G(A) = \begin{cases} v_1v_6|v_2v_4|v_3v_5|v_7 & \text{if } G = C_7, \\ v_1v_7|v_2v_4|v_3v_5|v_6 & \text{if } G = P_7. \end{cases}$$

4 Rough Approximations and Dependency of Graphs

In this section, we apply lower and upper approximations, rough membership,
positive region and degree of dependency to simple graphs. The case of complete
and bipartite complete graphs are fully developed.

4.1 Lower and Upper Approximations

For the A-lower and A-upper approximation functions we obtain the following
geometrical interpretation in the simple graph context.

Proposition 4.1. *[4] Let $G = (V(G), E(G))$ be a simple undirected graph and
let $\mathcal{I}[G]$ be the Boolean information system associated to G. Let A and Y be two
subsets of $V(G)$. Then:*

(i) $\mathbf{l}_A(Y) = \{v \in V(G) : (u \in V(G) \wedge N_G(u) \cap A = N_G(v) \cap A) \Longrightarrow u \in Y\}$.
(ii) $\mathbf{u}_A(Y) = \{v \in V(G) : \exists u \in Y : N_G(u) \cap A = N_G(v) \cap A\}$.

Therefore, $v \in \mathbf{l}_A(Y)$ iff all A-symmetric vertices of v are in Y. That is, the
lower approximation of a vertex set Y represents a subset of Y such that there
are no elements outside Y with the same connections of any vertex in $\mathbf{l}_A(Y)$
(relatively to A). Similarly, for the upper approximation, $v \in \mathbf{u}_A(Y)$ iff v is an
A-symmetric vertex of some $u \in Y$. That is, the upper approximation of Y is
the set of vertices with the same connections (w.r.t. A) of at least one element
in Y. So, it is natural to call $\mathbf{l}_A(Y)$ the A-*symmetry kernel* of Y and $\mathbf{u}_A(Y)$ the
A-*symmetry closure* of Y. In [4] it has been given a complete description for both
the A-lower and the A-upper approximations when $G = K_n$ and $G = K_{p,q}$.

Example 4.1. Let G be the graph in the Example 3.1. If $A = \{v_1, v_5\}$ and $Y = \{v_1, v_4\}$, it is immediate to verify that $l_A(Y) = \{v_4\}$ and $\mathbf{u}_A(Y) = \{v_1, v_4, v_5\}$.

4.2   Rough Membership and Dependency

For the rough-membership function and the rough-positive region we obtain the following geometrical interpretation.

Proposition 4.2. *Let* $G = (V(G), E(G))$ *be a simple undirected graph and let* $\mathcal{I}[G]$ *be the Boolean information system associated to G. Let A and Y be two subsets of $V(G)$. Then the rough-membership function is*

$$\mu_Y^A(v) = \frac{|\{v' \in Y : N_G(v) \cap A = N_G(v') \cap A\}|}{|\{v' \in V(G) : N_G(v) \cap A = N_G(v') \cap A\}|}.$$

If $B \subseteq V(G)$ *we have* $POS_A(B) = \{v \in V(G) : (v' \in V(G) \wedge N_G(v) \cap A = N_G(v') \cap A) \Longrightarrow (N_G(v) \cap B = N_G(v') \cap B)\}$.

Proof. It follows directly by (iii) of Theorem 3.1 and from the definitions of rough membership and positive region.   □

Example 4.2. Let G be the graph in the Example 3.1. If $A = \{v_1, v_5\}$ and $Y = \{v_1, v_4\}$, the corresponding rough membership function is

$$\mu_Y^A(v) = \begin{cases} 0 & \text{if } v = v_2 \vee v = v_3 \\ 1 & \text{if } v = v_4 \\ 1/2 & \text{if } v = v_1 \vee v = v_5 \end{cases}$$

Moreover, if $B = \{v_2, v_4\}$ then $\pi_G(B) = v_1 v_4 | v_2 v_3 v_5$, therefore $POS_A(B) = \{v_2, v_3, v_4\}$.

We consider now the complete graph $G = K_n$ and compute the rough membership function, the A-positive region of B (when A and B are any two vertex subsets) and the degree dependency function.

Proposition 4.3. *Let* $G = K_n$ *be the complete graph on n vertices and let A and Y be two subsets of $V(G)$. Then*

$$\mu_Y^A(v) = \begin{cases} 0 & \text{if } v \in A \setminus Y \\ 1 & \text{if } v \in A \cap Y \\ |A^c \cap Y|/|A^c| & \text{if } v \notin A \end{cases} \tag{5}$$

$$POS_A(B) = \begin{cases} A & \text{if } |A| < n - 1 \wedge B \nsubseteq A \\ V(G) & \text{otherwise} \end{cases} \tag{6}$$

$$\gamma_A(B) = \begin{cases} |A|/n & \text{if } |A| < n - 1 \wedge B \nsubseteq A \\ 1 & \text{otherwise} \end{cases} \tag{7}$$

Proof. 1. By (i) of Proposition 3.1 we deduce that

$$[v]_A \cap Y = \begin{cases} \emptyset & \text{if } v \in A \wedge v \notin Y \\ \{v\} & \text{if } v \in A \wedge v \in Y \\ A^c \cap Y & \text{if } v \notin A \end{cases}$$

Hence the thesis follows again by (i) of Proposition 3.1 and from the definition of rough membership function.

2. Let $\pi_{\mathcal{I}}(B) = Q_1|\ldots|Q_N$. We suppose at first that $|A| \geq n - 1$, then A^c is either empty or a singleton, therefore by (i) of Proposition 3.1 we have that $\pi_{K_n}(A) = v_1|\ldots|v_n$. This implies that $1_A(Q_i) = Q_i$ for $i = 1, \ldots, N$. Since $Q_1|\ldots|Q_N$ is a partition of $V(G)$, by (3) it follows that $POS_A(B) = V(G)$. We assume now that $|A| \leq n - 2$ and $B \subseteq A$. By (3), in order to obtain the first part of our thesis we must show that

$$\{v \in V(G) : [v]_A \subseteq [v]_B\} = V(G). \tag{8}$$

If $v \in A$, by (i) of Proposition 3.1 we have that $[v]_A = \{v\}$, therefore $[v]_A \subseteq [v]_B$. If $v \in A^c$, again by (i) of Proposition 3.1 we have that $[v]_A = A^c$. On the other hand, since $B \subseteq A$, then $v \in A^c \subseteq B^c$, so that, again by (i) of Proposition 3.1, it results that $[v]_B = B^c$. Hence also in this case $[v]_A \subseteq [v]_B$. This proves (8). We suppose now that $B \not\subseteq A$ and $|A| \leq n - 2$. In order to show the second identity for $POS_A(B)$, we prove that

$$\{v \in V(G) : [v]_A \subseteq [v]_B\} = A. \tag{9}$$

If $v \in A$, as previously $[v]_A = \{v\} \subseteq [v]_B$, therefore $A \subseteq \{v \in V(G) : [v]_A \subseteq [v]_B\}$. Let $v \in A^c$. As before we have $[v]_A = A^c$. Since $|A| < n - 1$, we have $|[v]_A| = |A^c| > 1$. We examine now $[v]_B$. By (i) of Proposition 3.1 we have that $[v]_B = \{v\}$ or $[v]_B = B^c$. In the first case, $[v]_A \not\subseteq [v]_B$ because $|[v]_A| > 1$. In the second case, $[v]_A = A^c \not\subseteq [v]_B = B^c$ because otherwise $B \subseteq A$, that is contrary to our assumption. This proves that $v \in A^c$ implies $[v]_A \not\subseteq [v]_B$, that is $A \supseteq \{v \in V(G) : [v]_A \subseteq [v]_B\}$. Hence we obtain (9).

3. The universe set for the information system $\mathcal{I}[G]$ is $U := V(G)$, therefore $|U| = n$ in our case. The result follows then directly by definition and the previous point. □

Finally, we analyze the case of bipartite graphs $G = K_{p,q}$.

Proposition 4.4. *Let $G = K_{p,q}$ be the complete bipartite graph on $p+q$ vertices with bipartition $(B_1|B_2)$. If A and B are two subsets of $V(G)$ we have that*

$$\mu_Y^A(v) = \begin{cases} |B_1 \cap Y|/|B_1| & \text{if } v \in B_1 \\ |B_2 \cap Y|/|B_2| & \text{if } v \in B_2 \end{cases} \tag{10}$$

$$POS_A(B) = \begin{cases} \emptyset & \text{if } A = \emptyset \wedge B \neq \emptyset \\ V & \text{otherwise} \end{cases} \tag{11}$$

$$\gamma_A(B) = \begin{cases} 0 & \text{if } A = \emptyset \wedge B \neq \emptyset \\ 1 & \text{otherwise} \end{cases} \tag{12}$$

Proof. 1. By (ii) of Proposition 3.1 we know that $[v]_A = B_i$ if $v \in B_i$, for $i = 1, 2$, therefore

$$[v]_A \cap Y = \begin{cases} B_1 \cap Y & \text{if } v \in B_1 \\ B_2 \cap Y & \text{if } v \in B_2 \end{cases}$$

and the thesis follows from the definition of rough membership function.

2. If $A = \emptyset$ and $B \neq \emptyset$ then $\pi_G(A) = V(G)$ and, by (ii) of Proposition 3.1, $\pi_B(G) = B_1|B_2$. Thus $[v]_A = V$ and $[v]_B = B_i$, for some $i = 1, 2$. This implies that $[v]_A \not\subseteq [v]_B$ for all $v \in V$, hence $POS_A(B) = \emptyset$. If $A = B = \emptyset$ then $\pi_G(A) = \pi_B(G) = V$, therefore $[v]_A = V \subseteq [v]_B = V$ for all $v \in V$. Hence $POS_A(B) = V$. If $A \neq \emptyset$ and $B = \emptyset$ then $\pi_G(A) = B_1|B_2$ by (ii) of Proposition 3.1, and $\pi_B(G) = V$. This implies that $[v]_A = B_i$ for some $i = 1, 2$ and $[v]_B = V$ for all $v \in V$. Hence $POS_A(B) = V$. Finally, if $A \neq \emptyset$ and also $B \neq \emptyset$, by (ii) of Proposition 3.1 we have that $\pi_G(A) = \pi_G(A) = B_1|B_2$. This implies that $[v]_A = [v]_B = B_i$ for all $v \in V$ and for some $i = 1, 2$. Hence also in this case $POS_A(B) = V$.

3. The result follows directly by definition and the above point.   □

5   Discernibility Matrix in Graph Theory

A fundamental investigation tool in rough set theory is the discernibility matrix of an information system [8]. In this section we discuss the important geometric role of the discernibility matrix in our graph theory context.

Let $\mathcal{I} = \langle U, Att, Val, F \rangle$ an information table such that $U = \{u_1, \ldots, u_m\}$ and $|Att| = n$. The discernibility matrix $\Delta[\mathcal{I}]$ of $\mathcal{I}$ is the $m \times m$ matrix having as (i, j)-entry the following attribute subset:

$$\Delta_{\mathcal{I}}(u_i, u_j) := \{a \in Att : F(u_i, a) \neq F(u_j, a)\}. \tag{13}$$

In the graph context we have the following result.

Proposition 5.1. *[3] If G is a simple undirected graph and $v_i, v_j \in V(G)$, then*

$$\Delta_G(v_i, v_j) := \Delta_{\mathcal{I}[G]}(v_i, v_j) = N_G(v_i) \, \triangle \, N_G(v_j). \tag{14}$$

In order to provide now a convenient geometrical interpretation of the identity (14), we can observe that if $v_i \neq v_j$ then $v \in \Delta_G(v_i, v_j)$ iff v is connected with exactly one vertex between v_i and v_j. This means that v is a *dissymmetry vertex* for v_i and v_j and therefore it is natural to call $\Delta_G(v_i, v_j)$ the *dissymmetry axis* of v_i and v_j and *dissymmetry number* of v_i and v_j the integer $\delta_{ij}(G) := |\Delta_G(v_i, v_j)|$. In graph terminology then the discernibility matrix $\Delta[G] := \Delta[\mathcal{I}[G]]$ becomes the *local dissymmetry matrix* of G. We also call *local numerical dissymmetry matrix* of G the numerical matrix $\Delta_{num}[G] := (\delta_{ij}(G))$. We use the term *local* because the dissymmetry is evaluated with respect to only two vertices.

Example 5.1. Let G be the graph in the Example 3.1. Then, it is easy to see that $\Delta_G(v_1, v_2) = \Delta_G(v_3, v_4) = \{v_1, v_2, v_3, v_4\}$, $\Delta_G(v_1, v_3) = \Delta_G(v_2, v_4) = V(G)$, $\Delta_G(v_1, v_4) = \Delta_G(v_2, v_3) = \{v_5\}$, $\Delta_G(v_1, v_5) = \{v_2, v_4\}$, $\Delta_G(v_2, v_5) = \{v_1, v_3\}$, $\Delta_G(v_3, v_5) = \{v_1, v_3, v_5\}$ and $\Delta_G(v_4, v_5) = \{v_2, v_4, v_5\}$.

We now introduce the following generalization of (13). If $Z \subseteq U$ we define the *generalized discernibility matrix* as

$$\Delta_{\mathcal{I}}(Z) := \{a \in Att \; : \; \exists \; z, z' \in Z : F(z, a) \neq F(z', a)\}, \tag{15}$$

and also set

$$\Delta_{\mathcal{I}}^c(Z) := Att \setminus \Delta_{\mathcal{I}}(Z) = \{a \in Att \; : \; \forall \; z, z' \in Z, F(z, a) = F(z', a)\} \tag{16}$$

It is clear then that

Proposition 5.2. *[3] $\Delta_{\mathcal{I}}^c(Z)$ is the unique attribute subset C of $\mathcal{I}$ such that :*
(i) $z \equiv_C z'$ for all $z, z' \in Z$;
(ii) if $A \subseteq Att$ and $z \equiv_A z'$ for all $z, z' \in Z$, then $A \subseteq C$.

In the graph context, the consequence of the Proposition 5.2 is that if $Z \subseteq V(G)$, then $\Delta_{\mathcal{I}[G]}^c(Z)$ is the maximum symmetry subset for Z, therefore we call $\Delta_G^c(Z)$ the *symmetry axis* of Z. Since $\Delta_{\mathcal{I}[G]}(Z)$ is the complementary subset of $\Delta_G^c(Z)$ in $V(G)$, it will be called the *dissymmetry axis* of Z. In particular, when $Z = \{v_i, v_j\}$ we re-obtain the previous definitions. It is also convenient to call *symmetry number* of v_i and v_j the integer $\delta_{ij}^c(G) := |\Delta_G^c(\{v_i, v_j\})|$. Let us note that for all i and j we have $\delta_{ij}(G) + \delta_{ij}^c(G) = |V(G)|$. By analogy with the dissymmetry case, we call *local symmetry matrix* of G the subset matrix $\Delta^c[G] := (\Delta_G^c(v_i, v_j))$ and *numerical local symmetry matrix* of G the subset matrix $\Delta_{num}^c[G] := (\delta_{ij}^c(G))$.

Example 5.2. Let G be the graph in the Example 3.1. Let $Z = \{v_2, v_3, v_5\}$, then $\Delta_{\mathcal{I}}(Z) = \{v_1, v_3, v_5\}$ and $\Delta_{\mathcal{I}}^c(Z) = \{v_2, v_4\}$.

By analogy with the case of the average degree of a graph (see [5]) it is natural to introduce both the corresponding averages of local symmetry and dissymmetry for G.

Definition 5.1. *We call* 2-local dissymmetry average *of G the average number $\delta(G)$ and, analogously,* 2-local symmetry average *of G the average number $\delta^c(G)$:*

$$\delta(G) := \frac{\sum_{1 \leq i < j \leq n} \delta_{ij}(G)}{\binom{n}{2}} \qquad \delta^c(G) := \frac{\sum_{1 \leq i < j \leq n} \delta_{ij}^c(G)}{\binom{n}{2}}.$$

The number $\delta(G)$ provides a global measure of the local dissymmetry level of G, whereas $\delta^c(G)$ measures globally the local symmetry level of G. Obviously we have that $\delta(G) + \delta^c(G) = n$. Let us consider at first a very extremal case. We denote by O_n the graph having n vertices $\{v_1, \ldots, v_n\}$ and no edge. Then it is clear that $\Delta_{num}[O_n]$ is the null-matrix and $\Delta_{num}^c[O_n]$ is the matrix having n in every place. So that we have $\delta(O_n) = 0$ and $\delta^c(O_n) = n$. This means that we consider O_n as the simple graph having the maximum level of local symmetry and the minimum level of local dissymmetry.

Example 5.3. Let G be the graph in the Example 3.1. By the computations made in Example 5.1 and by Definition 5.1, we have $\delta(G) = 3$ and $\delta^c(G) = 2$.

5.1   Cases of K_n and $K_{p,q}$

We consider now the 2-local dissymmetry averages for the graphs K_n and $K_{p,q}$.

Proposition 5.3. *(i) If $n \geq 2$ then $\Delta_{num}[K_n] = 2Adj(K_n)$ and $\delta(K_n) = 2$.*
(ii) $\Delta_{num}[K_{p,q}] = (p+q)Adj(K_{p,q})$ and $\delta(K_{p,q}) = \frac{2pq}{p+q-1}$.

Proof. (i): For any pair of distinct vertices v_i and v_j in K_n we have $\Delta_{K_n}(v_i, v_j) = N_{K_n}(v_i) \triangle N_{K_n}(v_j) = (V(K_n) \setminus \{v_i\}) \triangle (V(K_n) \setminus \{v_j\}) = \{v_i, v_j\}$, therefore $\Delta_{num}[K_n] = 2(J_n - I_n) = 2Adj(K_n)$ and obviously $\delta(K_n) = 2$.
(ii): Let $K_{p,q} = (B_1 | B_2)$, where $B_1 = \{x_1, \ldots, x_p\}$ and $B_2 = \{y_1, \ldots, y_q\}$. Then $N_G(x_i) \triangle N_G(x_{i'}) = N_G(y_j) \triangle N_G(y_{j'}) = \emptyset$ and $N_G(x_i) \triangle N_G(y_j) = B_1 \cup B_2 = V(K_{p,q})$ for all $i, i' \in \{1, \ldots, p\}$ and all $j, j' \in \{1, \ldots, q\}$ Hence, because of the block form of the adjacency matrix of K_{pq} and by definition of local numerical dissymmetry matrix, we deduce that $\Delta_{num}[K_{p,q}] = (p+q)Adj(K_{p,q})$. Thus we also obtain $\delta(K_{p,q}) = \frac{pq(p+q)}{\binom{p+q}{2}} = \frac{2pq}{p+q-1}$. □

5.2   Case of C_n

In the next result we determine the general form of all local dissymmetry subsets when $G = C_n$.

Proposition 5.4. *Let $n \geq 5$ and $G = C_n$. Then:*

$$\Delta_G(v_i, v_j) = \begin{cases} \{v_{i-1}, v_{j+1}\} & \text{if } j = i+2 \\ \{v_{i-1}, v_{i+1}, v_{j-1}, v_{j+1}\} & \text{if } i \leq j \leq n-1 \wedge j \neq i+2 \end{cases} \qquad (17)$$

where $1 \leq i < j \leq n$ and all sum indexes are taken $mod(n)$.

Proof. It is a direct consequence of (14), since $N_G(v_i) = \{v_{i-1}, v_{i+1}\}$, where $i = 1, \ldots, n$ and all the index sums are taken $mod(n)$. □

In order to give a compact description of the matrix $\Delta_{num}(P_n)$ we introduce some new type of matrices. Let n be a positive integer, the matrix Q_n is the matrix $Q_n = (q_{ij})$, where:

$$q_{ij} := \begin{cases} 1 & \text{if either } i = 1 \text{ or } j = 1 \\ 0 & \text{otherwise.} \end{cases}$$

Let now k be a positive integer such that $0 \leq k \leq n-1$. The matrix $Diag(n, k)$ is the matrix $Diag(n, k) = (x_{ij}^{(n,k)})$, where:

$$x_{ij}^{(n,k)} := \begin{cases} 1 & \text{if } |i - j| = k \\ 0 & \text{otherwise .} \end{cases}$$

At this point we can easily determine both the local numerical dissymmetry matrix and the 2-local dissymmetry average for C_n.

Proposition 5.5. *Let $n \geq 5$. Then:*

(i) $\Delta_{num}[C_n] = 4(J_n - I_n) - 2Diag(n,2) - 2Diag(n, n-2)$.

(ii) $\delta(C_n) = 4 - \frac{4}{n-1}$, *so that* $\lim_{n \to \infty} \delta(C_n) = 4$.

Proof. (i) : By Eq. (17) we obtain $\Delta[C_n] = (\delta_{ij})$, where $\delta_{ij} = 2$ if $j = i+2$ and $\delta_{ij} = 4$ otherwise, when $1 \leq i,j \leq n$ and where all the index sums are taken mod(n). It is easy to check that the previous values of δ_{ij} are equal to the correspondent entries of the matrix $4(J_n - I_n) - 2Diag(n,2) - 2Diag(n, n-2)$.

(ii) : By (17), there are exactly n couples of vertices $v_i, v_j \in V(C_n)$ such that $\delta_{ij}(C_n) = 2$ and $\binom{n}{2} - n$ couples such that $\delta_{ij}(C_n) = 4$. This means that

$$\sum_{1 \leq i < j \leq n} \delta_{ij}(C_n) = 2n + 4\left(\binom{n}{2} - n\right) = 4\binom{n}{2} - 2n.$$

By Definition (5.1) we finally obtain $\delta(C_n) = \frac{4\binom{n}{2} - 2n}{\binom{n}{2}} = 4 - \frac{4}{n-1}$. $\square$

5.3   Case of P_n

In the next result we describe the general form of all local dissymmetry subsets when $G = P_n$.

Proposition 5.6. *[3] Let $n \geq 4$ and $G = P_n$. Then:*

$$\Delta_G(v_1, v_j) = \begin{cases} \{v_1, v_2, v_3\} & \text{if } j = 2 \\ \{v_4\} & \text{if } j = 3 \\ \{v_2, v_{j-1}, v_{j+1}\} & \text{if } 4 \leq j \leq n-1 \\ \{v_2, v_{n-1}\} & \text{if } j = n \end{cases} \quad (18)$$

and

$$\Delta_G(v_i, v_j) = \begin{cases} \{v_{i-1}, v_i, v_{i+1}, v_{i+2}\} & \text{if } j = i+1 \\ \{v_{i-1}, v_{i+3}\} & \text{if } j = i+2 \\ \{v_{i-1}, v_{i+1}, v_{j-1}, v_{j+1}\} & \text{if } i+2 < j \leq n-1 \\ \{v_{i-1}, v_{i+1}, v_{n-1}\} & \text{if } j = n, i \neq n-2 \\ \{v_{n-3}\} & \text{if } j = n, i = n-2 \end{cases} \quad (19)$$

when $2 \leq i < j \leq n$.

By Proposition 5.6 we obtain the following result.

Proposition 5.7. *Let $n \geq 4$. Then:*

(i) $\Delta_{num}[P_n] = 4(J_n - I_n) - Q_n - 2Diag(n,2) - Diag(n, n-1)$.

(ii) $\delta(P_n) = \frac{4(n^2 - 3n + 3)}{n(n-1)}$, *so that* $\lim_{n \to \infty} \delta(P_n) = 4$.

Proof. See [3] for the proof of *(i)*. We now prove *(ii)*. By (18), $\sum_{j=2}^{n} \delta_{1j}(P_n) = 3+1+3(n-4)+2 = 3n-6$. Furthermore, by (19), we deduce that $\delta_{n-1,n}(P_n) = 3$, $\sum_{j=n-1}^{n} \delta_{n-2,j}(P_n) = 5$ and $\sum_{2 \leq i < j \leq n} \delta_{ij}(P_n) = \sum_{i=2}^{n-3}[4+2+4(n-i-3)+3]$, for $i = 2, \ldots, n-3$. With simple calculations we have $\sum_{1 \leq i < j \leq n} \delta_{ij}(P_n) = 2(n^2 - 3n + 3)$, and by Definition (5.1) we get the thesis. $\square$

6   Conclusion

We studied undirected simple graphs using rough set theory tools. In this way, we were able to introduce a new concept of symmetry in graphs (induced by the indiscernibility relation) and connect lower and upper approximations to it. Further, the discernibility matrix has a natural corresponding in graphs, since each entry is the set symmetric difference between the neighborhood of two vertices. We also introduce the new concept of generalized discernibility matrix whose role in other contexts has to be explored, as well as its relationship with formal concept analysis (FCA) operators. Indeed, the attribute subset $\Delta_{\mathcal{I}}^c(Z)$ can be compared and connected with the intension operator of FCA.

Other future works include extending this study to other category of graphs, for instance trees (which are bipartite but not complete) and to explore the important notion of reduct in the case of graphs.

References

1. Chen, J., Li, J.: An application of rough sets to graph theory. Inf. Sci. **201**, 114–127 (2012)
2. Cattaneo, G., Chiaselotti, G., Ciucci, D., Gentile, T.: On the connection of hypergraph theory with formal concept analysis and rough set theory. Submitted to Information Sciences (2015)
3. Chiaselotti, G., Ciucci, D., Gentile, T.: Simple graphs in granular computing. Submitted to Information Sciences (2015)
4. Chiaselotti, G., Ciucci, D., Gentile, T.: Simple undirected graphs as formal contexts. In: Baixeries, J., Sacarea, C., Ojeda-Aciego, M. (eds.) ICFCA 2015. LNCS, vol. 9113, pp. 287–302. Springer, Heidelberg (2015)
5. Diestel, R.: Graph Theory, Graduate Text in Mathematics, 4th edn. Springer, Heidelberg (2010)
6. Midelfart, H., Komorowski, J.: A rough set framework for learning in a directed acyclic graph. In: Alpigini, J.J., Peters, J.F., Skowron, A., Zhong, N. (eds.) RSCTC 2002. LNCS (LNAI), vol. 2475, p. 144. Springer, Heidelberg (2002)
7. Pawlak, Z.: Rough Sets: Theoretical Aspects of Reasoning About Data. Kluwer Academic Publisher, Dordrecht (1991)
8. Skowron, A., Rauszer, C.: The discernibility matrices and functions in information systems. In: Słowiński, R. (ed.) Intelligent Decision Support. Theory and Decision Library, vol. 11. Springer, Netherlands (1992)
9. Tang, J., She, K., William Zhu, W.: Matroidal structure of rough sets from the viewpoint of graph theory. J. Appl. Math. **2012** (2012). doi:10.1155/2012/973920
10. Wang, S., Zhu, Q., Zhu, W., Min, F.: Equivalent characterizations of some graph problems by covering-based rough sets. J. Appl. Math. **2013** (2013). doi:10.1155/2013/519173

The Connectivity of the Covering Approximation Space

Duixia Ma[✉] and William Zhu

Lab of Granular Computing, Minnan Normal University, Zhangzhou, China
maduixia@163.com, williamfengzhu@gmail.com

Abstract. As a covering approximation space, its connectivity directly reflects a relationship, which plays an important role in data mining, among elements on the universe. In this paper, we study the connectivity of a covering approximation space and give its connected component. Especially, we give three methods to judge whether a covering approximation space is connected or not. Firstly, the conception of the maximization of a family of sets is given. Particularly, we find that a covering and its maximization have the same connectivity. Second, we investigate the connectivity of special covering approximation spaces. Finally, we give three methods of judging the connectivity of a covering approximation space from the viewpoint of matrix, graph and a new covering.

Keywords: Covering approximation space · Connectivity · Granular computing

1 Introduction

At the Internet age, data collected and stored are enormous and inexact. Then, there are many puzzles in data intelligence processing, such as how to take effective ways to cope with the uncertainty of ubiquitous information. So how to solve such bottleneck problem becomes an important issue in computer science and industry. In order to deal with this issue, researchers have developed many techniques such as rough set theory [9], fuzzy set theory [21], computing with words [13], and granular computing [8].

Rough set theory was proposed by Pawlak [10] in 1982 as a tool for dealing with the vagueness and granularity in information system. It is built on equivalence relation [11]. However, the data in powerful computer systems and storage media are complex and redundant in the information-based society. Thus equivalence relation imposes restrictions and limitations on many practical applications. Therefore, Pawlak's rough sets has been extended to the covering-based rough set theory [22,23], relation-based rough sets [12,19,20], fuzzy rough sets [3,18]. That is to say, Pawlak's approximation space is extended to generalization approximation space.

However, covering approximation space is a class of generalization approximation space. In addition, covering is a common data structure and is used

© Springer International Publishing Switzerland 2015
D. Ciucci et al. (Eds.): RSKT 2015, LNAI 9436, pp. 435–445, 2015.
DOI: 10.1007/978-3-319-25754-9_38

to describe overlapping information blocks in information systems. Therefore, it is important to investigate the covering approximation space. For a covering approximation space $(U, \mathcal{C})$, where U is a universe and $\mathcal{C}$ is a covering of U, it is called connected if x is connected to y for each pair $x, y \in U$, namely, there exist $K_1, K_2, \cdots, K_n \in \mathcal{C}$ such that $x \in K_1$, $y \in K_n$ and $K_i \cap K_{i+1} \neq \emptyset$, for any $x, y \in U$, $i = 1, 2, \cdots, n$. That is to say, the relationship between elements of universe relates to connectivity of this covering approximation space.

In this paper, we mainly investigate the connectivity of the covering approximation space and give three methods to judge its connectivity in terms of matrix, graph and a new covering. In order to study expediently, we firstly give a new covering, namely, the maximization of a covering. Particularly, a covering and its maximization are proved to have the same connectivity. So we study the connectivity of a covering approximation space through maximization of this covering. Second, we explore the connectivity of some special covering approximation spaces. Finally, we give three approaches to judge the connectivity of a covering approximation space by matrix, graph and a new covering. It is the core of this paper.

The remainder of this paper is organized as follows. In Sect. 2, we review some basic definitions and related conclusions about covering approximation space and graph theory. The conception of maximization of a covering and its properties are given in Sect. 3. Section 4 studies the connectivity of the special covering approximation spaces. The most important work is Sect. 5. This section gives three methods to judge the connectivity of a covering approximation space. Finally, we conclude this paper in Sect. 6.

2   Preliminaries

Covering is often used to describe overlapping information blocks in information systems. Graph theory is an intuitive and visible mathematical model. In this section, we introduce the basic definitions and related results about covering and graph theory.

2.1   Covering Approximation Space

In this subsection, we give the basic concepts of the covering approximation space.

Definition 1 (*Covering [2]*). *Let U be a universe of discourse, $\mathcal{C}$ a family of subsets of U. If $\mathcal{C}$ is a family of nonempty subsets of U, and $\cup \mathcal{C} = U$, $\mathcal{C}$ is called a covering of U.*

Definition 2 (*Covering approximation space [2]*). *Let U be a nonempty set, $\mathcal{C}$ a covering of U. The pair $(U, \mathcal{C})$ is called a covering approximation space.*

The minimal description is an important concept in a covering approximation space.

Definition 3 (*Minimal description [24]*). *Let* $(U, \mathcal{C})$ *be a covering approximation space and* $x \in U$. *The family of sets* $Md_{\mathcal{C}}(x) = \{C \in \mathcal{C} \mid x \in C \wedge (\forall S \in \mathcal{C} \wedge x \in S \wedge S \subseteq C \Rightarrow C = S)\}$ *is called the minimal description of* x. *When there is no confusion, we omit the subscript* $\mathcal{C}$.

In the same way, we give the definition of maximal description.

Definition 4 (*Maximal description [14]*). *Let* $(U, \mathcal{C})$ *be a covering approximation space and* $x \in U$. *The family of sets* $Maxd_{\mathcal{C}}(x) = \{C \in \mathcal{C} \mid x \in C \wedge (\forall S \in \mathcal{C} \wedge C \subseteq S \Rightarrow C = S)\}$ *is called the maximal description of* x. *When there is no confusion, we omit the subscript* $\mathcal{C}$.

In the following definitions, we introduce two types of coverings.

Definition 5 (*Unary [25]*). $\mathcal{C}$ *is unary if* $|Md(x)| = 1$ *for all* $x \in U$.

Definition 6 (*Pointwise-covered [25]*). $\mathcal{C}$ *is called a pointwise-covered covering, if for any* $K \in \mathcal{C}$ *and* $x \in K$, $K \subseteq \cup Md(x)$.

The connectivity of covering approximation space has been used in medical diagnosis [5]. We will introduce the connected covering approximation space.

Definition 7 *[5]. Let* $(U, \mathcal{C})$ *be a covering approximation space.*
(1) Let $x, y \in U$. x *is called to be connected to* y *if there are* $K_1, K_2, \cdots, K_n \in \mathcal{C}$ *such that* $x \in K_1$, $y \in K_n$ *and* $K_i \cap K_{i+1} \neq \emptyset$ *for each* $i = 1, 2, \cdots, n - 1$.
(2) $(U, \mathcal{C})$ *is called connected if* x *is connected to* y *for each pair* $x, y \in U$.

2.2   Graph Theory

Graph is an important tool at the area of mathematics. It has been employed to find rough set reducts [6]. In this subsection, we present the basic concepts of graph theory.

Definition 8 (*Graph [17]*). *A set of elements and a relation between them is called a graph. Specifically, graph is a pair* (V, E), *there* V *is called the vertex set of the graph,* E *is a unordered pair from the element of* V, *is called the edge set of the graph. We say that* $G' = (V', E')$ *is a subgraph of* $G = (V, E)$ *if* $V' \subset V$ *and* $E' \subset E$.

Definition 9 *[1]. (1) Chain(Walk): Let* u *and* v *are two vertexes of the graph. The chain of the graph is a series of finite vertexes and edges* $u_0 e_1 u_1 \cdots u_{n-1} e_n u_n$ $(u = u_0, v = u_n)$, *is called* $u - v$ *chain, where* u_{i-1} *and* u_i, *are placed adjacent to* $e_i (1 \leq i \leq n)$, *are two endpoint of the* e_i.
(2) Path: The chain is called a path, if the internal point of chain is different.
(3) Connected graph: If there is a path between u *and* v, *there* u *and* v *are two different vertexes of the graph, then the graph is connected, otherwise, the graph is not connected. The maximum connected subgraph is called a connected component.*

Definition 10 (*Bigraph [17]*). *A bigraph (or bipartite graph)* $G = (V, E)$ *is a graph whose vertices can be divided into two disjoint sets* V_1 *and* V_2 *such that every edge connects a vertex in* V_1 *to one in* V_2. *We often use* $G = (V_1 \cup V_2, E)$ *to denote a bigraph.*

3   Maximization of a Covering

Let C be a family of sets on universe U. If there exist two elements K_1, K_2 of C, such that K_1 is a subset of K_2, then we omit K_1, and so on. Like that, we can get some new sets C'. Then the elements of C' do not have this containment relationships. We call C' the maximization of C. First, we give a symbol of set theory [7].

Let U be a finite universe and A be a family of subsets of U. Then
$$Max(A) = \{X \in A| \text{ For all } Y \in A. \text{ If } X \subseteq Y, \text{ then } X = Y\}.$$

Definition 11 (*Maximization of sets*). *Let C be a set of sets. C' is called the maximization of C if $C' = Max(C)$.*

Example 1. Let $C = \{\{a, c, d\}, \{a, b\}, \{c, d\}, \{f, d, b\}, \{a, b, f\}\}$ be a family of sets. Then the maximization of C is that $C' = \{\{a, c, d\} \{f, d, b\}, \{a, b, f\}\}$.

If a family of sets are a covering on the universe U, then the maximization of these sets is still a covering.

Proposition 1. *Let $C = \{K_1, K_2, \cdots, K_n\}$ be a covering of the universe U and C' the maximization of C. Then C' is also a covering of U.*

Proof. Since C is a covering of U, then $K_i \neq \emptyset (i = 1, 2, \cdots, n)$ and $\bigcup_{i=1}^{n} K_i = U$. According to Definition 11, $K \neq \emptyset$, $\forall K \in C'$. Since $\forall K_i \in C$ there must exist $K_j \in C'$, such that $K_i \subseteq K_j$, so $U = \bigcup_{i=1}^{n} K_i \subseteq \bigcup_{j \in I} K_j$, and $\bigcup_{j \in I} K_j \subseteq U$, then $\bigcup_{j \in I} K_j = U$. Therefore C' is a covering of U.

The following proposition indicates that the process of maximizing can not change the connectivity of the covering approximation space.

Proposition 2. *Let (U, C) be a covering approximation space and C' the maximization of C. (U, C) is connected if and only if (U, C') is connected.*

Proof. $\Rightarrow$) : If (U, C) is connected, then for any $x, y \in U$, $\exists K_1, \cdots, K_m \in C$ such that $x \in K_1$, $y \in K_m$ and $K_i \cap K_{i+1} \neq \emptyset$ $(i = 1, 2, \cdots, m)$. Since C' the maximization of C, thus there must exist $K'_1, \cdots, K'_m \in C'$ such that $K_i \subseteq K'_i$. Therefore $x \in K_1 \subseteq K'_1$, $y \in K_m \subseteq K'_m$ for any $x, y \in U$, and $K'_i \cap K'_{i+1} \supseteq K_i \cap K_{i+1} \neq \emptyset$. That is to say (U, C') is connected.

$\Leftarrow$) : It is straightforward.

Proposition 3. *Let C be a covering of the universe U and C' the maximization of C. Then $Maxd_C(x) = Maxd_{C'}(x)$ for all $x \in U$.*

Proof. On one hand, for any $K \in Maxd_C(x)$, then $x \in K$ and $\forall S \in C$, $x \in S \land K \subseteq S \Rightarrow S = K$, thus $K \in C'$. According to Definition 11, $S \not\subseteq K$ for all $S \in C'$. Therefore, $K \in Maxd_{C'}(x)$ i.e. $Maxd_C(x) \subseteq Maxd_{C'}(x)$. On the other hand, $\forall K' \in Maxd_{C'}(x)$, then $\forall S \in C$, $x \in S \land K' \subseteq S \Rightarrow S = K'$ based on Definition 11, that is to say $K' \in Maxd_C(x)$, i.e. $Maxd_C(x) \supseteq Maxd_{C'}(x)$. To sum up, $Maxd_C(x) = Maxd_{C'}(x)$.

Proposition 4. *Let $\mathcal{C}$ be a covering of the universe U and $\mathcal{C}'$ the maximization of $\mathcal{C}$. Then $Md_{\mathcal{C}'}(x) = Maxd_{\mathcal{C}'}(x)$.*

Proof. $\forall K_1, K_2 \in \mathcal{C}'$, if $K_1 \neq K_2$, then $K_1 \not\subseteq K_2$, hence $Md_{\mathcal{C}'}(x) = Maxd_{\mathcal{C}'}(x)$.

4   The Connectivity of Covering Approximation Space

The connected covering approximation space has been used in medical diagnosis [5]. Therefore, it is important and necessary to study the connectivity of a covering approximation space.

Proposition 5. *Let $(U, \mathcal{C})$ be a covering approximation space and $\mathcal{C}'$ the maximization of $\mathcal{C}$. If $\mathcal{C}'$ is unary on U and $\mathcal{C}' \neq \{U\}$, then $(U, \mathcal{C})$ is not connected.*

Proof. We only need to prove that $(U, \mathcal{C}')$ is not connected based on Proposition 2. If $\mathcal{C}'$ is unary on U and $\mathcal{C}' \neq \{U\}$, then for all $x \in U$, $|Md(x)| = 1$. Thus we suppose $(U, \mathcal{C}')$ is connected, then $\forall x, y \in U$, there exists $K_1, \cdots, K_m \in \mathcal{C}'$ such that $x \in K_1$, $y \in K_m$ and $K_i \cap K_{i+1} \neq \emptyset$ $(1 \leq i \leq m-1)$. Let $x' \in K_1 \cap K_2$, thus $|Md(x')| = 2$. It is contradictory with $|Md(x)| = 1$ for all $x \in U$. Therefore $(U, \mathcal{C}')$ is not connected. That is to say $(U, \mathcal{C})$ is not connected.

We only suppose that $\mathcal{C}' \neq \{U\}$ in the Proposition 5. It is because that $\mathcal{C}'$ is connected when $\mathcal{C}' = \{U\}$.

Proposition 6. *Let $(U, \mathcal{C})$ be a covering approximation space. If $\mathcal{C}'$ is unary on U, then $U|Md(x)$ $(x \in U)$ is a connected component of $(U, \mathcal{C})$.*

Proof. Since $\mathcal{C}'$ is unary, then $Md(x) = Md(y)$ for any $y \in Md(x)$, thus $x \sim y$, and if $z \in U - Md(x)$, then $Md(x) \cap Md(z) = \emptyset$, so $x \not\sim z$. That is to say $U|Md(x)$ $(x \in U)$ is a connected component of $(U, \mathcal{C})$.

Proposition 7. *Let $(U, \mathcal{C})$ be a covering approximation space and $\mathcal{C}'$ the maximization of $\mathcal{C}$. If $\mathcal{C}'$ is a partition on U and $\mathcal{C}' \neq \{U\}$, then $(U, \mathcal{C})$ is not connected.*

Proof. Since $\mathcal{C}'$ is a partition, thus $\forall K_1, K_2 \in \mathcal{C}'$, $K_1 \cap K_2 = \emptyset$, then $x \not\sim y$, for any $x \in K_1$, $y \in K_2$. Therefore $(U, \mathcal{C})$ is not connected.

Proposition 8. *Let $(U, \mathcal{C})$ be a covering approximation space and $\mathcal{C}'$ the maximization of $\mathcal{C}$. If $\mathcal{C}' = \{K_1, \cdots, K_m\}$ is a partition, then $U|K_i (i = 1, 2, \cdots, m)$ is connected component of $(U, \mathcal{C})$.*

Proposition 9. *Let $(U, \mathcal{C})$, a covering approximation space, be connected and $\mathcal{C}'$ the maximization of $\mathcal{C}$. Then there exists $x \in K$ for all $K \in \mathcal{C}'$, such that $|Md_{\mathcal{C}'}(x)| > 1$.*

Proof. If $(U, \mathcal{C}')$ is connected, then we suppose that $|Md_{\mathcal{C}'}(x)| = 1$ for all $x \in K$. Since $Md_{\mathcal{C}'} = Maxd_{\mathcal{C}'}$, hence $\forall x \in K$, $\forall y \in U - K$, thus x is not connected y, which is contradictory with $(U, \mathcal{C}')$ is connected. Therefore, there must exist $x \in K$ for all $K \in \mathcal{C}'$, such that $|Md_{\mathcal{C}'}(x)| > 1$.

The reverse of Proposition 9 may not hold from the following a counterexample.

Example 2. Let $U = \{a, b, c, d, e, f\}$ be a universe and $K_1 = \{a, b\}$, $K_2 = \{b, c\}$, $K_3 = \{f, d\}$, $K_4 = \{d, e\}$, $C = \{K_1, K_2, K_3, K_4\}$. There exist $b \in K_1, K_2$ such that $|Md(b)| > 1$, and $d \in K_3, K_4$ such that $|Md(d)| > 1$, but (U, C) is not connected.

5 Methods of Judging the Connectivity of a Covering Approximation Space

As a general covering approximation space, what we care is whether it is connected or not. In this section, we will explore the connectivity of the covering approximation space from the viewpoint of matrix, graph and a new covering.

5.1 From the Matrix Perspective

First, we give the matrix relating to the covering approximation space.

Definition 12. *[15] Let (U, C) be a covering approximation space and $U = \{u_1, u_2, \cdots, u_n\}$, $C = \{K_1, K_2, \cdots, K_m\}$. We define a matrix $A = (a_{ij})_{m \times n}$ as follows:*

$$a_{ij} = \begin{cases} 1, u_j \in K_i, \\ 0, u_j \notin K_i, \end{cases}$$

where $i = 1, 2, \cdots, m$; $j = 1, 2, \cdots, n$. Using u_j labels the jth column and K_i labels the ith row. A is called a matrix induced by C.

Example 3. Let $U = \{a, b, c, d\}$ be a universe. $K_1 = \{a, b\}$, $K_2 = \{a, c, d\}$, $K_3 = \{b, c\}$, $K_4 = \{b, d\}$, $K_5 = \{c, d\}$, $C = \{K_1, K_2, K_3, K_4, K_5\}$. Then $C' = \{K_1, K_2, K_3, K_4\}$, hence the matrix A_{ij} as follows:

$$A_{ij} = \begin{pmatrix} 1 & 1 & 0 & 0 \\ 1 & 0 & 1 & 1 \\ 0 & 1 & 1 & 0 \\ 0 & 1 & 0 & 1 \end{pmatrix}$$

If we denote all column vectors of the A as $\mathcal{A} = \{\alpha_1, \alpha_2, \cdots, \alpha_n\}$, then a family of vectors $\mathcal{A}$ can be regard as being composed of column vectors $\alpha_j (j = 1, 2, \cdots, n)$.

Remark 1. Let $\alpha_j = (a_{1j}, a_{2j}, \cdots, a_{mj})^\top$ be a column vector. Then we denote by $\|\alpha_j\| = a_{1j} + a_{2j} + \cdots + a_{mj}$, where $j = 1, 2, \cdots, n$.

Definition 13. *Let A be a matrix induced by C and $\mathcal{A}$ a set of all column vectors of A. Then we denote the maximization of the $\mathcal{A}$ by $\mathcal{A}_{max}$ as follows:*

$$\mathcal{A}_{max} = \{\alpha_i \in \mathcal{A} : \forall \alpha_j \in \mathcal{A}, \|\alpha_i\| \geq \|\alpha_j\|\}.$$

Example 4 (Continued from Example 3). $\mathcal{A}_{max} = \{\ \alpha_2\ \}$, where $\alpha_2 = (1\ 0\ 1\ 1)^{\top}$.

We take any one from $\mathcal{A}_{max}$, denoted as α. If a row of α is 1, then we omit column vectors whose values are 1 in the same row with α and the vectors we have omitted until there do not exist column vectors such that their values are 1 in the same row.

Example 5 (Continued from Example 3).

$$
A_{ij} = \begin{matrix} & \alpha_1 & \alpha_2 & \alpha_3 & \alpha_4 \\ & \begin{pmatrix} 1 & 1 & 0 & 0 \\ 1 & 0 & 1 & 1 \\ 0 & 1 & 1 & 0 \\ 0 & 1 & 0 & 1 \end{pmatrix} \end{matrix} \rightarrow \begin{matrix} \alpha_2 & \alpha_3 & \alpha_4 \\ \begin{pmatrix} 1 & 0 & 0 \\ 0 & 1 & 1 \\ 1 & 1 & 0 \\ 1 & 0 & 1 \end{pmatrix} \end{matrix} \rightarrow \begin{matrix} \alpha_2 & \alpha_4 \\ \begin{pmatrix} 1 & 0 \\ 0 & 1 \\ 1 & 0 \\ 1 & 1 \end{pmatrix} \end{matrix} \rightarrow \begin{matrix} \alpha_2 \\ \begin{pmatrix} 1 \\ 0 \\ 1 \\ 1 \end{pmatrix} \end{matrix}.
$$

Since $\|\alpha_2\|$ is the maximization of $\|\alpha_i\| (i = 1, \cdots, 4)$, hence we regard α_2 as α. Thus the first row of α is 1, so we omit α_1 whose the first row is 1. Because the values of α_3, α_4 and α_1 are 1 in the second row, we also omit α_3, α_4.

Definition 14. *Let (U, C) be a covering approximation space and A be a matrix induced by C. The new matrix through the above process is called the simplification of A, denoted by $Simp(A)$.*

Proposition 10. *Let (U, C) be a covering approximation space, A be a matrix induced by C and $N_{(U,C)}$ the number of connected component of (U, C). We denote the number of column vector of $Simp(A)$ by $N_{Simp(A)}$, then $N_{(U,C)} = N_{Simp(A)}$.*

Proof. We suppose $A = (\alpha_1\ \alpha_2\ \cdots\ \alpha_n)$, $Simp(A) = (\alpha_1\ \alpha_2\ \cdots\ \alpha_m)$, $(m \le n)$. On one hand, for any $\alpha_i (1 \le i \le n, i \ne 1, \cdots, m)$ there exist $\alpha_j (j = 1, \cdots, m)$ such that α_i, α_j have the same 1 in the same row, that is to say $a_i \sim a_j$ based on Definition 12. On the other hand, for any $\alpha_l, \alpha_k\ (l, k = 1, \cdots, m)$, then $a_l \not\sim a_k$ based on Definition 14 and Definition 12. Therefore $N_{(U,C)} = N_{Simp(A)}$.

Proposition 11. *Let (U, C) be a covering approximation space, A be a matrix induced by C. Then $Simp(A)$ only has one column vector if and only if (U, C) is connected.*

5.2 From the Graph Perspective

The following proposition is straightforward from the bigraph perspective [16].

Proposition 12. *Let (U, C) be a covering approximation space and G a bigraph induced by C. (U, C) is connected if and only if G is connected.*

The above proposition gives a useful way to judge the connectivity of a covering approximation space. We will introduce another method through a general graph.

Definition 15 (*Graph induced by K_i*). *Let $\mathcal{C} = \{K_1,\ K_2,\ \cdots,\ K_m\}$ be a covering. We define the graph $G(K_i) = (V, E)$ induced by $K_i = \{x_1,\ x_2,\ \cdots,\ x_l\}$ $(i = 1,\ \cdots,\ m)$ as follows:*
(1) $V = \{v_{K_i},\ v'_{K_i}\}$;
(2) $E = \{x_1,\ x_2,\ \cdots,\ x_l\}$, where x_j connects v_{K_i} with v'_{K_i}, $j = 1, 2, \cdots, l$.

Example 6. Let $K = \{u_1, u_2, u_3, u_4\}$, then the graph induced by K as shown in Fig. 1:

Fig. 1. The graph induced by K.

Remark 2. Let $G_1 = (V_1, E_1)$, $G_2 = (V_2, E_2)$ be two graphs induced by K_1, K_2 respectively, where $V_1 = \{v_{K_1}, v'_{K_1}\}$, $V_2 = \{v_{K_2}, v'_{K_2}\}$. If $K_1 \cap K_2 \neq \emptyset$, then we denote as $v_{K_1} = v_{K_2}$ and $v'_{K_1} = v'_{K_2}$.

Definition 16 (*Graph union [4]*). *Let $G(K_1) = (V_1, E_1)$, $G(K_2) = (V_2, E_2)$ be two graphs induced by K_1 and K_2. Then the union of $G(K_1)$ and $G(K_2)$ is defined as a graph $(V_1 + V_2, E_1 + E_2)$, denoted by $G(K_1) \cup G(K_2)$.*

It is clearly that a covering approximation space can induce a graph from the following definition.

Definition 17. *Let $(U, \mathcal{C})$ be a covering approximation space. The graph induced by $\mathcal{C}$ is called the graph induced by $(U, \mathcal{C})$, denoted by $G(\mathcal{C})$.*

Example 7. Let $(U, \mathcal{C})$ be a covering approximation space and $U = \{u_1, u_2, u_3, u_4, u_5\}$, $\mathcal{C} = \{\{u_1, u_3\}, \{u_3, u_5\}, \{u_2, u_4\}\}$. Then the graph induced by $\mathcal{C}$ as shown in the Fig. 2:

Proposition 13. *Let $(U, \mathcal{C})$ be a covering approximation space and $G(\mathcal{C})$ a graph induced by $\mathcal{C}$. $G(\mathcal{C})$ is connected if and only if $(U, \mathcal{C})$ is connected.*

Proof. It is straightforward based on the Definitions 15 and 16 and Remark 2.

Proposition 14. *Let $(U, \mathcal{C})$ be a covering approximation space and $G(\mathcal{C})$ a graph induced by $\mathcal{C}$. Then the numbers of connected component of $(U,\ \mathcal{C})$ and the graph $G(\mathcal{C})$ are equal.*

Example 8 (Continued from Example 7). Since the graph induced by $\mathcal{C}$ has two connected component from the Fig. 2, thus $N_{(U,\mathcal{C})} = 2$.

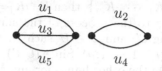

Fig. 2. The graph induced by $\mathcal{C}$.

5.3 From Covering Perspective

In the following discussion, we will continue to simplify a covering in terms of a covering for herself. First, we give the notion of friends of an element.

Definition 18 *[26]. Let $(U, \mathcal{C})$ be a covering approximation space. For any $x \in U$, $\cup\{K : x \in K \wedge K \in \mathcal{C}\}$ is called the friends of x, denoted by $Friends(x)$.*

According to Definition 18, it is clear that $\{Friends(x) : x \in U\}$ is a covering on the U, we denote by $\mathcal{C}_1$. We already know that $\mathcal{C}_1'$ is also a covering and both of them have the same connectivity in Sect. 3. Now, for every element x on U, we can get the friends of x in the covering approximation space $(U, \mathcal{C}_1')$. Then $\{Friends(x) : x \in U\}$ is also a covering, denoted by $\mathcal{C}_2 = \{Friends(x) : x \in U\}$, then we get $\mathcal{C}_2'$. Following it until we get a partition $\mathcal{C}_N'$.

Example 9. Let $(U, \mathcal{C})$ be a covering approximation space. $\mathcal{C} = \{C_1, C_2, C_3\}$, $C_1 = \{a, c\}$, $C_2 = \{a, b, d\}$, $C_3 = \{b, e, f\}$. Since $Friends(a) = \{a, b, c, d\}$, $Friends(b) = \{a, b, d, e, f\}$, $Friends(c) = \{a, c\}$, $Friends(d) = \{a, b, d\}$, $Friends(e) = Friends(f) = \{b, e, f\}$. Thus $\mathcal{C}_1' = \{\{a, b, c, d\}, \{a, b, d, e, f\}\}$. In the same way, we get $\mathcal{C}_2 = \{U, \{a, b, c, d\}, \{a, b, d, e, f\}\}$, then $\mathcal{C}_2' = \{U\}$.

If $\mathcal{C}_N'$ is a partition induced by above process, then this partition is called the simplification of $\mathcal{C}$. It is provided in the following Definition 19.

Definition 19. *Let $(U, \mathcal{C})$ be a covering approximation space. If $\mathcal{C}_N'$ is a partition induced by above process, then we call $\mathcal{C}_N'$ is the simplification of $\mathcal{C}$, denote by $\mathcal{C}_N'$.*

The notion of simplification of a covering can help us to judge the connectivity of a covering approximation space from the following proposition.

Proposition 15. *Let $(U, \mathcal{C})$ be a covering approximation space and $\mathcal{C}_N'$ the simplification of $\mathcal{C}$. Then $(U, \mathcal{C})$ is connected if and only if $\mathcal{C}_N' = \{U\}$.*

Proof. It is straightforward based on the process of the simplification of a covering.

Proposition 16. *Let $(U, \mathcal{C})$ be a covering approximation space and $\mathcal{C}_N'$ the simplification of $\mathcal{C}$. Then an equivalence classes of $\mathcal{C}_N'$ is a connected component of $(U, \mathcal{C})$.*

Proof. We suppose $\mathcal{C}'_N = \{K_1, \cdots, K_m\}$, then $K_i \cap K_j = \emptyset (i, j = 1, \cdots, m)$ based on Definition 19. According to the process of the simplification of a covering, there must exist a covering $\mathcal{C}'_n$ and $z \in U$ such that $x \in Md_{\mathcal{C}'_n}(z)$ and $y \in Md_{\mathcal{C}'_n}(z)$ for any $x, y \in K_i (i = 1, \cdots, m)$. Since $(U, \mathcal{C})$ and $(U, \mathcal{C}'_n)$ have the same connectivity, thus $x \sim y$. On the other hand, for any $x \in K_i$ and $y \in K_j (i \neq j)$, since $K_i \cap K_j = \emptyset$, thus $x \not\sim y$. Therefore an equivalence classes of $\mathcal{C}'_N$ is a connected component of $(U, \mathcal{C})$.

6   Conclusions

In order to improve the high-efficiency in data mining, we have studied the connectivity of the covering approximation space and given three methods to judge its connectivity. First, we have given a conception of maximization of a family of sets, so the maximization of a covering has been given. Then we have investigated the relationship between covering and its maximization. Especially, we give that a covering and its maximization have the same connectivity. Second, we have studied the connectivity of special covering approximation spaces. Finally, we have given three methods to judge the connectivity of a covering approximation space with the aid of matrix, graph and a new covering.

Acknowledgments. This work is in part supported by The National Nature Science Foundation of China under Grant Nos. 61170128, 61379049 and 61379089, the Key Project of Education Department of Fujian Province under Grant No. JA13192, the Project of Education Department of Fujian Province under Grant No. JA14194, the Zhangzhou Municipal Natural Science Foundation under Grant No. ZZ2013J03, and the Science and Technology Key Project of Fujian Province, China Grant No. 2012H0043.

References

1. Bollobás, B.: Modern Graph Theory. Springer, New York (1998)
2. Bonikowski, Z., Bryniarski, E., Wybraniec-Skardowska, U.: Extensions and intentions in the rough set theory. Inf. Sci. **107**, 149–167 (1998)
3. Dubois, D., Prade, H.: Rough fuzzy sets and fuzzy rough sets. Int. J. Gen. Syst. **17**, 191–209 (1990)
4. Gao, S.: Graph Theory and Network Flow Theory. Higher Education Press, Beijing (2009)
5. Ge, X.: Connectivity of covering approximation spaces and its applications on epidemiological issue. Appl. Soft. Comput. **25**, 445–451 (2014)
6. Jensen, R., Shen, Q.: Finding rough set reducts with ant colony optimization. In: Proceedings of the 2003 UK Workshop on Computational Intelligence, pp. 15–22 (2003)
7. Lai, H.: Matroid Theory. Higher Education Press, Beijing (2001)
8. Lin, T.Y.: Granular computing on binary relations. In: Alpigini, J.J., Peters, J.F., Skowron, A., Zhong, N. (eds.) RSCTC 2002. LNCS (LNAI), vol. 2475, pp. 296–299. Springer, Heidelberg (2002)

9. Pawlak, Z.: Rough Sets: Theoretical Aspects of Reasoning About Data. Kluwer Academic Publishers, Boston (1991)

10. Pawlak, Z.: Rough sets. Int. J. Comput. Inf. Sci. **11**, 341–356 (1982)

11. Pawlak, Z., Skowron, A.: Rudiments of rough sets. Inf. Sci. **177**, 3–27 (2007)

12. Skowron, A., Stepaniuk, J.: Tolerance approximation spaces. Fundamenta Informaticae **27**, 245–253 (1996)

13. Wang, F.: Outline of a computational theory for linguistic dynamic systems: toward computing with words. Int. J. Intell. Control Syst. **2**, 211–224 (1998)

14. Wang, Z., Shu, L., Ding, X.: Minimal description and maximal description in covering-based rough sets. Fundamenta Informaticae **128**, 503–526 (2013)

15. Wang, S., Zhu, W., Zhu, Q., Min, F.: Characteristic matrix of covering and its application to boolean matrix decomposition. Inf. Sci. **263**, 186–197 (2014)

16. Wang, S., Zhu, W., Zhu, Q., Min, F.: Four matroidal structures of covering and their relationships with rough sets. Int. J. Approximate Reasoning **54**, 1361–1372 (2013)

17. West, D., et al.: Introduction to Graph Theory. Pearson Education, Singapore (2002)

18. Wu, W., Leung, Y., Mi, J.: On characterizations of (I, T) -fuzzy rough approximation operators. Fuzzy Sets Syst. **154**, 76–102 (2005)

19. Yao, Y.Y.: On generalizing pawlak approximation operators. In: Polkowski, L., Skowron, A. (eds.) RSCTC 1998. LNCS (LNAI), vol. 1424, p. 298. Springer, Heidelberg (1998)

20. Yao, Y.: Relational interpretations of neighborhood operators and rough set approximation operators. Inf. Sci. **111**, 239–259 (1998)

21. Zadeh, L.A.: Fuzzy sets. Inf. Control **8**, 338–353 (1965)

22. Zakowski, W.: Approximations in the space (u, π). Demonstratio Mathematica **16**, 761–769 (1983)

23. Zhu, W., Wang, F.: Reduction and axiomization of covering generalized rough sets. Inf. Sci. **152**, 217–230 (2003)

24. Zhu, W.: Relationship between generalized rough sets based on binary relation and covering. Inf. Sci. **179**, 210–225 (2009)

25. Zhu, W., Wang, F.: On three types of covering-based rough sets. IEEE Trans. Knowl. Data Eng. **19**, 1131–1144 (2007)

26. Zhu, W., Wang, F.: A new type of covering rough sets. In: 2006 3rd International IEEE Conference on Intelligent Systems, pp. 444–449. IEEE (2006)

Detecting Overlapping Communities with Triangle-Based Rough Local Expansion Method

Zehua Zhang[1]([✉]), Nan Zhang[2], Caiming Zhong[3], and Litian Duan[1]

[1] College of Computer Science and Technology, Taiyuan University of Technology,
Shanxi 030024, China
{zehua_zhang,zhangnan0851}@163.com, dlt19890203@126.com
[2] School of Computer and Control Engineering, Yantai University,
Shandong 264005, China
[3] College of Science and Technology, Ningbo University, Zhejiang 315211, China
zhongcaiming@nbu.edu.cn

Abstract. Overlapping communities structures could effectively reveal the internal relationships in real networks, especially on the ownership problems on the nodes in overlapping areas between communities. Hence, the overlapping community detection research becomes a hotspot topic on graph mining in the decade, while the local expansion methods based on structural fitness function could simultaneously discover overlapped and hierarchical structures. Aimed at community drift and redundant calculation problems on general local expansion methods, the paper presents a novel local expansion method based on rough neighborhood that carries out the heuristic technology by the community seed inspiration and the triangle optimization on the boundary domain. The method could directly generate natural overlaps between communities, reduce the computational complexity and improve the detection on the overlapped boundary area. Finally, the experimental results on some real networks also show that the rough expansion method based on triangle optimization could be more effective in detecting the overlapping structures.

Keywords: Rough set theory · Overlapping community detection · Rough neighborhood expansion · Triangle optimization

1 Introduction

The communities on networks as a common phenomenon in actual complex systems are the nodes sets with some common properties, such as the proteins structures with some dissimilar functions in biological networks, the various interest groups in social networks and the different research fields on the scientific collaboration networks, etc. The community depicts close connections within the community and loose aggregation between communities in actual networks [1]. The community structures used to mean some statistical properties on networks,

© Springer International Publishing Switzerland 2015
D. Ciucci et al. (Eds.): RSKT 2015, LNAI 9436, pp. 446–456, 2015.
DOI: 10.1007/978-3-319-25754-9_39

which could effectively reveal the topological characteristics about network structures, functions and communications. In the past decade, community detection methods have been widely applied in the research field of computer science, statistical physics, biology, sociology and so on [2]. However, the nodes of complex structures existing in many applications could simultaneously belong to many communities. For instance, some people could participate in the different topics in the forum, and some protein in protein-protein interaction networks could possess many functions and belong to some different motifs. The hard partition in traditional community detection methods cannot solve the Overlapping Community Detection Problem (OCDP) [4], which the overlap nodes could belong to several communities. The OCDP focuses on the overlap structures, especially on the nodes ownerships in overlapping area.

Recently, many researchers devote into solving OCDP from different perspectives of node densities, edge features, structure measurements. At present, the overlapping community detection algorithms mainly include clique percolation algorithm (CPM) [4], local fitness methods [5], and edge-graph based partition methods [6], the hybrid methods [7–11], etc. Due to many links may exist between dense overlapping communities, the current OCDP research has no consistent definition on the overlapping community as the strong or weak connected communities. The discovered overlapping communities mostly depends on the given community metrics and the arbitrary parameters, while the local fitness methods [5] effectively generate the natural overlapping community structures, and avoids the setup on the measurement about the overlapped area. Besides, the local expansion methods show the hierarchial structures by the expansion process of nodes. The local fitness maximization method (LFM) [5] is based on local structure fitness function to generate communities with multi-resolution strategy, which consider the features on local structures to gradually expand the local communities and form natural overlapping fields between communities. However, it is noticed that a great deal of replicated computation exists in the update process of structure fitness, especially in the dense overlapping network. In addition, the LFM methods are sensitive to the initial communities.

Greedy clique expansion algorithm (GCE) [21] adopts to merge the generated duplicate communities to reduce redundant calculation rather than the node exclusion as community generated process in LFM methods. When the initial nodes are located in sparse overlapped area, the community drift problems are likely to occur. The community drift problem that the initial nodes would respectively attach to various communities at different times brings a lot of redundant calculation. The Merging of Overlapping Natural Communities (MONC) [9] regards the new joining nodes within community should not reduce the resolution of the original community module, so the structure fitness function could be transformed into the resolution function about local extended structures. And further studies [11] has proven the local expansion method effective based on ground-truth. So our previous work [12] presents the local expansion method based on rough neighborhood and introduces the stable rough graph to analyze the inherent structural features of communities.

The paper presents an improved Rough Neighborhood based overlapping community detection Method (RNM), further aimed at community drift and redundant calculation problems from the heuristic seeds in boundary area between communities. Firstly, a novel heuristic strategy based on triangle optimization about community seeds is adopted to reduce the computational complexity and improve the detection capability in local expansion process. On condition of the structural fitness maximization, the cliques with the nodes of great degree are expanded as local seed communities based on the heuristic triangle optimization. Then the proposed local expansion method merges the adjacent duplicate communities by the stability measurement and finally generates the natural overlapping communities.

The paper is arranged as follows. The Sect. 2 introduces the preliminaries. The next section elaborates the community discovery with triangle-based rough expansion method, and analyzes the heuristic technology by community seed inspiration and the triangle optimization strategy in the boundary domain. In the Sect. 4 the experiment results on some actual networks with known structures depict that the proposed method is feasible. The paper ends in the Sect. 5.

2   Preliminaries

The global topological structure of the whole network sometimes is hard to be completely obtained on some dynamic networks. And especially for the large scale network, the whole network is hardly loaded into the internal memory. Therefore, Clauset (2005) [3] proposed a community generative model based on the local modularity measurement. The calculation process of the local expansion method does not load the entire network, which can effectively deal with some large-scale networks, and greatly reduce the time cost and space complexity. Furthermore, each local expansion process is relatively independent, so the local generative method can also be accelerated by parallel processing.

The general view on the network community is the node sets of dense internal connections and sparse external connections. The mathematical notations and their descriptions in the paper are shown as Table 1. The letter d associates with the nodes, while the letter D associates with the graph structure. And the

Table 1. Notations for the community detection algorithms

Symbols	Description
$d(v_i) = \sum_{v_j \in V} M_{ij}$	The total number of links related to node v_i
$d_{v_i}^{in}(S) = \sum_{v_j \in S} M_{ij}$	The number of links from node v_i into insides of subgraph S
$d_{v_i}^{out}(S) = \sum_{v_j \notin S} M_{ij}$	The number of links from node v_i to outsides of subgraph S
$D(S)$	The total number of links related to subgraph S
$D_{in}^{S} = \sum_{v_i, v_j \in S} M_{ij}$	The number of links are totally contained in subgraph S
$D_{out}^{S} = \sum_{v_i \in S, v_j \notin S} M_{ij}$	The number of links from insides to outsides in subgraph S

symbols without supreme label and the subscript are concerned with the total number of connections.

3   Triangle-Based Rough Local Expansion Method

The overlapping community detection problems pay more attention to the uncertainty on overlapped regions between communities. Rough set theory (1982) [13] is an effective method of analysis on uncertain problems, by the advantage of the existing knowledge to describe imprecise or vague concepts. The research related to rough sets has been widely used to solve the problems with inaccurate and incomplete information [14,15,18,19]. In the paper, we propose the rough neighborhood expansion method based on triangular information granulation. The rough upper and lower approximation is used to analyze the search progress on community and local structure extension. the uncertain measurement on communities based on the connected triangles is introduced. The triangular strategy could be combined with other heuristic methods to effectively improve the local search progress, and avoid the community drifts on the local extension.

3.1   Community Detection Method Based on Rough Neighborhood

As a matter of fact, the neighborhood operator on the approximation space of a graph shows the cover progress on the domain to describe the concept of community. Yao [14] and Wu [17]studied the k-neighborhood operators and the related properties of neighborhood systems. Hu [15] discussed the mixed feature extraction problem in neighborhood information system. Lin and Qian [18] further studied the neighborhood based multi-granulation rough sets. The neighborhood relationship R defined on the graph structure could be expressed as adjacency matrix $M(R) = (r_{ij})_{|V| \times |V|}$, and the neighborhood of the node v_i is shown as $r(v_i) = \{v_j \in V | r_{ij} = 1\}$.

Definition 1. *Neighborhood approximation on graph [12]: Given the neighborhood approximate space $U = (G, \Gamma, R)$ on the finite and nonempty graph $G(V, E)$, and the edge set $E \subseteq V \times V$. While Γ is the set of the connected subgraphs on G, called the community candidates. For any subgraph $S \subseteq \Gamma$ on the neighborhood space, its upper and lower approximation could be defined:*

$$\underline{R}(S) = \{v_i \in V | r(v_i) \subseteq S\}; \overline{R}(S) = \{v_i \in V | r(v_i) \cap S \neq \emptyset\}. \tag{1}$$

By the above definition, the lower approximation $\underline{R}(S)$ is also called the positive region of S, denoted as $POS(S)$, which shows the node sets must belong to the subgraph S. And the boundary region $BN(S) = \overline{R}(S) - \underline{R}(S)$ implies nodes that is not sure whether belong to S. In other words, the nodes in the overlapped area do not assure their ownerships for communities. Besides, the nodes in the negative region which satisfies $NEG(S) = G - \overline{R}(S)$ cannot be included in S by the current knowledge. So we could draw a conclusion that the boundary

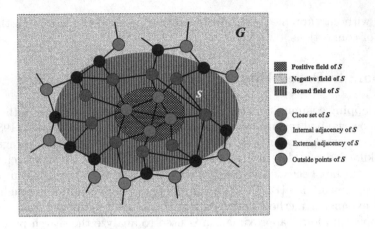

Fig. 1. The rough expansion process on the local neighborhood

region mainly leads to uncertainty of overlapping communities. With the larger boundary area, the certainty on networks is lower.

The rough expansion process on the local neighborhood is shown as Fig. 1. It is remarkable that the nodes in the close sets which must belong to the positive region have no links with the external adjacent nodes of S. Besides, the internal and the external adjacent nodes are combined into the boundary region.

Definition 2. *The stable rough graph [12]: Given the graph $G(V, E, f)$, $E \subseteq V \times V$, and $S \subseteq G$. $f : V \times S \to \mathbf{R}$ is the structure information function on G, and R is the real number. For $\forall V_A \in V$ and $V_B \in S$, if $f(S \cup V_A) \leq f(S)$ and $f(S - V_B) \leq f(S)$ are both satisfied, then the subgraph S is called the stable rough subgraph on G, denoted as $\Phi(S)$. When $G = \cup_{i=1}^{k} \Phi(S_i)$, the stable rough graph could be indicated as $\Phi(G)$, and k is the number of the stable rough subgraphs.*

The stable rough graph could effectively describe the expansion progress on communities generation and the fitness maximization on structures.

3.2   Neighborhood Expansion Strategy with Triangle Optimization

To begin with, some heuristic methods based on cliques such as CPM and GCE have high computation comlexity $O(kn^2 \log n)$ in average situation [21], especially in some dense networks with the nested structures and the uneven distribution. Besides, cliques structures seldom exist in the sparse network, so the clique methods do not well in the seeds initialization. In order to avoid the initial seeds falling in the overlapped area, the proposed method adopts the local triangle structure measurement to reduce the global computation as much as possible, and improve the search efficiency on the community seeds of initialization.

Definition 3. *The connected triangle optimization: Given $G(V, E)$, $S \subseteq G$. $T \subseteq S$ is a triangle structure on S. If $d_{v_i}^{in}(S) > d_{v_i}^{out}(S)$ is satisfied for*

$v_i \in T$, T is called the strongly connected triangle on the subgraph S, denoted as Δ_S^{T}. While $\sum_{v_i \in S} d_{v_i}^{in}(S) > \sum_{v_i \in S} d_{v_i}^{out}(S)$ is satisfied for $v_i \in T$, T is called the weakly connected triangle on the subgraph S, denoted as ∇_S^{T}.

Obviously, the structure with the strongly connected triangle is the subset of the strongly connected graph. Especially, the strongly connected graph is equal to strongly connected triangle on the complete graph, similarly in the weakly connected triangle. The number of strongly and weakly connected triangles to some extent describes the link aggregation between nodes. Therefore, the method based on the connected triangles could be used to analyze the the uncertainty measurement of the nodes belonging to boundaries. The connected triangles measurement on the graph G is drawn as the conclusion:

$$M(G) = pN(\Delta_G^{\mathrm{T}}) + (1-p)N(\nabla_G^{\mathrm{T}}) \tag{2}$$

The parameter $p \in [0,1]$ depicts the empirical effect on the graph between the strongly and weakly connected triangles. When $p = 1$, the strong impact about the network is uniquely considered. $N(\Delta_G^{\mathrm{T}})$ and $N(\nabla_G^{\mathrm{T}})$ separately denote the triangle number of Δ_G^{T} and ∇_S^{T}.

3.3 The Heuristic Algorithm Based on Triangle Optimization

Because LFM methods are sensitive to the initial seeds, the inspired strategy with the greatest node degree is easy to fall into the dense local networks. And the methods with the maximum clique such as GCE, has a high computational cost. Especially, the effective cliques are hard to be obtained in the unbalance networks, like the unbalance star networks with some sparse local structures. So the triangle based heuristic method combines the feature of node degree and the local triangle structure to improve the search on the unbalance networks.

Algorithm 1: the triangle based heuristic progress
Input: a network $G(V, E)$, the parameter p.
Output: community seed S_0.
Begin:
Step1: Initialize Seed $v_o = \emptyset$, and sort $v_0 \leftarrow \arg\max_{v_i}(D(v_i))$;
Step2: Construct the community seed S_0, $v_i \in r(v_0)$ and $\forall v_i \in S_0$;
Step3: Calculate the uncertainty of S_0, $M(S_0) = pN(\Delta_{S_0}^{\mathrm{T}}) + (1-p)N(\nabla_{S_0}^{\mathrm{T}})$;
Step4: For $\forall v_j \in S_0'$, $i \neq j$, the community seed S_0', $v_k \in r(v_j)$ and $\forall v_k \in S_0'$.
Step5: Calculate $M(S_0')$, the uncertainty of S_0'.
Step6: If $M(S_0) \geq M(S_0')$, then output community seed S_0.
Step7: Otherwise, return the previous Step1 and $v_o = v_j$.

In the worst case at the complete network, the time cost on the Step1 is $O(N^2)$, N denotes the number of the nodes. From Step2 to Step6, the community heuristic algorithm needs to be compared up to $k * (k-1)/2$ times at most, and loops the N times, where k means the number of the neighbors on v_i. So the total computation complexity is $O(k^2 N^3)$ in the worst case. In fact, the most case only needs search on the local space, and $k \ll N$. In the average case, the total computation complexity is $O(N^2 \log N)$ like GCE. However, the proposed method has better performance both on the dense and unbalanced networks.

3.4   The Heuristic Local Expansion Algorithm

By the stable rough graph, the formalized definition of the overlapping communities detection problem could be described as follow:

Definition 4. *Given the graph $G(V, E, f, \alpha)$, f is the structure information function on G and α depicts the resolution ratio. For the local community candidate sets $S \subseteq G$ and $S_i \subseteq S$, $S_i = \arg\max_{S} f_{(S,\alpha)}(S)$ could be found, so that it satisfies $\Phi(G) = \cup_{i=1}^{k} S_i$.*

$$f_{(S,\alpha)} = \frac{\sum_{v_i, v_j \in S} M_{ij}}{\left(\sum_{v_i, v_j \in S} M_{ij} + \sum_{v_i \in S, v_j \notin S} M_{ij}\right)^{\alpha}} \tag{3}$$

Algorithm 2: the local expansion progress on networks
Input: the network $G(V, E)$, seeds S_0, the resolution ratio α.
Output: the communities partition $S_i \subseteq S$.
Begin
Step1: Initialize the communities, $S_i \leftarrow S_0$.
Step2: Calculate the structure fitness $f_{(S_i,\alpha)}^{u}$, for all neighbors u on S_i, $u \notin S_i$;
Step3.1: If $f_{(S_i,\alpha)}^{u} \geq 0$, then the node $u \leftarrow \arg\max_{u}(f_{(S_i,\alpha)}^{u})$.
Step3.2: Update $S_i \leftarrow S_i \cup \{u\}$, and return the Step1 in the algorithm 2.
Step3.3: Until $f_{(S_i,\alpha)}^{u} < 0$, update $G \leftarrow G - S_i$, $i = i + 1$, and return the Step1 of the algorithm1.
Step4.1: When G is empty, the candidates are merged into the natural overlaps.
Step4.2: For all $S_i \subseteq S$, calculate the similarity between the community candidates: $\delta(S_i, S_j) = \frac{|S_i \cap S_j|}{\min(|S_i|, |S_j|)}$;
Step4.3: If $\delta(S_i, S_j) \geq 0.5$, $S_i \leftarrow S_i \cup S_j$, and update $S \leftarrow S - S_i$.
Step5: Until $S = \emptyset$, the algorithm is terminated, and all the S_i are output as the partition about overlapping communities.

The methods based on merging of natural community could avoid the reduplicative computation on the structure fitness of the nodes in the LFM method, but produce a large number of community candidates with repeated structures. Therefore, the proposed method could choose the suitable community seeds to reduce the redundant candidates and improve the local search progress.

4   Experiments and Analysis

In the section, the proposed algorithm would be evaluated with other OCDP methods on some real networks with the known communities, in the machine environment with Intel Dual-core 2.26GHz, RAM 4GB and Matlab 2014b.

Table 2. Actual network data sets with the known communities

Network	N_n	N_e	D_a	N_c
$D_1 : KarateClub$	34	78	4.588	2
$D_2 : DolphinNet$	62	159	5.129	2
$D_3 : FootballLeague$	115	613	10.661	12

The data sets about the actual network are shown as the Table 2. The network data sets respectively show several common evaluating networks, D_1: Zacharys karate club, D_2:Dolphin's associations and D_3:NCAA college football league.

The proposed algorithm would be respectively compared with the classical CDPs algorithm FN [1], and three OCDPs methods:CPM [4], LFM [6], and GCE [21]. As for the experiments settings, the all resolution parameter is set as $\alpha = 1$ and the merging community threshold as $\varepsilon = 0.5$. Due to the random initialization, the calculation results are the average of ten test runs. The performance are evaluated by the modularity measurement Q function and $F1Score$.

The Fig. 2 as follow depicts that the proposed RNM method on the karate club networks is able to find the ideal communities as same as the known data set. The red nodes No.3,9,10,31 are the overlapping region, while the overlap nodes indicate the hesitant people in line with relationships in the karate club. Besides, the proposed method with the higher Recall in Fig. 3 shows the nodes partition on overlapping areas are more accurate, except that the Recall of LFM with a mass of redundant computation is slightly higher than RNM on D3.

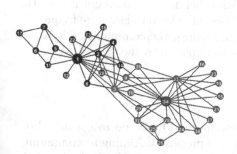   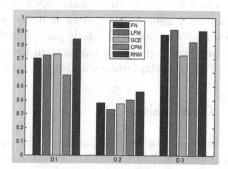

Fig. 2. RNM on karate club network   **Fig. 3.** The comparison results on recall rate

The higher precision in Fig. 4 and F1-score in Fig. 5 prove the RNM are more effective in the most time. The precision on RNM is relatively lower on the karate club, and our method considers the nodes partition in the overlapping area. Algorithm comparison results on the Table 3 show the number of the detected communities always is conform with the known communities.

It is noticed that the modularity Q are not the best result, but the precision on RNM is the highest except on the NCAA college football league. The above reason may come from the resolution limit that lead to discover the communities

Table 3. Performance evaluation with the modularity Q

Networks	FN	LFM	GCE	CPM	RNM
D_1	0.4155(4)	0.4126(2)	0.3955(2)	0.4020(2)	0.3993(2)
D_2	0.5273(4)	0.4793(2)	0.4903(2)	0.5113(2)	0.4925(2)
D_3	0.6051(13)	0.5962(13)	0.6028(12)	0.6085(11)	0.5947(12)

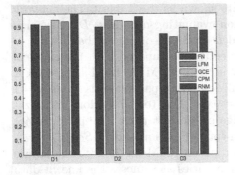

Fig. 4. The comparison results on precision

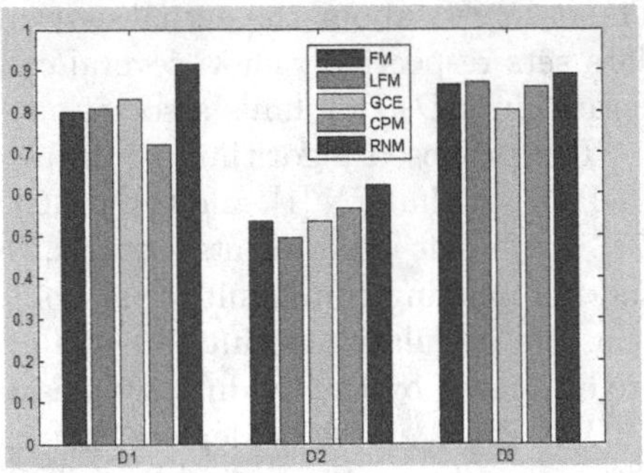

Fig. 5. The comparison results on F1-score

with the similar scale. Besides, on the college football league, the lower precision comes from the several points divided into other wrong communities. The union structure is loose in the community named *Independents* and the teams in the community usually have more matches with other league communities, so the community is easy to be falsely partitioned. Because the matches more frequently happen in few groups of teams than the other teams, the communities *Sun belt* and *Western Athletic* easily lead to be broken apart into the other leagues.

5 Conclusions

In the paper, the local community expansion method based on rough neighborhood with the heuristic triangle optimization is proposed. Aiming at community drift happening on boundary area in the local expansion method, the heuristic strategies on the community seeds are taken with the nodes of generous degree and triangle optimization. The community candidates could be expanded and merged into some natural overlapping communities. Finally the experiments on real networks with other OCDP algorithms also show that the proposed method based on triangle heuristic optimization is more effective and could reveal the more details on the overlap regions. Correspondingly, the real networks also confirm the approach is more consistent with the actual network structure. The future work will in view of the actual problems further combine the triangle optimization strategy and adaptive resolution selection as the focus of the study.

Acknowledgments. The research is supported by the Creative Research Groups and the Qualified Personnel Foundation of Taiyuan University of Technology of China (Grant No. 2014TD056), the National Natural Science Foundation of China (No. 61503273 and 61403329) and General Program (No. 61175054), and the Natural Science Foundation of Shandong Province (No. ZR2013FQ020).

References

1. Girvan, M., Newman, M.E.J.: Community structure in social and biological networks. Proc. Natl. Acad. Sci. **99**(12), 7821–7826 (2002)
2. Fortunato, S.: Community detection in graphs. Phys. Rep. **486**(3), 75–174 (2010)
3. Clauset, A.: Finding local community structure in networks. Phys. Rev. E **72**(2), 026132 (2005)
4. Dernyi, I., Palla, G., Vicsek, T.: Clique percolation in random networks. Phys. Rev. Lett. **94**(16), 160202 (2005)
5. Evans, T.S., Lambiotte, R.: Line graphs, link partitions, and overlapping communities. Phys. Rev. E **80**(1), 016105 (2009)
6. Lancichinetti, A., Fortunato, S.: Detecting the overlapping and hierarchical community structure in complex networks. New J. Phys. **11**(3), 033015 (2009)
7. Ahn, Y.Y., Bagrow, J.P., Lehmann, S.: Link communities reveal multiscale complexity in networks. Nature **466**(7307), 761–764 (2010)
8. Ball, B., Karrer, B., Newman, M.E.J.: An efficient and principled method for detecting communities in networks. Phys. Rev. E **84**(3), 036103 (2011)
9. Havemann, F., Heinz, M., Struck, A., et al.: Identification of overlapping communities and their hierarchy by locally calculating community-changing resolution levels. J. Stat. Mech. Theory Experiment **2011**(01), P01023 (2011)
10. Yang, J., Leskovec, J.: Defining and evaluating network communities based on ground-truth. Knowl. Inf. Syst. **42**(1), 181–213 (2015)
11. Gregory, S.: Fuzzy overlapping communities in networks. J. Stat. Mech. Theory Experiment **2011**(02), P02017 (2011)
12. Zhang, Z., Miao, D., Qian, J.: Detecting overlapping communities with heuristic expansion method based on rough neighborhood. Chin. J. Comput. **36**(10), 2078–2086 (2013)
13. Pawlak, Z.: Rough sets. Int. J. Comput. Inf. Sci. **11**(5), 341–356 (1982)
14. Yao, Y.Y.: Relational interpretations of neighborhood operators and rough set approximation operators. Inf. Sci. **111**(1), 239–259 (1998)
15. Hu, Q., Yu, D., Liu, J., et al.: Neighborhood rough set based heterogeneous feature subset selection. Inf. Sci. **178**(18), 3577–3594 (2008)
16. Radicchi, F., Castellano, C., Cecconi, F., Loreto, V., Parisi, D.: Defining and identifying communities in networks. Proc. Natl. Acad. Sci. **101**, 2658–2663 (2004)
17. Wu, W., Zhang, W.: Neighborhood operator systems and approximations. Inf. Sci. **144**(1), 201–217 (2002)
18. Lin, G., Qian, Y., Li, J.: NMGRS: neighborhood-based multigranulation rough sets. Int. J. Approximate Reasoning **53**, 1080–1093 (2012)
19. Lusseau, D.: The emergent properties of a dolphin social network. Proc. R. Soc. Lond. Ser. B Biol. Sci. **270**, S186–S188 (2003)
20. Newman, M.E.J.: Finding community structure in networks using the eigenvectors of matrices. Phys. Rev. E **74**(3), 036104 (2006)

21. Lee, C., Reid, F., McDaid, A., et al.: Detecting highly overlapping community structure by greedy clique expansion. In: Proceedings of the 4th SNA-KDD Workshop, pp. 33–42 (2010)
22. Lin, T.Y.: Granular computing on binary relations I: data mining and neighborhood systems. In: Proceedings of Rough sets in knowledge discovery (1998)

Modeling and Learning

Rough Sets for Finite Mixture Model Based HEp-2 Cell Segmentation

Abhirup Banerjee$^{(\boxtimes)}$ and Pradipta Maji

Biomedical Imaging and Bioinformatics Lab, Machine Intelligence Unit,
Indian Statistical Institute, Kolkata, India
{abhirup_r,pmaji}@isical.ac.in

Abstract. Automatic extraction of HEp-2 cells from an image is one key component for the diagnosis of connective tissue diseases. The gradual transition between cell and surrounding tissue renders this process difficult for any computer aided diagnostic systems. In this regard, the paper presents a new approach for automatic HEp-2 cell segmentation by incorporating a new probability distribution, called stomped normal (SN) distribution. The proposed method integrates judiciously the concept of rough sets and the merit of the SN distribution into an finite mixture model framework to provide an accurate delineation of HEp-2 cells. The intensity distribution of a class is represented by SN distribution, where each class consists of a crisp lower approximation and a probabilistic boundary region. Finally, experiments are performed on a set of HEp-2 cell images to demonstrate the performance of the proposed algorithm, along with a comparison with related methods.

Keywords: Rough sets · HEp-2 cells · Segmentation · Stomped normal distribution · Expectation-maximization

1 Introduction

The connective tissue disease (CTD) refers to a group of disorders that has the connective tissues of the body as a target of pathology. Many CTDs feature abnormal immune system activity with inflammation in tissues as a result of an immune system that is directed against ones own body tissues (autoimmunity). The CTDs such as systemic lupus erythematosus (SLE), Sjögrens syndrome (SS), systemic sclerosis (SSc), and rheumatoid arthritis (RA) [1] are characterized by the presence of antinuclear autoantibodies (ANAs) in the blood of patients. Using this property, immunofluorescent antinuclear antibody test has become the gold standard in the diagnosis of these disorders [1] for many decades.

Indirect immunofluorescent antinuclear antibody (IIF-ANA) test is a technique that is becoming increasingly important for the diagnosis of the CTDs, because of its simplicity, inexpensiveness, and high sensitivity and specificity [2]. In case of IIF-ANA tests, the most used substrate is the human epithelial type 2 (HEp-2) cells. These cells have larger nuclei and nucleoli than rodent tissue

© Springer International Publishing Switzerland 2015
D. Ciucci et al. (Eds.): RSKT 2015, LNAI 9436, pp. 459–469, 2015.
DOI: 10.1007/978-3-319-25754-9_40

cells, which facilitates their microscopic observation and provides greater sensitivity in the detection of small quantities of antibodies with better resolution of some antigens. However, despite the advantages, the IIF approach is labour intensive, time consuming, and also dependent on the experience and expertise of the physician. Each ANA specimen must be examined by at least two scientists, which renders the test result subjective, and leads to large variabilities across personnel and laboratories [3]. As a consequence of this limitation, there is a strong demand for a complete automation of the procedure that would result in increased test repeatability and reliability, easier and faster result reporting, and lower costs. In recent years, there has been a growing interest towards the realization of computer aided diagnostic (CAD) systems for the analysis of IIF images [2,4–7]. But, these methods execute manual segmentation of HEp-2 cells before applying staining pattern recognition. To make the analysis of IIF images automatic, the automatic segmentation of HEp-2 cells is necessary.

Segmentation is the process of partitioning an image space into some non-overlapping meaningful homogeneous regions. To process each cell individually and automatically, one needs to employ some segmentation method to separate the image from microscope into individual cells. Unfortunately, the process of automatically extracting cells from IIF images is a challenging process due to the gradual transition between cell, artifacts, mitotic cells, and surrounding tissue. This results in the ambiguity of the structural boundaries. Hence, one important problem in cell segmentation from IIF images is uncertainty, which includes imprecision in computations and vagueness in class definitions. In this background, the possibility concept introduced by the theory of probability, fuzzy set, and rough sets have gained popularity in modeling and propagating uncertainty. The segmentation of images using fuzzy c-means has been reported in [8], while rough-fuzzy c-means is used in [9] for image segmentation. One of the most popular framework to model classes for segmentation is the probabilistic model, which labels the pixels according to their probability values, calculated based on the intensity distribution of the image. With a suitable assumption about the distribution, these approaches attempt to solve the problem of estimating the associated class label, given only the intensity value of each pixel. In this regard, finite mixture (FM) model, more specifically, the finite Gaussian mixture (FGM) model, is one of the mostly used model for image segmentation [10,11].

However, all the FM based statistical frameworks generally assume that each class is normally distributed around its mean. Now, normal or Gaussian distribution is a unimodal distribution, which attains highest probability density value at a single point in its range of variation, which is mean; and probability density decreases symmetrically as it traverses from the mean. So, in that case, only a single intensity value (mean of the distribution) ensures the belongingness of a pixel to the class. In this regard, to define each class of images, the paper incorporates a new multimodal probability distribution, termed as stomped normal (SN) distribution [15], that can include multiple intensity values to represent a specific class. Integrating the concept of rough sets and the merit of SN distribution, the paper presents a novel approach for HEp-2 cell segmentation. Each class consists of two regions, namely, a crisp lower approximation and a probabilistic boundary. Integration of rough sets and SN distribution deals with uncertainty, vagueness, and incompleteness in class definition and enables efficient handling of overlapping classes. Finally, the effectiveness of the proposed algorithm, along with a comparison with related algorithms, is demonstrated on a set of HEp-2 cell images both qualitatively and quantitatively.

2   Stomped Normal Distribution

Gaussian distribution is a unimodal distribution. It attains highest probability density at its mean; and probability density decreases symmetrically towards its both ends. If a class follows normal distribution, it indicates that any pixel in the class, has highest probability of belongingness at the mean value of the distribution and the probability decreases with its deviation from mean. Hence, in case of normal distribution, only a single intensity value of mean ensures the association of a pixel to the class. However, there exists multiple intensity values in an image that ensure the association of a pixel to a specific class. A pixel attaining any of those intensity values will surely be considered as an element of the class. But this property is nullified when one uses unimodal Gaussian distribution to model a class, which leads to an inaccurate segmentation performance.

To overcome the shortcomings of Gaussian distribution, in the proposed method, each class is considered as the union of two disjoint parts, namely, lower approximation and boundary region. Let $\underline{A}(\Gamma_l)$ and $B(\Gamma_l)$ be the lower approximation and boundary region of class Γ_l, and $\overline{A}(\Gamma_l) = \{\underline{A}(\Gamma_l) \cup B(\Gamma_l)\}$ denotes the upper approximation of class Γ_l. Hence, each class Γ_l is represented by the tuple $< \underline{A}(\Gamma_l), \overline{A}(\Gamma_l) >$. The lower approximation influences the overlapping characteristics of the class. According to the definitions of lower approximation and boundary region of rough sets [14], if a pixel $i \in \underline{A}(\Gamma_l)$, then $i \notin B(\Gamma_l), \forall l \in \mathcal{L}$, that is, the pixel i is contained in class Γ_l definitely. On the other hand, if $i \in B(\Gamma_l)$, then the pixel possibly belongs to Γ_l and potentially belongs to another class.

In this regard, a new probability distribution, termed as SN distribution [15], is incorporated to fit the intensity distribution of each class. It attains fixed (highest) probability density in a region (lower approximation) around its mean and the density decreases (boundary region) while moving away from the uniform region. The probability density function (pdf) of SN distribution is defined as:

$$f(y; \mu, \sigma, k) = \begin{cases} \frac{\frac{1}{\sigma}\phi(k)}{2(1-\Phi(k)+k\phi(k))}, & \text{if } |\frac{y-\mu}{\sigma}| < k \\ \frac{\frac{1}{\sigma}\phi(\frac{y-\mu}{\sigma})}{2(1-\Phi(k)+k\phi(k))}, & \text{otherwise} \end{cases} \tag{1}$$

where $\phi()$ and $\Phi()$ are the probability density and distribution functions of standard normal distribution respectively. Alternatively, the pdf can be written as:

$$f(y; \mu, \sigma, k) = \frac{1}{D}\frac{1}{\sigma}\phi(z), \quad y \in \mathbb{R}; \quad \text{where} \quad z = \begin{cases} k, & \text{if } |\frac{y-\mu}{\sigma}| < k \\ \frac{y-\mu}{\sigma}, & \text{otherwise} \end{cases} \tag{2}$$

and $D = 2(1 - \Phi(k) + k\phi(k))$. In case $k = 0$, the SN distribution is reduced to normal distribution with mean μ and variance σ^2. Additional properties of SN distribution are discussed in [15].

Combining the concept of rough sets and SN distribution, the lower approximation and boundary region of lth class are defined as follows:

$$\underline{A}(\Gamma_l) = \left\{ i \in \mathcal{S} : |\frac{y_i - \mu_l}{\sigma_l}| < k \right\} ; \tag{3}$$

$$B(\Gamma_l) = \{ i \in \mathcal{S} : i \notin \underline{A}(\Gamma_p), \forall p \in \mathcal{L} \} . \tag{4}$$

Hence, each class Γ_l is represented by a crisp lower approximation $\underline{A}(\Gamma_l)$ and a probabilistic boundary $B(\Gamma_l)$.

3   SNFM: Proposed Segmentation Algorithm

In the proposed segmentation method, the intensity distribution of the image is modeled as a finite mixture of SN distributions. Let y_i be the intensity value of the ith pixel, where $i \in \mathcal{S} = \{1, 2, \cdots, N\}$ and x_i denotes its corresponding label, $x_i \in \mathcal{L} = \{1, 2, \cdots, L\}$. Hence, the image can be represented as a mixture of finite number of SN distributions as follows:

$$p(y_i|\theta) = \sum_{l \in \mathcal{L}} p(y_i|x_i = l)\, \omega_l \quad \forall\, i \in \mathcal{S} \tag{5}$$

where $\quad p(y_i|l) = \frac{1}{D}\frac{1}{\sigma_l}\phi(z_{il})$, $y_i \in \mathbb{R}$, $\quad z_{il} = \begin{cases} k, & \text{if } i \in \underline{A}(\Gamma_l) \\ \frac{y_i - \mu_l}{\sigma_l}, & \text{otherwise} \end{cases}$,

$\omega_l = p(x_i = l)\ \forall i \in \mathcal{S}$, and Γ_l is the class having label l.

Assuming the pixel intensities are statistically independent, the probability density of the entire image can be written as

$$p(\underline{y}|\theta) = \prod_{i \in \mathcal{S}} p(y_i|\theta) = \prod_{i \in \mathcal{S}} \sum_{l \in \mathcal{L}} p(y_i|l)p(l). \tag{6}$$

As the estimation of parameters $\theta = \{\omega_l, \mu_l, \sigma_l; l \in \mathcal{L}\}$ from the above expression using either ML or MAP principles is computationally infeasible, the EM algorithm can be used to solve the above problem. The standard EM algorithm consists of two parts: first it tries to estimate a set of latent variables based on the given data in its E-step; and then in the M-step, it tries to find the optimum estimates of the parameters of the distribution based on the original variables and the new set of latent variables. Iteratively optimizing these two steps, the EM algorithm converges to its local optimum solution. The latent variables for the EM algorithm are defined in this problem as

$$\delta_{il} = \begin{cases} 1, & \text{if } x_i = l \\ 0, & \text{otherwise.} \end{cases}$$

In the E-step, the latent variables are estimated, given the observed variables and the current estimate of the parameters:

$$E(\delta_{il}|y_i, \theta^{(t)}) = \frac{p^{(t)}(y_i|l)w_l^{(t)}}{\sum_{m \in \mathcal{L}} p^{(t)}(y_i|m)w_m^{(t)}} = W_{il}^{(t)} \tag{7}$$

where $\theta^{(t)}$ is the estimate of the parameters at tth iteration. The expression of W_{il} in (7) calculates the belongingness of the ith pixel to Γ_l. Hence, it can be considered as the membership value of pixel i to class Γ_l, and the corresponding expression as the membership function.

As per the definitions of lower approximation and boundary region of a class, based on rough set theory and SN distribution, if a pixel belongs to the lower approximation of a specific class, it definitely belongs to that class. So, the membership value of the pixel to that class should be 1 and to other classes should be 0. On the other hand, the pixels in boundary regions should have different membership values to different classes as there exists ambiguity in its belongingness to a particular class. So, the membership function is modified as:

$$W_{il} = \begin{cases} 1, & \text{if } i \in \underline{A}(\Gamma_l) \\ \dfrac{p(y_i|l)w_l}{\sum_{m \in \mathcal{L}} p(y_i|m)w_m}, & \text{else if } i \in B(\Gamma_l) \\ 0, & \text{otherwise.} \end{cases} \tag{8}$$

In the M-step, the Q-function, that is, expected complete data log-likelihood, is calculated, given the current estimate of the parameters:

$$Q(\theta|\theta^{(t)}) = E_X[\log p(\underline{x}, \underline{y}|\theta)|\underline{y}, \theta^{(t)}]$$

$$= \sum_{i \in \mathcal{S}} \sum_{l \in \mathcal{L}} W_{il}^{(t)} \left\{ \log \omega_l - \log \sigma_l - \frac{z_{il}^2}{2} + C \right\}.$$

Optimizing the Q-function with respect to parameters, we get:

$$\hat{\mu}_l^{(t+1)} = \frac{\sum\limits_{i \in B(\Gamma_l)} W_{il}^{(t)} y_i}{\sum\limits_{i \in B(\Gamma_l)} W_{il}^{(t)}}; \tag{9}$$

and

$$(\hat{\sigma}_l^2)^{(t+1)} = \frac{\sum\limits_{i \in B(\Gamma_l)} W_{il}^{(t)} (y_i - \hat{\mu}_l^{(t+1)})^2}{\sum\limits_{i \in \mathcal{S}} W_{il}^{(t)}}. \tag{10}$$

The derived estimate of μ_l considers only intensity of the boundary region. It puts higher weightage in the non-uniform region, which in presence of noise and outliers degrades the parameter estimation procedure. Hence, computation of the μ_l is modified to include the effects of both lower approximation and boundary region. The final estimate of μ_l, calculated based on the weighting average of the crisp lower approximation and probabilistic boundary, is as follows:

$$\hat{\mu}_l^{(t+1)} = \alpha \, \frac{\sum\limits_{i \in \underline{A}(\Gamma_l)} y_i}{|\underline{A}(\Gamma_l)|} + (1 - \alpha) \, \frac{\sum\limits_{i \in B(\Gamma_l)} W_{il}^{(t)} y_i}{\sum\limits_{i \in B(\Gamma_l)} W_{il}^{(t)}}; \qquad (11)$$

where the parameter α corresponds to the relative importance of lower approximation. Similarly, in the estimate of σ_l^2, the weighted squared deviation of intensity values from their mean in lower approximation region is included in the numerator. But, in this case, equal weightage is given to both lower approximation and boundary region, because assigning higher weightage to any of them will either reduce the effect of non-uniform variation of the data, or increase the effect of noise and outliers. So, the final estimate of σ_l^2 is expressed as follows:

$$(\hat{\sigma}_l^2)^{(t+1)} = \frac{\sum\limits_{i \in \underline{A}(\Gamma_l)} (y_i - \hat{\mu}_l^{(t+1)})^2 + \sum\limits_{i \in B(\Gamma_l)} W_{il}^{(t)} (y_i - \hat{\mu}_l^{(t+1)})^2}{|\underline{A}(\Gamma_l)| + \sum\limits_{i \in B(\Gamma_l)} W_{il}^{(t)}}. \qquad (12)$$

The mixing parameters, that is, ω_l, $l \in \mathcal{L}$, are estimated by using a constrained optimization technique on Q-function,

$$\frac{\partial}{\partial \omega_l} \left[Q(\theta | \theta^{(t)}) + \lambda \left(\sum_{l \in \mathcal{L}} \omega_l - 1 \right) \right] = 0. \qquad (13)$$

where λ is the Lagrange multiplier. Solving (13), the estimate of the mixing parameters is obtained as follows:

$$\hat{\omega}_l^{(t+1)} = \frac{1}{N} \sum_{i \in \mathcal{S}} W_{il}^{(t)}. \qquad (14)$$

4   Experimental Results and Discussion

The performance of the proposed SN distribution and finite mixture (FM) model based segmentation method (SNFM) is studied and compared with that of finite Gaussian mixture (FGM) model [10] and several c-means algorithms: fuzzy c-means (FCM) [12], rough-fuzzy c-means (RFCM) [9], and robust rough-fuzzy c-means (rRFCM) [13] algorithms. To analyze the performance of different algorithms, the experimentation is done on some HEp-2 cell images obtained from "MIVIA HEp-2 Images Dataset" [2]. The comparative performance analysis is studied with respect to three segmentation metrics, namely, Dice coefficient, sensitivity, and specificity.

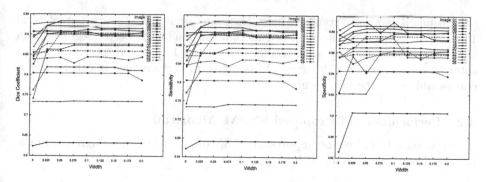

Fig. 1. Variation of several segmentation evaluation indices for different values of width parameter k of SN distribution

Table 1. Performance of proposed SNFM algorithm for HEp-2 cell images

Img.	Dice			Sensitivity			Specificity		
No.	SNFM	FGM	NRS	SNFM	FGM	NRS	SNFM	FGM	NRS
1	**0.933**	0.893	0.893	**0.932**	0.891	0.891	**0.983**	0.973	0.973
2	0.819	0.762	**0.820**	0.823	0.778	**0.823**	0.965	0.956	**0.965**
4	**0.804**	0.804	0.800	**0.804**	0.804	0.802	**0.942**	0.942	0.941
6	0.912	0.907	**0.912**	**0.913**	0.906	0.912	0.978	0.977	**0.978**
9	**0.900**	0.826	0.842	**0.903**	0.834	0.849	**0.989**	0.982	0.983
13	**0.844**	0.839	0.842	**0.849**	0.837	0.846	**0.962**	0.961	0.960
14	**0.903**	0.855	0.855	**0.903**	0.858	0.858	**0.976**	0.964	0.964
15	0.873	0.741	**0.883**	0.874	0.754	**0.882**	0.959	0.922	**0.962**
18	**0.921**	0.869	0.883	**0.919**	0.866	0.880	**0.986**	0.977	0.979
19	0.632	**0.653**	0.633	0.673	**0.689**	0.669	0.903	**0.907**	0.901
20	0.734	**0.753**	0.733	0.753	**0.767**	0.750	0.943	**0.946**	0.941
21	**0.906**	0.897	0.903	**0.908**	0.900	0.905	**0.974**	0.971	0.972
22	0.929	0.925	**0.931**	**0.930**	0.923	0.929	0.971	0.968	**0.971**
23	**0.905**	0.845	0.860	**0.905**	0.847	0.862	**0.980**	0.967	0.971
25	**0.876**	0.853	0.853	**0.884**	0.877	0.877	**0.962**	0.955	0.955
28	**0.859**	0.847	0.858	0.864	0.869	**0.874**	0.958	0.956	**0.958**
p-value		0.0021	0.0125	-	0.0046	0.0078	-	0.0021	0.0055

4.1 Importance of the Width of SN Distribution

The width parameter k of the SN distribution is an important parameter in this problem. Varying the width of SN distribution in the range of 0.0 to 0.2 with common difference 0.025, the proposed algorithm is applied on several HEp-2 cell images; and the results, after comparing using segmentation evaluation indices, are presented in Fig. 1. From the results reported in Fig. 1, it is observed that the proposed method attains its best Dice coefficient in 14 cases out of total 16 cases at width 0.1. The proposed method also achieves better segmentation with respect to sensitivity and specificity in 13 and 14 cases, respectively, out of total 16 cases each, at width $k = 0.1$; whereas its performance in other cases is comparable with the best result.

4.2 Performance of Proposed SNFM Algorithm

This section establishes the importance of SN distribution and rough sets in the proposed segmentation algorithm.

Importance of SN Distribution: From the results reported in Table 1, it can be seen that the proposed SNFM algorithm provides better segmentation than FGM in 14, 13, and 14 cases, out of total 16 cases each, on MIVIA HEp-2 cell images dataset with respect to Dice coefficient, sensitivity, and specificity, respectively. The comparative performance analysis is also reported in terms of p-value computed through Wilcoxon signed-rank test (one-tailed). The SNFM method attains lower p-values for all quantitative indices with respect to FGM, which are statistically significant considering 0.05 as the level of significance.

Importance of Rough Sets: From the results reported in Table 1, it is easily observed that the proposed segmentation method using rough sets produces better segmentation results than its nonrough counter part (NRS) in 11, 13, and 11

Table 2. Comparative performance analysis of SNFM and different clustering algorithms on HEp-2 cell images

Img. No.	Dice				Sensitivity				Specificity			
	SNFM	FCM	RFCM	rRFCM	SNFM	FCM	RFCM	rRFCM	SNFM	FCM	RFCM	rRFCM
1	0.933	**0.942**	0.941	0.942	0.932	**0.942**	0.941	0.942	0.983	**0.986**	0.985	0.985
2	0.819	**0.820**	0.817	0.819	0.823	**0.823**	0.818	0.823	0.965	**0.965**	0.965	0.965
4	**0.804**	0.787	0.787	0.787	**0.804**	0.788	0.788	0.788	**0.942**	0.938	0.938	0.938
6	**0.912**	0.865	0.834	0.872	**0.913**	0.869	0.840	0.875	**0.978**	0.967	0.960	0.969
9	0.900	0.907	0.901	**0.907**	0.903	0.907	0.900	**0.907**	0.989	0.990	0.989	**0.990**
13	**0.844**	0.797	0.755	0.837	**0.849**	0.809	0.773	0.843	**0.962**	0.951	0.942	0.960
14	**0.903**	0.843	0.843	0.843	**0.903**	0.846	0.846	0.846	**0.976**	0.961	0.961	0.961
15	**0.873**	0.864	0.873	0.873	**0.874**	0.866	0.874	0.874	**0.959**	0.957	0.959	0.959
18	**0.921**	0.909	0.906	0.914	**0.919**	0.913	0.910	0.917	**0.986**	0.984	0.983	0.985
19	**0.632**	0.613	0.613	0.613	**0.673**	0.660	0.660	0.660	**0.903**	0.899	0.899	0.899
20	0.734	0.753	**0.762**	0.692	0.753	**0.768**	0.765	0.708	0.943	0.946	**0.948**	0.936
21	**0.906**	0.832	0.832	0.857	**0.908**	0.832	0.832	0.856	**0.974**	0.954	0.954	0.960
22	**0.929**	0.868	0.847	0.889	**0.930**	0.874	0.854	0.893	**0.971**	0.946	0.938	0.954
23	**0.905**	0.896	0.890	0.901	**0.905**	0.897	0.890	0.901	**0.980**	0.978	0.977	0.979
25	**0.876**	0.872	0.862	0.872	**0.884**	0.867	0.853	0.867	**0.962**	0.960	0.957	0.960
28	**0.859**	0.807	0.807	0.831	**0.864**	0.796	0.796	0.824	**0.958**	0.944	0.944	0.950
p-value		3.8E-3	4.1E-3	4.2E-3	-	3.1E-3	2.1E-3	2.9E-3	-	3.1E-3	4.1E-3	2.4E-3

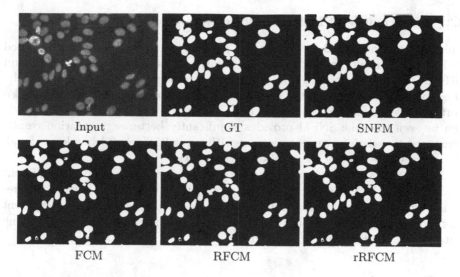

Fig. 2. Input image of HEp-2 cells (image no. 18), ground truth, and segmented images

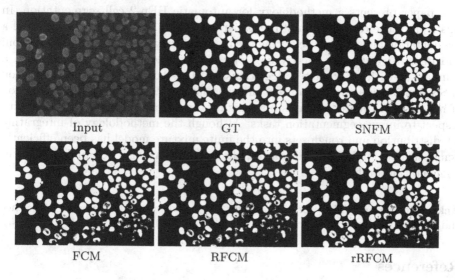

Fig. 3. Input image of HEp-2 cells (image no. 22), ground truth, and segmented images

cases, out of total 16 cases each, with respect to Dice coefficient, sensitivity, and specificity, respectively. Hence, all the results reported in Table 1 establish the importance of using rough sets in terms of highest segmentation accuracy. The comparative performance analysis is also reported in terms of p-value computed through Wilcoxon signed-rank test (one-tailed). The SNFM method attains statistically significant segmentation results than its non rough counter part with respect to lower p-values (< 0.05) for all segmentation evaluation indices.

4.3   Performance of Different Clustering Algorithms

This section presents the comparative performance analysis of the proposed SNFM method and several clustering algorithms such as FCM, RFCM, and rRFCM. Results are reported in Table 2 with respect to three indices, namely, Dice coefficient, sensitivity, and specificity. The Wilcoxon signed-rank test is also performed for significance analysis. From all the results reported in Table 2, it can be seen that the SNFM provides significantly better segmentation results compared to existing segmentation algorithms at 95 % confidence level. In all the cases, the p-value of the test is less than 0.05.

Figs. 2 and 3 depict the comparative segmentation performance of different algorithms on HEp-2 cell images, along with their original images and corresponding ground truth images. The segmented outputs generated by different methods establish the fact that the proposed method generates more promising outputs that do the existing algorithms.

5   Conclusion

The paper presents a methodology for automatic HEp-2 cell segmentation. In this regard, the contribution of the paper is two-fold, namely, development of a segmentation algorithm, integrating the merits of rough sets and SN distribution; and demonstrating the effectiveness of the proposed algorithm, along with a comparison with other related algorithms, on HEp-2 cell images. The integration of SN distribution and the concept of lower approximation and boundary region of rough sets is geared towards maximizing the utility of rough clustering with respect to image segmentation tasks. Although the methodology of integrating SN distribution and rough sets into the finite mixture model has been efficiently demonstrated for HEp-2 cell images, the concept can also be applied to other segmentation problems.

Acknowledgment. This work is partially supported by the Department of Science and Technology, Government of India, New Delhi (grant no. SB/S3/EECE/050/2015).

References

1. Meroni, P.L., Schur, P.H.: ANA screening: an old test with new recommendations. Autoimmun. Rev. **69**(8), 1420–1422 (2010)
2. Foggia, P., Percannella, G., Soda, P., Vento, M.: Benchmarking HEp-2 cells classification methods. IEEE Trans. Med. Imaging **32**(10), 1878–1889 (2013)
3. Soda, P., Iannello, G.: Aggregation of classifiers for staining pattern recognition in antinuclear autoantibodies analysis. IEEE Trans. Inf. Technol. Biomed. **13**(3), 322–329 (2009)
4. Stoklasa, R., Majtner, T., Svoboda, D.: Efficient k-NN based HEp-2 cells classifier. Pattern Recogn. **47**(7), 2409–2418 (2014)

5. Theodorakopoulos, I., Kastaniotis, D., Economou, G., Fotopoulos, S.: HEp-2 cells classification via sparse representation of textural features fused into dissimilarity space. Pattern Recogn. **47**(7), 2367–2378 (2014)

6. Wiliem, A., Sanderson, C., Wong, Y., Hobsone, P., Minchin, R.F., Lovell, B.C.: Automatic classification of human epithelial type 2 cell indirect immunofluorescence images using cell pyramid matching. Pattern Recogn. **47**(7), 2315–2324 (2014)

7. Yang, Y., Wiliem, A., Alavi, A., Lovell, B.C., Hobson, P.: Visual learning and classification of human epithelial type 2 cell images through spontaneous activity patterns. Pattern Recogn. **47**(7), 2325–2337 (2014)

8. Gong, M., Liang, Y., Shi, J., Ma, W., Ma, J.: Fuzzy C-means clustering with local information and kernel metric for image segmentation. IEEE Trans. Image Proces. **22**(2), 573–584 (2013)

9. Maji, P., Pal, S.K.: Rough set based generalized fuzzy C-means algorithm and quantitative indices. IEEE Trans. System, Man, and Cybern., Part B: Cybern. **37**(6), 1529–1540 (2007)

10. Liang, Z., MacFall, J.R., Harrington, D.P.: Parameter estimation and tissue segmentation from multispectral MR images. IEEE Trans. Med. Imaging **13**(3), 441–449 (1994)

11. Nguyen, T.M., Wu, Q.M.J.: Fast and robust spatially constrained gaussian mixture model for image segmentation. IEEE Trans. Circuits Syst. Video Technol. **23**(4), 621–635 (2013)

12. Hall, L.O., Bensaid, A.M., Clarke, L.P., Velthuizen, R.P., Silbiger, M.S., Bezdek, J.C.: A comparison of neural network and fuzzy clustering techniques in segmenting magnetic resonance images of the brain. IEEE Trans Neural Netw. **3**(5), 672–682 (1992)

13. Maji, P., Paul, S.: Rough-fuzzy clustering for grouping functionally similar genes from microarray data. IEEE/ACM Trans. Comput. Biol. Bioinf. **10**(2), 286–299 (2013)

14. Pawlak, Z.: Rough Sets: Theoretical Aspects of Reasoning about Data. Kluwer Academic, Dordrecht (1991)

15. Banerjee, A., Maji, P.: Rough sets and stomped normal distribution for simultaneous segmentation and bias field correction in brain MR images. IEEE Trans. Image Proces. (Accepted)

Water Quality Prediction Based on an Improved ARIMA- RBF Model Facilitated by Remote Sensing Applications

Jiying Qie[✉], Jiahu Yuan, Guoyin Wang, Xuerui Zhang,
Botian Zhou, and Weihui Deng

Big Data Mining and Applications Center, Chongqing Institute of Green
and Intelligent Technology, Chinese Academy of Sciences,
Chongqing 400714, China
qiejiying@cigit.ac.cn

Abstract. Remote sensing technique are great used to assess and monitor water quality. An efficient and comprehensive method in monitoring water quality is of great demand to prevent water pollution and to mitigate the adverse impact on the livestock and crops caused by polluted water. This study focused on a typical water area, where eutrophication is the main problem, and thus, the total nitrogen was chosen as an important parameter for this study. The research contains two parts. The first part is the methodology development, an algorithms was proposed to inverse the total nitrogen (TN) concentrations from the field imagery acquisition. The squared correlation coefficient between the inversion values and measured values was 0.815. The second part is the deduction of water quality parameter (TN) from upstream to downstream. An improved hybrid model of Autoregressive Integrated Moving Average (ARIMA) model and Radial basis function neural network (RBF-NN) was developed to simulate and forecast variation trend of the water quality parameter. We evaluated our method using data sets from satellite. Our method achieved the competing predicting performance in comparison with the state-of-the-art method on missing data completion and data predicting. Generally, the evaluation results indicated that the developed methods were successfully applied in forecasting the water quality parameters and filling in missing data which cannot be inversed in space by satellite images due to the cloud and mist interference, and were of promising accuracy.

Keywords: HJ-1 · Water quality prediction · Total nitrogen · ARIMA · Radial basis function · Hybrid model

1 Introduction

Water quality using remote sensing has been studied since the 1980s with different sensors in almost every inland water. Remote sensing of lakes using satellite images offers a good spatial and temporal coverage while some variables of water quality can only be assessed up to several times per year such as chlorophyll-a (Chl-a), total suspended sediment (TSS), suspended minerals (SM), turbidity, Secchi disk depth (SDD), particulate organic carbon, and colored dissolved organic matter (CDOM) [1].

© Springer International Publishing Switzerland 2015
D. Ciucci et al. (Eds.): RSKT 2015, LNAI 9436, pp. 470–481, 2015.
DOI: 10.1007/978-3-319-25754-9_41

Many studies used multi-spectral remote sensing image to acquire nitrogen, phosphorus, chlorophyll- a concentrations and other water quality parameters and then evaluate eutrophication, which achieved good results [2, 3]. Eutrophication is one of the common stressor with obvious symptoms (e.g., cyanobacterial blooms) such as changes in biotic communities, food web disturbances, and degradation of water quality leading to biodiversity loss [4]. Studies have shown that, nitrogen and phosphorus concentration of the water body are important factors affecting algal blooms and eutrophication [5–10].

In this study, in order to enhance the reliability of sampling monitoring samples, an effort has been made in assessing total nitrogen concentration and determining the environmental quality of a typical water area through a remote sensing application and particularly the HJ-1 satellite imagery of same dates as the sampling campaigns.

In recent years, due to convergence of environmental concerns and availability of innovative computational intelligence approaches, the level of interest in the analysis and prediction of water quality has increased substantially [11, 12]. ARIMA models are one of the most popular model based on mathematic theory for time series forecasting over the past two decades, which have enjoyed useful applications in forecasting economic [13] and environment [14, 15]. Although ARIMA models are quite flexible in that they can represent several different types of time series, e.g., pure autoregressive (AR), pure moving average (MA) and combined AR and MA (ARMA) series, their major limitation is the pre-assumed linear form of the model. Therefore, no nonlinear patterns can be captured by the ARIMA models. Over the past several years, some nonlinear models based on Data Driven for time series data forecasting have been proposed as alternative techniques. The most widely used nonlinear models for time series forecasting is the artificial neural networks (ANNs) model [16–21]. Zhang and Hu [22] summarized the different applications of neural networks for predictions. Hatzikos et al. [23] utilized ANNs for the prediction of seawater quality indictors like PH and dissolved oxygen (DO). Chang [19] presented a comparative study of the ANNs for digital game content stocks price prediction. Another ANNs models applied for time series prediction are self-organizing RBF model [24], adaptive BP neural networks model [25], and dynamic evolving neural-fuzzy model [26], etc.

As many problems are composed of both linear and nonlinear correlation structures, using the hybrid models has become a common practice to overcome the limitations of components and improve the forecasting accuracy. Zhang [27] developed a hybrid model of ARIMA and ANN, and applied the model to some real data sets. Faruk [28] used the hybrid model of ARIMA and back propagation neural network for water quality prediction of the Büyük Menderes basin located in southwest Turkey area. However, all the hybrid models mentioned above are based on two assumptions [29]. One assumption is that the relationship between the linear and nonlinear components is additive and this may underestimate the relationship between the components and degrade performance. Another is that one may not guarantee that the residuals of the linear component may comprise valid nonlinear patterns. In this paper, on the basis of Zhang's model, a novel hybrid model combining ARIMA model and RBF-NN model was applied in the remote sensing of water quality data prediction.

The outline of this paper is as follows. In Sect. 2, we describe the materials and methods including the basic concepts and the form of the proposed model and its components. In Sect. 3, we show and analyze experiment results. And in the last section we draw the conclusion.

2   Materials and Methods

2.1   Study Area

Since the adverse impact from eutrophication in recent years, a typical water area was selected as a focal pilot area for water quality in this study. It is a tributary of the Yangtze River, originated in the junction of Hunan, Hubei and Jiangxi province. The study area is located approximately between 29 28'-29 55' N and 113 32'-114 13' E, covers an area of 57 km^2, and the average annual runoff is 2.8 billion m^3 (Fig. 1). Climate in the area is typical subtropical and monsoonal. It has hot and humid summer but cold and dry winters, with 1600 mm annual average rainfall. The reservoir has played an important role in flood control, power generation, and navigation.

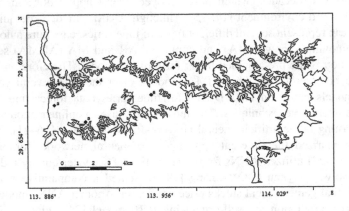

Fig. 1. Distribution of water quality sampling points

Field measurements of total nitrogen has been conducted at six sampling stations on October 10, 2011. The values are as shown in Table 1. The sampling network was established in order to cover the lake spatially, taking into account authropogenic pressures, the different habitats, and the hydromorphological conditions of the lake. Water samples were collected from the upper layer of the surface (approximately 50 cm) and transported to the laboratory for analysis. In the study area, the total nitrogen (TN) level is high.

2.2   Data Sources

The HJ-1A/B satellite image of the water area on 10 October 2011 was used for this study. It was acquired from China Centre For Resources Satellite Data and Application. Due to the small size of the area, the number of satellite data and sampling stations

Table 1. Statistical comparison of observed and predicted data from the proposed model and other three modes

Model	TN		
	R	MAPE (%)	RMSE (%)
ARIMA	0.9056	0.27	0.45
RBF-NN	0.9172	0.25	0.38
Zhang's model	0.9126	0.26	0.39
Our proposed model	0.9269	0.23	0.36

were considered to be adequate for monitoring variation of total nitrogen concentration. Previous studies indicated that the middle infrared bands(TM5 and TM7) showed low correlations and independent relationships with water quality parameters, which could be due to the low water depth penetration of these wavelengths [30], since they are absorbed in just a few centimeters of water. Accordingly, these bands were omitted from further consideration; thus, the analysis was restricted to TM bands 1 to 4 and their combinations for total nitrogen estimation.

2.3 Digital Data Processing

After selecting the study area scenes and the appropriate dates (near simultaneous remote sensing data collection at the time of water samples), the digital data were submitted to the pre-processing including radiometric calibration, data clipping, geometric correction, atmospheric correction, and band extraction. A patch for reading the data from HJ-1 satellite was used to calibrate the radiance of the raw images. To increase the speed of computation, the raw images were clipped to obtain the images for the study area. Furthermore, atmospheric correction was conducted using the ENVI software with a spectral curve based on the spectral response function provided by the HJ-1 satellite data system.

2.4 Fundamental Theory

The patterns and changing trend of the water quality parameters data may not easily be captured by stand-alone models since water quality data could include a variety of characteristics like a non-Gaussian error or heteroskedasticity. Appling the ARIMA model to complex nonlinear problems may not adequate and using the ANNs to model linear problems may have obtained unsatisfactory results. Therefore, the hybrid techniques that decompose a time series into its linear and nonlinear components are a compromise and effective approach. Zhang [27] have successfully applied the hybrid model to the three well-known data sets and yielded satisfactory results. However, there are two assumptions mentioned in Sect. 1 with both Zhang's model and the other hybrid models, which have been applied for water quality prediction. A novel hybrid model, which can overcome the above-mentioned assumptions of traditional hybrid models and also limitations of linear and nonlinear models by using the unique advantages of ARIMA and ANNs models, will be proposed in the following part.

As mentioned in the linear and nonlinear hybrid models literature [26–29], the water quality time series also can be considered to be decomposed into its linear autocorrelation structure and its nonlinear component. In the proposed model, the water quality time series is considered as a function of a linear and nonlinear component. Thus, the predicted value for time t can be described by

$$y_t = f(L_t, N_t),$$ (1)

where L_t denotes the linear component and N_t denotes the nonlinear component. Both of the two components have to be estimated from the water quality time series data.

In the first stage, a ARIMA model was used to model the linear component, then the residuals from the linear model will contain only the nonlinear relationship. We can obtain the following expression:

$$L_t = \left[\sum_{i=1}^{p} \phi_i z_{t-i} - \sum_{j=1}^{q} \theta_j \alpha_{t-j} \right] + e_t = \hat{L}_t + e_t,$$ (2)

Where L_t is the predicted value at time t from the Eq. (1), $zt = (1 - B^d)(yt - \mu)$, and e_t is the residual at time t from the linear model. Residuals are very important in diagnosis of the sufficiency of the linear model. A linear model is not sufficient if there are still linear correlation structures left in the residuals. However, any nonlinear patterns cannot be detected by residual analysis. For this reason, even if a model has passed diagnostic checking, the model may still not be adequate in that nonlinear relationships have not been appropriately modeled. Any significant nonlinear pattern in the residuals will indicate the limitation of the ARIMA model. The results of the first stage are the predicted values and residuals of the linear model. Then these values are used as input for the next stage.

In the second stage, a Radial Basis Function Neural Network (RBF-NN) model is used in order to model the nonlinear and probable linear relationships existing in residuals of the linear model and the observed data. Thus, the combined predictive model can be as follows:

$$\begin{aligned} y_t &= f(N_t^1, \hat{L}_t, N_t^2) \\ &= f(z_{t-1}, \dots, z_{t-\hat{m}}, \hat{L}_t, e_{t-1}, \dots, e_{t-\hat{n}}) \\ &= \sum_{t=1}^{N} \omega_i \phi_i(Xt) \end{aligned}$$ (3)

$$N_t^1 = f^1(z_{t-1}, \dots, z_{t-m}),$$ (4)

$$N_t^2 = f^2(e_{t-1}, \dots, e_{t-m}),$$ (5)

where f^1, f^2, f are the nonlinear function determined by the RBF-NN. N is the number of the hidden neuron. N and m are integers and often referred to as orders of the model, respectively, and $\hat{n} \le n$ and $\hat{m} \le m$ are integers that are determined in design process of final RBF neural network. We should note that anyone of the above-mentioned variables $z_i = (i = t - 1, \ldots, t - m)$, $\hat{L}_t$ and $e_i (i = t - 1, \ldots, t - n)$ may be deleted in the design process of final neural network.

3   Experiments and Analysis

3.1   Calibration of Models Relating HJ-1 and Total Nitrogen Data

The HJ-1A/B satellite data in conjunction with in situ water sampling provide the means in establishing a relationship between satellite-derived reflectance values and total nitrogen concentration. Initially, attempts were made to find combinations, transformations, or logarithmical transformations of bands which would provide more information about total nitrogen concentration in the reservoir than only one band. Such combinations and band transformations concern ratios of B1/B2, B2/B3, B1/B4, B2/B4, and B3/B4; multiplications of B1*B4 and B2*B4; and the logarithmical transformations of log(B1/B2), log(B1/B3), etc. Subsequently, digital numbers of each transformed image were retrieved from those regions where the sampling stations are located. Linear regression models were developed between reflectivity values and in situ measurements for all transformed images and sampling stations. Dependent variables were total nitrogen concentration, and independent variables were image bands, band ratios, and other band combinations (Fig. 2).

This regression analysis resulted in models that predict total nitrogen concentration, and the selection of the best applicable model was based on the value of the correlation of determination between the reflectance and the values of total nitrogen. Predictive model with high coefficient of determination (R2), which was developed based on field sampling of 10 October 2011, was applied to assess water quality and the deduction of the water quality parameter from upstream to downstream. As the experience shows, a strong association was expected between total nitrogen and nitrate (Fig. 3).

3.2   Forecasting of Water Quality

The water quality parameter(TN) has totally 578 samples. We used the first 2/3 samples to train the model and the rest 1/3 samples were used for model verification and comparison. In each model, the input and output of the data set were firstly normalized to the range of [0,1].

The result of the ARIMA modeling process was shown in Fig. 4(a). The Pearson product moment correlation coefficient (R) values between the observed values and the predicted values for total nitrogen (TN) was 0.9056, the mean absolute percentage errors (MAPE) was 0.27 %, and the root mean square errors (RMSE) was 0.0045.

We applied the RBF-NN models to the one-step-ahead prediction of the total nitrogen (TN). In this paper, the Gaussian kernel function was selected as the Radial Basis Function and the learning rate was set to be 2.25. Through repeated experiments, the best structure of RBF-NN was $N^{(7-9-1)}$ for the water quality parameter (TN) prediction. In the model verification phase, the last third samples were used to predict the

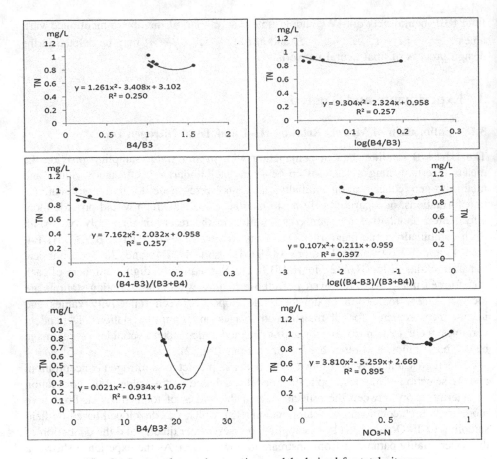

Fig. 2. Scatter plots and regression models derived for total nitrogen

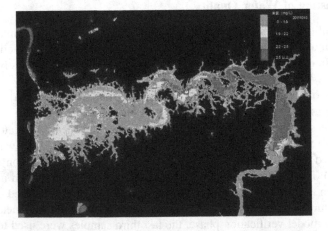

Fig. 3. Total nitrogen map regression model

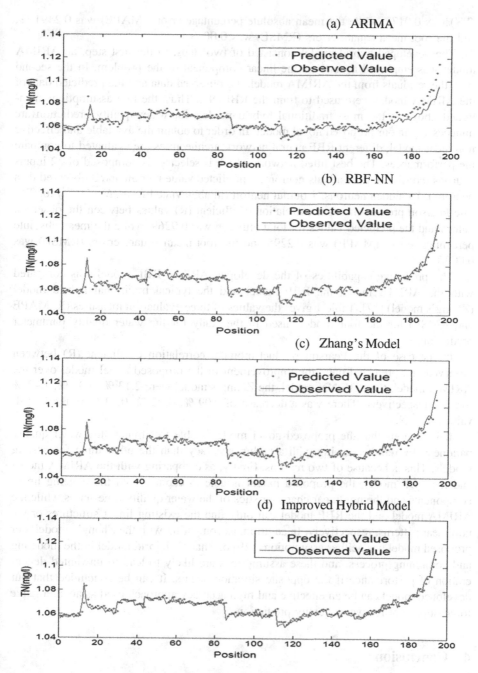

Fig. 4. Observed versus predicted data for TN

water quality parameters. Figure 4(b) compares the model predictions for water quality parameters with the observations. The Pearson product moment correlation coefficient (R) value between the observed values and the predicted values for total nitrogen

(TN) was 0.9172, while the mean absolute percentage errors (MAPE) was 0.2491 %, and the root mean square errors (RMSE) was 0.0038.

The developed hybrid model consisted of two steps. In the first step, an ARIMA model was employed to analyze the linear component of the problem. In the second step, the residuals from the ARIMA model, the observed data and the predicted data of the ARIMA model were used to train the RBF-NN. Thus, the two assumptions mentioned above in the most traditional hybrid models for water quality prediction are unnecessary in our proposed novel model. In order to obtain the available and effective nonlinear model, different RBF neural network architectures are evaluated to compare the performances. The best fitted network which is selected, is composed of 22 inputs neurons (consist of 18 residuals neurons, 1 predicted value neuron and 3 observed data neurons), 15 hidden neurons, 1 output neuron (in abbreviated form, $N^{(22-15-1)}$) for TN. The Pearson product-moment correlation coefficient (R) values between the observed values and the predicted values for total nitrogen was 0.9269, while the mean absolute percentage errors (MAPE) was 0.2295 and the root mean square errors (RMSE) was 0.0036.

The predictive capabilities of the developed ARIMA-RBF model was compared with the ARIMA mode, the RBF model and the typical traditional hybrid model (Zhang's model) [27]. Table 1 gives the values of three evaluation measures (R, MAPE and RMSE) for the four models used in the study for the water quality parameter prediction.

In the case of the Pearson product-moment correlation coefficient (R) between observed and predicted data, the improvement of the proposed novel model over the ARIMA model, the RBF-NN model, the Zhang's model were 2.13 %, 0.97 %, 1.43 % for TN, respectively. There was a decrease of 0.09 %, 0.02 %, 0.03 % in the RMSE values for TN.

It is obvious that the proposed novel model is able to simulate the water quality parameter by remote sensing with a higher accuracy than the other three forecasting models. This is because of two reasons. Firstly, as comparing with the ARIMA model and the RBF model, the proposed model is able to capture both the existing linear components and the nonlinear characteristics of the water quality time series while the ARIMA model and the RBF model can only find the existing linear structures or the nonlinear patterns, respectively. In addition, as comparing with the Zhang's model, our proposed model has neither assumptions [29] of Zhang's hybrid model in the modeling and combining process, and these assumptions are likely to lead to unwanted degeneration of performance if the opposite situation occurs. It can be concluded that our developed model can be an effective and high accuracy approach used as an alternative forecasting tool for water quality prediction.

4   Conclusions

Remote sensing has been proven to offer good spatial and temporal coverage while some variables of water quality can potentially be assessed up to several times per year. The results of this study showed that HJ-1 data can be used effectively to determine total nitrogen concentrations and that typical ground reference data can be used in combination with satellite imagery to establish regression models, when continuous or frequent in sit measurements are impossible to obtain at a long-term basis, which is usually the case. However, this particular algorithm should be improved since it has not reached high levels of accuracy. Possible error sources include the geo-referencing of satellite data and conversion of the sampling stations coordinates, the influence of aquatic plants, and the fact that regression analysis was based on only six sampling sites. In order to increase predicting accuracy and produce more widely applicable models, more frequent water samples, a denser network of stations, and satellite data of greater resolution shall be focused on future.

Improving the forecasting accuracy for the water quality parameter is an important and difficult task for decision makers of water resource management. In recent years, hybrid techniques that decompose a water quality time series into its linear and non-linear form are gradually becoming a mainstream in water quality prediction. In this paper, the proposed novel hybrid model can overcome the above-mentioned limitations and obtain the more general and accurate forecasting outcome than traditional hybrid ARIMA-ANN models. In the ARIMA-RBF model, we used the unique capability of ARIMA model to identify and magnify the existing linear component, and then a RBF neural network was used to capture the underlying data generating process and predict changes on future, by using the residuals from the linear model, the observed data and the predicted value of the ARIMA model. The experiment results have successfully shown that the prediction conducted by the developed model performed much better than either single model or the traditional hybrid model.

Acknowledgements. This work is supported by National Science and Technology Major Project (2014ZX07104-006), the Hundred Talents Program of CAS (NO. Y21Z110A10).

References

1. Allan, M.G., Hamilton, D.P., Hicks, B.J., Brabyn, L.: Landsat remote sensing of chlorophyll a concentrations in central North Island lakes of New Zealand. Int. J. Remote Sens. **32**, 2037–2055 (2011)
2. Torbick, N., Hu, F., Zhang, J., Qi, J., Zhang, H., Becker, B.: Mapping chlorophyll-a concentrations in West Lake, China using Landsat 7 ETM+. J. Great Lakes Res. **34**, 559–565 (2008)
3. Qiao, P.L., Zhang, J.X., Lin, Z.J.: The application of remote sensing technique to monitoring and evaluating water pollution in the Shiyang river valley. Remote Sens. Land Resour. **4**, 39–45 (2003)
4. Moustaka-Gouni, M., Vardaka, E., Michaloudi, E., Kormas, K.A., Tryfon, E., Mihalatou, H., et al.: Plankton food web structure in a eutrophic polymictic lake with a history in toxic cyanobacterial blooms. Limnol. Oceanogr. **51**, 715–727 (2006)
5. Ji, W., Wu, Y.: Jiangxi Poyang Lake National Natural Reserve Study. China Forestry Press, Beijing (2002)
6. Ji, W., Lin, W., Huang, X.: Wuhan East Lake Water Column Floating Particulate Organic Carbon, Nitrogen, Phosphorous Decade Dynamic. Science Press (1995)

7. Lin, W., Wang, J.: Research of Wuhan East Lake Phosphorous Nutritional Status, pp. 108–128. Science Press (1995)
8. Wang, Y., Jiao, N.Z.: Research progresses in nutrient bottom-up effect on phytoplankton growth. Mar. Sci. Qingdao Chin. Ed. **24**, 30–32 (2000)
9. Qiong, C.: The influence to water bloom by nitrogen, phosphorous. Bull. Biol. **41**(5), 12–14 (2006)
10. Yang, L., Qin, B., Chen, F., et al.: Eutrophication Mechanisms and control technology and its applications. Chin. Sci. Bull. **51**(16), 1857–1866 (2006)
11. Chau, K.W.: A review on integration of artificial intelligence into water quality modelling. Mar. Pollut. Bull. **52**, 726–733 (2006)
12. Zou, X., Wang, G., Gou, G., Li, H.: A divide-and-conquer method based ensemble regression model for water quality prediction. In: Lingras, P., Wolski, M., Cornelis, C., Mitra, S., Wasilewski, P. (eds.) RSKT 2013. LNCS, vol. 8171, pp. 397–404. Springer, Heidelberg (2013)
13. Kumar, M., Anand, M.: An application of time series arima forecasting model for predicting sugarcane production in India. Stud. Bus. Econ. **9**, 81–94 (2014)
14. Kavasseri, R.G., Seetharaman, K.: Day-ahead wind speed forecasting using f-ARIMA models. Renewable Energy **34**, 1388–1393 (2009)
15. Sun, H., Koch, M.: Time series analysis of water quality parameters in an estuary using Box-Jenkins ARIMA models and cross correlation techniques. Comput. Methods Water Resour. **11**, 230–239 (1996)
16. Tang, Z., de Almeida, C., Fishwick, P.A.: Time series forecasting using neural networks vs. Box-Jenkins methodology. Simulation **57**, 303–310 (1991)
17. Tang, Z., Fishwick, P.A.: Feedforward neural nets as models for time series forecasting. ORSA J. Comput. **5**, 374–385 (1993)
18. Zhang, G.P., Patuwo, B.E., Hu, M.Y.: A simulation study of artificial neural networks for nonlinear time-series forecasting. Comput. Oper. Res. **28**, 381–396 (2001)
19. Chang, T.S.: A comparative study of artificial neural networks, and decision trees for digital game content stocks price prediction. Expert Syst. Appl. **38**, 14846–14851 (2011)
20. Al-Saba, T., El-Amin, I.: Artificial neural networks as applied to long-term demand fore-casting. Artif. Intell. Eng. **13**, 189–197 (1999)
21. Hwarng, H.B.: Insights into neural-network forecasting of time series corresponding to ARMA (p, q) structures. Omega **29**, 273–289 (2001)
22. Zhang, G., Hu, M.Y.: Neural network forecasting of the British Pound/US Dollar exchange rate. Omega **26**, 495–506 (1998)
23. Hatzikos, E., Anastasakis, L., Bassiliades, N., Vlahavas, I.: Simultaneous prediction of multiple chemical parameters of river water quality with tide. In: Proceedings of the Second International Scientific Conference on Computer Science, IEEE Computer Society, Bulgarian Section (2005)
24. Han, H.G., Chen, Q.L., Qiao, J.F.: An efficient self-organizing RBF neural network for water quality prediction. Neural Networks **24**, 717–725 (2011)
25. Yu, S., Zhu, K., Diao, F.: A dynamic all parameters adaptive BP neural networks model and its application on oil reservoir prediction. Appl. Math. Comput. **195**, 66–75 (2008)
26. Kasabov, N.K., Song, Q.: DENFIS: dynamic evolving neural-fuzzy inference system and its application for time-series prediction. IEEE Trans. Fuzzy Syst. **10**, 144–154 (2002)
27. Zhang, G.P.: Time series forecasting using a hybrid ARIMA and neural network model. Neurocomputing **50**, 159–175 (2003)
28. Faruk, D.Ö.: A hybrid neural network and ARIMA model for water quality time series prediction. Eng. Appl. Artif. Intell. **23**, 586–594 (2010)

29. Taskaya-Temizel, T., Casey, M.C.: A comparative study of autoregressive neural network hybrids. Neural Networks **18**, 781–789 (2005)
30. Lathrop, R.G.: Use of Thematic Mapper data to assess water quality in Green Bay and central Lake Michigan. Photogramm Eng. Remote Sens. **52**, 671–680 (1986)
31. Khashei, M., Bijari, M.: A novel hybridization of artificial neural networks and ARIMA models for time series forecasting. Appl. Soft Comput. **11**, 2664–2675 (2011)

Roughness in Timed Transition Systems Modeling Propagation of Plasmodium

Andrew Schumann[1] and Krzysztof Pancerz[1,2]([✉])

[1] University of Information Technology and Management, Sucharskiego Str. 2,
35–225 Rzeszów, Poland
andrew.schumann@gmail.com
[2] Chair of Computer Science, Faculty of Mathematics and Natural Sciences,
University of Rzeszów, Prof. S. Pigonia Str. 1, 35-310 Rzeszów, Poland
kkpancerz@gmail.com

Abstract. In the paper, we propose to use rough sets to describe some ambiguities in anticipation of states in propagation of plasmodium modeled by timed transition systems. A *Physarum* machine, that is a biological computing device implemented in the plasmodium of *Physarum polycephalum*, is considered as a modeled system. The plasmodial stage of *Physarum polycephalum* can be treated as a natural transition system. In the presented approach, both a standard definition of rough sets and the Variable Precision Rough Set Model (VPRSM) are used.

Keywords: Rough sets · Variable precision rough set model · Timed transition systems · *Physarum* machines

1 Introduction

Rough sets proposed by Z. Pawlak [10] are an appropriate tool to deal with rough (ambiguous, imprecise) concepts in the universe of discourse. Therefore, we propose to apply rough sets to model some ambiguities in anticipation of states that can be noticed in propagation of plasmodium. As an example of the modeled system, a *Physarum* machine is considered. This machine is a programmable amorphous biological computer, experimentally implemented in the vegetative state of *Physarum polycephalum* (also called slime mould) [1]. *Physarum polycephalum* is a one-cell organism belonging to the species of order *Physarales*. The plasmodium of *Physarum polycephalum* spread by networks can be programmable. In propagating and foraging behavior of the plasmodium, we can perform useful computational tasks. This ability was noticed by T. Nakagaki et al. in [5]. There has been, and still is, a lot of research on *Physarum polycephalum*. In *Physarum Chip Project: Growing Computers from Slime Mould* [2] funded by the Seventh Framework Programme (FP7), we are going to construct an unconventional computer on programmable behavior of *Physarum polycephalum*. The *Physarum* machine comprises an amorphous yellowish mass with networks of protoplasmic veins, programmed by spatial configurations of

© Springer International Publishing Switzerland 2015
D. Ciucci et al. (Eds.): RSKT 2015, LNAI 9436, pp. 482–491, 2015.
DOI: 10.1007/978-3-319-25754-9_42

attracting and repelling stimuli. The plasmodium looks for attractants, propagates protoplasmic veins towards them, feeds on them and goes on. As a result, a transition system is built up. Therefore, *Physarum* propagation can be treated as a kind of a natural transition system with states presented by attractants and events presented by plasmodium transitions between attractants [1].

To program computational tasks for *Physarum* machines, we are developing a new object-oriented programming language (see [7,9,11]) called the *Physarum* language. The proposed language can be used for developing programs for *Physarum polycephalum* by the spatial configuration of stimuli (attracting and repelling). In the *Physarum* language, we use some high-level models of propagation of plasmodium, e.g., ladder diagrams [14], Petri nets [13], transition systems [7], and timed transition systems [12].

In the presented approach, we use timed transition systems [4] to model behavior of *Physarum* machines. In timed transition systems, the quantitative lower-bound and upper-bound timing constraints are imposed on transitions between states. This ability of modeling behavior of *Physarum* machines is important because attracting and repelling stimuli can be activated and/or deactivated for proper time periods to perform given computational tasks.

In [8], we considered transition system models of behavior of *Physarum* machines in terms of rough set theory. As it was mentioned earlier, in the behavior of *Physarum* machines, one can notice some ambiguities in *Physarum* propagation that influence exact anticipation of states of machines in time. To model these ambiguities, rough set models created over transition systems were proposed. Models described in [8] were based on the original definition of rough sets proposed by Z. Pawlak [10]. In fact, this definition is rigorous in terms of set inclusion. Therefore, in Sect. 3, we extend the model described in [8] to a model based on the more relaxed and generalized rough set approach, called the Variable Precision Rough Set Model (VPRSM), proposed by W. Ziarko in [16]. The VPRSM approach was defined on the basis of the notion of majority set inclusion instead of the standard set inclusion. The majority set inclusion is parameterized. Therefore, a model becomes more flexible. Moreover, in Sect. 3, we adapt the rough set model proposed in [8] for timed transition systems.

2   Basic Definitions

In this section, we recall necessary definitions, notions and notation concerning both rough sets and transition systems.

2.1   Rough Sets

The idea of rough sets (see [10]) consists of the approximation of a given set by a pair of sets, called the lower and the upper approximation of this set. Some sets cannot be exactly defined. If a given set X is not exactly defined, then we employ two exact sets (the lower and the upper approximation of X) that define X roughly (approximately). Let $U \neq \emptyset$ be a finite set of objects we are interested in. U stands for the universe. Any subset $X \subseteq U$ of the universe is called a concept in U. Let R be any equivalence relation over U. We denote an equivalence class of any $u \in U$ by $[u]_R$. With each subset $X \subseteq U$ and any equivalence relation R over U, we associate two subsets:

- $\underline{R}(X) = \{u \in U : [u]_R \subseteq X\}$,
- $\overline{R}(X) = \{u \in U : [u]_R \cap X \neq \emptyset\}$,

called the R-lower and R-upper approximation of X, respectively. An R-boundary region of X is a set $BN_R(X) = \overline{R}(X) - \underline{R}(X)$. If $BN_R(X) = \emptyset$, then X is sharp (exact) with respect to R. Otherwise, X is rough (inexact). Roughness of a set can be characterized numerically. To this end, the accuracy of approximation of X with respect to R is defined as:

$$\alpha_R(X) = \frac{card(\underline{R}(X))}{card(\overline{R}(X))},$$

where $card$ denotes the cardinality of the set and $X \neq \emptyset$.

The definitions given earlier are based on the standard definition of the set inclusion. Let U be the universe and $A, B \subseteq U$. The standard set inclusion is defined as

$$A \subseteq B \text{ if and only if } \underset{u \in A}{\forall} \ u \in B.$$

In some situations, the application of this definition seems to be too restrictive and rigorous. W. Ziarko proposed in [16] some relaxation of the original rough set approach. His proposition was called the Variable Precision Rough Set Model (VPRSM). The VPRSM approach is based on the notion of majority set inclusion. Let U be the universe, $A, B \subseteq U$, and $0 \leq \beta < 0.5$. The majority set inclusion is defined as

$$A \overset{\beta}{\subseteq} B \text{ if and only if } 1 - \frac{card(A \cap B)}{card(A)} \leq \beta,$$

where $card$ denotes the cardinality of the set. $A \overset{\beta}{\subseteq} B$ means that a specified majority of elements belonging to A belongs also to B. One can see that, if $\beta = 0$, then the majority set inclusion becomes the standard set inclusion. By replacing the standard set inclusion with the majority set inclusion in definitions of approximations, we obtain the following two subsets:

- $\underline{R}^\beta(X) = \{u \in U : [u]_R \overset{\beta}{\subseteq} X\}$,
- $\overline{R}^\beta(X) = \{u \in U : \frac{card([u]_R \cap X)}{card([u]_R)} > \beta\}$,

called the R_β-lower and R_β-upper approximation of X, respectively.

2.2 Transition Systems

Transition systems are used to describe behavior of systems with distinguished states and transitions between states. Formally, a transition system is a quadruple $TS = (S, E, T, I)$ (cf. [6]), where:

- S is the non-empty set of states,
- E is the set of events,
- $T \subseteq S \times E \times S$ is the transition relation,
- $I \subseteq S$ is the set of initial states.

Usually transition systems are based on actions which may be viewed as labeled events. If $(s, e, s') \in T$ then the idea is that TS can go from s to s' as a result of the event e occurring at s. A single element $(s, e, s') \in T$ is called shortly a transition. Any transition system $TS = (S, E, T, I)$ can be presented in the form of a labeled graph with nodes corresponding to states from S, edges representing the transition relation T, and labels of edges corresponding to events from E. It is assumed, in transition systems mentioned earlier, that all events happen instantaneously. In timed transition systems, timing constraints restrict the times at which events may occur (see [4]). The timing constraints are classified into two categories: lower-bound and upper-bound requirements. Let N be a set of nonnegative integers. Formally, a timed transition system $TTS = (S, E, T, I, l, u)$ consists of:

- an underlying transition system $TS = (S, E, T, I)$,
- a minimal delay function (a lower bound) $l : E \to N$ assigning a nonnegative integer to each event,
- a maximal delay function (an upper bound) $u : E \to N \cup \{\infty\}$ assigning a nonnegative integer or infinity to each event.

3   Roughness in Timed Transition Systems

Behavior of systems can be characterized by some ambiguities of transitions between states. Such ambiguities influence anticipation of next states. We present an approach to model ambiguities of anticipation of next states based on rough sets. Models based on rough sets are built over timed transition systems describing behavior of systems. In the presented approach, both a standard definition of rough sets and VPRSM (Variable Precision Rough Set Model) are used.

Remark 1. In the remaining part of the paper, we assume, for timed transition systems, that the events may occur only at discrete time instants. Therefore, whenever time instant t is used, it means that $t = 0, 1, 2, \ldots$.

Let $TTS = (S, E, T, I, l, u)$ be a timed transition system. For each state $s \in S$ in TTS and each $t \in \{0, 1, 2, \ldots\}$, we can determine its direct successors and predecessors at the time instant t. Let

$$Post_t(s, e) = \{s' \in S : (s, e, s') \in T \land l(e) \leq t \leq u(e)\}$$

and

$$Pre_t(s, e) = \{s' \in S : (s', e, s) \in T \land l(e) \leq t \leq u(e)\},$$

then the set $Post_t(s)$ of all direct successors of the state $s \in S$ at t is given by

$$Post_t(s) = \bigcup_{e \in E} Post_t(s, e)$$

and the set $Pre_t(s)$ of all direct predecessors of the state $s \in S$ at t is given by

$$Pre_t(s) = \bigcup_{e \in E} Pre_t(s, e).$$

For a given state $s \in S$, all states included in $Post_t(s)$ are called the states directly reachable from the state s at t. If $Post_t(S) = \emptyset$, then s is said to be a goal state at t in the timed transition system TTS. The predecessors of the predecessors (and so on) of a given state $s \in S$ are said to be indirect predecessors of s.

If there exists the state $s \in S$ in the timed transition system TTS at the time instant t such that $card(Post_t(s)) > 1$, where $card$ is the cardinality of the set, then TTS is called a non-deterministic timed transition system at t. One can see that, in non-deterministic timed transition systems, we deal with ambiguity of direct successors of some states, i.e., at some time instants, there exist states having no uniquely determined direct successors. In the presented approach, we propose to manage this ambiguity using rough set theory.

Analogously to rough approximation of sets defined in rough set theory, we can define rough anticipation of states over transition systems. In the proposed approach, the anticipation of states is made via direct predecessor states of the anticipated ones. Therefore, we call this anticipation the predecessor anticipation. Analogously to rough approximation, we can distinguish two kinds of anticipations, called the lower predecessor anticipation and the upper predecessor anticipation.

Let $TTS = (S, E, T, I, l, u)$ be a timed transition system and $X \subseteq S$. The lower predecessor anticipation $\underline{Pre}_t(X)$ of X at the time instant t is given by

$$\underline{Pre}_t(X) = \{s \in S : Post_t(s) \neq \emptyset \land Post_t(s) \subseteq X\}.$$

The lower predecessor anticipation $\underline{Pre}_t(X)$ consists of all states from which TTS surely goes to the states in X as results of any events occurring at these states at the time instant t.

Let $TTS = (S, E, T, I, l, u)$ be a timed transition system and $X \subseteq S$. The upper predecessor anticipation $\overline{Pre}_t(X)$ of X at the time instant t is given by

$$\overline{Pre}_t(X) = \{s \in S : Post_t(s) \cap X \neq \emptyset\}.$$

The upper predecessor anticipation $\overline{Pre}_t(X)$ consists of all states from which TTS possibly goes to the states in X as results of some events occurring at these states at the time instant t. It means that TTS can also go at t to the states from outside X.

The set $BN_{Pre,t}(X) = \overline{Pre}_t(X) - \underline{Pre}_t(X)$ will be referred to as the boundary region of predecessor anticipation of X at the time instant t. If $BN_{Pre,t}(X) = \emptyset$, then the anticipation of the set X of states on the basis of their direct predecessors, at the time instant t, is exact. In the opposite case (i.e., $BN_{Pre,t}(X) \neq \emptyset$), the anticipation of X at t is rough (inexact). The accuracy of anticipation can be defined analogously to the accuracy of approximation in rough set theory, i.e.:

$$\alpha_t(X) = \frac{card(\underline{Pre_t}(X))}{card(\overline{Pre_t}(X))}.$$

Specifically, we can determine the lower and the upper predecessor anticipations of a single state $s \in S$ at the time instant t, denoted as $\underline{Pre_t}(s)$ and $\overline{Pre_t}(s)$, respectively. Moreover, the boundary region of the predecessor anticipation of s at t is denoted as $BN_{Pre,t}(s)$.

By replacing the standard set inclusion with the majority set inclusion in the original definition of the lower predecessor anticipation of a set of states in a timed transition system, we obtain the generalized notion of the β-lower predecessor anticipation.

Let $TTS = (S, E, T, I, l, u)$ be a timed transition system, $X \subseteq S$, and $0 \leq \beta < 0.5$. The β-lower predecessor anticipation $\underline{Pre_t^\beta}(X)$ of X at the time instant t is given by

$$\underline{Pre_t^\beta}(X) = \{s \in S : Post_t(s) \neq \emptyset \wedge Post_t(s) \overset{\beta}{\subseteq} X\}.$$

The β-lower predecessor anticipation of X at t consists of each state from which TTS goes, in most cases (i.e., in terms of the majority set inclusion) to the states in X as results of events occurring at these states at the time instant t. A rough set description of a given transition system enables us to determine whether a given state anticipates, unambiguously or nearly unambiguously, the next state that is one of the states from a distinguished set.

Let $TTS = (S, E, T, I, l, u)$ be a timed transition system, $X \subseteq S$, and $0 \leq \beta < 0.5$. We can use the following nomenclature for each state $s \in S$:

- If

$$\underset{t \in \{0,1,2,\dots\}}{\forall} \quad s \in \underline{Pre_t}(X),$$

then s is said to be a continuous strict anticipator of states from X. It means that s always anticipates (i.e., at each time instant) states from X.
- If s is not a continuous strict anticipator of states from X, but

$$\underset{t \in \{0,1,2,\dots\}}{\exists} \quad s \in \underline{Pre_t}(X),$$

then s is said to be an interim strict anticipator of states from X. It means that s sometimes (not always) anticipates states from X.
- If s is not a continuous and interim strict anticipator of states from X, but

$$\underset{t \in \{0,1,2,\dots\}}{\forall} \quad s \in \underline{Pre_t^\beta}(X),$$

then s is said to be a continuous quasi-anticipator of states from X.

– If s is not a continuous and interim strict anticipator and continuous quasi-anticipator of states from X, but

$$\underset{t \in \{0,1,2,\dots\}}{\exists} \ s \in \underline{Pre}_t^\beta(X),$$

then s is said to be an interim quasi-anticipator of states from X.

A set of all strict continuous anticipators of X will be denoted by $\overline{Ant}(X)$, a set of all interim strict anticipators of X by $\overleftrightarrow{Ant}(X)$, a set of all continuous quasi-anticipators of X by $\widetilde{Ant}(X)$, and a set of all interim quasi-anticipators of X by $\overset{\leftrightarrow}{Ant}(X)$. One can see that for a given X, $\overline{Ant}(X) \subseteq \widetilde{Ant}(X)$.

4   Modeling Propagation of Plasmodium

As it was mentioned in Sect. 1, in *Physarum* machines, the plasmodium looks for attractants, propagates protoplasmic veins towards them, feeds on them and goes on. Activated repellents cause annihilation of the protoplasmic veins. In a real-life implementation of *Physarum* machines, attractants are sources of nutrients or pheromones, on which the plasmodium feeds. In case of repellents, the fact that plasmodium of *Physarum* avoids light and some thermo- and salt-based conditions is used. As a result of computations, the plasmodium forms a network of protoplasmic veins connecting attractants and original points of the plasmodium. Original points of the plasmodium and attractants occupied by the plasmodium are called active points in the *Physarum* machines.

Formally, a structure of the *Physarum* machine is defined as a triple $\mathcal{PM} = \{P, A, R\}$, where:

– $P = \{ph_1, ph_2, \dots, ph_k\}$ is a set of original points of plasmodium.
– $A = \{a_1, a_2, \dots, a_m\}$ is a set of attractants.
– $R = \{r_1, r_2, \dots, r_n\}$ is a set of repellents.

In a standard case, positions of original points, attractants, and repellents are considered in the two-dimensional space.

We can also take into consideration a dynamics of the *Physarum* machine describing the behavior of the *Physarum* machine in time. One can see that a dynamics of the *Physarum* machine (i.e., establishing or annihilating the protoplasmic veins) can be controlled by means of attractants or repellents, more specifically, by means of their activation or deactivation in time.

A dynamics of the plasmodium causes that the set A of attractants can be divided, at each time instant t, into two disjoint subsets:

– $A_\bullet^t$ - a set of attractants occupied by the plasmodium at t,
– $A_\circ^t$ - a set of unoccupied attractants at t.

The set Π^t of all active points at time instant t consists of a set P of original points of the plasmodium as well as a set $A_\bullet^t$ of all attractants occupied by the plasmodium at t.

Formally, a dynamics of the *Physarum* machine over time can be described by the family of protoplasmic veins propagating by plasmodium. Let $\{ \Pi^t \}_{t=0,1,2,\ldots}$ be a family of the sets of all active points at time instants $t = 0, 1, 2, \ldots$ in the *Physarum* machine $\mathcal{PM}$. A dynamics of $\mathcal{PM}$ over time is defined by the family $V = \{ V^t \}_{t=0,1,2,\ldots}$ of the sets of protoplasmic veins propagated by the plasmodium, where $V^t = \{ v_1^t, v_2^t, \ldots, v_{r_t}^t \}$ is the set of all protoplasmic veins of the plasmodium present at time instant t in $\mathcal{PM}$. Each vein $v_i^t \in V^t$, where $i = 1, 2, \ldots, r_t$, is an unordered pair $\{ \pi_j^t, \pi_k^t \}$ of two adjacent active points $\pi_j^t \in \Pi^t$ and $\pi_k^t \in \Pi^t$, connected directly by v_i^t.

Let $\Theta = \{ 0, 1, \ldots \}$ be a space of discrete time instants when states of the *Physarum* machine are noticed. Moreover, let a structure $\mathcal{PM} = \{ P, A, R \}$ of the *Physarum* machine and the family V of protoplasmic veins propagating by its plasmodium be given. We assume that all attractants from A are activated and each repellent from R can be activated/deactivated only once if it is inactive/active at $t = 0$. On the basis of V, we can built a model of $\mathcal{PM}$ in the form of a timed transition system $TTS = (S, E, T, I, l, u)$. The model may be created in the following way:

- S consists of states corresponding both to all original points of plasmodium from the set P and to all attractants from the set A.
- I consists of initial states corresponding to all original points of plasmodium from the set P.
- E, T, l, and u are defined on the basis of V. A transition $(s, e, s') \in T$ and an associated event $e \in E$ are established if there exists $t \in \Theta$ such that the vein $\{ \pi_j^t, \pi_k^t \} \in V^t$, where π_j^t corresponds to s and π_k^t corresponds to s'. Minimal and maximal delay functions, i.e., l and u, respectively, are determined on the basis of minimal and maximal t for which the vein exists.

A rough set model described in Sect. 3 may be created over a timed transition system model.

Let us consider an exemplary *Physarum* machine $\mathcal{PM} = \{ P, A, R \}$, where $P = \{ ph_1 \}$, $A = \{ a_1, a_2, a_3, a_4, a_5, a_6, a_7 \}$, and $R = \{ r_1 \}$, shown in Fig. 1(a). We assume a situation that, at the beginning, r_1 is inactive, and it is activated at $t = 5$. Protoplasmic veins created in $\mathcal{PM}$ at $t < 5$ and $t \geq 5$ are shown in Figs. 1(b) and

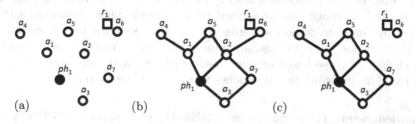

(a)   (b)   (c)

Fig. 1. An exemplary *Physarum* machine: (a) a structure, (b) protoplasmic veins for $t < 5$, (c) protoplasmic veins for $t \geq 5$.

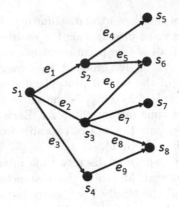

Fig. 2. A timed transition system model of $\mathcal{PM}$.

(c), respectively. A timed transition system model $TTS = (S, E, T, I, l, u)$ of the *Physarum* machine $\mathcal{PM}$ is shown in Fig. 2.

In this model, among others, $S = \{s_1, s_2, s_3, s_4, s_5, s_6, s_7, s_8\}$, where s_1 corresponds to ph_1, s_2 corresponds to a_1, s_3 corresponds to a_2, ..., s_8 corresponds to a_7, $I = \{s_1\}$, $l(e_1) = l(e_2) = l(e_3) = l(e_4) = l(e_5) = l(e_6) = l(e_7) = l(e_8) = l(e_9) = 0$, $u(e_1) = u(e_2) = u(e_3) = u(e_4) = u(e_5) = u(e_6) = u(e_8) = u(e_9) = \infty$, and $u(e_7) = 4$.

Let us assume that we are interested in the set $X = \{s_5, s_6, s_8\}$ of goal states and $\beta = 0.5$. For X and $t < 5$, we obtain $\underline{Pre_t}(X) = \{s_2, s_4\}$, $\overline{Pre_t}(X) = \{s_2, s_3, s_4\}$, but $\underline{Pre_t}^{0.5}(X) = \{s_2, s_3, s_4\}$. For X and $t \geq 5$, we obtain $\underline{Pre_t}(X) = \{s_2, s_3, s_4\}$ and $\overline{Pre_t}(X) = \{s_2, s_3, s_4\}$. It means that s_2 and s_4 are continuous strict anticipators of states from X, s_3 is an interim strict anticipator of states from X but also s_3 is a continuous quasi-anticipator of states from X.

5   Conclusions

We have shown that rough set theory can be used to describe some ambiguities in behavior of systems described by transition systems. A special attention has been focused on the predecessor anticipation of states. It is worth noting that the presented approach can be applied to all systems whose behaviors can be modeled by timed transition systems, for example, systems based on flows of some phenomenon. More interesting approaches for further research on ambiguity modeling are those based on combined rough sets and fuzzy sets (cf. [3]) or those based on probabilistic approaches to rough set theory (cf. [15]).

Acknowledgment. This research is being fulfilled by the support of FP7-ICT-2011-8.

References

1. Adamatzky, A.: Physarum Machines: Computers from Slime Mould. World Scientific (2010)
2. Adamatzky, A., Erokhin, V., Grube, M., Schubert, T., Schumann, A.: Physarum chip project: growing computers from slime mould. Int. J. Unconventional Comput. **8**(4), 319–323 (2012)
3. Dubois, D., Prade, H.: Rough fuzzy sets and fuzzy rough sets. Int. J. Gen. Syst. **17**(2–3), 191–209 (1990)
4. Henzinger, T.A., Manna, Z., Pnueli, A.: Timed transition systems. In: de Bakker, J., Huizing, C., de Roever, W., Rozenberg, G. (eds.) Real-Time: Theory in Practice. LNCS, vol. 600, pp. 226–251. Springer, Heidelberg (1992)
5. Nakagaki, T., Yamada, H., Toth, A.: Maze-solving by an amoeboid organism. Nature **407**, 470–470 (2000)
6. Nielsen, M., Rozenberg, G., Thiagarajan, P.: Elementary transition systems. Theor. Comput. Sci. **96**(1), 3–33 (1992)
7. Pancerz, K., Schumann, A.: Principles of an object-oriented programming language for Physarum polycephalum computing. In: Proceedings of the 10th International Conference on Digital Technologies (DT'2014), pp. 273–280. Zilina, Slovak Republic (2014)
8. Pancerz, K., Schumann, A.: Rough set models of Physarum machines. Int. J. Gen. Syst. **44**(3), 314–325 (2015)
9. Pancerz, K., Schumann, A.: Some issues on an object-oriented programming language for Physarum machines. In: Bris, R., Majernik, J., Pancerz, K., Zaitseva, E. (eds.) Applications of Computational Intelligence in Biomedical Technology, Studies in Computational Intelligence, vol. 606, pp. 185–199. Springer International Publishing, Switzerland (2016)
10. Pawlak, Z.: Rough Sets: Theoretical Aspects of Reasoning about Data. Kluwer Academic Publishers, Dordrecht (1991)
11. Schumann, A., Pancerz, K.: Towards an object-oriented programming language for Physarum polycephalum computing. In: Szczuka, M., Czaja, L., Kacprzak, M. (eds.) Proceedings of the Workshop on Concurrency, Specification and Programming (CS&P'2013), pp. 389–397. Warsaw, Poland (2013)
12. Schumann, A., Pancerz, K.: Timed transition system models for programming Physarum machines: extended abstract. In: Popova-Zeugmann, L. (ed.) Proceedings of the Workshop on Concurrency, Specification and Programming (CS&P'2014), pp. 180–183, Chemnitz, Germany (2014)
13. Schumann, A., Pancerz, K.: Towards an object-oriented programming language for Physarum polycephalum computing: a petri net model approach. Fundam. Informaticae **133**(2–3), 271–285 (2014)
14. Schumann, A., Pancerz, K., Jones, J.: Towards logic circuits based on Physarum polycephalum machines: the ladder diagram approach. In: Cliquet Jr., A., Plantier, G., Schultz, T., Fred, A., Gamboa, H. (eds.) Proceedings of the International Conference on Biomedical Electronics and Devices (BIODEVICES'2014), pp. 165–170. Angers, France (2014)
15. Yao, Y.: Probabilistic rough set approximations. Int. J. Approximate Reasoning **49**(2), 255–271 (2008)
16. Ziarko, W.: Variable precision rough set model. J. Comput. Syst. Sci. **46**(1), 39–59 (1993)

A Study on Similarity Calculation Method for API Invocation Sequences

Yu Jin Shim[1], TaeGuen Kim[1], and Eul Gyu Im[2]([✉])

[1] Department of Computer Software, Hanyang University, Seoul, Korea
{luvtdw,cloudio17,imeg}@hanynag.ac.kr
[2] Division of Computer Science and Engineering, Hanyang University, Seoul, Korea

Abstract. Malware variants have been developed and spread in the Internet, and the number of new malware variants is increases every year. Recently, malware is applied with obfuscation and mutation techniques to hide its existence, and malware variants are developed with various automatic tools that transform the properties of existing malware to avoid static analysis based malware detection systems. It is difficult to detect such obfuscated malware with static-based signatures, so we have designed a detection system based on dynamic analysis. In this paper, we propose a dynamic analysis based system that uses the API invocation sequences to compare behaviors of suspicious software with behaviors of existing malware.

Keywords: Malware detection · API invocation sequence · Dynamic analysis · Similarity calculation method

1 Introduction

Since Morris worm emerged, the number of malware has steadily increased. AV-TEST [1] reported that more than 220,000 malware was discovered every day in 2014. One of main reasons that the number of malware has continuously increased is that attackers can have financial gains through attacks with malware. Recent malware is used by criminals to achieve specific goals, such as stealing confidential data, harvesting accounts, or identity thefts. In order to defend against malware, many researches on the malware detection has been performed.

The previous researches can be classified into two categories: static analysis based detections [15,16] and dynamic analysis based detections [10–14,17–19]. Static analysis based detections use static features that can be extracted from malware. Static features represent syntatic properties of malware, and they can be extracted without executing the malware. In contrast, dynamic analysis based detections use dynamic features, such as API (Application Programming Interface) invocation sequences, instruction traces, changes in the infected systems. To extract these kinds of features, it is necessary to execute the target malware in a controlled environment. Dynamic analysis is more resistant to code obfuscation techniques because dynamic features represent semantic information of

© Springer International Publishing Switzerland 2015
D. Ciucci et al. (Eds.): RSKT 2015, LNAI 9436, pp. 492–501, 2015.
DOI: 10.1007/978-3-319-25754-9_43

malware. This is the reason why we choose to use dynamic analysis to detect malware variants.

In this paper, dynamic anaysis based detection framework is proposed, and the framework uses two kinds of behavior similarity calculation methods. Behaviors are defined as API invocation sequences of software, and an n-gram-based comparison method and a sequence alignment-based comparison method are used to calculate the similarities of API invocation sequences. Each similarity calculation method has different advantages and disadvantages, so these methods can be used selectively by malware analyzers.

The rest of the paper is organized as follows: Sect. 2 discusses related work. Section 3 presents the proposed framework for malware detection in detail. Section 4 shows the experimental results which show possibilities of malware variant detection. Section 5 summarizes our research and addresses the contributions and limitations.

2   Related Work

Wu et al. [4] performed a research on the malware detection system. The system uses API sequences to define the behaviors of malware, and behaviors are classified into 4 categories; file related behaviors, process related behaviors, network related behaviors and registry related behaviors. Risk scores of behaviors of malware are measured, and the risk scores are used in malware detection.

Apel et al. [5] analyzed the similarity calculation methods that is used in malware detection. They compared four similarity calculation methods: edit distance, approximated edit distance, normalized compression distance, manhattan distance. They learned that manhattan distance is most accurate and has least time complexity.

Rieck et al. [6] proposed the system that classifies malware variants using behavior patterns. They monitored malware behaviors using CWSandbx. They extracted strings that represent malicious behaviors, and frequencies of the strings are counted. A series of frequencies of the string is expressed as a vector, and the density of the vector is used to classify the malware. SVM algorithm is used as a classificaion algorithm. According to their experimental results, the accuracy of malware classification was about 70 %.

Alazab et al. [7] analyzed malicious behaviors that are frequently used by malware. They suggested a method to learn malware behaviors by the extraction of API invocations. API invocation frequencies and API invocation sequences are utilized in order to analyse malware behaviors, and they defined six kinds of malicious behaviors that can be used for attacks. Consequently, It is found out that searching files to infect and read/write files are most frequently used behaviors.

Bayer et al. [8] investigated statistics to find out what kinds of malicious behaviors are. Anubis which is a service for dynamic analysis was used to collect the information related with malicious behaviors. The known samples were used in experiments to specify the behaviors of each malware family.

Xu et al. [9] suggested a malware detection system. They used a PE binary parser to extract API sequences from PE binary files, and the extracted API invocation sequences were used to define signatures of malware, then they detected polymorphic malware by measuring the similarities between signatures of existing malware and suspicious program. They used a sequence alignment method in similiarity calculation.

3   The Similarity Measurement System

Malware invokes APIs to perform its malicious behaviors, so we utilized the API invocation sequences to detect malware variants. We used Cuckoo sandbox [3] to extract the dangerous API invocation sequences from malware. The dangerous API invocation sequences are used to calculate the similarities of two binary executable files.

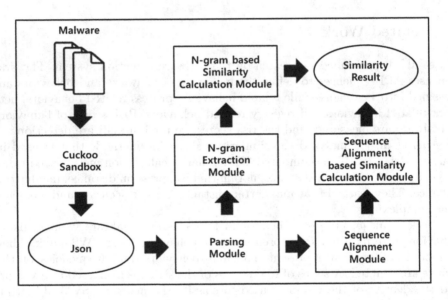

Fig. 1. The Processing Flow of the Framework

3.1   Overall System Architecture

Figure 1 shows the overall process of our proposed framework. The processing flows have four steps:

(1) Two binary executable files are inputted to the proposed framework.
(2) Two given binary files are executed, and the dangerous API invocation sequences of them are extracted.
(3) Dangerous API invocation sequences are transformed into API code sequences of which elements are all integer values, not string values.
(4) Final similarity of two API code sequences is computed and reported.

3.2   Extraction of API Calling Sequence and Transformation to API Code Sequence

Cuckoo sandbox is utilized to automatically analyze suspicious files. This tool monitors the API invocation sequences of the malicious processes while running in an isolated environment. Cuckoo sandbox specifies 182 dangerous APIs, and the specified dangerous APIs are only logged. Each element of the API invocation sequence provided by Cuckoo sandbox is expressed as a string of an API label. There are many comparison operations in the similarity calculation methods. The API invocation sequences should be transformed to API code sequences of which elements are represented integer values because it is more convenient to compare integer values than string values. Mapping table is used to map a set of the dangerous APIs to integer values.

3.3   Similarity Calculation of API Code Sequences

Two kinds of methods are used to calculate the similarity of given two binary files. N-gram based similarity calculation and sequence alignment based similarity calculation are performed to compare accuracy of results of the two methods.

N-Gram Based Similarity Calculation. n-gram is a sub-sequence of n elements from a given sequence, and N-gram model is one of a representative statistical language model. In our case, N-grams of API code sequence are extracted and their frequencies are used to calculate the similarity score. The method to extract the N-grams is straightforward. The whole sequence is scanned with a fixed sliding window, and each scanned sub-sequence in the sliding window is counted. For example, if there is a string, "SIGNATURE", 5-grams in the string are "SIGNA", "IGNAT", "GNATU", "NATUR", "ATURE". and each 5-gram's frequency is one. This example is explained with a string, but the API code sequences are used in the actual N-gram extraction process of the framework.

After the N-gram frequencies are extracted, each N-gram frequency of the API code sequence is used to calculate the cosine similarity. Cosine similarity calculation is a measure that computes the similarity score between two vectors of n dimensions, and it is widely used in the data mining field as a distance measure. The angle of given vectors is calculated using the Euclidean dot product, and it ranges from 0 to 1. In case of the comparison of API code sequences, the similarity results will range from 0 to 1 because there is no negative frequency of N-grams. Cosine similarity equation is explained in the below. Two series of frequency values of extracted N-grams are used as input values of *equation*.

$$Similarity = \cos(\theta) = \frac{A \cdot B}{\|A\|\|B\|} = \frac{\sum_{i=1}^{n} A_i \times B_i}{\sqrt{\sum_{i=1}^{n} (A_i^2)} \times \sqrt{\sum_{i=1}^{n} (B_i^2)}}. \tag{1}$$

Sequence Alignment Based Similarity Calculation. Sequence alignment is a way of arranging sequences, which is widely used in the bioinformatics field. In bioinformatics, sequence alignment is used to find common patterns in sequences, such as DNA sequences or peptide sequences. The sequences in bioinformatics may have some noisy tokens, and when common patterns are extracted without any arrangements, the common patterns would be missed due to some noisy tokens. It is necessary to align the sequences to extract the common patterns more accurately. API code sequences share this property, so a sequence alignment algorithm arranges the API code sequences to enhance the accuracy of similarity results. Among various sequence alignment algorithms, Smith-Waterman algorithm [20] is chosen to arrange the API code sequences. Smith-Waterman algorithm inserts gaps into sequences to make them aligned properly.

Two given API code sequences are arranged by Smith-Waterman algorithm, and then the common sub-sequences of aligned sequences are extracted. The lengths of common sub-sequences are used to calculate the similarity scores using the equation represented in below.

$$Similarity = \frac{(Length\ of\ Max\ length\ substring)}{\frac{(sum\ of\ two\ strings)}{2}}. \tag{2}$$

4   Experiment

We conducted several experiments to figure out the accuracy and to measure the performance of similarity calculation methods. Experimental results are explained in next subsections.

4.1   Similarity Results Among the Malware in Different or Same Family

Similarities of malware in the same family and different families are measured to check whether the proposed framework can be used in malware classifications. Malware samples were downloaded from VxHeaven [2], and one hundred fifty malware samples from ten malware families were used. The number of malware samples in each family was fifteen. Figures 2 and 3 show the results of the similarity calculations. The graph in Fig. 2 shows similarity results that are computed with the N-gram based similarity calculation method, and the graph in Fig. 3 shows similarity results that are computed with the sequence alignment based similarity calculation method. The black bar represents the average of the similarities among malware samples in the same family and the gray bar represents the average of the similarities between malware samples from different families. Most of the similarities of samples from the same malware family were higher than those from different malware families. These results show that our proposed framework can be used as a distance measure to classify the malware variants.

4.2   Similarity Results Between Malware and Benign Programs

Similarities between malware samples and benign programs are measured to check whether the proposed framework can detect the malware. Malware samples that were used in these experiments were same with the samples that are

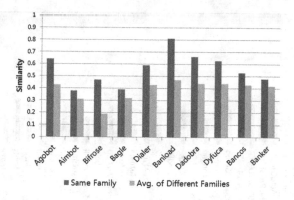

Fig. 2. Similarity calcuation result with N-gram based method

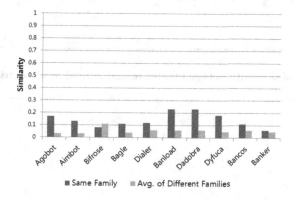

Fig. 3. Similarity calcuation result with sequence alignment based method

explained in the previous subsection. Table 1 shows the list of benign samples that were used in these experiments.

The similarities of samples from same malware families had already measured, so we only calculated the similarities between malware samples and benign samples. Figures 4 and 5 show the experimental results, and each figure represents the similarities that were measured by the N-gram based method and the sequence alignment based method respectively. The average of the similarity results of the N-gram based method was 0.558 and that of the sequence alignment based method was 0.142. It means that malware and benign programs can be distinguished because most of the similarities between malware and benign programs were lower than the similarities of samples from the same malware family.

4.3   Measurement of Time Performance

We randomly generated API code sequences to measure computational overheads of two similarity calculation methods. The number of elements of the

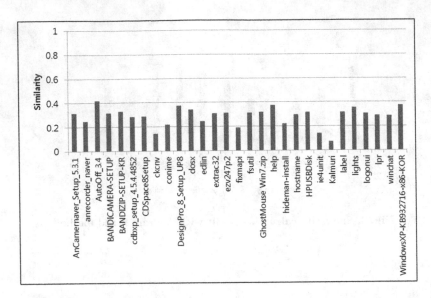

Fig. 4. Result of similarity between malware and normal program using N-gram

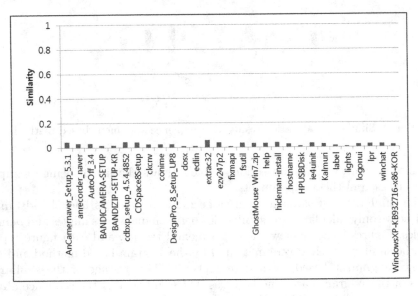

Fig. 5. Result of similarity between malware and normal program using Smith-Waterman

generated sequences ranges from one thousand to five thousands, and the length of each sequence is increased by the length of one thousand. Figure 6 shows the time measurement of two similarity calculation methods. It is noted that the computation time for the extraction of the API code sequence was excluded. As shown in Fig. 6, the computation time of the sequence alignement based

Table 1. Benign Programs

AnCamernaverSetup5.3.1.exe	anrecordernaver.exe	ckcnv.exe
AutoOff3.4.exe	BANDICAMERA-SETUP.exe	dosx.exe
BANDIZIP-SETUP-KR.exe	cdbxpsetup4.5.4.4852.exe	edlin.exe
conime.exe	DesignPro8SetupUP8.exe	ie4uinit.exe
EveryonTV1.7.00installer.zip	extrac32.exe	winchat.exe
ezv247p2.exe	label.exe	fixmapi.exe
WindowsXP-x86-KOR.exe	lpr.exe	fsutil.exe
GhostMouse Win7.zip	lights.exe	help.exe
hideman-install.exe	CDSpace8Setup.exe	hostname.exe
HPUSBDisk.exe	Kalmuri.exe	logonui.exe

method was increased exponentially, and that of the n-gram based method was linearly increased. There is huge difference between two time measurements. For example, when the sequences of which length is one thousand are compared, time measurement of the n-gram based method was about 0.7 s and that of the sequence alignment based method was about 12 s.

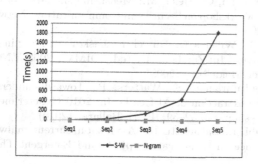

Fig. 6. Time performance of two similarity calculation methods

5   Discussion

In this paper, we used two kinds of similarity calculation methods to analyze the API invocation sequences of binary programs, and the accuracy and time performance were measured to figure out which method is more accurate and has less overheads.

We expected that the sequence alignment based method will be more accruate in similarity calculation, but there was no big difference in similarity results. Both of methods can be utilized as similarity calculation for malware detections.

Computational overheads of the similarity calculation methods were also measured. As a result, the n-gram based method has much less time overheads than the sequence alignment based method. The sequence alignment method uses a dynamic programming algorithm, and the method finds the optimal solution to maximize the number of common patterns in two sequences. In summary, the n-gram based similarity calculation method is more suitable when the detection should be processed in real-time, and the sequence alignement based similarity calculation method is more suitable when the detailed analysis should be conducted.

Acknowledgments. This research was supported by Next-Generation Information Computing Development Program through the National Research Foundation of Korea (NRF) funded by the Ministry of Science, ICT & Future Planning (2011-0029923)

References

1. The Independent IT-Security Institute. http://www.av-test.org/en/
2. The site for providing information about computer viruses. http://vxheaven.org/
3. Cuckoo Sandbox. http://www.cuckoosandbox.org/
4. Wu, L., Ping, R., Ke, L., Hai-xin, D.: Behavior-based Malware analysis and detection. In: First International Workshop on Complexity and Data Mining, pp. 39–42. IEEE, Nanjing (2011)
5. Apel, M., Bockermann, C., Meier, M.: Measuring similarity of malware behavior. In: The 5th LCN Workshop on Security in Communications Networks, pp. 891–898. IEEE, Zurich (2009)
6. Rieck, K., Holz, T., Willems, C., Düssel, P., Laskov, P.: Learning and classification of malware behavior. In: Zamboni, D. (ed.) DIMVA 2008. LNCS, vol. 5137, pp. 108–125. Springer, Heidelberg (2008)
7. Alazab, M., Venkataraman, S., Watters, P.: Towards understanding malware behaviour by the extraction of API calls. In: 2010 Cybercrime and Trustworthy Computing Workshop, pp. 52–59. IEEE, Ballarat (2010)
8. Bayer, U., Habibi, I., Balzarotti, D.: A view on current malware behaviors. In: USENIX conference on Large-scale Exploits and Emergent Threats, p. 8. ACM, Boston (2009)
9. Xu, J.-Y., Sung, A.H., Chavez, P., Mukkzmala, S.: Polymorphic malicious executable scanner by API sequence analysis. In: Hybrid Intelligent Systems, pp. 378–383. IEEE, Kitakyushu (2004)
10. Natani, P., Vidyarthi, D.: Malware detection using API function frequency with ensemble based classifier. In: Security in Computing and Communications, pp. 379–388. IEEE, Mysore (2004)
11. Soo, H.K., Kyoung, K.I., Gyu, I.E.: Malware family classification method using API sequential characteristic. In: The International Conference on IT Convergence and Security, pp. 613–626. Springer, Huangshi (2011)
12. De Huang, H., Lee, C.-S., Kao, H.-Y., Tsai, Y.L., Chang, J.-G.: Malware behavioral analysis system: twman. In: Intelligent Agent, pp. 1–8. IEEE, Paris (2011)
13. Rieck, K., Holz, T., Willems, C., Düssel, P., Laskov, P.: Learning and classification of malware behavior. In: Zamboni, D. (ed.) DIMVA 2008. LNCS, vol. 5137, pp. 108–125. Springer, Heidelberg (2008)
14. Purui, S., Lingyun, Y., Dengguo, F.: Exploring malware behaviors based on environment constitution. In: Computational Intelligence and Security, pp. 320–325. IEEE, Suzhou (2008)

15. Moser, A., Kruegel, C., Kirda, E.: Exploring multiple execution paths for malware analysis. In: Security and Privacy, pp. 231–245. IEEE, Berkeley (2008)
16. Moser, A., Kruegel, C., Kirda, E.: Byte level nGram analysis for malware detection. In: 5th International Conference on Information Processing, pp. 51–59. Bangalore (2011)
17. Jian, L., Ning, Z., Ming, X., YongQing, S., JiouChuan, L.: Malware behavior extracting via maximal patterns. In: The 1st International Conference on Information Science and Engineering, pp. 1759–1764. IEEE, Nanjing (2009)
18. Moser, A., Kruegel, C., Kirda, E.: Analysis of machine learning techniques used in behavior-based malware detection. Advances in Computing. Control and Telecommunication Technologies, pp. 201–203. IEEE, Jakarta (2010)
19. Bayer, U., Moser, A., Kruegel, C., Kirda, E.: Dynamic analysis of malicious code. J. Comput. Virology **2**, 67–77 (2006)
20. Smith, T.F., Waterman, M.S.: Identification of common molecular subsequences. J. Mol. Biol. **147**(1), 195–197 (1981)

Fast Human Detection Using Deformable Part Model at the Selected Candidate Detection Positions

Xiaotian Wu[1](✉), KyoungYeon Kim[2], Guoyin Wang[1], and Yoo-Sung Kim[2]

[1] Key Laboratory of Chongqing Computation and Intelligence,
Chongqing University of Posts and Telecommunications,
Chongqing 400065, People's Republic of China
williamxtwu@gmail.com
[2] Department of Information and Communication Engineering, Inha University,
Incheon 402-751, Korea
yskim@inha.ac.kr

Abstract. We integrate the classic deformable part models (DPM) with the object proposal approaches to achieve a fast and accurate human detection system. The proposed method avoids exhaustive sliding window search, which accelerating the detection speed and reducing the incorrect false positives. In this paper, EdgeBoxes and BING are selected as the candidate object proposal methods to generate the candidate detection positions for the DPM, because their good performance and fast speed. The DPM is only carried on the candidate locations selected by EdgeBoxes and BING for fast human detection. Experiments on PASCAL 2007 dataset for human detection show that the proposed method accelerates the detection speed and reduces the incorrect detections effectively, and EdgeBoxes is better than BING.

Keywords: Human detection · Deformable part model · Object proposals · Sliding windows · Candidate positions

1 Introduction

Human detection is a key problem and a hotspot in the field of computer vision. The goal of human detection is to determine whether an image include a human or not, and where the human is in the image. Human detection has been widely used in many fields including surveillance, robotics, and automatic navigation.

Overall, the human detection detectors can be divided into two parts, global detectors and part-based detectors. The global approaches using a single feature description for the complete person. Typically simple, i.e. we train a discriminative classification on top of the feature descriptions. For the part-based approaches, the individual feature descriptors are utilized for body parts. Recently, human detection has a great progress utilizing part-based approaches which is also utilized in the field of object detection [1]. Appearances of the object' parts such as persons'

© Springer International Publishing Switzerland 2015
D. Ciucci et al. (Eds.): RSKT 2015, LNAI 9436, pp. 502–512, 2015.
DOI: 10.1007/978-3-319-25754-9_44

head, torso, foot are represented by Histograms of Gradients (HOG) [2], which is the most commonly used. Various spatial models have been proposed to represent the spatial relationships of parts. Among them, the deformable part models (DPM) [1] adopting a star model where each part is independent of all other part locations and is connected to a central reference part only, has a great success against the PASCAL VOC 2007 benchmarks [3] and has become the framework of the later excellent detection methods. It is worth mentioning that the PASCAL VOC 2007 dataset [4] are widely acknowledged as difficult testbeds for detection tasks, especially for human detection, because people in this dataset usually with different postures, complex backgrounds and even occlusion.

However, it is a pity that the DPM is pretty slow and generates many false positives in the results. It is well-known that the speed is significant for the human detection. With the development of high computing hardware, Graphics Processing Unit(GPU) implementation is popular. Gadeski, et al. [5] utilizes a GPU implementation of the DPM and achieves a good result. However, in this paper, we focus on improving the detection algorithm to accelerate the runtime, rather than just implementation. We argue that the low speed is due to partly the fact that the DPM using sliding window scheme at all possible scales and positions in the image. The exhaustive searching is low efficiency and some background locations that get a high score may be regarded as the correct detections. In this paper, we propose a new method to solve this problem. We first utilize object proposal method to find some candidate locations that may exist objects, in this stage many locations that possible become the false positives be filtered. Then the DPM only carried on these selected candidate locations. By this way, we accelerate the detection speed. For the object proposal methods, according to [6], we select EdgeBoxes [7] and BING [8] as our candidate methods, since their fast speed. The recommended one will be gotten by the experiments. Two comparisons are conducted in our work, one is between the proposed method and the original DPM, the other one is between EdgeBoxes and BING. Experiments carried on the challenging PASCAL VOC 2007 dataset show that our proposed method is faster and more accurate than the original DPM, and the EdgeBoxes is better than the BING.

Two comparisons are conducted in our work, one is between the proposed method and the original DPM, the other one is between EdgeBoxes and BING. Experiments carried on the challenging PASCAL VOC 2007 dataset show that our proposed method is faster and more accurate than the original DPM, and the EdgeBoxes is better than the BING.

The rest of this paper is organized as follows. Section 2 reviews the related works. Our proposed detection method will be presented in Sect. 3. We show experiments in Sect. 4 and conclude the paper in Sect. 5.

2   Related Work

There are extensive methods for accelerating the speed of the DPM, but here we mentation just a few relevant works using detection proposals. Among them, [9] uses hashing technique to identify promising locations for reducing search positions. [10] uses two segmentation methods to get the detection proposals and adopts the DPM to detecting objects, which is similar to ours. But, the difference is [10] constrains the search locations of part filters which is bad for human detection, because people often have many deformations. While segmentation has found limited useful in object detection [11], generally it cannot provide the accurate object regions. For the object proposals methods, the research focuses currently on the field of detection, see [6] for a survey. Uijlings, et al. [12] and Manen, et al. [13] generate proposals by merging the super-pixels. Cheng, et al. [8] and Alexe, et al. [14] use the multiple foreground-background segmentations to get the object proposals. Then, we introduce the related methods in detail which play important roles in our works.

2.1   Deformable Part Model

The original DPM adopts a sliding window detector scheme. The model can include multi-components, every component detects the dissimilar angles (e.g. the side and frontage of a person), every component consists of two kinds of filters, a root filter and the part filters. A coarse root filter that approximately covers an entire object and higher resolution part filters that cover smaller parts of the object.

The detected image is represented as a group of feature maps. A filter F is a rectangular template defined by an array of d-dimensional weight vector. The score of F at a position (x,y) in a feature map G is the dot product of the filter and a sub-window of the feature map, with the top-left corner at (x,y):

$$\sum_{x',y'} F[x',y'] \cdot G[x+x',y+y'], \tag{1}$$

We define a score at dissimilar positions and scales in an image using an N levels image pyramid constructed via repeated smoothing and subsampling. Next we compute a feature map from each level of the pyramid. Then, in every level we compute M kinds of feature maps, which adjacent scales differ twice. The $M*N$ feature maps called feature pyramid H. In the M image octave, the feature will increasingly better with it lower. In the stage of detection, the part filter is placed at the twice lower scale than the root filter.

The model is defined by an $(n+2)$-tuple $(F_0, P_1, ... P_n, b)$, where F_0 is a root filter, P_i is a model for the ith part, and b is a real-valued bias term. P_i is defined by (F_i, v_i, d_i), F_i is a part filter, v_i is a two-demensional vector specifying an "anchor" position for F_i, and d_i is a four-dimensional vector specifying coefficients of a quadratic function defining a deformation cost.

An object hypothesis specifies the location of each filter in the model in a feature pyramid. $z=(p_0,...,p_n)$, where $p_i = (x_i, y_i, l_i)$ is the position and the level number of the ith filter placed in the feature pyramid. The score of detection is defined:

$$score(p_0, ..., p_n) = \sum_{i=0}^{n} F_i' \phi(H, p_i) - \sum_{i=0}^{n} d_i \cdot \phi(dx_i, dy_i) + b, \tag{2}$$

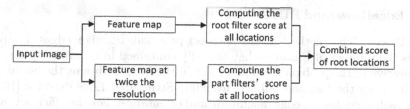

Fig. 1. The matching processing of the original DPM. We compute the response of root filter at all locations. The final score is yielded by combination of the results of root and part filters.

where
$$(dx_i, dy_i) = (x_i, y_i) - (2(x_i, y_i) + v_i) \tag{3}$$
denotes the displacement of the ith part relative to its anchor position.

$$\phi(dx, dy) = (dx, dy, dx^2, dy^2) \tag{4}$$

are deformation features.

A hypothesis will be determined as a correct detection if its score is greater than the pre-learned threshold. Figure 1 describes the matching processing flow chart of the original DPM in one scale.

2.2   Drawbacks and Corresponding Reasons of the DPM

By evaluating the results of the DPM conducting on the PASCAL VOC 2007 dataset for detecting human, the multiple object detection accuracy(MODA) [15], the missed and false positives rate, is -0.107 which is low, and the detection speed is pretty slow. Hence, we have to admit the fact that the DPM is outstanding but exists some drawbacks too. Firstly, the computational efficiency for detection is low which results in the slow speed, and numerous false positives exist in the results. Secondly, context information, the relationship of objects occurred in one image, is not to be taken into consideration in the DPM.

Then, we analyze the reason why these disadvantages exist in the DPM. For the low computational efficiency, it is because the DPM adopts sliding window paradigm that searching targets at all scales and positions. The merit of it is never missing every possible position and scale in the image, however, the demerits is also obvious. Firstly, the general weakness of exhaustive search methods is low computational efficiency, the sliding window approach is no exception. Secondly, some background places that gets a high score in the detection stage but not including human, which causes a large number of incorrect detections. To the context information, the reason why the DPM does not considers it is that including it is too complex and more time-consuming. However, the speed is important for the detection task.

2.3   EdgeBoxes and BING

EdgeBoxes. This method generates object proposals based on edges. The main idea is the number of contours that are fully contained in an arbitrary size box is indicative of the probability of an object in the box. It obtains the initial edge map by using the fast and publicly available Structured Edge detector [16,17], where each pixel has an edge magnitude and orientation. For the efficiency, adjacent edge pixels of similar orientation are clustered together to form a group. It computes the affinities between the edge groups based on their relative positions and orientations such that groups that have high affinity forming long continuous contours. It computes the score of a bounding box by summing the edge strength of all edge groups within the box, then minus the strength of edge groups that are part of a contour that straddles the boundary of the box. It adopts a sliding window paradigm to evaluate proposals in every scale and position and generates a score which denoting the likelihood of an object exists in a sub-window of the image. Then, it finds the top proposals from millions of possible candidates.

BING. Binarized Normed Gradients for the possibility of object generates the detection proposals by using the idea that generic object with well-defined boundary looks alike strongly when we observe the norm of the gradient, after resizing the image windows to a small fixed size. Hence, BING resizes a sub-window which includes an object to $8 * 8$ and learns a generic possibility of the object using the norm of the gradients, a simple 64D feature vector, in a cascaded SVM(support vector machine) framework. It predefines quantized window sizes $\{(w_0, h_0)\}$, where $w_0, h_0 \in \{10, 20, 40, 80, 160, 320\}$, the image is resized to all of these 36-size, then it employs a $8 * 8$ size window to scan over these 36-size images and extracting NG (Normed Gradients) features from them for detection.

The training consists of two stages. In stage 1, we train a single model w using linear SVM [18] for evaluating the score of the confidence of whether a bounding box contain an object or not. It extracts the NG features of the ground truth windows and random sampled background windows as positive and negative samples respectively. For some sizes (e.g. $10 * 500$) of sub-windows are unlike to contain an object. So, in stage 2, it learns a coefficient and a bias terms for each quantized size i. For the detection, it extracts a NG feature f from a $8 * 8$ window to scan over all the 36-size images. Then, the inner product between f and w is defined to the response s'. The final score s of a hypothesis is defined:

$$s = v_i \cdot s' + t_i \tag{5}$$

where, v_i, t_i are separately learned coefficient and a bias terms learned for each quantized size i in training stage 2, which considering the confidence of the size of a hypothesis includes an object.

Particularly worth mentioning is that Cheng et al. [8] adopt the model binary approximation method and a series of optimized mechanisms make the speed achieves 300 frames per second on CPU. There is no doubt that it is the fastest object proposal method so far.

3 DPM at Selected Candidate Detection Positions

Integrated with the object proposal approaches, our work is based on the DPM and further improving it for detecting human. As described in Sect. 2.2, the original DPM itself is pretty complex and utilizing a sliding window scheme which results in the slow detection speed and even gives rise to many unnecessary false positives. So, if we have some candidate detection locations guide the searching process, we need not to take an exhaustive searching, it will improve the detection speed absolutely.

Then, how can we get the candidate detection locations? The current mainstream methods are object proposals [7,8,12–14,19–22], proposals based on segment [10,23]. Between them, the object proposal methods are predominant in this field for their good performance. In our work, we select EdgeBoxes [7] and BING [8], two of the best object proposals methods, as our candidate methods for generating the candidate detection positions. The reason why we select them is that they have high detection rate and fast speed. According to Hosang et al. [6], BING is the fastest method and EdgeBoxes is the best compromise in speed versus quality among all the state-of-the-art object proposals methods. They are the only two methods that the runtime is less than 1 second in [6]. Given just 1,000 proposals, EdgeBoxes achieves over 96 % detection rate and about 86 % recall at overlap threshold of 0.5, 93 % and 86 % to BING, the detection rate and recall of them are very high. At the same time, the running time of them is 0.25 s and 0.003 s per image respectively, which are very fast.

In our proposed method, for a detected image, we first adjust the parameters of EdgeBoxes or BING to get at most 1,000 candidate bounding boxes, then convey the top-left corners set T of the bounding boxes to the DPM as candidate root locations, in this way, we reduce the detection positions from $\sim 10^6$ to $\sim 10^3$ or less. Then, we construct a feature pyramid consisted of several levels feature maps, which is same to the DPM. For a proposal position $P\ (x_i,\ y_i) \in T$ in the image, we convert it to $P^{'}\ (\ x_i^{'},\ y_i^{'},\ l\ ) \in T^{'}$ according to a scaling factor pre-defined

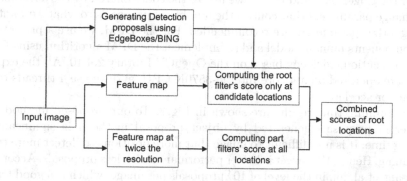

Fig. 2. The processing of the proposed method. We compute only the response of root filter in proposal positions. The final score is yielded by combination of the results of root and part filters.

by the DPM, where (x_i, y_i) is the position of the pixel space in the image and (x'_i, y'_i, l) denotes the position and level of the block space in the feature map H, respectively. T' indicates the candidate locations set in H. In every level of the feature pyramid, we put the root filter only in the locations of T', then we calculate the root score which is the inner product between the root filter and the sub-window of the feature map. The part filters' scores are calculated by the part models at the twice resolution of the feature map relative to its root filter computed. We combine the root and parts scores together according to (2). Next, we search the positions in a feature map that the value of integrated score is greater than the threshold which is considered as the potential correct detections. For example, if the hypothesis in $P'(x'_i, y'_i, l)$ is considered as a correct detection, we convert this detection location back to the pixel space for generating the object bounding box. Finally, the NMS algorithm will be utilized here for eliminating the overlapped detections. The flowchart of the proposed method is shown in Fig. 2.

4   Experimental Results

In this section, we evaluate our proposed method with proposals generated by EdgeBoxes and BING respectively, with the DPM on the PASCAL VOC 2007 [4] dataset for detecting human using the MODP (Multiple Object Detection Precision) and the MODA (Multiple Object Detection Accuracy) metrics [15]. We compare the running time and the detection results between them on the 4,952 test images of the PASCAL VOC 2007 dataset. By the way, the experimental results also show the comparison between the EdgeBoxes and the BING.

It is worth noting that the evaluation metrics MODP and MODA measure the degree of the precision of detected objects and the ratio of incorrect detections (miss and false) over the total detections, respectively. The range of MODA is $(-\infty, 1)$. And the MODA value 1 denotes that there are no incorrect detections, which is an ideal situation. For the MODP, according to the PASCAL VOC challenge convention, we set the overlap threshold to 0.5.

For the EdgeBoxes and BING, we utilize the codes offered by the authors. We just change parameters that control the number of proposals to what we want for testing, other parameters are recommended. For the DPM, we use a pre-trained two components human model and re-implement the DPM algorithm using C++ for the execution efficiency based on the OpenCV Library 2.4.10. All the experiments are operated on a 4-core Intel i5-3570K CPU, and we use 4 threads to all of them for accelerating.

The running time results are shown in Fig. 3. To our proposed method, the number of proposals generated by object proposal methods is significant to running time. It is not difficult to image that more positions to detect more time-consuming. Hence, the target is better performance with less proposals. According to Hosang et al. [6], in the level of 10^3 proposals per image, which is a good trade-off between the speed and proposals quality. At the same time, if we utilize just 1000 proposals get a better results than the DPM, it will enough prove that the proposed method improved the DPM. In addition, to explore the influence of the

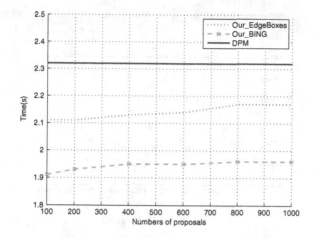

Fig. 3. The running time results.

number of proposals to the running time, we generate different number of proposals from 100 to 1000 using our proposed method with EdgeBoxes and BING, respectively.

As shown in Fig. 3, Our_EdgeBoxes means we firstly generate proposals using EdgeBoxes, then using our proposed method to detect, the same to Our_BING. As a whole, the proposed method is faster than the DPM. Notice that the running time of BING and EdgeBoxes is 0.003 s and 0.25 s per image when the proposals number is one thousand. Hence, excepting the running time of EdgeBoxes and BING, the proposed method 0.4 s faster than the DPM approximately. To Our_EdgeBoxes and Our_BING, the latter is faster than the former because of the running time of EdgesBox is slow than the BING. And, with the number of proposals grows, the running time have a slight increase, because more positions to detect.

The Fig. 4 (a) and (b) shows the MODP and MODA results respectively. As is shown in Fig. 4 (a), the MODP of Our_EdgesBoxes almost equal to the DPM and even better than the DPM. It is obvious that the proposals' quality of EdgeBoxes is better than that of BING.

Generally, we will hold the view that the precision of the proposed method at best equal to the DPM, because it detects human only in some positions. But when the proposals number is 400, Our_EdgeBoxes exceeds the DPM. This is due partly to the fact that the DPM generates many overlap windows when detecting, because the exhaustive searching. Many positions adjacent to a person will also get a high score. Hence, the DPM need NMS to reduce the overlap windows. But the NMS eliminates overlap windows just according to an overlap criterion without considering the score, which results in some uncertainties in the NMS. For our proposed method, we detecte the human at selected candidate positions, so we do not have many overlap windows.

In Fig. 4 (b), both the MODA of Our_EdgesBoxes and Our_BING have a considerable improvement than the DPM, and their curves are pretty similar.

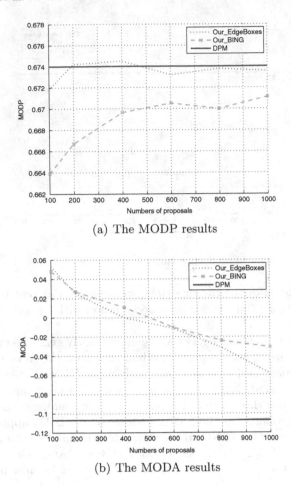

(a) The MODP results

(b) The MODA results

Fig. 4. The detection results of the experiment.

It indicates that the proposed method decrease the incorrect detections effectively. And with the number of proposals grows, the MODA of the proposed method decreases.

As a whole, comparing with the DPM, the proposed method accelerates the detection speed and decreases the incorrect detections effectively with little MODP drops. And, the proposals quality of EdgeBoxes is better than BING. Hence, we recommend EdgeBoxes as the object proposal method in our proposed method.

5   Conclusion

By using detection proposals generated by the object proposal methods, we improve the speed and accurate of the original DPM for detecting human. By combining the excellent object proposal methods we reduced the search positions from $\sim 10^6$ of exhaustive searching to $\sim 10^3$. Experimental results suggest that the proposed method is faster than the original DPM and reducing the false positive detections effectively at the same time. An intuitionistic comparison between the two of the state-of-the-art object proposal methods is done in our work. And we recommend the EdgeBoxes. For the next work, we will try to take the context information into account for improving the detection precision and take more verification experiments on the ImageNet Large Scale Visual Recognition Challenge (ILSVRC) [24,25] dataset.

References

1. Felzenszwalb, P., Girshick, R., McAllester, D., Ramanan, D.: Object detection with discriminatively trained part-based models. J IEEE Trans. Pattern Anal. Mach. Intell. **32**(9), 1627–1645 (2010)
2. Dalal, N., Triggs, B.: Histograms of oriented gradients for human detection. In: 2005 IEEE Conference on Computer Vision and Pattern Recognition, vol. 1, pp. 886–893. IEEE Press, San Diego (2005)
3. Everingham, M., Van Gool, L., Williams, C.K.I., Winn, J., Zisserman, A.: The PASCAL Visual Object Classes Challenge (VOC2007) (2007) Results http://www.pascalnetwork.org/challenges/VOC/voc2007/
4. Everingham, M., Van Gool, L., Williams, C.K.I., Winn, J., Zisserman, A.: The pascal visual object classes (VOC) challenge. J. Int. J. Comput. Vis. **88**, 303–338 (2010)
5. Gadeski, E., Fard, H.O., Borgne, H.L.: GPU deformable part model for object recognition. J. Real-Time Image Proc. (2014). doi:10.1007/s11554-014-0447-5
6. Hosang, J., Benenson, R., Schiele, B.: How good are detection proposals, really?. In: 2014 British Machine Vision Conference, Nottingham (2014). ArXiv:1406.6962
7. Zitnick, C.L., Dollár, P.: Edge boxes: locating object proposals from edges. In: Fleet, D., Pajdla, T., Schiele, B., Tuytelaars, T. (eds.) ECCV 2014, Part V. LNCS, vol. 8693, pp. 391–405. Springer, Heidelberg (2014)
8. Cheng, M.M., Zhang, Z., Lin, W.Y., Torr, P.H.S.: BING: Binarized normed gradients for objectness estimation at 300fps. In: 2014 IEEE Conference on Computer Vision and Pattern Recognition, pp. 3286–3293. IEEE Press, Columbus, OH (2014)
9. Sadeghi, M.A., Forsyth, D.: 30HZ Object detection with DPM V5. In: 13th European Conference on Computer Vision, pp. 65–79. Zurich (2014)
10. Yang, Y., Li, S.-P.: Fast object detection with deformable part models and segment location' hint. J. Acta Automatica Sinica **38**, 540–548 (2012)
11. Gu, C., Lim, J.J., Arbelaez, P., Malik, J.: Recognition using regions. In: 2009 IEEE Conference on Computer Vision and Pattern Recognition, pp. 1030–1037. IEEE Press, Miami, FL (2009)
12. Uijlings, J.R.R., van de Sande, K.E.A., Gevers, T., Smeulders, A.W.M.: Selective search for object recognition. J. Int. J. Comput. Vis. **104**, 154–171 (2013)
13. Manen, S., Guillaumin, M., Van Gool, L., Leuven, K.: Prime object proposals with randomized prims algorithm. In: 2013 IEEE International Conference on Computer Vision, pp. 2536–3543. IEEE Press, Sydney, NSW (2013)
14. Alexe, B., Deselaers, T., Ferrari, V.: Measuring the objectness of image windows. J IEEE Trans. Pattern Anal. Mach. Intell. **34**, 2189–2202 (2012)
15. Kasturi, R., Goldgof, D., Soundararajan, P., et al.: Framework for performance evaluation of face, text, and vehicle detection and tracking in video: Data, metrics, and protocol. J IEEE Trans. Pattern Anal. Mach. Intell. **31**(2), 319–336 (2009)

16. Dollr, P., Zitnick, C.L.: Structured forests for fast edge detection. In: 2013 IEEE International Conference on Computer Vision, pp. 1841–1848. IEEE Press, Sydney, NSW (2013)
17. Dollr, P., Zitnick, C.L.: Fast edge detection using structured forests. J. IEEE Trans. Pattern Anal. Mach. Intell. (2014)
18. Fan, R.-E., Chang, K.-W., Hsieh, C.-J., Wang, X.-R., Lin, C.-J.: Liblinear: A library for large linear classification. J. The Journal of Machine Learning Research **9**, 1871–1874 (2008)
19. Carreira, J., Sminchisescu, C.: CPMC: automatic object segmentation using constrained parametric min-cuts. J IEEE Trans. Pattern Anal. Mach. Intell. **34**, 1312–1328 (2012)
20. Rahtu, E., Kannala, J., Blaschko, M.: Learning a category independent object detection cascade. In: 2011 IEEE International Conference on Computer Vision, pp. 1052–1059. IEEE Press, Barcelona (2011)
21. Endres, I., Hoiem, D.: Category-independent object proposals with diverse ranking. J IEEE Trans. Pattern Anal. Mach. Intell. **36**, 222–234 (2014)
22. Rantalankila, P., Kannala, J., Rahtu, E.: Generating object segmentation proposals using global and local search. In: 2014 IEEE Conference on Computer Vision and Pattern Recognition, pp. 2417–2424. IEEE Press, Columbus, OH (2014)
23. Girshick, R., Donahue, J., Darrell, T., and Malik, J.: Rich feature hierarchies for accurate object detection and semantic segmentation. In: 2014 IEEE Conference on Computer Vision and Pattern Recognition, pp. 580–587. IEEE Press, Columbus, OH (2014)
24. Deng, J., Wei, D., Socher, R., Berg, A., Satheesh, S., Su, H., Khosla, A., Fei-Fei, L.: ImageNet Large Scale Visual Recognition Competition (ILSVRC2012) (2012). http://www.image-net.org/challenges/LSVRC/2012/
25. Deng, J., Wei, D., Socher, R., Li, L.-J., Li, K., Fei-Fei, L.: ImageNet: a large-scale hierarchical image database. In: 2009 IEEE International Conference on Computer Vision, pp. 248–255. IEEE Press, Miami, FL (2009)

A New Method for Driver Fatigue Detection Based on Eye State

Xinzheng Xu[1](✉), Xiaoming Cui[1], Guanying Wang[1], Tongfeng Sun[1],
and Hongguo Feng[2]

[1] School of Computer Science and Technology, China University of Mining
and Technology Xuzhou, Jiangsu 221116, China
xuxinzh@163.com
[2] 77626 Troops, Tibet Autonomous Region, Tibet 851400, China

Abstract. Fatigue driving is one of main problems threatening driving safety. Therefore, it attracts numerous researchers interests. This paper introduces a new method based on eye feature to research the fatigue driving. Firstly, the face is detected by the model of skin-color in the YCbCr color space, which extracted face region from complex background quickly and accurately. Secondly, eye detection includes extracting eye region and detecting eye two steps. Specifically, the proposed method extracts eye region in face image based on gray-scale projection and then detect eye using Hough transform. Finally, calculate the area of the eye profile after dilation and use it as the parameter to analysis eye state. Put forward the standard to recognize fatigue base on the PER-CLOS. The experiment results illustrate the efficiency and accurately of the proposed method, especially, detected face as well as extracted eye region with a high accuracy.

Keywords: Fatigue detection · Face detection · Extracting eye region · Eye detection

1 Introduction

In recent years, more and more researchers have been engaged in fatigue driving recognition and many important achievements have been made, such as lane-track alarm system [1], EEG method [2], heart rate detector [3], etc. Although there are many ways to realize driver fatigue detection to a certain extent, the performances of these methods are various. These methods can be divided into three categories. The first way is to detect fatigue through the analysis of the vehicle status. For example, American Ellison Research Labs monitored lane track to achieve fatigue detection in 2004. While, some researchers detected fatigue through other vehicle status such as the speed of vehicle running and the turning of the steering wheel. But those ways are hard to get a standard to judge whether the drivers are dozing. The second method is based on the physiological property such as EEG and heart rates. According to the research of Japan Pioneer Company, the heart rate will be slow when drivers are drowsy. So through

© Springer International Publishing Switzerland 2015
D. Ciucci et al. (Eds.): RSKT 2015, LNAI 9436, pp. 513–524, 2015.
DOI: 10.1007/978-3-319-25754-9_45

physical characteristics like heart rate, EEG can make an accurate detection. However, these techniques are usually more expensive and will annoy the drivers, so they are not easily accepted by the drivers. In general, driver fatigue monitoring system should consider that whether it is easy to accept for the driver, timeliness, reliability, scalable, and the cost of these. With the development of digital image processing and computer technology, more mature techniques are provided to analyze videos and images. Using digital image processing technology to find out fatigue in time and then prevent it which is the third approach. Fatigue recognition according to facial features has following advantages: real time, reliability and little interference to the driver. Driver fatigue recognition based on facial feature is a trend. Many researchers proposed different methods in this area [4,5]. The eye contains a lot of information, so the research and analysis of the human eye has become a very hot issue.

The paper is organized as follows. System design is presented in Sect. 2. Section 3 shows the design of fatigue detection algorithm. Section 4 contains the results of experiment experiments. The paper is concluded in Sect. 5.

2   System Design

This driver fatigue monitoring system includes Pre-treatment (video acquisition and image preprocessing), face detection, eye detection, eyes feature extraction and driving fatigue recognition. The driver fatigue monitoring system is shown in Fig. 1 and we will introduce these in detail in the following sections.

3   Fatigue Detection Algorithm Design

This driver fatigue monitoring system detects driver drowsiness based on eye features. Getting the picture from the video and using skin-color to detect face. Using skin-color way to detect face has many advantages, but it is influenced by many factors (light, clothes similar to skin color, etc.), so choosing YCbCr color space and add restricted conditions are necessary. Gray-scale projection method combines with Circular Hough transform method to extract the eye region accurately. Finally, according to eye status we can recognize driving fatigue.

3.1   Face Detection

Face detection should be done before eye detection, so the first step is face detection. The face detection ways currently used include Eigenface [6], neural networks [7], Gabor transform [8], skin-color [9–11]. Compared with mentioned method above, the face image based on skin-color method can extract face from complex background quickly and accurately. It achieves good real-time performance and strong practicability. The skin-color model is established should choose appropriate color space. RGB, NTSC, YCbCr and HSV are used commonly color space. YCbCr color space put the RGB color space divided into three

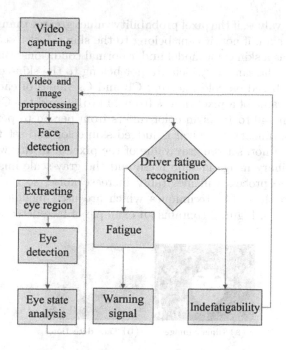

Fig. 1. The system framework

components, a brightness component(Y) and the two color component(Cr and Cb). YCbCr model is commonly used in color digital video model. In this color space, Y contains brightness information, Chromaticity information is stored in the Cr and Cb. Cb shows green component relative reference value, Cr shows red component relative reference value, and they are independent relationship. YCbCr model data can be described by double precision. As brightness component and color component are separated in YCbCr color space, reduce the effects of light. Cr and Cb contain chrominance information and fit to establish skin-color model14. So we chose YCbCr color space. The RGB components and the YCbCr components can be converted by the following formulas.

$$\begin{cases} Y = 0.299R + 0.578G + 0.114 \\ Cb = 0.1482(B - R) + 0.2910(B - G) + 128 \\ Cr = 0.3678(B - R) - 0.0714(B - R) + 128 \end{cases} \quad (1)$$

The skin color distribution similar to Gaussian distribution, Gaussian model is proposed base on the theoretical. Based on this, calculate the probability of each pixel belongs to skin color which use two-dimensional Gaussian in the color image and get the probability value of skin. Those values make skin probability graph. The higher probability of the pixel color values of skin area in figure, it is the candidate skin region.

The values of the pixel probability are calculated by the principle of two-dimensional Gaussian model of skin color detection. Selecting a threshold base on those probability values, if the pixel probability values greater than the threshold, it belongs to the skin, if not, it cant belong to the skin. Most researchers choose Gaussian model as a skin color model under normal conditions. But this method needs to consider the samples which are not belong to the skin-color.

In the model based on color, getting Cb and Cr values of each pixel of the image. If the Cr value of a pixel ranges from 140 to160 and the Cb value of this pixel ranges from 140 to 195(skin color differs from person to person, the data base on this experiment), the pixel is judged skin-color and set the gray value of the pixel 255. If not, set the gray value of the pixel 0. In this way, we get the binary image. Binary image include noise and the gray-scale images converted into binary image process will inevitably increase noise, so the binary image needs to be smoothed. The techniques which are used include corrosion and expansion of images. Figure 2 contains an example of face detection procedure.

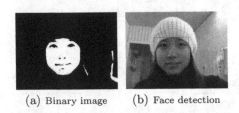

(a) Binary image   (b) Face detection

Fig. 2. Illustration of face detection algorithm

If the image background is complex, we may get multiple candidate face region. Those are candidate skin region need to judgment. And it will be affected by realistic environment, such as clothes, skin-color background, so we will get some candidate skin region as described earlier in this paper, it must be added some conditions to limit:

(1) The limitation of length and width of the skin region, these values are set according to the proportion of face in the image.
(2) Height and width ratio, in reality, the ratio is set range from 0.6 to 2.
(3) The region should contain the eyes, in other words there are two black areas in real face region.

Determine the face region accurately by adding the three restrictions. Figure 3 shows the illustration of face detection adding the restrictions mentioned above from complicated background.

The procedures of recognizing face based on skin-color are as follows:

Step 1: The color images from RGB color space converted into the YCbCr color space. The function we used is YCbCr = rgb2ycbcr(RGB).
Step 2: The color images change into binary image base on the values of Cb and Cr which get according to the skin in YCbCr model distribution range.
Step 3: Binary image includes noise and the gray-scale images converted into binary image process will inevitably increase noise, so this step we eliminate the noise.

(a) Candidate skin region   (b) Face detection

Fig. 3. Illustration of face detection with restrictions

Step 4: Corrosion and fill hole processing.
Step 5: We will get some candidate skin region, we choose true face region from them. According to the three restrictions which have already been mentioned we determine the face region accurately.

3.2 Extraction the Eye Area

The extracting face region is the basis for the recognize fatigue driving. Recognize fatigue driving by the eye state. The human eye is one of vital organ to reflect the fatigue driving or not, it contains a lot of information. The researches show that recognize fatigue and eye state have a close relationship, eyes are closed or almost closed state in a long time when people tired, at the same time the blink rate is accelerated and the closed time will become long. So this system through the eye state to recognize fatigue driving.

The gray level around the eye part is lower than other parts in the face, so we can detect the human eye using this characteristic. Gray-scale projection method is used widely in image processing. Using this method to process the original image directly, the noise is relatively large and it is difficult to achieve desired results. But with the development of technology, the gray projection method are also improved, this method in the human eye feature extraction is widely applied [12,13]. The system is based on previous studies on the application of this theoretical knowledge, Combination with the knowledge of the digital image. In order to improve the efficiency and reduce calculation, we use the binary image instead of the original image to gain the horizontal and vertical integral projection. Binary image of the face can clearly find eyebrow, eyes, nose and mouth etc. The binary image with gray-scale projection can provide a more determined accurately area of the human eye.

The binary image of the human face can clearly find eyebrows, eyes and mouth. Using gray-scale projection algorithm to deal with the binary image can get the position of eye roughly. $G(x,y)$ represents the gray value of the pixel in(x,y), $H(x)$ stands for the value of binary image horizontal integral projection, $V(y)$ stands for the value of binary image vertical integral projection, which are shown as follows:

$$H(x) = \frac{1}{x_2 - x_1} \sum_{x_1}^{x_2} G(x, y) \tag{2}$$

$$V(y) = \frac{1}{y_2 - y_1} \sum_{y_1}^{y_2} G(x, y) \tag{3}$$

We have got coordinate of the face area in the section A and extracted face area base on the coordinate. In Fig. 4, (a) shows the extracted face region, (b) shows the binary image of the face and (c) shows the extracted eye region. It is gained by the gray-scale projection.

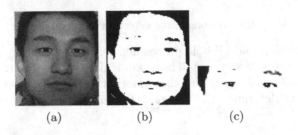

(a)   (b)   (c)

Fig. 4. Extraction the eye area

We get the gray value from the binary image of the face with horizontal and vertical integral projection. Figure 5 shows the gray value of horizontal integral. Figure 6 shows the gray value of vertical integration.

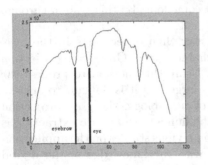
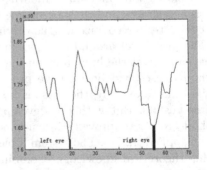

Fig. 5. The gray value of horizontal integral

Fig. 6. The gray value of horizontal integral

The black region which closes to forehead shows eyebrows and eyes. It corresponds to the Horizontal integral projection curve are two minimum.

It is easy to roughly get the ordinate of eyebrows and eyes from the figure, so that we can choose the appropriate width. The same method, two valleys in vertical gray projection curve represent left and right eyes. So we extract the eye region.

3.3 Eye Detection

Hough transform which describes the border of region, is often used to detect geometry, such as circular, oval or line in the image. In this paper, we use this method to detect eye accurately. Before using Hough transform we should detect the borders of the image. By comparing the different edge detection algorithms, we find Prewitt algorithm in this system performs best. It is easy to find that the eye is similar to oval in binary figure. But using the oval model to judge human eyes need to determine the center axislength and minor axis length and steer of the oval which similar to eye. Detecting an oval needs five parameters [14], it is a great amount of calculation. The human iris is circular. Locating the iris circle model by Hough transform to determine the human eye is open or close, if it is close at that time we can't detect the circle [15]. It is only requires three parameters, the center (x, y) and radius r.

The basic idea of the Hough transform is according to the majority points of the boundary to determine the curve. So through describing the boundary of the curve, image space changes into curves space. So it is good tolerance and robustness to some possible noise of area boundary.

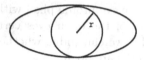

Fig. 7. Eye model

The appearance of eye model is shown in Fig. 7. We know the circle can be expressed with the following formula:

$$(x - x_0)^2 + (y - y_0)^2 = r^2 \tag{4}$$

From Eq. 4 we can see that a circle requires three parameters. Judging a circle with the three parameters still exists a certain difficulty. So researchers limit the center (x, y) and radius r within a certain range, and then calculate the parameters. In this way, the parameters have reduced. It reduces much calculating obviouslly. The radius of the circle is calculated first for a very small and sealed area, then judgment all the points on the edge. Consequently, we can quickly identify the edge of the circle.

Set the parameter of the circle is one pixel, the step of radians change is 0.2 Pixel, the minimum Radius $r_{min} = 5$, the maximum Radius, $r_{max} = 8$, the threshold value is 0.685. These parameters are set according to the system. Figure 8 shows the illustration of Hough transform.

This paper presents the gray-scale projection and the Hough transform to quickly detect eye. It includes two steps which are extracting the eye area and eye detection accurately. This section is very important for the driver fatigue monitoring system, we describes two algorithms in section B and C. The procedures of recognizing eye are follows:

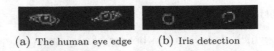

(a) The human eye edge   (b) Iris detection

Fig. 8. Illustration of Hough transform

Step 1: Segmentation of face region and change it to binary image.

Step 2: Binary image of face with horizontal integral projection get the distribution of the vertical graph. We get the value of y that the minimum gray values which stand for the eye and eyebrow.

Step 3: The width adds to 2d with y values as the center. Get out the region.

Step 4: The region with horizontal integral projection and get two lows from the graph which can determine the eye part. The eye area can be extracted.

Step 5: Detect the border of the image by Prewitt algorithms.

Step 6: Set the parameters of the circle with $r_{min} = 5$, $r_{max} = 8$ and threshold value is 0.685. Then eye detection by Hough transform.

3.4 Eye State Analysis

We calculate the quantity of the eye profile pixel with dilation. Under the same conditions, the pixels number of the open eyes certainly more than the eyes closed pixels. So based on this we can judge whether human eyes are open or closed, and judge how much eyes are open.

PERCLOS is a classical method to determine whether human eyes are fatigue or not, the fatigue recognition model based P80 criterion [16]. The 100 % open is the eyes largest area in all images during a period of time. If the eye closed degree more than 80 % is determined closed state. In this paper, the pixels number of the eye largest is the eyes largest area.

To deal with each frame from the human eye features respectively, the number of eyes closed frames is $CloseFrame\_Num$ and the total number of frames that deal with is $SumFrame\_Num$. According to the following formula the value of PERCLOS can be calculated.

$$PERCLOS = \frac{CloseFrame\_Num}{SumFrame\_Num} \times 100\% \tag{5}$$

If the value of PERCLOS in experiment is greater than the threshold that we set the value of 20 %, we think the driver is fatigue, then the alarm system start warning.

4   Experimental Results

In this section, we take the video which acquired by camera in simulation driving condition as an example. Video processing using MATLAB converted to images. The video in the system is 30 frames /s and $640 * 480$ pixels per a frame. Based on the fact that movement is continuous and not too fast, there is no need for testing each frame. First of all, we extract one image per 5 frames. In order to not only accurately judge the results but also improve the efficiency of the system in this way. So we extract 6 frames per second [17].

Lots of experiments show that this method is better than other methods. It can identify the face more accurately and quickly which is the foundation for later processing. Figures 9 and 10 show the original image and the framed face image recognize face by this method.

(a) Original image   (b) The framed face image

Fig. 9. Detected face 1

(a) Original image   (b) The framed face image

Fig. 10. Detected face 2

The experiments illustrate that the face detection method can detect the faces accurately even though in different distances and positions, and the accuracy is over 95 %. It also can meet the real-time face detection. So the face detection method is the basis of the fatigue state judgment.

After extracting the face and then extract eyes from the face image based on the described algorithms for eye detection. Some examples of the detected eyes are shown in Figs. 11 and 12.

Numerous experiments demonstrate the viability of the proposed method. In this section we extract the eye region and detect eye better.

Figure 13 shows the eye edge after dilation of person who is not drowsy. And Fig. 14 illustrates the eye edge of a sleepy driver. According to Figs. 13 and 14, the area of eye edge in Fig. 13 is bigger than what in Fig. 14. Two testers with four 30 s videos which simulation of driving situations, according to the algorithm, each video includes 180 frames, Table 1 shows the experimental results.

From Table 1, the system can detect the faces accurately and the extracted eye region is exact, but using Hough transform to detect eye is imperfect. From the experimental results, the proposed algorithm is faster and has higher

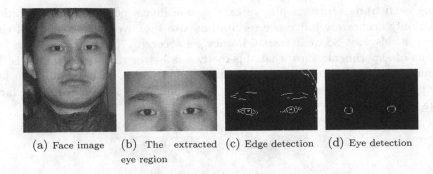

(a) Face image   (b) The extracted eye region   (c) Edge detection   (d) Eye detection

Fig. 11. Eye detection 1

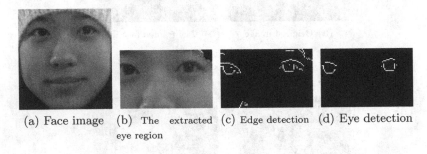

(a) Face image   (b) The extracted eye region   (c) Edge detection   (d) Eye detection

Fig. 12. Eye detection 2

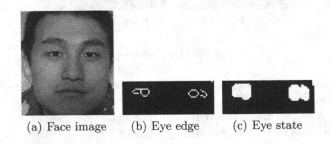

(a) Face image   (b) Eye edge   (c) Eye state

Fig. 13. Results when person not in drowsy state

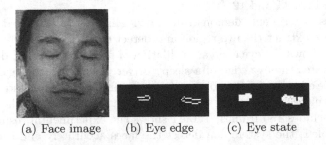

(a) Face image   (b) Eye edge   (c) Eye state

Fig. 14. Results when person is in drowsy state

accuracy. The proposed algorithm can reach correct detection rate of 86.4 % on average. On the other side, the correct rate of video 4 is lower than video1, the main reason is that the system is easily influenced by illumination.

Table 1. Experimental results

	video 1	video 2	video 3	video 4
Total frames	180	180	180	180
Face Detection	180	178	178	177
Extracted eye region	176	174	172	172
Eye Detection	166	163	151	150
Detected Fatigue	165	161	149	147
Correct Rate	91.7 %	89.4 %	82.8 %	81.7 %

5   Conclusion

In this paper, we propose a method for fatigue driving detection base on eye feature. As can be seen from the experimental results the system detected face as well as extracted eye region with a high accuracy of more than 95 %. Using the last system, fatigue detection is not as well as expected. But the system can fully meet the real-time and accurate requirements. We will realize driver fatigue monitoring system on the hardware platform and improve the accuracy in the future study, besides, we will study more evaluation indexes or compare its performance with other similar methods, to demonstrate the effectiveness of the proposed approach.

Acknowledgments. This work is supported by the Basic Research Program (Natural Science Foundation) of Jiangsu Province of China (No.BK20130209), the Fundamental Research Funds for the Central Universities (No.2013QNA24), the Project Funded by China Postdoctoral Science Foundation (No.2014M560460), the Project Funded by Jiangsu Postdoctoral Science Foundation (No.1302037C).

References

1. Pei, Z., Song, Z.H., Zhou, Y.M.: Research status and development trend of motor vehicle driver fatigue evaluation method. J. China Agric. Univ. **6**(6), 101–105 (2001)
2. Lampetch, S., Punsawad, Y., Wongsawat, Y.: EEG-based mental fatigue prediction for driving application. In: Biomedical Engineering International Conference (BMEICON), pp. 1–5 (2012)
3. Vicente, J., Laguna, P., Bartra, A., Bailon, R.: Detection of driver's drowsiness by means of HRV analysis. In: Computing in Cardiology, pp. 89–92 (2011)

4. Wang, P., Shen, L.: A method detecting driver drowsiness state based on multi-features of face. In: 2012 5th International Congress on Image and Signal Processing (CISP 2012), pp. 1171–1175 (2012)
5. Lee, B.G., Chung, W.Y.: Driver alertness monitoring using fusion of facial features and bio-signals. IEEE Sens. J. **12**(7), 2416–2422 (2012)
6. Watta, P., Gandhi, N., Lakshmanan, S.: An Eigenface approach for estimating driver pose. In: 2000 Proceedings Intelligent Transportation Systems, pp. 376–381. IEEE (2000)
7. Ni, Q.K., Guo, C., Yang, J.: Research of face image recognition based on probabilistic neural networks. In: 2012 24th Chinese Control and Decision Conference (CCDC), pp. 3885–3888 (2012)
8. Shan, D., Ward, R.K.: Improved face representation by nonuniform multilevel selection of gabor convolution features. IEEE Trans. Sys. Man Cybern. Part B Cybern. **39**(6), 1408–1419 (2009)
9. Zhao, Y.L., Gao, Z., Wu, W.X.: The detection algorithm of locomotive driverss fatigue based on vision. In: 2010 3rd International Congress on Image and Signal Processing (CISP2010), pp. 2686–2690 (2010)
10. Devi, M.S., Choudhari, M.V., et al.: Driver drowsiness detection using skin color algorithm and circular hough transform. In: 2011 Fourth International Conference on Emerging Trends in Engineering and Technology, pp. 129–134 (2009)
11. Wu, C.D., Zhang, C.B.: Detecting and locating method of human face in driver fatigue surveillance. J. Shenyang Jianzhu Univ. Nat. Sci. **25**(2), 386–389 (2009)
12. Lu, L., Yang, Y., Wang, L., Tang, B.: Eye location based on gray projection. In: 2009 Third International Symposium on Intelligent Information Technology Application, pp. 58–60 (2009)
13. Feng, J.Q., Liu, W.B., Yu, S.L.: Eyes location based on gray-level integration projection. Comput. Simul. **22**(4), 75–76 (2005)
14. Yang, Q.F., Gui, W.H., et al.: Eye location novel algorithm for fatigue driver. Comput. Eng. Appl. **44**(6), 20–24 (2008)
15. Qu, P.S., Dong, W.H.: Eye states recognition based on eyelid curvature and fuzzy logic. Comput. Eng. Sci. **29**(8), 50–53 (2007)
16. Pan, X.D., Li, J.X.: Eye state-based fatigue drive monitoring approach. J. Tongji Univ. Nat. Sci. **39**(2), 231–235 (2011)
17. Wang, Y., Hu, J.W.: A method for detection of driver eye fatigue state based on 3G video. Electron. Sci. Tech. **24**(10), 84–85 (2011)

Facial Expression Recognition Based on Quaternion-Space and Multi-features Fusion

Yong Yang[1,2(✉)], Shubo Cai[1], and Qinghua Zhang[1]

[1] Chongqing Key Laboratory of Computational Intelligence, Chongqing University of Posts and Telecommunications, Chongqing 400065, People's Republic of China
{yangyong,zhangqh}@cqupt.edu.cn, yongyang@inha.ac.kr,
caishubo1991@163.com
[2] School of Information and Communication Engineering,
Inha University, Incheon 402-751, Korea

Abstract. There is an increasing trend of using feature fusion technique in facial expression recognition. However, when traditional serial or parallel feature fusion methods are used, the problem of highly dimensional features and insufficient fusion of possible feature categories always exist. In order to solve these problems, a novel facial expression recognition method based on quaternion-space and multi-features fusion is proposed. Firstly, four different kinds of expression features are extracted such as Gabor wavelet, LBP, LPQ and DCT features, then PCA+CCA framework is proposed and used to reduce the dimensions of the four original features. Secondly, quaternion is used to construct the combinative features. Thirdly, a novel quaternion-space HDA method is proposed and used as the dimensional reduction method of the combinative features. Finally, SVM is used and set as the classifier. Experimental results indicate that the proposed method is capable of fusing four kinds of features more effectively while it achieves higher recognition rates than the traditional feature fusion methods.

Keywords: Facial expression recognition · Multi-features fusion · Quaternion · Dimensional reduction · Quaternion-space HDA

1 Introduction

Psychologists consider that human emotions mainly consist of 7 % of linguistic information, 38 % of vocal information and 55 % of facial expression information [1]. Besides, facial expression can reflect one's true feelings while people can judge the emotional status from the subtle expression changes on the face. Therefore, facial expression is crucial for non-verbal communication. Since automatic

This paper is partially supported by The National Natural Science Foundation of China under Grant No.61472056 and No.61300059, and the Ministry of Science, ICT & Future Planning(MSIP) of Korea in the ICT R & D Program 2013 under Grant No.10039149.

D. Ciucci et al. (Eds.): RSKT 2015, LNAI 9436, pp. 525–536, 2015.
DOI: 10.1007/978-3-319-25754-9_46

facial expression recognition technology was proposed in the 1970s, it has always been the research hotspot of pattern recognition and it has been widely applied in the intelligent surveillance systems, intelligent terminals, medical treatments, educations, et al.

Most of the traditional facial expression recognition methods are based on at most two categories of features. However, pattern description will be incomprehensive under this circumstance. In order to eliminate this limitation, data fusion technique was proposed in the 1990s [2]. It mainly consists of three levels: pixel-level fusion, feature-level fusion and decision-level fusion. Feature fusion is capable of reserving valid distinguishing information of multi-features while eliminating data redundancy, and it can be divided as the serial feature fusion methods and the parallel feature fusion methods [3].

Traditionally, serial feature fusion simply connects multiple feature vectors end to end to form a supervector in the real space [4–6]. However, parallel feature fusion is usually more effective and efficient than serial feature fusion, which is relatively complicated and uses complex vector to fuse two kinds of feature vectors in the unitary space [7]. In 2009, Luo proposed a modifying parallel feature fusion method and applied it into expression recognition. This method took advantage of the maximum scatter difference discrimination analysis method with different weighs and achieved 85.4 % recognition rate [8]. In 2009, Bai fused Gabor wavelet and LBP features of facial expressions parallelly, and by proposing the null-space LDA method to fuse the parallel combinative feature, this method achieved 87.57 % recognition rate [9]. In 2015, Yang proposed a novel facial expression recognition method based on PCA+unitary-space HDA two-step dimensional reduction and parallel feature fusion which fuses Gabor wavelet and LBP features, and this method achieved 87.17 % recognition rate [10]. Although the parallel feature fusion methods conquer the probable dimensional disaster problem which always exists in the serial feature fusion methods, the shortcoming of parallel feature fusion methods still exist that feature categories for fusion can't exceed two, and it may leads to incomprehensive facial expression description. In 2007, Lang proposed a newly parallel feature fusion method which is based on quaternion theory and generalized Fisher discriminant analysis method from real space to quaternion space, and it achieved remarkable results in face detection [11]. Therefore, in order to solve the problems of incomplete description of facial expression which is derived from the traditional feature fusion methods, a facial expression recognition method based on quaternion-space and multi-features fusion is proposed and the main contributions are as follows: 1. The PCA+CCA framework is proposed and used as the first-step dimensional reduction method; 2. Quaternion is used to fuse four kinds of facial expression features; 3. The quaternion-space HDA method is proposed and used as the second-step dimensional reduction method.

The rest of paper is organized as follows: In Sect. 2, basic concepts of quaternion theory are introduced. In Sect. 3, the framework of a facial expression recognition method based on multi-features fusion is proposed. Simulation experiments and results analysis are presented in Sect. 4. Conclusions and future works are discussed in the last section.

2 Introduction of Quaternion Theory

Quaternions are a number system that extend the traditional complex numbers and a quaternion variable q is defined as follows [12]:

$$q = q_r + iq_i + jq_j + kq_k, \tag{1}$$

where $q_r, q_i, q_j, q_k \in \mathbf{R}$, i, j and k are imaginary units:

$$i^2 = j^2 = k^2 = -1, \tag{2}$$

$$ij = -ji = k, jk = -kj = i, ki = -ik = j. \tag{3}$$

The conjugate operation and modular arithmetic are defined as follows:

$$q^H = q_r - iq_i - jq_j - kq_k, \tag{4}$$

$$\|q\| = \sqrt{qq^H} = \sqrt{q_r^2 + q_i^2 + q_j^2 + q_k^2}, \tag{5}$$

where H denotes conjugate transpose. Besides, other basic operational rules of quaternion are defined as follows:

$$q_1 + q_2 = (q_{r1} + q_{r2}) + i(q_{i1} + q_{i2}) + j(q_{j1} + q_{j2}) + k(q_{k1} + q_{k2}), \tag{6}$$

$$\begin{aligned} q_1 \cdot q_2 = &(q_{r1}q_{r2} - q_{i1}q_{i2} - q_{j1}q_{j2} - q_{k1}q_{k2}) \\ &+ i(q_{r1}q_{i2} + q_{i1}q_{r2} + q_{j1}q_{k2} - q_{k1}q_{j2}) \\ &+ j(q_{r1}q_{j2} + q_{r2}q_{j1} + q_{i2}q_{k1} - q_{k2}q_{i1}) \\ &+ k(q_{r1}q_{k2} + q_{k1}q_{r2} + q_{i1}q_{j2} - q_{j1}q_{i2}), \end{aligned} \tag{7}$$

$$q_1 = q_2 \Leftrightarrow q_{r1} = q_{r2}, q_{i1} = q_{i2}, q_{j1} = q_{j2}, q_{k1} = q_{k2}. \tag{8}$$

The relationship between dualistic complex variate and quaternary hyper-complex variate can be described as follows:

$$q = q_r + iq_i + jq_j + kq_k = (q_r + iq_i) + (q_j + iq_k)j = q_{c1} + q_{c2}j, \tag{9}$$

where q_{c1} and q_{c2} are two complex variables.

The matrix which consists of quaternion elements is called the quaternion matrix and there're two equivalent representation which are second order equivalent complex matrix and fourth order equivalent real matrix. Assume that a $m \times n$ order quaternion matrix M_q can be described as the sum of two complex matrix: $M_q = M_{c1} + M_{c2}j$, where M_{c1} and M_{c2} are all $m \times n$ order complex matrix. Therefore, the equivalent complex matrix M_q^c is defined as follows:

$$M_q^c = \begin{pmatrix} M_{c1} & M_{c2} \\ -\overline{M_{c2}} & \overline{M_{c1}} \end{pmatrix} \in C^{2m \times 2n}, \tag{10}$$

where M_q^c is a $2m \times 2n$ order complex matrix, and $\overline{M_{c1}}$ and $\overline{M_{c2}}$ are conjugate forms of M_{c1} and M_{c2}. $\mathbf{C}$ denotes the complex field.

Assume that a $m \times n$ order quaternion matrix M_q can be described as the sum of four real matrix: $M_q = M_r + M_i i + M_j j + M_k k$, where M_r, M_i, M_j and M_k are $m \times n$ order real matrix. Therefore, the equivalent real matrix M_q^r is defined as follows:

$$M_q^r = \begin{pmatrix} M_r & M_i & M_j & M_k \\ -M_i & M_r & -M_k & M_j \\ -M_j & M_k & M_r & -M_i \\ -M_k & -M_j & M_i & M_r \end{pmatrix} \in \mathbf{R}^{4m \times 4n}, \tag{11}$$

where M_q^r is a $4m \times 4n$ order real matrix. $\mathbf{R}$ denotes the real field.

3   Facial Expression Recognition Method Based on Multi-features Fusion

3.1   Framework of the Proposed Method

The pipeline of the proposed method is as follows: 1. Image pre-processing methods which include size cropping, gray-scale pretreatment and histogram equalization are applied on each expression images; 2. Four kinds of facial features are extracted such as Gabor wavelet [13], LBP [14], LPQ [15] and DCT [16]; 3. The PCA+CCA framework is proposed and used to reduce the dimensions of the four features above; 4. Quaternion is used to construct the combinative feature [11]; 5. The quaternion-space HDA method is proposed to reduce the dimensions of the combinative feature; 6. SVM is used as the classifier. Figure 1 shows the process of the proposed method.

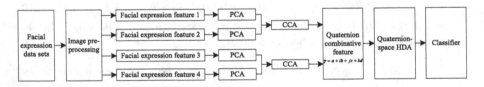

Fig. 1. Flow chart of the proposed facial expression recognition method based on quaternion-space and multi-features fusion

3.2   PCA+CCA Framework

The dimensions of the original facial expression features are relatively high, when fusing them directly by quaternion may results dimensional disaster problem. PCA is a dimensional reduction method which uses the covariance matrix to analyze the distribution of the sample data while reducing feature dimensions [17]. However, PCA only analyze the principal component distribution of single feature category and ignores the correlation between different feature categories.

On the other hand, CCA is also a dimensional reduction method which analyzes the canonical correlation between different feature categories.

CCA method aims at finding a pair of projection axis to acquire maximum correlation between two feature vectors and by maximizing Pearson correlation coefficient to get CCA optimal identification function [18]:

$$
\begin{aligned}
\rho &= \frac{\alpha^T E\left[xy^T\right]\beta}{\sqrt{\alpha^T E\left[xx^T\right]\cdot\alpha\beta^T E\left[yy^T\right]\beta}} \\
&= \frac{\alpha^T C_{xy}\beta}{\sqrt{\alpha^T C_{xx}\alpha\cdot\beta^T C_{yy}\beta}},
\end{aligned}
\tag{12}
$$

where ρ is the canonical correlation coefficient, x and y are the two original feature vectors respectively, C_{xy} is the cross-covariance matrix of the two original features, C_{xx} and C_{yy} are the auto-covariance matrix of the two original features respectively. E denotes the expectation symbol, T denotes the transpose symbol.

Then, the canonical correlations between x and y can be found by solving the eigenvalue equations below:

$$
\begin{cases}
C_{xx}^{-1}C_{xy}C_{yy}^{-1}C_{yx}\alpha = \rho^2\alpha \\
C_{yy}^{-1}C_{yx}C_{xx}^{-1}C_{xy}\beta = \rho^2\beta
\end{cases},
\tag{13}
$$

where ρ^2 can be seen as the eigenvalues, α and β are the eigenvectors.

Therefore, in order to synthesize the advantages of both methods mentioned above, the PCA+CCA dimensional reduction framework is proposed: Firstly, PCA is applied on each sample matrix of the four original features respectively. Then CCA is applied on each two of the output features which have been already processed by PCA. After processed by PCA+CCA framework, the dimensions and redundancy of the expression features will be largely reduced meanwhile maintaining the comprehensive descriptions of facial expressions.

3.3   Quaternion-Space Combinative Feature

Suppose the sample pattern space is Ω, and the four feature vector space are A, B, C and D respectively. If sample $\xi \in \Omega$, and its four feature vectors are $\alpha \in A$, $\beta \in B$, $\tau \in C$ and $\sigma \in D$. After data pre-processing: $\alpha' = \alpha/\|\alpha\|$, $\beta' = \beta/\|\beta\|$, $\tau' = \tau/\|\tau\|$, $\sigma' = \sigma/\|\sigma\|$, the quaternion-space combinative feature γ is defined as follows [11]:

$$
\gamma = \{\alpha' + i\beta' + j\tau' + k\sigma', \alpha \in A, \beta \in B, \tau \in C, \sigma \in D\}.
\tag{14}
$$

3.4   Quaternion-Space HDA Method

After feature fusion, the dimension of the combinative feature is still high. Therefore, a hybrid discriminant analysis method which is based on quaternion-space(quaternion-space HDA) is proposed. This method is the further extension of the real-space HDA method [19] and it's not only adaptive to the dimensional reduction of the hypercomplex data, but also gives both consideration of between-class discriminant information and total descriptive information in the quaternion space.

The three scatter matrix in the quaternion space are defined as follows [11]:

$$S_{q-b} = \sum_{i=1}^{C} P(w_i)(m_i - m_0)(m_i - m_0)^H, S_{q-b} \in SC_n^{\geq}(Q), \qquad (15)$$

$$S_{q-w} = \sum_{i=1}^{C} P(w_i)E\{(x - m_i)(x - m_i)^H / w_i\}, S_{q-w} \in SC_n^{\geq}(Q), \qquad (16)$$

$$S_{q-t} = S_{q-b} + S_{q-w} = E\{(x - m_0)(x - m_0)^H\}, S_{q-t} \in SC_n^{\geq}(Q), \qquad (17)$$

where S_{q-b}, S_{q-w} and S_{q-t} are the between-class scatter matrix, the within-class scatter matrix and the total-class scatter matrix which are all defined in the quaternion space and belong to semidefinite quaternion matrix space $SC_n^{\geq}(Q)$. Besides, C is the number of pattern categories, x is the sample matrix, $P(w_i)$ is the prior probability of the i-th sample category, m_i is the mean value of the i-th sample category and m_0 is the mean value of the whole sample.

The optimal identification function of the hybrid discriminant analysis method based in quaternion space can be described as follows:

$$J(W)_{opt}^{quaternion-spaceHDA} = \underset{W}{\text{argmax}} \frac{|W^H((1 - \lambda)S_{q-b} + \lambda S_{q-t})W|}{|W^H((1 - \eta)S_{q-w} + \eta I)W|}, \qquad (18)$$

where W is the projection axis and I is the unit matrix. Eq. (18) is the further extension of the generalized Raleigh entropy in the quaternion space and forming the optimal identification function which contains two kinds of optimal problem in the quaternion space. By adjusting the tradeoff coefficient (λ, η) Eq. (18) can balance the problem analysis advantage between the between-class discriminant problem and the total-class descriptive problem in the quaternion space. Besides, the constraint conditions should be met are as follows when solving Eq. (18):

$$w_i^H((1 - \lambda)S_{q-b} + \lambda S_{q-t})w_i = \begin{cases} \lambda_i, i = j \\ 0, \ i \neq j \end{cases}, \qquad (19)$$

$$w_i^H((1 - \eta)S_{q-w} + \eta I)w_i = \begin{cases} 1, i = j \\ 0, i \neq j \end{cases}, \qquad (20)$$

where λ_i is the i-th eigenvalue and w_i is the corresponding i-th eigenvector. By using w_i to form the projection axis while each two of the eigenvectors are orthogonal.

However, if the optimal identification function of the quaternion-space HDA method is calculated directly, the computational complexity will be extremely high. Therefore, the equivalent real matrix of the quaternion matrix in Eq. (11) is used to solve the projection axis of Eq. (18). Suppose w_i^r and w_i^q are the i-th eigenvectors and form the projection axis of both matrix mentioned above respectively. The relationship of them can be described as follows:

$$w_i^r = (w_{i1}, w_{i2}, w_{i3}, w_{i4})^T, \tag{21}$$

$$w_i^q = (w_{i1} - w_{i2}i - w_{i3}j - w_{i4}k), \tag{22}$$

where w_{in} is the n-th component of w_i^r and the dimension of w_i^r is four times as big as w_i^q.

4   Experiments

4.1   Experimental Settings

In the following experiments, six basic facial expressions are used as the classification label [20]: anger, disgust, fear, happy, sad and surprise. Meanwhile, 10-fold cross validation method is applied on both JAFFE [21] and CK+ [22] facial expression data sets which contains 183 and 600 images selected as samples respectively.

Image cropping method is as follows: Suppose the length between two pupils is d, so the cropping width and height is $1.8d$ and $2.2d$ respectively as pupils are set as the symmetric points. After cropping, pure facial expression images are obtained. After each images are converted to gray-scale and histograms are equalizing, then they are normalized to the size of 100×120. Finally, each images are divided into 3×3 non-overlapping blocks.

Four kinds of features are extracted on each blocks of the images in this paper: Gabor wavelet, LBP, LPQ and DCT feature. Gabor kernel with 5 scales and 8 orientations is applied, and by calculating the mean value and the standard deviation to form the $9 \times 5 \times 8 \times 2 = 720$ dimensional Gabor wavelet statistical features. $LBP_{P,R}^{U2}$ operator with $P=8$ and $R=2$ forming the $9 \times 59 = 531$ dimensional LBP features. The gray-scale histogram is calculated when LPQ and DCT are applied and the feature dimensions are $9 \times 256 = 2304$ repectively. After PCA+CCA processing, the quaternion combinative features are constructed by these features sequently.

4.2   Experimental Results Analysis

Experiment #1. In order to testify the validity of the proposed method, different feature fusion experiments are taken as follows: The quaternion-space multi-features fusion method which fuses Gabor, LBP, LPQ and DCT features is used in experiment 1 and (λ, η) is $(0.5, 0.5)$; The parallel feature fusion method which fuses Gabor+LBP, Gabor+LPQ, LBP+LPQ, Gabor+DCT, LBP+DCT and LPQ+DCT are used in experiment 2~7 respectively and (λ, η) is $(0.5, 0.5)$; The null-space LDA parallel feature fusion method [9] is used in experiment 8; The serial feature fusion method [6] is used in experiment 9; The traditional serial feature fusion method which fuses Gabor+LBP, Gabor+LPQ, LBP+LPQ,

Table 1. Experimental performances of different feature fusion methods(%)

No	Feature fusion methods	JAFFE data sets			CK+ data sets		
		Avg	Max	Min	Avg	Max	Min
1	Quaternion-space multi-features fusion	84.44	88.89	77.78	90.17	98.33	83.33
2	Gabor+LBP parallel feature fusion [10]	80.55	94.44	50.00	87.17	95.00	78.33
3	Gabor+LPQ parallel feature fusion	80.00	94.44	72.22	88.33	100.00	73.33
4	LBP+LPQ parallel feature fusion	81.11	100.00	61.11	87.50	96.67	71.67
5	Gabor+DCT parallel feature fusion	78.89	94.44	38.89	87.67	93.33	83.33
6	LBP+DCT parallel feature fusion	75.56	88.89	44.44	89.00	96.67	83.33
7	LPQ+DCT parallel feature fusion	74.45	88.89	55.56	85.00	100.00	76.67
8	Null-space LDA parallel feature fusion [9]	76.11	88.89	61.11	80.00	88.33	75.00
9	PCA+LBP serial feature fusion [6]	74.44	83.33	55.56	83.17	88.33	71.67
10	Gabor+LBP serial feature fusion	75.00	88.89	50.00	73.50	86.67	60.00
11	Gabor+LPQ serial feature fusion	72.22	88.89	55.56	78.89	88.89	72.22
12	LBP+LPQ serial feature fusion	75.56	88.89	55.56	78.33	90.00	50.00
13	Gabor+DCT serial feature fusion	75.56	88.89	66.67	79.33	90.00	71.67
14	LBP+DCT serial feature fusion	76.11	88.89	66.67	67.00	86.67	33.33
15	LPQ+DCT serial feature fusion	68.33	83.33	27.78	82.17	96.67	71.67
16	Gabor single feature	65.56	77.78	50.00	66.50	73.33	60.00
17	LBP single feature	68.33	83.33	50.00	71.83	75.00	65.00
18	LPQ single feature	63.33	88.89	33.33	67.28	78.33	61.11
19	DCT single feature	61.67	66.67	55.56	67.16	78.33	35.00

Gabor+DCT, LBP+DCT and LPQ+DCT are used in experiment 10~15 respectively; The single feature method which includes Gabor, LBP, LPQ and DCT are used in experiment 16~19 respectively. Table 1 shows the experimental results.

Experimental results in Table 1 reflect that recognition rates of the proposed method are much higher than the traditional feature fusion methods and single feature methods on both data sets. It indicates that the expression description is more accurate and comprehensive when using four kinds of features, and the proposed two-steps dimensional reduction framework can effectively avoid the dimensional disaster problem which always exists in the serial feature fusion method. When comparing with the null-space LDA method [9] and PCA+LBP serial feature fusion method [6], recognition rate of the proposed method is 8.33 %, 10.17 % and 10.00 % and 7.00 % higher on both data sets. It also indicates that the proposed method doesn't have to consider whether the null space of sample sets exists while avoiding the dimensional disaster more effectively. Furthermore, stable recognition rates have been achieved by the proposed method, so it's more robust than other algorithms.

Experiment #2. In order to verify the validity and necessity of the proposed two-step dimensional reduction framework, several contrast experiments are taken as follows: The proposed two-steps dimensional reduction method is used in experiment 1 and (λ, η) is $(0.5, 0.5)$; The quaternion-space LDA and PCA

Table 2. Necessity verification of the proposed two-steps dimensional reduction framework(%)

No	Dimensional reduction methods	JAFFE data sets			CK+ data sets		
		Avg	Max	Min	Avg	Max	Min
1	PCA+CCA+quaternion-space HDA	84.44	88.89	77.78	90.17	98.33	83.33
2	PCA+CCA+quaternion-space LDA	67.22	83.33	38.89	77.67	90.00	66.67
3	PCA+CCA+quaternion-space PCA	65.56	83.33	50.00	79.00	95.00	63.33
4	PCA+quaternion-space HDA	63.33	77.78	44.44	60.50	81.67	38.33
5	PCA+CCA only	76.67	94.44	50.00	85.33	93.33	81.67
6	quaternion-space HDA only	-	-	-	-	-	-

which are used in the second-step dimensional reduction are used in experiment 2~3 respectively; PCA+quaternion-space HDA method is used in experiment 4; PCA+CCA and quaternion-space HDA are only used in experiment 5~6 respectively. Table 2 shows the experimental results.

Experimental results in Table 2 reflect that recognition rates of the proposed two-steps dimensional reduction method are much higher than the one-step methods. It verifies the necessity and effectiveness of the PCA+CCA+quaternion-space HDA framework. By changing of the tradeoff coefficients, recognition rates are much lower due to the less optimal problems which have been considered. When CCA isn't used, low recognitions have been achieved due to the ignorance of the correlation analysis between multi-features of PCA. However, no results will be achieved by only using quaternion-space HDA and the reason is that highly dimensional features are directly fused by quaternion so that dimensional disaster is triggered and leads to the overflow of the computational memory.

Experiment #3. In order to testify the differentiation and the validity of the proposed quaternion-space HDA method with alternative tradeoff coefficients, several contrast experiments within different (λ, η) are taken as follows: (0.5, 0.5), (0.0, 0.0), (0.0, 1.0), (1.0, 0.0) and (1.0, 1.0). Table 3 shows the experimental results.

Table 3. Experimental performances of quaternion-space HDA with different tradeoff coefficients(%)

(λ, η)	JAFFE data sets			CK+ data sets		
	Avg	Max	Min	Avg	Max	Min
(0.5, 0.5)	84.44	88.89	77.78	90.17	98.33	83.33
(0.0, 0.0)	67.22	83.33	38.89	77.67	90.00	66.67
(0.0, 1.0)	75.00	88.89	55.56	81.67	98.33	73.33
(1.0, 0.0)	67.77	83.33	44.44	75.50	90.00	58.33
(1.0, 1.0)	65.56	83.33	50.00	79.00	95.00	63.33

Experimental results in Table 3 reflect that experimental performance is better when two optimal problems are considered than that when only one optimal problem is considered. Moreover, experimental performance of the optimal identification problem which only considers quaternion-space between-class scatter matrix is better than other problems, so it indicates that when only considering one optimal problem, the between-class discriminant information is more distinguishing than the total-class descriptive information in data discrimination.

Experiment #4. In order to measurement the real-time performance of the proposed method, several contrast experiments are taken by calculating the average testing time consumptions of 10 facial expression images. Table 4 shows the experimental results.

Table 4. Real-time performances of different feature fusion methods(s)

Feature fusion methods	Time
Parallel feature fusion(Gabor+LBP) [10]	2.897
Parallel feature fusion(Gabor+LBP)	3.063
Parallel feature fusion(LBP+LPQ)	3.164
Quaternion-space multi-features fusion	5.299

Experimental results in Table 4 show that average time consumptions of the parallel feature fusion methods [10] and the proposed method are approximately 3 seconds and 5 seconds respectively. However, real-time requirement of pattern recognition system is generally in a time scale of millisecond. It reflects the computational complexity of the proposed method is relatively high. Therefore, how to improve the real-time performance and computational complexity is the point of the further research.

5   Conclusion

A novel facial expression recognition method which is based on quaternion-space and multi-features fusion is proposed in this paper. By fusing four kinds of feature to construct the quaternion combinative feature and several contrast experiments are completed. The experimental results show that comparing to the traditional feature fusion methods and single feature methods, recognition rates have been greatly improved by the proposed method. Besides, contrast experiments such as necessity verification of the two-steps dimensional reductions and the quaternion-space HDA method with different tradeoff coefficients are set. Although the amount of the facial expression feature categories which are being fused is up to four, it is still limited and the computational complexity is relatively high. Therefore, the next steps of research will mainly focus on how to fuse more than four kinds of facial expression features while reducing the algorithm complexity.

References

1. Mehrabian, A.: Silent Messages: Implicit Communication of Emotions and Attitudes. Wadsworth, Belmont (1981)
2. Abidi, M.A., Gonzalez, R.C.: Data Fusion in Robotics and Machine Intelligence. Academic Press, San Diego (1992)
3. Yang, J., Yang, J.Y., Zhang, D., et al.: Feature fusion: parallel strategy vs. serial strategy. J Pattern Recogn. **36**(6), 1369–1381 (2003)
4. Liu, C.J., Wechsler, H.: A shape- and texture-based enhanced Fisher classifier for face recognition. J. IEEE Trans. Image Process. **10**(4), 598–608 (2001)
5. Kotsia, I., Nikolaidis, N., Pitas, I.: Fusion of geometrical and texture information for facial expression recognition. In: 2006 IEEE International Conference on Image Processing, pp. 2649–2652. IEEE Press, Atlanta (2006)
6. Luo, Y., Wu, C.M., Zhang, Y.: Facial expression feature extraction using hybrid PCA and LBP. J. China Univ. Posts Telecommun. **20**(2), 120–124 (2013)
7. Yang, J., Yang, J.Y., Wang, Z.Q., et al.: A novel method of combined feature extraction. J. Chinese J. Comput. **25**(6), 570–575 (2002)
8. Luo, F., Wang, G.Y., Yang, Y., et al.: Facial expression recognition based on improved parallel features fusion. J. Guangxi Univ. Nat. Sci. Ed. **34**(5), 700–703 (2009)
9. Bai, G., Jia, W.H., Jin, Y.: Facial expression recognition based on fusion features of lbp and gabor with lda. In: 2nd International Congress on Image and Signal Processing(CISP2 009), pp. 1–5. IEEE Press, Tianjin (2009)
10. Yang, Y., Cai, S.B.: Facial expression recognition method based on two-steps dimensionality reduction and parallel feature fusion. J. Chongqing Univ. Posts Telecomm. Nat. Sci. Ed. **27**(3), 377–385 (2015)
11. Lang, F.N., Zhou, J.L., Zhong, F., et al.: Quaternion based image information parallel fusion. J. Acta Automatica Sinica **33**(11), 1136–1143 (2008)
12. Hamilton, W.R.: Elements of Quaternions. Longmans Green and Company, London (1866)
13. Ruan, J.X., Yin, J.X., Chen, Q., et al.: Facial expression recognition based on gabor wavelet transform and relevance vector machine. J. Inf. Comput. Sci. **11**(1), 295–302 (2014)
14. Verma, R., Dabbagh, M.Y.: Fast facial expression recognition based on local binary patterns. In: 2013 26th Annual IEEE Canadian Conference on Electrical and Computer Engineering (CCECE), pp. 1–4. IEEE Press, Regina (2013)
15. Wang, Z., Ying, Z.L.: Facial Expression Recognition Based on Local Phase Quantization and Sparse Representation. In: 2012 Eighth International Conference on Natural Computation (ICNC), pp. 222–225. IEEE Press, Chongqing (2012)
16. Kharat, G.U., Dudul, S.V.: Neural network classifier for human emotion recognition from facial expressions using discrete cosine transform. In: First International Conference on Emerging Trends in Engineering and Technology (ICETET 2008), pp. 653–658. IEEE Press, Nagpur (2008)
17. Xanthopoulos, P., Pardalos, P.M., Trafalis, T.B.: Principal component analysis. Robust Data Mining, pp. 21–26. Springer Press, New York (2013)
18. Hardoon, D., Szedmak, S., Shawe-Taylor, J.: Canonical correlation analysis: an overview with application to learning methods. Neural Comput. **16**(12), 2639–2664 (2004)
19. Yu, J., Tian, Q., Rui, T., et al.: Integrating discriminant and descriptive information for dimension reduction and classification. IEEE Trans. Circuits Sys. Video Technol. **17**(3), 372–377 (2007)

20. Ekman, P., Friesen, W.V.: Constants across cultures in the face and emotion. J. Pers. Soc. Psychol. **17**(2), 124 (1971)
21. Lyons, M., Akamatsu, S., Kamachi, M., et al.: Coding facial expressions with gabor wavelets. In: Third IEEE International Conference on Automatic Face and Gesture Recognition, pp. 200–205. IEEE Press, Nara (1998)
22. Lucey, P., Cohn, J.F., Kanade, T., et al.: The Extended cohn-kanade dataset (CK+): a complete dataset for action unit and emotion-specified expression. In: 2010 IEEE Computer Society Conference on Computer Vision and Pattern Recognition Workshops (CVPRW), pp. 94–101. IEEE Press, San Francisco (2010)

Correction to: Interactive Granular Computing

Andrzej Skowron and Andrzej Jankowski

Correction to:
Chapter "Interactive Granular Computing"
in: D. Ciucci et al. (Eds.): *Rough Sets*
and Knowledge Technology, **LNAI 9436,**
https://doi.org/10.1007/978-3-319-25754-9_5

The acknowledgement section of this paper originally referred to grant DEC-2013/09/B/ST6/01568. The reference to this grant has been removed from the acknowledgement section at the request of one of the authors.

The updated version of this chapter can be found at
https://doi.org/10.1007/978-3-319-25754-9_5

Author Index

Printed in the United States
by Baker & Taylor Publisher Services

Printed in the United States
by Baker & Taylor Publisher Services